AF683973

ICL Contribution to Landslide Disaster Risk Reduction

Series Editor

Kyoji Sassa, The International Consortium on Landslides, ICL, Kyoto, Japan

The ICL Contribution to Landslide Disaster Risk Reduction book-series publishes integrated research on all aspects of landslides. The volumes present summaries on the progress of landslide sciences, disaster mitigation and risk preparation. The contributions include landslide dynamics, mechanisms and processes; volcanic, urban, marine and reservoir landslides; related tsunamis and seiches; hazard assessment and mapping; modeling, monitoring, GIS techniques; remedial or preventive measures; early warning and evacuation and a global landslide database.

More information about this series at http://www.springer.com/series/16332

ICL and Springer created a new book series "ICL Contribution to Landslide Disaster Risk Reduction" in 2019 which is registered as ISSN 2662-1894 (print version) and ISSN 2662-1908 (electronic version). The first books in this series are six volume of books "Understanding and Reducing Landslide Disaster Risk" containing the recent progress of landslide science and technologies from 2017 to 2020.

Editor-in-Chief: Kyoji Sassa
Assistant Editor-in-Chief: Željko Arbanas

Organizational Structure of the Fifth World Landslide Forum

Organizers

International Consortium on Landslides (ICL)
Global Promotion Committee of International Programme on Landslides (IPL-GPC), including: United Nations Educational, Scientific and Cultural Organization (UNESCO), World Meteorological Organization (WMO), Food and Agriculture Organization (FAO), United Nations Office for Disaster Risk Reduction (UNDRR), United Nations University (UNU), International Science Council (ISC), World Federation of Engineering Organizations (WFEO), International Union of Geological Sciences (IUGS), International Union of Geodesy and Geophysics (IUGG) Kyoto University (KU), Japan Landslide Society (JLS), Japanese Geotechnical Society (JGS), Japan Society for Natural Disaster Science (JSNDS) and Japan Association for Slope Disaster Management (JASDiM)

Co-sponsors

Cabinet Office (Disaster Management Bureau) of Japan; Ministry of Foreign Affairs of Japan (MOFA); Ministry of Education, Culture, Sports, Science and Technology-Japan (MEXT); Ministry of Land Infrastructure, Transport and Tourism (MLIT); Ministry of Agriculture, Forestry and Fisheries (MAFF); Science Council of Japan (SCJ); Japan International Cooperation Agency (JICA); Japan Society of Civil Engineers (JSCE); Japanese Society of Irrigation, Drainage and Rural Engineering (JSIDRE); Japan Society of Erosion Control Engineering; Japan Society of Engineering Geology.

Supporting Organizations with Finance

Tokyo Geographical Society
International Union of Geological Sciences (IUGS)
Association for Disaster Prevention Research, Kyoto, Japan

Organizing Committee

Honorary Chairpersons
Audrey Azoulay, Director-General of UNESCO*
Mami Mizutori, Special Representative of the United Nations Secretary-General for Disaster Risk Reduction*

Petteri Taalas, Secretary-General of WMO*
Qu Dongyu, Director-General of FAO*
David Malone, Under-Sectretary General of the Unitred Nations and Rector of UNU
Daya Reddy, President of ISC
Gong Ke, President of WFEO
Qiuming Cheng, President of IUGS
Kathryn Whaler, President of IUGG
Qunli Han, Executive Director of Integrated Research on Disaster Risk (IRDR)
Walter Ammann, President and CEO of Global Risk Forum GRF Davos, Switzerland
Juichi Yamagiwa, President of Kyoto University, Japan
Angelo Borrelli, Head of the National Civil Protection Department, Italian Presidency of the Council of Ministers, Italy
Darko But, Director General of the Administration for Civil Protection and Disaster Relief of the Republic of Slovenia, Slovenia
Akifumi Nakao, Director, International Cooperation Division, Disaster Management Bureau, Cabinet Office, Japan
Kazuyuki Imai, Director General of Sabo Department, Ministry of Land Infrastructure, Transport and Tourism, Japan*
Chungsik Yoo, President of the International Geosynthetics Society
Rafig Azzam, President of the International Association for Engineering Geology and the Environment
(*to be confirmed)

Chairpersons
Kyoji Sassa, Professor Emeritus, Kyoto University; Secretary General of ICL
Peter T. Bobrowsky, Geological Survey of Canada; President of ICL
Kaoru Takara, Kyoto University, Japan; Executive Director of ICL

Members
Željko Arbanas (University of Rijeka, Croatia)
Snježana Mihalić Arbanas (University of Zagreb, Croatia)
Nicola Casagli (University of Firenze, Italy)
Fausto Guzzetti (Department of Civil Protection, Italy)
Matjaž Mikoš (University of Ljubljana, Slovenia)
Paola Reichenbach (Research Institute for Geo-Hydrological Protection, National Research Council, Italy)
Shinji Sassa (Port and Airport Research Institute, Japan)
Alexander Strom (Geodynamics Research Center LLC, Russia)
Binod Tiwari (California State University, Fullerton, USA)
Veronica Tofani (University of Firenze, Italy)
Vít Vilímek (Charles University in Prague, Czech Republic)
Fawu Wang (Tongji University, China)

Chairpersons of Local Organizing Committee
Kaoru Takara (Kyoto University)
Daisuke Higaki (Japan Landslide Society)
Ikuo Towhata (Japanese Geotechnical Society)

Secretary Generals
Ryosuke Uzuoka (Disaster Prevention Research Institute, Kyoto University)
Kazuo Konagai (International Consortium on Landslides)
Khang Dang (International Consortium on Landslides)

International Scientific Committee

Beena Ajmera, North Dakota State University, USA
Snježana Mihalić Arbanas, University of Zagreb, Croatia
Željko Arbanas, Faculty of Civil Engineering, University of Rijeka, Croatia
Amin Askarinejad, Technische Universiteit Delft, Delft, The Netherlands
Peter T. Bobrowsky, Geological Survey of Canada, Sidney, Canada
Michele Calvello, University of Salerno, Italy
Giovanna Capparelli, Universita degli Studi della Calabria, Rende, Italy
Nicola Casagli, University of Florence, Italy
Yifei Cui, Tsinghua University, Beijing, China
Sabatino Cuomo, University of Salerno, Fisciano, Italy
Khang Dang, International Consortium on Landslides, Kyoto, Japan
Elias Garcia-Urquia, National Autonomous University of Honduras, Tegucigalpa, Honduras
Stefano Luigi Gariano, Research Institute for Geo-Hydrological Protection, CNR, Perugia, Italy
Daniele Giordan, Research Institute for Geo-Hydrological Protection, CNR, Italy
Fausto Guzzetti, Department of Civil Protection, Italy
Baator Has, Asia Air Survey, Tokyo, Japan
Hans-Balder Havenith, Universite de Liege, Liege, Belgium
D. P. Kanungo, Central Building Research Institute (CBRI), Roorkee, Uttarakhand, India
Oded Katz, Geological Survey of Israel, Jerusalem, Israel
Kazuo Konagai, International Consortium on Landslides, Kyoto, Japan
Doan Huy Loi, International Consortium on Landslides, Kyoto, Japan
Ping Lu, Tongji University, Shanghai, China
Olga Mavrouli, University of Twente, Enschede, The Netherlands
Matjaž Mikoš, Faculty of Civil and Geodetic Engineering, University of Ljubljana, Slovenia
Alessandro C. Mondini, Research Institute for Geo-Hydrological Protection, CNR, Italy
Veronica Pazzi, Department of Earth Science, University of Florence, Florence, Italy
Dario Peduto, Department of Civil Engineering, University of Salerno, Fisciano, Italy
Paola Reichenbach, Research Institute for Geo-Hydrological Protection, CNR, Italy
Paola Salvati, Research Institute for Geo-Hydrological Protection, CNR, Italy
Katsuo Sasahara, Kochi University, Japan
Kyoji Sassa, International Consortium on Landslides, Kyoto, Japan
Shinji Sassa, Port and Airport Research Institute, Japan
Andrea Segalini, University of Parma, Italy
Hendy Setiawan, Universitas Gadjah Mada, Yogyakarta, Indonesia
Alexander Strom, Geodynamics Research Center LLC, Moscow, Russia
Kaoru Takara, Kyoto University, Japan
Faraz Tehrani, Deltares, Delft, The Netherlands
Binod Tiwari, California State University, Fullerton, California, USA
Veronica Tofani, University of Florence, Italy
Ryosuke Uzuoka, Kyoto University, Kyoto, Japan
Vít Vilímek, Faculty of Science, Charles University, Prague, Czech Republic
Fawu Wang, College of Civil Engineering, Tongji University, Shanghai, China
Gonghui Wang, Kyoto University, Kyoto, Japan
Mike Winter, Winter Associates Limited, Kirknewton, UK
Hiromitsu Yamagishi, Hokkaido Research Center of Geology (HRCG), Sapporo, Japan

Local Organizing Committee

Shinro Abe, Okuyama Boring Co., Ltd.
Kiminori Araiba, Fire and Disaster Management College
Shiho Asano, Forestry and Forest Products Research Institute
Has Baator, Asia Air Survey Co., Ltd.

Hiromu Daimaru, Forestry and Forest Products Research Institute
Khang Dang, International Consortium on Landslides
Mitusya Enokida, Japan Conservation Engineers & Co., Ltd.
Kumiko Fujita, International Consortium on Landslides
Kazunori Hayashi, Okuyama Boring Co., Ltd.
Daisuke Higaki, The Japan Landslide Society
Kiyoharu Hirota, Kokusai Kogyo Co., Ltd.
Kazuo Konagai, International Consortium on Landslides
Taketoshi Marui, MARUI & Co., Ltd.
Satoshi Nishikawa, Nagoya University
Keisuke Oozone, OYO Corporation
Katsuo Sasahara, Kochi University
Kyoji Sassa, International Consortium on Landslides
Shinji Sassa, Port and Airport Research Institute
Go Sato, Teikyo Heisei University
Nobuyuki Shibasaki, Nippon Koei Co., Ltd.
Nobuo Sugiura, Japan Association for Slope Disaster Management
Kaoru Takara, Kyoto University
Keisuke Takimoto, GODAI KAIHATSU Corporation
Yoko Tomita, Public Works Research Institute
Ikuo Towhata, The Japanese Geotechnical Society
Kenichi Tsukahara, Kyushu University
Ryosuke Tsunaki, Sabo & Landslide Technical Center
Taro Uchida, Saitama University
Mie Ueda, International Consortium on Landslides
Ryosuke Uzuoka, Kyoto University
Fawu Wang, Tongji University
Hiroshi Yagi, Yamagata University
Hiromitsu Yamagishi, Shin Engineering Consultants Co., Ltd.
Maki Yano, OSASI Technos Inc.

Foreword by Mami Mizutori

More landslides can be expected as climate change exacerbates rainfall intensity. The long-term trend of the last 40 years has seen the number of major recorded extreme weather events almost double, notably floods, storms, landslides, and wildfires.

Landslides are a serious geological hazard. Among the host of natural triggers are intense rainfall, flooding, earthquakes or volcanic eruption, and coastal erosion caused by storms that are all too often tied to the El Niño phenomenon. Human triggers including deforestation, irrigation or pipe leakage, and mine tailings, or stream and ocean current alteration can also spark landslides. Landslides can also generate tsunamis, as Indonesia experienced in 2018.

Globally, landslides cause significant economic loss and many deaths and injuries each year. Therefore, it is important to understand the science of landslides: why they occur, what factors trigger them, the geology associated with them, and where they are likely to happen.

Landslides with high death tolls are often a result of failures in risk governance, poverty reduction, environmental protection, land use and the implementation of building codes. Understanding the interrelationships between earth surface processes, ecological systems, and human activity is the key to reducing landslide risk.

The Sendai Framework for Disaster Risk Reduction, the global plan to reduce disaster losses adopted in 2015, emphasizes the importance of tackling these risk drivers through improved governance and a better understanding of disaster risk.

One important vehicle for doing that is the Sendai Landslide Partnerships 2015–2025 for global promotion of understanding and reduction of landslide risk facilitated by the International Consortium on Landslides (ICL) and signed by the leaders of 22 global stakeholders, including the UN Office for Disaster Risk Reduction (UNDRR), during the Third UN World Conference on Disaster Risk Reduction in Sendai, Japan.

The Sendai Landslide Partnerships—featured on the Sendai Framework Voluntary Commitments online platform—helps to provide practical solutions and tools, education, and capacity building, to reduce landslide risks.

The work done by the Sendai Partnerships can be of value to many stakeholders including civil protection, planning, development and transportation authorities, utility managers, agricultural and forest agencies, and the scientific community.

UNDRR fully supports the work of the Sendai Landslide Partnerships and ICL and looks forward to an action-oriented outcome from the 5th World Landslide Forum to be held in November 2020 in Kyoto, Japan. Successful efforts to reduce disaster losses are a major contribution to achieving the overall 2030 Agenda for Sustainable Development.

Mami Mizutori
United Nations Special Representative of the Secretary-General for Disaster Risk Reduction

Foreword by the Assistant Director-General for the Natural Sciences Sector of UNESCO for the Book of the 5th *World Landslide Forum*

As the world slowly recovers from the COVID-19 global pandemic, and looking back at the way this crisis developed, it becomes evident that as a global community we were not prepared for an event of this scale. Although not commonly perceived as such, biological hazards such as epidemics are included in the Sendai Framework for Disaster Risk Reduction 2015–2030. In that sense, the preparedness approach for a pandemic is very similar to that of a geophysical natural hazard such as landslides.

Although natural hazards are naturally occurring phenomena, the likelihood of their occurrence and of associated disasters is rising. Climate change, urban pressure, under-development and poverty and lack of preparedness are increasingly transforming these natural hazards into life-threatening disasters with severe economic impacts. Therefore, Disaster Risk Reduction (DRR) is gaining momentum on the agenda of the UN system of Organizations including UNESCO. While the Sendai Framework for Disaster Risk Reduction 2015–2030 is the roadmap for DRR, other global agendas including the Sustainable Development Goals, the Paris Climate Agreement and the New Urban Agenda have targets which cannot be attained without DRR.

In shaping its contribution to those global agendas, UNESCO is fully committed in supporting its Member States in risk management, between its different mandates and disciplines and with relevant partners. The International Consortium on Landslides (ICL) is UNESCO's key partner in the field of landslide science. The Organization's support to the Consortium is unwavering. Since ICL was established in 2002, the two organizations have a long history of cooperation and partnership and UNESCO has been associated with almost all of ICL activities. I am very glad that ICL and UNESCO are mutually benefitting from their collaboration.

The 5th World Landslide Forum (WLF5) is expected to represent a milestone in the history of landslide science particularly for scientists and practitioners. One of the major outcomes of WLF5 will be the Kyoto 2020 Commitment for global promotion of understanding and reducing landslide disaster risk (KLC2020). This commitment is expected to strengthen and expand the activities of the Sendai Landslide Partnership 2015–2025. With UNESCO already engaged as a partner, the adoption of this international commitment will raise global awareness on landslide risk and mobilize wider partnerships that draw together stakeholders from all levels of society, across different regions, sectors and disciplines.

It is my great pleasure to congratulate the organizers for holding this event and assure you that UNESCO is fully committed in contributing to its success. As part of that contribution, our Organization is proud to host a session on landslides and hazard assessment at UNESCO-designated sites such as natural World Heritage sites, biosphere reserves and UNESCO Global Geoparks. This session aims to assess landslide impacts on our shared cultural and natural heritage, providing the best opportunity to generate public awareness and capacity development for landslide disaster reduction.

I am confident that WLF5 will contribute to further advance the knowledge of both scientists and practitioners regarding landslide disaster risk reduction. This book paves the way for the science, knowledge and know-how which will feature in the deliberations of the Forum. UNESCO commends all of the contributors to this publication. I look forward to an enhanced collaboration between UNESCO and ICL in future activities and undertakings.

Shamila Nair-Bedouelle
Assistant Director-General for Natural Sciences
UNESCO

Preface I

Understanding and Reducing Landslide Disaster Risk

Book Series: ICL Contribution to Landslide Disaster Risk

The International Consortium on Landslides (ICL) was established in pursuance of the 2002 Kyoto Declaration "Establishment of an International Consortium on Landslides," with its Statutes adopted in January 2002. The Statutes define the General Assembly of ICL as follows: in order to report and disseminate the activities and achievements of the Consortium, a General Assembly shall be convened every 3 years by inviting Members of the International Consortium on Landslides, individual members within those organizations, and all levels of cooperating organizations and individual researchers, engineers, and administrators. The General Assembly developed gradually prior to, during and after its first meeting in 2005. In the light of the 2006 Tokyo Action Plan, the Assembly was further facilitated at, and following the First World Landslide Forum held in November 2008. On the occasion of each of its triennial forums, ICL publishes the latest progress of landslide science and technology for the benefit of the whole landslide community including scientists, engineers, and practitioners in an understandable form. Full color photos of landslides and full color maps are readily appreciated by those from different disciplines. We have published full color books on landslides at each forum. In 2019, ICL created a new book series "ICL Contribution to Landslide Disaster Risk Reduction" ISSN 2662-1894 (print version) and ISSN 2662-1908 (electronic version). Six volumes of full color books *Understanding and Reducing Landslide Disaster Risk* will be published in 2020 as the first group of books of this series.

The Letter of Intent 2005 and the First General Assembly 2005

The United Nations World Conference on Disaster Reduction (WCDR) was held in Kobe, Japan, 18–22 January 2005. At this Conference, ICL organized session 3.8 "New international Initiatives for Research and Risk Mitigation of Floods (IFI) and Landslides (IPL)" on 19 January 2005 and adopted a "Letter of Intent" aimed at providing a platform for a holistic approach in research and learning on 'Integrated Earth System Risk Analysis and Sustainable Disaster Management'. This Letter was agreed upon and signed, during the first semester of 2005, by heads of seven global stakeholders including the United Nations Educational, Scientific and Cultural Organization (UNESCO), the World Meteorological Organization (WMO), the Food and Agriculture Organization of the United Nations (FAO), the United Nations International Strategy for Disaster Risk Reduction (UNISDR-currently UNDRR), the United Nations University (UNU), the International Council for Science (ICSU-Currently ISC), and the World Federation of Engineering Organizations (WFEO).

The first General Assembly of ICL was held at the Keck Center of the National Academy of Sciences in Washington D.C., USA, on 12–14 October 2005. It was organized after the aforementioned 2005 World Conference on Disaster Reduction (WCDR). ICL published the first full color book reporting on Consortium activities for the initial 3 years, 2002–2005 titled "Landslides-Risk analysis and sustainable disaster management". In the preface of this book, the Letter of Intent for Integrated Earth System Risk Analysis and Sustainable Disaster Management was introduced. Results of the initial projects of the International Programme on Landslides (IPL) including IPL C101-1 Landslide investigation in Machu Picchu World Heritage, Cusco, Peru and previous agreements and MoU between UNESCO, ICL and the Disaster Prevention Research Institute of Kyoto University including UNESCO/KU/ICL UNITWIN Cooperation programme were published as well in this book.

The 2006 Tokyo Action Plan and the First World Landslide Forum 2008

Based on the Letter of Intent, the 2006 Tokyo Round-Table Discussion—"Strengthening Research and Learning on Earth System Risk Analysis and Sustainable Disaster Management within UN-ISDR as Regards Landslides"—towards a dynamic global network of the International Programme on Landslides (IPL) was held at the United Nations University, Tokyo, on 18–20 January 2006. The 2006 Tokyo Action Plan—Strengthening research and learning on landslides and related earth system disasters for global risk preparedness—was adopted. The ICL exchanged Memoranda of Understanding (MoUs) concerning strengthening cooperation in research and learning on earth system risk analysis and sustainable disaster management within the framework of the United Nations International Strategy for Disaster Reduction regarding the implementation of the 2006 Tokyo action plan on landslides with UNESCO, WMO, FAO, UNISDR (UNDRR), UNU, ICSU (ISC) and WFEO, respectively in 2006. A set of these MoUs established the International Programme on Landslides (IPL) as a programme of the ICL, the Global Promotion Committee of IPL to manage the IPL, and the triennial World Landslide Forum (WLF), as well as the concept of the World Centres of Excellence on Landslide Risk Reduction (WCoE).

The First World Landslide Forum (WLF1) was held at the Headquarters of the United Nations University, Tokyo, Japan, on 18–21 November 2008. 430 persons from 49 countries/regions/UN entities were in attendance. Both Hans van Ginkel, Under Secretary-General of the United Nations/Rector of UNU who served as chairperson of the Independent Panel of Experts to endorse WCoEs, and Salvano Briceno, Director of UNISDR who served as chairperson of the Global Promotion Committee of IPL, participated in this Forum. The success of WLF1 paved the way to the successful second and third World Landslide Forum held in Italy and China respectively.

The Second World Landslide Forum 2011 and the Third World Landslide Forum 2014

The Second World Landslide Forum (WLF2)—Putting Science into Practice—was held at the Headquarters of the Food and Agriculture Organization of the United Nations (FAO) on 3–9 October 2011. It was jointly organized by the IPL Global Promotion Committee (ICL, UNESCO, WMO, FAO, UNDRR, UNU, ISC, WFEO) and two ICL members from Italy: the Italian Institute for Environmental Protection and Research (ISPRA) and the Earth Science Department of the University of Florence with support from the Government of Italy and many Italian landslide-related organizations. It attracted 864 participants from 63 countries.

The Third World Landslide Forum (WLF3) was held at the China National Convention Center, Beijing, China, on 2–6 June 2014. A high-level panel discussion on an initiative to create a safer geoenvironment towards the UN Third World Conference on Disaster Risk Reduction (WCDRR) in 2015 and forward was moderated by Hans van Ginkel, Chair of Independent Panel of Experts for World Centers of Excellence (WCoE). In a special address to this high-level panel discussion, Irina Bokova, Director-General of UNESCO, underlined that countries should be united to work against natural disasters and expressed commitment that UNESCO would like to further deepen cooperation with ICL. Ms. Bokova awarded certificates to 15 World Centres of Excellence.

The Sendai Landslide Partnerships 2015 and the Fourth World Landslide Forum 2017

The UN Third World Conference on Disaster Risk Reduction (WCDRR) was held in Sendai, Japan, on 14–18 March 2015. ICL organized the Working Session "Underlying Risk Factors" together with UNESCO, the Japanese Ministry of Land, Infrastructure, Transport and Tourism (MLIT) and other competent organizations. The session adopted ISDR-ICL Sendai Partnerships 2015–2025 (later changed to Sendai Landslide Partnerships) for global promotion of understanding and reducing landslide disaster risk as a Voluntary Commitment to the World Conference on Disaster Risk Reduction, Sendai, Japan, 2015 (later changed to Sendai Framework for Disaster Risk Reduction). After the session on 16 March 2015, the Partnerships was signed by Margareta Wahlström, Special Representative of the UN Secretary-General for Disaster Risk Reduction, Chief of UNISDR (UNDDR), and other representatives from 15 intergovernmental, international, and national organizations. Following the Sendai Landslide Partnerships, the Fourth World Landslide Forum was held in Ljubljana, Slovenia from 29 May to 2 June in 2017. On that occasion, five volumes of full color books were published to disseminate the advances of landslide science and technology. The high-level panel discussion on 30 May and the follow-up round table discussion on 31 May adopted the 2017 Ljubljana Declaration on Landslide Risk Reduction. The Declaration approved the outline of the concept of "Kyoto 2020 Commitment for global promotion of understanding and reducing landslide disaster risk" to be adopted at the Fifth World Landslide Forum in Japan, 2020.

The Fifth World Landslide Forum 2020 and the Kyoto Landslide Commitment 2020

The Fifth World Landslide Forum was planned to be organized on 2–6 November 2020 at the National Kyoto International Conference Center (KICC) and the preparations for this event were successfully ongoing until the COVID-19 pandemic occurred over the world in early 2020. The ICL decided to postpone the actual Forum to 2–6 November 2021 at KICC in Kyoto, Japan. Nevertheless, the publication of six volumes of full color books *Understanding and Reducing Landslide Disaster Risk* including reports on the advances in landslide science and technology from 2017 to 2020 is on schedule. We expect that this book will be useful to the global landslide community.

The Kyoto Landslide Commitment 2020 will be established during the 2020 ICL-IPL Online Conference on 2–6 November 2020 on schedule. Joint signatories of Kyoto Landslide Commitment 2020 are expected to attend a dedicated session of the aforementioned Online Conference, scheduled on 5 November 2020 which will also include and feature the Declaration of the launching of KLC2020. Landslides: Journal of the International Consortium on Landslides is the common platform for KLC2020. All partners may contribute and publish news and reports of their activities such as research, investigation, disaster reduction administration in the category of News/Kyoto Commitment. Online access or/and hard copy of the Journal will be sent to KLC2020 partners to apprise them of the updated information from other partners. As of 21 May 2020, 63 United Nations, International and national organizations have already signed the KLC2020.

Call for Partners of KLC2020

Those who are willing to join KLC2020 and share their achievements related to understanding and reducing landslide disaster risk in their intrinsic missions with other partners are invited to inform the ICL Secretariat, the host of KLC2020 secretariat (secretariat@iclhq.org). The ICL secretariat will send the invitation to the aforementioned meeting of the joint signatories and the declaration of the launching of the KLC2020 on 5 November 2020.

Eligible Organizations to be Partners of the KLC2020

1. ICL member organizations (full members, associate members and supporters)
2. ICL supporting organization from UN, international or national organizations and programmes
3. Government ministries and offices in countries having more than 2 ICL on-going members
4. International associations /societies that contribute to the organization of WLF5 in 2021 and WLF6 in 2023
5. Other organizations having some aspects of activities related to understanding and reducing landslide disaster risk as their intrinsic missions.

Kyoji Sassa
Chair of WLF5/
Secretary-General of ICL
Kyoto, Japan

Peter T. Bobrowsky
President of ICL
Sidney, Canada

Kaoru Takara
Executive Director of ICL
Kyoto, Japan

Appendix: World Landslide Forum Books

WLF	Place/participants	Title	Editors	Publisher/pages
WLF0 (1st General Assembly) 2005	Washington D.C., USA 59 from 17 countries/UNs	Landslides-Risk Analysis and Sustainable Disaster Management	Kyoji Sassa, Hiroshi Fukuoka, Fawu Wang, Goghui Wang	Springer/377 pages ISBN: 978-3-540-2864-6
WLF1 2008	Tokyo, Japan 430 from 49 countries/regions/UNs	Landslides-Disaster Risk Reduction	Kyoji Sassa, Paolo Canuti	Springer/649 pages ISBN: 978-3-540-69966-8
WLF2 2011	Rome, Italy 864 from 63 countries	Landslide Science and Practice Vol. 1 Landslide inventory and Sustainability and Hazard Zoning	Claudia Margottini, Paolo Canuti, Kyoji Sassa	Springer/607 pages ISBN: 978-3-642-31324-0
		Vol. 2 Early Warning, Instrumentation and Monitoring		Springer/685 pages ISBN: 978-3-642-31444-5
		Vol. 3 Spatial Analysis and Modelling		Springer/440 pages ISBN: 978-3-642-31309-7
		Vol. 4 Global Environmental Change		Springer/431 pages ISBN: 978-3-642-31336-3
		Vol. 5 Complex Environment		Springer/354 pages ISBN: 978-3-642-31426-1
		Vol. 6 Risk Assessment, Management and Mitigation		Springer/789 pages ISBN: 978-3-642-31318-9
		Vol. 7 Social and Economic Impact and Policies		Springer/333 pages ISBN: 978-3-642-31312-7
WLF3 2014	Beijing, China 531 from 45 countries/regions/UNs	Landslide Science for a Safer Geoenvironment Vol. 1 The International Programme on Landslides (IPL)	Kyoji Sassa, Paolo Canuti, Yueping Yin	Springer/493 pages ISBN: 978-3-319-04998-4
		Vol. 2 Methods of Landslide Studies		Springer/851 pages ISBN: 978-3-319-05049-2
		Vol. 3 Targeted Landslides		Springer/717 pages ISBN: 978-3-319-04995-3
WLF4 2017	Ljubljana, Slovenia 588 from 59 countries/regions/UNs	Advancing Culture of Living with Landslides Vol. 1 ISDR-ICL Sendai Partnerships 2015-2025	Kyoji Sassa, Matjaž Mikoš, Yueping Yin	Springer/585 pages ISBN: 978-319-53500-5

(continued)

WLF	Place/participants	Title	Editors	Publisher/pages
		Vol. 2 Advances in Landslide Science	Matjaž Mikoš, Binod Tiwari, Yueping Yin, Kyoji Sassa	Springer/1197 pages ISBN: 978-319-53497-8
		Vol. 3 Advances in Landslide Technology	Matjaž Mikoš, Željko Arbanas, Yueping Yin, Kyoji Sassa	Springer/621 pages ISBN: 978-3-319-53486-2
		Vol. 4 Diversity of Landslide Forms	Matjaž Mikoš, Nicola Casagli, Yueping Yin, Kyoji Sassa	Springer/707 pages ISBN: 978-3-319-53484-8
		Vol. 5 Landslides in Different Environments	Matjaž Mikoš,Vít Vilímek,Yueping Yin, Kyoji Sassa	Springer/557 pages ISBN: 978-3-319-53482-4
WLF5	2020 (publication) 2021 (Forum)	Understanding and Reducing Landslide Disaster Risk Vol. 1 Sendai Landslide Partnerships and Kyoto Landslide Commitment	Kyoji Sassa, Matjaž Mikoš, Shinji Sassa, Peter T. Bobrowsky, Kaoru Takara, Khang Dang	Springer In Process
		Vol. 2 From mapping to hazard and risk zonation	Fausto Guzzetti, Snježana Mihalić Arbanas, Paola Reichenbach, Kyoji Sassa, Peter T. Bobrowsky, Kaoru Takara	
		Vol. 3 Monitoring and early Warning	Nicola Casagli, Veronica Tofani, Kyoji Sassa, Peter T. Bobrowsky, Kaoru Takara	
		Vol. 4 Testing, modelling and risk assessment	Binod Tiwari, Kyoji Sassa, Peter T. Bobrowsky, Kaoru Takara	
		Vol. 5 Catastrophic landslides and Frontier of Landslide Science	Vit Vilimek, Fawu Wang, Alexander Strom, Kyoji Sassa, Peter T. Bobrowsky, Kaoru Takara	
		Vol. 6 Specific topics in landslide science and applications	Željko Arbanas, Peter T. Bobrowsky, Kazuo Konagai, Kyoji Sassa, Kaoru Takara	

Preface II

Landslides represent a major cause of loss of life, injuries, property damage, socio-economic disruption and environmental degradation. Froude and Petley (2018) determined that, in the period from January 2004 to December 2016, 55.997 people were killed in 4.862 landslide events, with Asia representing the dominant geographical area in terms of distribution.

A contribution to the implementation of effective disaster risk reduction strategies is provided by landslide monitoring and early warning, which are tools to forecast the potential occurrence of disasters.

Landslide monitoring means comparing landslide conditions (e.g. areal extent, rate of movement, surface topography or soil moisture) over time, in order to assess a landslide's activity. The measurement of the superficial displacement of a slope often represents the most effective method for defining its behaviour, allowing for the observation of the response to triggering factors and for the assessment of the effectiveness of the mitigation measures (Tofani et al., 2013).

The retrieval over time of superficial ground displacements can be based on conventional geotechnical techniques, topographical techniques and remote sensing. In particular, remote sensing images represent a powerful tool to measure landslide displacement, as they offer a synoptic view that can be repeated at different time intervals and can be available at various scales.

Earth Observation (EO) techniques are very effective for landslide detection, mapping, monitoring and hazard assessment. Applications are originating from nearly all types of sensors available today (Tofani et al., 2013). Rapid developments in this field are fostered by the very high spatial resolution obtained by optical systems (currently in the order of tens of centimetres) and by the launching of SAR sensors, purposely built for interferometric applications with revisiting times of few days, such as TerraSAR-X, COSMO-SkyMed and Sentinel-1.

Synthetic Aperture Radar (SAR) techniques have been demonstrated to be highly valuable in measuring land motion. Unlike conventional geodetic monitoring systems, SAR-based applications permit the measurement of surface deformation over vast areas with millimetre to centimetre accuracy and at a frequency varying between 1 month to several days with the earliest satellites. In 2014, the launch of the Sentinel-1 mission provided a new opportunity for InSAR (Interferometric SAR) monitoring applications due to increased acquisition frequency and the regularity of acquisitions (Raspini et al., 2019).

Advanced terrestrial remote sensing technologies, such as Ground-Based SAR (GB-InSAR), Terrestrial Laser Scanner (TLS), Infrared Thermography (IRT) and Digital Photogrammetry (DP) are nowadays applied in the field of slope instability detection, mapping and monitoring, for short-/long-term landslide management (real-time, near real-time and deferred time) (Casagli et al., 2017). They are characterized by operational efficiency and accuracy of data not reached by traditional methods: high-resolution acquisition, multifunction versatility, device portability, low-cost sensors, easy and fast data processing. Such equipment allows for systematic and easily updatable acquisitions of data that may also enhance the implementation of effective early warning systems (EWSs)at slope scale (Casagli et al. 2017).

According to United Nations International Strategy for Disaster Reduction (UNISDR, 2009), an EWS is defined as "the set of capacities needed to generate and disseminate timely and meaningful warning information to enable individuals, communities and organizations threatened by a hazard to prepare and to act appropriately and in sufficient time to reduce the possibility of harm or loss". This definition encompasses both technical-scientific and social-economic elements, that altogether mark the important, and often neglected, difference between a mere monitoring system and an EWS.

The definition above also includes the concept that EWSs must be people-centred, that is the objective must be to empower individuals and communities threatened by hazards to act and reduce the possibility of loss (UNISDR, 2006), meaning that the population is not a passive recipient of authorities' instructions but is an active participant of the civil protection system.

Two categories of EWSs can be defined on the basis of their scale of analysis: local systems for single slopes and regional systems. Local-scale landslide EWSs are based on reliable continuous monitoring of relevant indicators (e.g. displacements, rainfall, groundwater level) that are assumed to be precursory of landslides triggering or reactivations and that can be different depending on landslide typology. There are two main approaches for forecasting aimed at early warning (Intrieri et al., 2019): one makes use of thresholds, i.e. quantitative signals whose exceeding suggests a probable failure but no time frame for such an occurrence can be provided; the other approach employs empirical or semi-empirical forecasting methods, typically based on creep models, that implement equations and/or graphical solutions to obtain the time of failure.

In a regional-scale EWS two approaches are mainly used: empirical approaches based on rainfall thresholds and physically based deterministic models. While the first is widely used for operational applications at the regional scale, the latter are more commonly applied in smaller areas, since the main constraint to their application at the regional scale is the poor knowledge of the spatial organization of the hydrological and geotechnical parameters due to the extreme heterogeneity and inherent variability of the soil when considered at the large scale (Tofani et al., 2016).

The importance of EWSs for Disaster Risk Reduction (DRR) is highlighted in several risk reduction policies proposed by international organizations such as the Sendai Framework for Disaster Risk Reduction 2015–2030 that was adopted at the Third UN World Conference on Disaster Risk Reduction in Sendai, Japan, on 18 March 2015.

Also the ISDR-ICL Sendai Partnership 2015–2025 (Sassa, 2015), that proposes tools for implementing and monitoring the post-2015 framework for disaster risk reduction and the sustainable development goals, agree on the development of fields of cooperation in research and capacity building focused on the development of people-centred early warning technology for landslides with increased precision and reliable prediction both in time and location, especially in a changing climate context. Furthermore, the 2020 Kyoto Commitment (KC2020) for global promotion of understanding and reducing landslide disaster risk that is duty to the Sendai Landslide Partnerships 2015–2025, the Sendai Framework for Disaster Risk Reduction 2015–2030, the 2030 Agenda Sustainable Development Goals, the New Urban Agenda and the Paris Climate Agreement (Sassa, 2019), clearly highlights the promotion of people-centred early warning technologies for landslides.

This volume compiles the results of the studies on landslide monitoring and early warning conducted all over the world. The volume collects one Theme lecture, one Keynote paper and 37 research papers from 17 countries (Australia, China, Czech republic, France, Germany, Greece, India, Indonesia, Italy, Japan, New Zealand, Norway, Switzerland, Taiwan, United Kingdom, Uzbekistan, Vietnam).

The volume starts with the Theme lecture entitled "Monitoring and Early Warning Systems: Applications and Perspectives" which illustrates and discusses several case studies in which different risk mitigation strategies based on EWSs are adopted. Each of the cases presented shows a peculiarity that can help in the definition of a modern and reliable landslide EWS, while the recent and upcoming technological and scientific advancements are the premise of even more accurate and meaningful landslide EWSs.

The volume is then divided into two parts: Part I: Monitoring and Remote Sensing for Landslide Risk Mitigation which collects 18 papers and Part II: Landslide Early Warning Systems, Forecasting Models and Time Prediction of Landslides which includes 20 papers.

Part I includes the Keynote lecture entitled "Defining Kinematic and Evolutive Features of Earth Flows Using Integrated Monitoring and Low-Cost Sensors" in which an integrated monitoring system based on traditional monitoring techniques and specifically developed low-cost sensors has allowed for the reconstruction of the relationship among basal-slip surface geometry, deformation styles and pattern, geomorphic structures, movement velocity and sediment discharge during ordinary and extraordinary movements for three earthflows in southern Italy. Part I includes also manuscripts on the application of different monitoring techniques for landslide investigation and risk management, comprising conventional geotechnical techniques, remote sensing images with special reference to PSI techniques and geophysical techniques.

Part II contains papers concerning the development and application of EWSs, the forecasting of landslides at the regional scale and papers on the time prediction of the landslide occurrence. The majority of the papers of this part focuses on the development and application of regional-scale EWSs based on the use of rainfall thresholds, some papers focus on the application of innovative monitoring techniques for the set-up of local EWSs while few others propose innovative methods for time prediction of landslides at the regional and local scale.

Florence, Italy

Nicola Casagli
Veronica Tofani

References

Casagli N, Frodella W, Morelli S, Tofani V, Ciampalini A, Intrieri E, Raspini F, Rossi G, Tanteri L, Lu P (2017) Spaceborne, UAV and ground-based remote sensing techniques for landslide mapping, monitoring and early warning. Geoenvironmental Disasters. 4(1), 9

Froude M J, Petley D (2018) Global fatal landslide occurrence from 2004 to 2016. Natural Hazards and Earth System Sciences. 18: 2161–2181

Intrieri E, Carlà T, Gigli G (2019) Forecasting the time of failure of landslides at slope-scale: A literature review. Earth-science reviews. 193: 333–349

Raspini F, Bianchini S, Ciampalini A, Del Soldato M, Montalti R, Solari L, Tofani V, Casagli N (2019) Persistent Scatterers continuous streaming for landslide monitoring and mapping: the case of the Tuscany region (Italy). Landslides. 16(10): 2033–2044

Sassa K (2015) ISDR-ICL Sendai Partnerships 2015–2025 for global promotion of understanding and reducing landslide disaster risk. Landslides, 12 (4), 631–640.

Sassa K (2019) The Fifth World Landslide Forum and the final draft of the Kyoto 2020 Commitment. Landslides, 16(2), 201–211.

Tofani V, Raspini F, Catani F, Casagli N (2013) Persistent Scatterer Interferometry (PSI) technique for landslide characterization and monitoring. Remote Sensing. 5(3): 1045–1065.

Tofani V, Bicocchi G, Rossi G, Segoni S, D'Ambrosio M, Casagli N, Catani F (2017) Soil characterization for shallow landslides modeling: a case study in the Northern Apennines (Central Italy). Landslides. 14: 755–770.

UNISDR (2006) Developing an early warning system: a checklist. The Third International Conference on Early Warning (EWC III). Available at: https://www.unisdr.org/2006/ppew/info-resources/ewc3/checklist/English.pdf, Last access: March 2016, 2006.

UNISDR (2009) Terminology on Disaster Risk Reduction. Available at: https://www.undrr.org/publication/2009-unisdr-terminology-disaster-risk-reduction. Last access: March 2020

Contents

Monitoring and Early Warning Systems: Applications and Perspectives

2020 Kyoto Japan

Nicola Casagli, Emanuele Intrieri, Tommaso Carlà, Federico Di Traglia, William Frodella, Giovanni Gigli, Luca Lombardi, Massimiliano Nocentini, Federico Raspini, and Veronica Tofani

Abstract

One of the most efficient and cost-effective tools for landslide risk mitigation is often the setup of an early warning system (EWS). Even if the latter encompass both technical-scientific and social-economic topics, the focus of this note is on the monitoring and forecasting components of a slope-scale landslide EWS. In this framework, a landslide monitoring system is required to provide reliable and continuously updated data for quantitatively catching the scenario evolution, thus allowing for correct forecasting analyses and prompt actions for risk mitigation. Landslide monitoring systems based on remote sensing techniques represent efficient and robust tools for risk mitigation, allowing for a low environmental and economic impact and high operator safety in difficult environments. Among these techniques, radar interferometry is one of the most widely adopted and reliable methods, whose advantages include very high accuracy, operation during all weather conditions, and high spatial and temporal coverage. Radar interferometry output data, due to their high accuracy and acquisition frequency (which is getting higher and higher for satellite applications too), perfectly fit in the prediction activity, enabling very often to make accurate and prompt time of failure or scenario evolution forecasts. In this note, a number of case studies are presented, describing the employed monitoring systems and the associated techniques adopted for risk mitigation. In particular, an integrated EWS for rockslide risk mitigation, a landslide EWS in a volcanic environment, a landslide failure prediction using satellite InSAR and a rockfall monitoring and associated time of failure prediction are presented. Each of the cases presented shows a peculiarity that can help in the definition of the characteristics and potential of a modern and reliable landslide EWS, while the recent and upcoming technological and scientific advancements are the premise of even more accurate and meaningful landslide EWSs.

Keywords

Risk management • Disaster risk reduction • Remote sensing • Forecasting • Landslides • Interferometry

Introduction

Among natural hazards, landslides play a relevant role in terms of life loss. Varnes and IAEG Commission on Landslides (1984) found out that ca. 14% of total casualties from 1971 to 1975 were due to slope failure, although accurate estimations are difficult since landslides are frequently considered secondary effects and deaths are often attributed to the primary trigger (storms, earthquakes) (Petley 2012). Focusing on more recent years, Froude and Petley (2018) determined that, in the period from January 2004 to December 2016, 55.997 people were killed in 4.862 landslide events, with Asia representing the dominant geographical area in terms of distribution. On the other hand, the global annual cost related to landslides is about $19.8 billion, which is about 17% of the average annual global natural disaster losses (Haque et al. 2016).

These numbers highlight the importance of continuous work from both researchers and practitioners in the field of landslide risk mitigation. Among the many possible approaches, one that is generally considered cost-effective and complementary is the setup of early warning systems (EWS) (Intrieri et al. 2013).

According to United Nations International Strategy for Disaster Reduction (UNISDR 2009), an EWS is defined as "the set of capacities needed to generate and disseminate

N. Casagli · E. Intrieri (✉) · T. Carlà · F. Di Traglia · W. Frodella · G. Gigli · L. Lombardi · M. Nocentini · F. Raspini · V. Tofani
Department of Earth Science, Università Degli Studi Di Firenze, Via G. La Pira 4, 50121 Florence, Italy
e-mail: emanuele.intrieri@unifi.it

© Springer Nature Switzerland AG 2021
N. Casagli et al. (eds.), *Understanding and Reducing Landslide Disaster Risk*, ICL Contribution to Landslide Disaster Risk Reduction, https://doi.org/10.1007/978-3-030-60311-3_1

timely and meaningful warning information to enable individuals, communities and organizations threatened by a hazard to prepare and to act appropriately and in sufficient time to reduce the possibility of harm or loss".

This definition encompasses both technical-scientific and social-economic elements, that altogether mark the important, and often neglected, difference between a mere monitoring system and an EWS. These elements can be resumed as:

- knowledge of the risks;
- monitoring and analysis of the hazards;
- forecasting;
- communication or dissemination of alerts and warnings;
- local capabilities to respond to the warnings received (which includes the education of the population in "normal times").

These and other components of an EWS have been further detailed by Intrieri et al. (2012) and Fathani et al. (2016).

The expression "end-to-end warning system" is also used to emphasize that warning systems need to span all steps from hazard detection through to community response (UNISDR 2009).

The definition above also includes the concept that EWSs must be people-centred, that is the objective must be to empower individuals and communities threatened by hazards to act and reduce the possibility of loss (UNISDR 2006), meaning that the population is not a passive recipient of authorities' instructions but is an active participant of the civil protection system.

The importance of EWSs for Disaster Risk Reduction (DRR) is highlighted in several Risk reduction policies proposed by international organizations such as the Sendai Framework for Disaster Risk Reduction 2015–2030 that was adopted at the Third UN World Conference on Disaster Risk Reduction in Sendai, Japan, on March 18, 2015.

The Sendai Framework for Disaster Risk Reduction 2015–2030 outlines seven clear targets and four priorities for action to prevent new and reduce existing disaster risks: (i) Understanding disaster risk; (ii) Strengthening disaster risk governance to manage disaster risk; (iii) Investing in disaster reduction for resilience and; (iv) Enhancing disaster preparedness for effective response, and to "Build Back Better" in recovery, rehabilitation and reconstruction.

It aims to achieve the substantial reduction of disaster risk and losses in lives, livelihoods and health and in the economic, physical, social, cultural and environmental assets of persons, businesses, communities and countries over the next 15 years.

Also the ISDR-ICL Sendai Partnership 2015–2025 (Sassa, 2015) that proposes Tools for implementing and Monitoring the Post-2015 Framework for Disaster risk Reduction and the Sustainable Development Goals agree on the development of fields of cooperation in research and capacity building focused on the development of people-centered early warning technology for landslides with increased precision and reliable prediction both in time and location, especially in a changing climate context.

The 2020 Kyoto Commitment (KC2020) for global promotion of understanding and reducing landslide disaster risk is a duty to the Sendai Landslide Partnerships 2015–2025, the Sendai Framework for Disaster Risk Reduction 2015–2030, the 2030 Agenda Sustainable Development Goals, the New Urban Agenda and the Paris Climate Agreement (Sassa 2019). The KC2020 clearly highlights as Action 1 among all the priority actions the promotion of people-centered early warning technologies for landslides.

The focus of this paper is on the monitoring and forecasting components of a slope-scale landslide EWS, which will be discussed through the description of some real cases of early warning experiences, each one peculiar in its own right. However, by no means it is intended that the social and cultural side of EWSs are secondary. On the contrary, activities such as communication campaigns, information about the risk and the EWS, education about the safe behaviours, drills, questionnaires to verify the knowledge, the risk perception and gather feedback are often underrated but necessary and relatively not expensive activities, as the Hong Kong experience clearly shows (Mak et al. 2007; Kong et al. 2020); otherwise, if the operational systems do not meet the expectations of the end-users, they might not trust the system (Thiebes and Glade 2016). In fact, monitoring and forecasting are but two steps in a longer process that starts from engaging the community in "normal times" and finishes with communicating an easily understandable warning using the proper media (Intrieri et al. 2020).

As far as the technical aspect are concerned, the first stage is probably defining the most representative parameters for assessing the stability of the slope (e.g. rainfall, water level, pore pressures, vibrations, surficial movements, displacements at some depth and so on). Generally, regional-scale EWSs rely on rainfall (Guzzetti et al. 2020). At slope-scale, which is the focus of this paper, the choice of the indicators depends on the type of landslide. For debris flows, pendulums or trip-wires detecting the passage of the material can be used (Bossi et al. 2015), while for creep landslides the monitoring of displacement and its derivatives is usually preferred because they are direct indicators of the stability of a slope (Intrieri and Gigli 2016; Pecoraro et al. 2019).

What is invariable is that the monitoring frequency must keep up with the presumed activation of the landslide (which means that near real-time monitoring is not always necessary, in some cases daily measurements or less can be enough, as shown in Gigli et al. 2011). Furthermore, robustness should be preferred over accuracy. In fact, pre-failure accelerations follow power-law trends (usually with a high exponent, which makes them very similar to an exponential law; Intrieri et al. 2019) which typically cause very high signal-to-noise ratios. On the other hand, communication failures or damages to the instrumentation (maybe caused by the landslide's movements) can compromise the whole system.

Once the key indicators are identified, the appropriate technology for monitoring must be selected.

Then monitoring data must then be elaborated so that they can provide input for the prediction activity. There are two main approaches for forecasting aimed at early warning (Intrieri et al. 2019): one makes use of thresholds, i.e. quantitative signals whose exceeding suggests a probable failure but no time frame for such an occurrence can be provided; the other approach employs empirical or semi-empirical forecasting methods, typically based on creep models, that implement equations and/or graphical solutions to obtain the time of failure.

Concerning the first approach, the major problem is the specifying thresholds that are both reliable and effective. Usually this is done based on past (personal) experience (the so called "expert judgment"), past trends in the measurements and the knowledge of the rupture mechanism(s), rather than in objective data (Nadim and Intrieri 2011). The scientific literature has many cases where expert judgement is implemented for thresholds definition (Blikra 2012; Iovine et al. 2006; Lombardi et al. 2017; Zavodni and Broadbent 1978). This is because, in most cases, even for reactivated landslides, there is no measured record of the indicators preceding failure.

Thresholds can also be defined using more general, not site-specific methods, with a certain degree of exportability, such as those proposed by Brox and Newcomen (2003), Carlà et al. (2017c), Crosta and Agliardi (2003), Xu et al. (2011). A more detailed dissertation on this topic can be found on Intrieri et al. (2019).

Whether the case, warning thresholds should be used carefully, and there is no justification to define too many warning levels or too precise limits. For a slope scale EWS, thresholds derived from kinematic indicators (displacement, velocity, acceleration) and pore pressure/water table measurements are preferable to rainfall intensity since they are more directly correlated with the stability conditions of the slope. Rainfall thresholds could be implemented as well, but mainly as a support to decision makers or as simple pre-warnings. In any case, the design of the early warning system must be flexible such that the threshold parameters can be changed as more information is gathered on the performance of the monitoring system and on the behaviour of the slope or the region being monitored (Nadim and Intrieri 2011).

The second approach is based on forecasting methods such as those invented by Azimi et al. (1989), Fukuzono (1985), Hao et al. (2016), Mufundirwa et al. (2010), Saito (1969), Voight (1989). Among these, the most used is by far Fukuzono's (1985) method, mostly because of its intuitive graphical solution (Atzeni et al. 2015; Carlà et al. 2017a, b; Dick et al. 2015; Intrieri et al. 2019; Kothari and Momayez 2018; Rose and Hungr 2007); its rationale is that, if the displacement velocity of a landslide during the tertiary creep stage increases (theoretically toward infinite), its inverse number will tend to decrease to zero. Therefore, extrapolating the inverse velocity until when it intercepts the time axis will give the time of failure. A common simplification when applying this method and that can lead to mistakes is that the interpolation is generally assumed to be linear.

More details on the forecasting methods and their applications have been discussed by Intrieri et al. (2019).

Of the four case histories presented in this paper, the first two (Quincinetto landslide and Stromboli volcano) use the first approach and thresholds defined through expert judgement. The last two (Xinmo and Gallivaggio landslides) employed the second approach; in particular, Fukuzono's method (1985) has been used with hindsight for Xinmo landslide while for Gallivaggio landslide it ensured the clearance of the area and the lack of injuries or victims.

Landslide Monitoring

The aim of this section is to describe the landslide monitoring systems and the adopted techniques of the following presented case studies.

The purpose of a landslide monitoring system is to provide reliable data for catching the characteristic information of collapses in time, and make a correct analysis, evaluation, prediction and control of the kinematics and evolutionary process (Pinggen 2004). The current availability of advanced remote sensing technologies in the field of landslide analysis allows for rapid and easily updatable data acquisitions, improving the monitoring traditional capabilities (Guzzetti et al. 2012). Landslide monitoring systems based on remote sensing techniques represent cost-effective means to reduce the risk, allowing for a low environmental and economic impact (Intrieri et al. 2012, 2013). Amongst the remote sensing techniques adopted in an integrated real-time monitoring and warning system, radar interferometry represents one of the most widely adopted and reliable method for the remote monitoring of landslide displacements (Casagli et al.

2017). In this framework ground displacements and slope deformations represent the relevant parameter for the analysis of slope instability, especially in the field of failure prediction (Carlà et al. 2016, 2019a), emergency management (Lombardi et al. 2017), long-term stabilization works (Ferrigno et al. 2017) and residual reactivations (Confuorto et al. 2017; Frodella et al. 2018). In particular, the ability to make numerous point measurements of displacement over the landslide body allows for the detection and mapping of the actively deforming slopes (Bardi et al. 2014; Bianchini et al. 2015a, b; Ciampalini et al. 2014, 2015; Nolesini et al. 2016; Raspini et al. 2015; Intrieri et al. 2019), the characterization and monitoring of landslide mechanisms (Tofani et al. 2013; Solari et al. 2018) and, through the analysis of time series of deformation, the identification of velocity changes in the landslide evolution (Berti et al. 2013; Del Soldato et al. 2018), as well as the modeling of large slope instability (Berardino et al. 2003).

Rapid developments in this field are fostered by the very high spatial resolution obtained by the launching of new Synthetic Aperture Radar (SAR) satellite sensors, purposely built for interferometric applications with revisiting times of few days, such as Sentinel-1 (Raspini et al. 2018, 2019; Del Soldato et al. 2019; Solari et al. 2019).

The family of SAR satellite sensors orbits the Earth at an altitude ranging from 500 to 800 km, following sun-synchronous, near-polar orbits, slightly inclined with respect of Earth meridians. The most commonly used bands in SAR applications are C-band (5–6 GHz, ~5.6 cm wavelength), X-band (8–12 GHz, ~3.1 cm wavelength) and L-band (1–2 GHz ~ 23 cm wavelength) with a temporal resolution depending on the satellite revisiting time.

A SAR image is composed of pixels characterized by amplitude and phase values. Phase values of a single SAR image partly depend on the sensor-target distance and are the key element to detect ground displacement. SAR interferometry is the technique focused on measuring changes of signal phase over time through the analysis of at least two SAR images (Fruneau et al. 1996).

A suitable approach to exploit phase variation between two consecutive radar images acquired over the same target is the Differential Interferometric SAR (D-InSAR) (Bamler and Hartl 1998). Geometrical and temporal decorrelation and atmospheric effects caused by the variation of the phase reflectivity value of some radar targets reduce the reliability of the D-InSAR technique (Berardino et al. 2003). In order to overcome these limitations InSAR-based information can be enhanced through multi-temporal interferometric techniques (MT-InSAR), based on analysis of long stacks of coregistered SAR imagery (Ferretti et al. 2001, 2011). An extensive review of the several developed MT-InSAR approaches can be found in Crosetto et al. (2016) and Casagli et al. (2017). MT-InSAR analysis is designed to generate time series of ground deformation for individual measurement points, assuming different types of deformation models (e.g., linear, nonlinear or hybrid). Signal analysis of a network of coherent radar targets (Permanent Scatterers, PS) enables to estimate the occurred displacement, acquisitions by acquisition.

Line Of Sight (LOS) deformation rate can be estimated with an accuracy theoretically better than 0.1 mm/year. Each measurement is referred temporally and spatially to a unique reference image and to a stable reference point. In the field of landslide investigations the potential of SAR data has been exploited at different scales: from national (Adam et al. 2011) to regional (Meisina et al. 2013, 2016; Ciampalini et al. 2016a, b) basin (Lu et al. 2012) slope and building scale (Ciampalini et al. 2014; Bianchini et al. 2015a, b; Nolesini et al. 2016), as well as in different phases of landslide response (Canuti et al. 2007) and public safety (Farina et al. 2008).

The terrestrial application of this technique is carried out through ground-based radar interferometric systems (GB-InSAR) and is applied in the field of the monitoring of slope deformations in the framework short- and long-term landslide monitoring (Bardi et al. 2017; Carlà et al. 2019b; Frodella et al. 2016, 2017; Intrieri et al. 2015; Pratesi et al. 2015). These systems are characterized by operational efficiency and accuracy of data not reached by traditional methods: high-resolution acquisition, multifunction versatility, fast acquisition and data processing (Casagli et al. 2017, 2018).

GB-InSAR systems consist of a computer-controlled microwave transceiver, characterized by a transmitting and receiving antennas, which by moving along a mechanical linear rail is capable to synthesize a linear aperture along the azimuth direction (Tarchi et al. 1997; Pieraccini et al. 2002). The obtained SAR image contains amplitude and phase information of the observed scenario backscattered echo in the acquiring time interval (than 1 min with the most modern systems) (Luzi et al. 2004, 2010; Monserrat et al. 2014).

The displacement obtained from the phase difference calculation can be represented in 2D maps, in which the chromatic scale covers a total value corresponding to half of the wavelength used. However, since the phase is periodic, it cyclically assumes the same values crating image interpreting problems. This issue, known as phase ambiguity, can be solved through interpretation based on field geological knowledge or by adopting apposite phase unwrapping algorithms (Ghiglia and Romero 1994), which count the number of cycles performed by the wave obtaining cumulated displacement maps. Given the relative short distances at which GB-InSAR apparatuses usually operate (typically less than 3 km), they work in Ku band (1.67–2.5 cm).

In landslide studies the main research applications of GB-InSAR soon became focused on slope monitoring

(Tarchi et al. 2003; Pieraccini et al. 2002, 2003), civil protection purposes (Del Ventisette et al. 2011; Bardi et al. 2014, 2017; Lombardi et al. 2017; Frodella et al. 2016, 2017, 2018) and more recently for mining safety (Carlà et al. 2018).

Wireless Sensor Networks (WSNs) are another modern monitoring tool that are attracting an increasing interest due to their potentials for applicability in hazardous and inaccessible scenarios such as active landslides (Rosi et al. 2011). After deployment in the landslide environment, wireless sensors create a network by inter-connecting to each other; this fulfils an important need for real-time monitoring, especially in hazardous or remote conditions (Mucchi et al. 2018). WSN technology has the capability to quickly capture, process, and transmit data in real-time (Intrieri et al. 2018a). The WSNs in literature (Hill and Sippel 2002; Terzis et al. 2006; Fernández-Steeger et al. 2009; Ramesh et al. 2009; Rosi et al. 2011) mainly exploit radiofrequency signals to provide connectivity to the sensor nodes not to measure the distance between nodes.

Integrated Landslide EWS

Quincinetto Landslide

The Quincinetto landslide is a large, bowl-shaped slope instability in hard metamorphic rocks located on a steep slope in the Dora Baltea valley, at the border between the Piedmont and Aosta Valley regions (Northern Italy). There are many unknowns about the exact kinematic and geometrical features of the rockslide, as sub-surface investigations are virtually impossible due to the difficult site accessibility; as of today, the site can be accessed only by walking an arduous mountain trail. The rockslide, whose volume can be very roughly estimated in about 500 000 m^3, appears as a highly chaotic mass of large, heavily disjointed blocks and fragments of eclogitic micaschists, which constitute the main lithology in the area; rock blocks vary from few tens to few thousands of m^3 in size. This “debris field” is distributed throughout the entire rockslide area, and it is split in two parts by a middle scarp, represented by a sub-vertical rock cliff of few tens of meters in height. The main head scarp at the top of the slope is identified by an even higher sub-vertical cliff ($\approx$60 m).

A large talus cone extends from the toe of the rockslide down to the valley bottom, where several giant remnants from past rockfall events are found (Fig. 1). In the 1960s, as the risk related to the Quincinetto landslide had not been identified, the path of the newly constructed A5 highway was traced so as to pass right next to the distal edge of the talus cone. This means that the highway, one of the most important infrastructures in Northern Italy, is critically exposed to future rockfalls and rockslide reactivations. Highway closures would have enormous socio-economic repercussions, as the road is the main thoroughfare leading to the Mont Blanc and Grand Saint Bernard tunnels at the Italy-France and Italy-Switzerland borders, respectively.

The Monitoring Activities

Satellite InSAR monitoring performed by the Sentinel-1 constellation in the period November 2014–March 2020 revealed widespread movements within the rockslide area, with velocities in the order of 10–30 mm/year (Fig. 2; the section of the A5 highway at risk can be seen on the upper right edge of the image). Velocities appear to be higher close to the top of the talus cone (areas 1 and 3), and slightly lower in the area between the top and middle scarps (area 2).

The persistent deformation activity is a predisposing factor for the destabilization and detachment of the blocks at the toe of the rockslide, some of which exceed 1000 m^3 in size. Due to their size and geometry, these massive blocks lie in apparent precarious equilibrium (Fig. 3). Figure 2 also indicates that no measuring points are available between areas 1 and 3, behind the apex of the talus cone. This is arguably due to the toe of the rockslide being locally affected by velocities that are too high for being captured by means of the InSAR technique.

In November 2018, a GBInSAR campaign was thus started in order to improve the level of understanding with regards to the movements of the frontal blocks; the instrument was installed in the valley bottom, at a distance of about 1 km from the rockslide. The installed GB-InSAR system operates in the Ku band (frequency of about 17 GHz) and moves on a 4 m long track to create the so-called

Fig. 1 Rockfall remnant at the toe of the talus cone

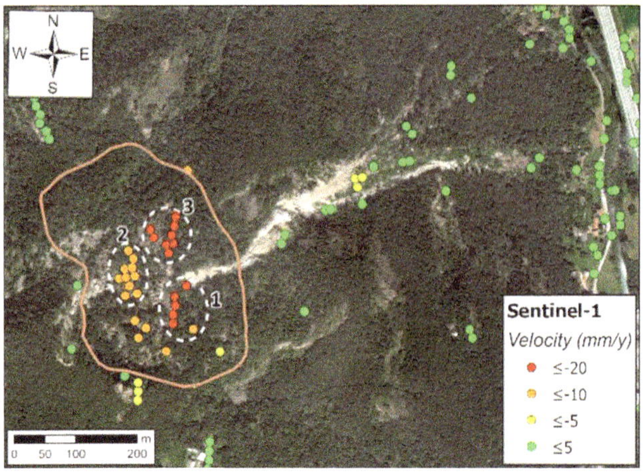

Fig. 2 Satellite InSAR data collected by the Sentinel-1 constellation in the period from October 2014 to March 2020

Fig. 3 Frontal image of the Quincinetto landslide, with detail of the head scarp, middle scarp, and frontal unstable blocks

synthetic aperture, in order to improve image resolution and at the same time maintain a compact and portable size. Figure 4 shows the resulting cumulative displacements measured between the start of the monitoring campaign and March 2020: while poor coverage is obtained over a large part of the debris field, excellent coverage is obtained over the two sub-vertical scarps and over the frontal blocks at the toe of the rockslide, areas that are not visible with the satellite InSAR monitoring. The frontal blocks indeed appear to represent by far the most unstable sector, with cumulative displacements exceeding 20 cm over a 15-month period. This observation prompted the decision to complement the GBInSAR monitoring with a WSN of extensometers and tiltmeters installed over the two largest blocks lying in close proximity to the apex of the talus cone. While no significant tilting was detected, the trends of displacement measured by the extensometers were in full agreement with the GBInSAR data. Displacement rates were observed not to be constant over time, but rather to be sensitive to pulses of acceleration; Fig. 5 details the displacements of one of the unstable blocks near the apex of the talus cone, as measured by the GBInSAR and by an extensometer in the last three months of 2019. These are also compared with the values of 7-day cumulative rainfall for the same period. It appears that the deformation activity is significantly controlled by rainfall, which dictates the slow-to-fast transition of the displacement rates. Then, these enter a slow decelerating trend as soon as the phase of heavy precipitation ends.

Results

The results of the monitoring campaigns thus made it possible to gain critical insights into the complex kinematics of the Quincinetto landslide and the type of risk potentially affecting the exposed section of the A5 highway on the valley bottom. In particular, satellite InSAR data suggested that there are currently no signs of impending paroxysmal reactivations of the entire rockslide mass. The major source of risk is rather associated with the presence of few very large blocks lying in precarious equilibrium at the forefront of the slope instability. These appear to be moving according to a simple sliding mechanism triggered by heavy rainfall, hence suggesting a high degree of kinematic freedom as well as a poor amount of cohesion with the bedrock. Continued movement could eventually lead to toppling of the blocks over the slope change where the apex of the talus cone is located, with subsequent rapid fall and rolling towards the toe of the slope. All these observations allow for the implementation of a robust early warning strategy, under which a series of displacement and/or rainfall alert

Fig. 4 Map of cumulative displacements measured by the GBInSAR in the period from November 2018 to March 2020

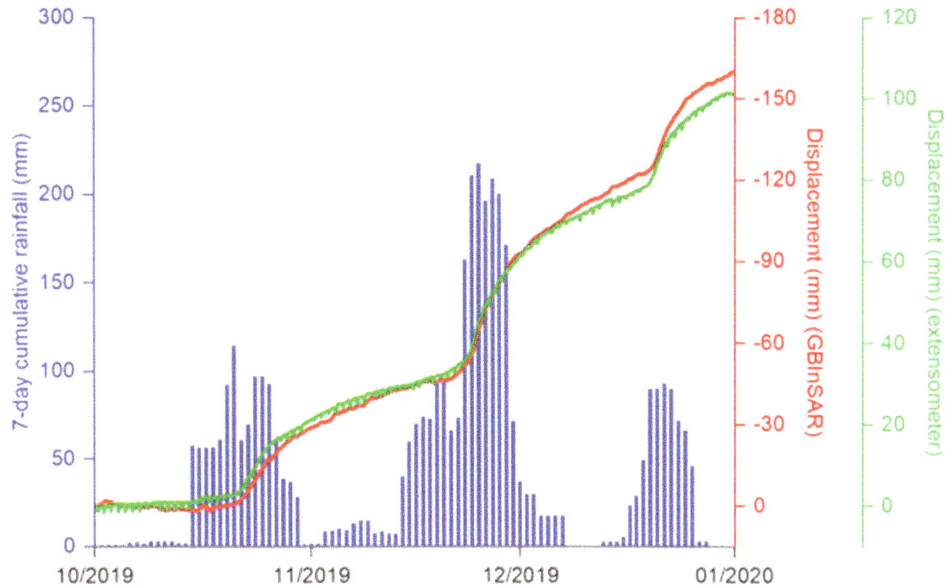

Fig. 5 Displacement time series measured by the GBInSAR and by one of the extensometers in the last three months of 2019

thresholds can be related to a set of predetermined response actions regulating the management of the traffic on the highway and the possible enforcement of temporary road closures. In particular, the EWS was established under a 3 level-scale, with the separation from one level to the other being defined by a predetermined velocity threshold measured by the GBInSAR. The three warning levels were associated with a state of:

1. ordinary alert: no significant movements;
2. moderate alert: beginning of an acceleration phase;
3. high alert: the acceleration phase could potentially lead to the failure of one of the frontal unstable blocks.

Under ordinary alert, traffic on the highway is regularly open. If the first warning threshold is surpassed, a partial road closure is undertaken; this allows technical personnel to prepare for a possible complete closure of the road section at risk, while local officials are also informed. At this stage, the GBInSAR data are also cross-checked with the extensometer and tiltmeter data, in order to verify the accuracy of the displacement measurements and exclude the presence of instrumental error of any kind. If a high alert stage is then entered, a complete closure of the section of highway at risk is enforced; at the same time, selected detour routes and informative panels for the drivers are activated. The procedure thus makes it possible to coordinate actions and countermeasures in a consistent way, and to regulate traffic over the A5 highway based on objective criteria relating the observed displacement rates of the rockslide with the actual level of risk posed to public safety.

Landslide EWS in a Volcanic Environment

Stromboli (Italy) is a volcanic island located in the Tyrrhenian Sea, that has experienced several large mass wasting phenomena, forming two depressions on its NW (Sciara del Fuoco; SdF) and SE flank, showing bilateral flank instability. Nowadays, the activity mainly occurs from a summit crater terrace located at ca. 750 m a.s.l., and from ephemeral vents within the SdF (Di Traglia et al. 2018a). The distinctive persistent Strombolian activity is characterized by intermittent explosions, showing intensity and frequency fluctuations over time. It is often punctuated by lava overflows from the crater terrace (Calvari et al. 2016), and/or by flank eruptions, with the outpouring of lava flows from lateral vents (Di Traglia et al. 2020). The most hazardous phenomena that can hit Stromboli Island are mass flow-induced tsunamis (Fornaciai et al. 2019), which make it one of the tsunami sources in the Mediterranean Sea (Cerase et al. 2019). Mass flows can be of two types:

- intrusion-related landslides from the NW unstable flank of the volcano (Sciara del Fuoco; SdF), as occurred on 30th December 2002 (Tinti et al. 2006);
- by the entry into the sea of pyroclastic density currents (PDCs) produced during paroxysmal, as occurred on 3rd July 2019 and 28th August 2019 (Turchi et al. 2020).

Tsunamis occurred in recent times as in 1879, 1916, 1919, 1930, 1944, and 1954, 2002, accounting for an average of 1 tsunami every 20 years (Maramai et al. 2005), and affecting the coast of Stromboli (Fornaciai et al. 2019), and secondly, the coasts of the other Aeolian islands and the Tyrrhenian coasts of southern Italy (Fornaciai et al. 2019). Deposits of potentially tsunamigenic landslides have been discovered in marine sediments offshore of the island of Stromboli (Di Roberto et al. 2010). Recent findings revealed the occurrence of three tsunamis likely related to repeated flank collapses struck during the Late Middle Ages (Rosi et al. 2019). Present-day volcano slope instability comprises mobilization of coarse-grained and fine-grained sediments directly or indirectly related to the eruptive activity (Di Traglia et al. 2018b). Slope instability phenomena at Stromboli are classified, on the base of their size and movement, into three types (Schaefer et al. 2019): (1) "deep-seated gravitational slope deformations" evolving to "rock or debris avalanches" from the SdF (volumes > 10^6 m^3); (2) "rock (rotational or planar) slides" evolving to "rock avalanches" from the SdF (volumes ≈ 10^6 m^3); 3) "rock falls" or "gravel/debris slides" evolving to " gravel/debris flows" (volumes ≈ 10^5 m^3).

The Stromboli GBInSAR Monitoring System

The NE portion of the summit crater terrace and the northern portion of the SdF are monitored by two GBInSAR devices, located in a stable area N of the SdF (Fig. 6). The first GBInSAR (GBInSAR NE400; Model: GBInSAR LiSALab; Revisiting time: 11 min; Antonello et al. 2004) was installed in February 2003, during the 2002–03 flank eruption, while the second device (GBInSAR NE190; Model: GBInSAR LiSAmobile k09; Revisiting time: 2 min) was installed on 14 December 2014, after the 2014 flank eruption (Di Traglia et al. 2018a).

Negative and positive values of displacement indicate, respectively, a movement towards or away from the sensor. Since the GBInSAR systems are located in a stable area north of the SdF, and its LOS allows us to detect the N-S components of the movements in all direction; negative displacement may represent inflation of the crater area of the volcano or inflation and sliding of the SdF, while positive displacement may represent deflation of the crater area (Casagli et al. 2009). Range and cross-range resolution are on average 2 m × 2 m, with a measurement precision referred to the displacement of less than 1 mm (Casagli et al. 2009). The capability of InSAR to detect ground displacement depends on the persistence of phase coherence over appropriate time intervals (Lu et al. 2002). Loss in coherence mainly depends on chaotic ground movements (Antonello et al. 2004), e.g. grain avalanches. A coherence mask (threshold = 0.8) is set to mask the noisy areas of the interferogram (Luzi et al. 2010). The phase values can be affected by ambiguity (unwrapped phase) but, due to the short elapsed time (11 min) between two subsequent measurements on Stromboli volcano, the interferometric displacements are usually smaller than half wavelength and unwrapping procedures (as described by Ghiglia and Romero 1994) are not necessary.

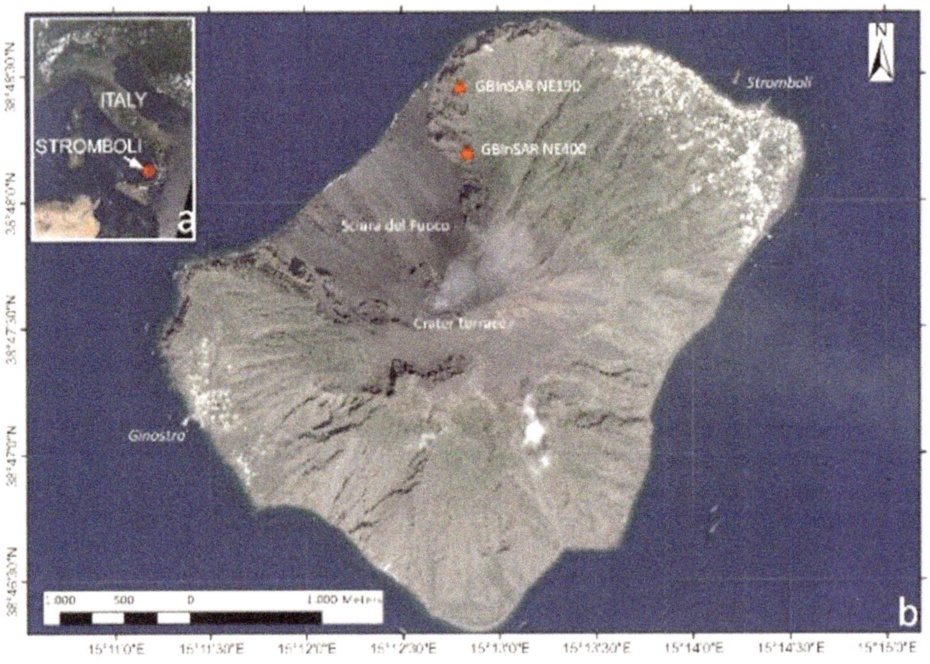

Fig. 6 Geographic location of (a) of the island of Stromboli. (C) PLEAIDES-1 orthorectified panchromatic image (true colour, collected on September 2018) of the island of Stromboli. The locations of Stromboli and Ginostra villages, and the Sciara del Fuoco and the crater terrace are reported. Interferometric analysis allows us to derive the displacement field of the observed portion of the SdF and of the crater area in the elapsed time

Results

The GBInSAR devices usually record (Fig. 7):

- the inflation/deflation of the summit plumbing system (Di Traglia et al. 2018a);
- the gravitational re-adjustment of the lava breccia, sometimes evolving into rockfalls (Schaefer et al. 2019);
- thermal contraction of the lava field, mainly in area of lower pre-effusive slope angle (Schaefer et al. 2019);
- persistent flank motion (Schaefer et al. 2019).

The method proposed by Di Traglia et al. (2014) for early warning considered only the displacement rate as proxy for the warning levels for mitigating the risk connected with the Sciara del Fuoco dynamics and the increase in eruptive activity. The approach was applied during the 2014 flank eruption, allowing for the anticipation of the beginning of the eruption 11 h in advance (Di Traglia et al. 2017). However, the "displacement rate thresholds" approach was not enough to discriminate the intensity of the instability phenomena, given that very high displacement rates can also be measured in very small rock

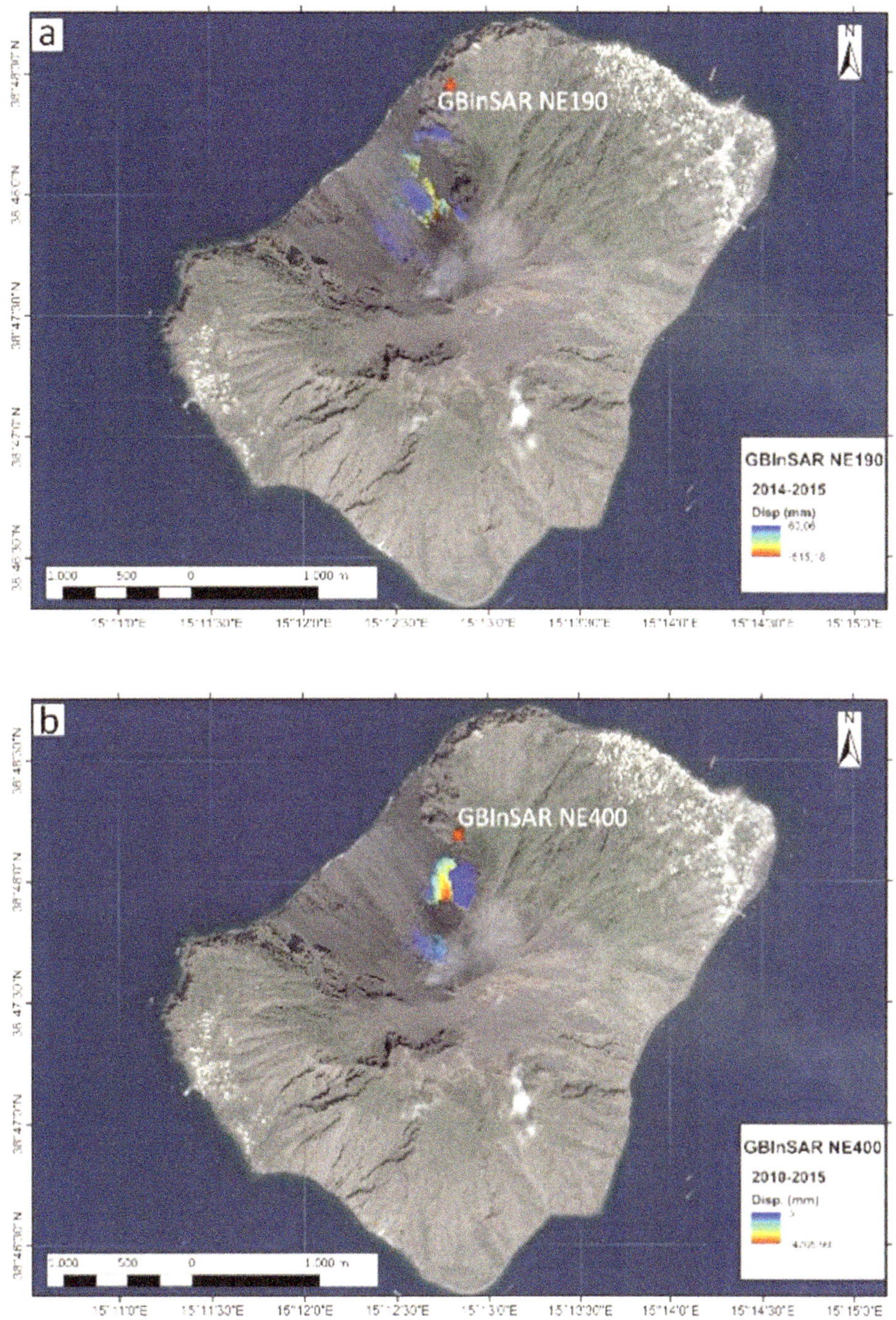

Fig. 7 Displacement maps based on the **a** GBInSAR NE190 (December 2014–December 2015), **b** GBInSAR NE400 (January 2010–December 2015)

volumes. For this reason, a new approach was proposed considering both the displacement rate thresholds and the magnitude, i.e. the volume characterized by anomalous displacement rates (Fig. 7). Based on the integration of GBInSAR displacement rate and different instability scenarios (Schaefer et al. 2019), the level of hazard has been estimated, taking into account the ability to generate tsunamis (based on results of Fornaciai et al. 2019). This coupled method enables to estimate the intensity and to define levels of criticality for the instability phenomena in the Sciara del Fuoco area (Fig. 8).

Landslide Forecasting from Satellite

The Xinmo landslide (Ngawa Prefecture, Sichuan Province, Mao County, China, Fig. 9) occurred in the early morning of the 24 June 2017. The landslide can be classified as a rock avalanche: it originated as a large rockslide from the ridge of an alpine slope of about 55–60° and evolved in an extremely rapid flow-like movement of fragmented rock. The main type of the bedrock in this area is quartzite with cataclastic texture, then highly susceptible to fracture and breakup during high velocity movement. About 4.5 million m^3 of rock detached from the top of the mountain crest. During its travel along the slope, the landslide entrained a large amount of pre-existing debris. The landslide hit the Xinmo village with a velocity of 250 km/h (Fan et al. 2017), burying 62 houses and killing more than 80 people.

The course of the Songping gully, which flows at an elevation of 2280 m from WNW to SSE in the area, was dammed for more than 1 km (Scaringi et al. 2018) by the sliding mass, with a maximum thickness of debris of more than 10 m and a total area of 1.5 km^2.

The mean elevation of the source area of the landslide is about 3400 m a.s.l., while the foot of the landslide, where the Xinmo village is located, has an elevation of about 2300 m a.s.l.; considering the runout distance (L) of about 2600 m, the angle of reach (*fahrböschung*) is equal to 22°. This value fits quite well the empirical graphs for rock avalanches plotting travel angle against volume (Rickenmann 2005) and area against volume (Li 1983), obtained with historical data from the literature. At regional scale (Fig. 9), the landslide area is situated at the eastern margin of the tectonically active Tibetan Plateau. The landslide area is placed within a north trending earthquake zone and it is affected by frequent earthquakes. Historically, all these events are located along major faults in the boundary zone between the Tibetan Plateau and the Sichuan Basin. Although no evidence shows that the Xinmo landslide was triggered by the repeated earthquakes, these seismic events may have caused intensive neotectonic deformations, weakening and partially destabilizing the slope from where the rock avalanche detached.

Landslide Failure Prediction Using Sentinel-1

Immediately after the landslide an InSAR analysis on a stack of 45 C-band SAR images (frequency 5.4 GHz, wavelength 5.6 cm) acquired by the ESA (European Space Agency) Sentinel-1 satellites from 9 October 2014 to 19 June 2017 was performed. The analysis was carried out on Sentinel-1 images acquired with the Interferometric Wide (IW) swath mode, the main acquisition mode over land which acquires data with a 250-km swath at 5 m × 14 m spatial resolution. Images have been captured along satellite track n.62 in descending orbit and with an incidence angle of 40.78°.

Sentinel-1 archive was processed through the SqueeSAR approach (Ferretti et al. 2011), one of the most advanced

		Volume range			
		Small (10^3-10^4)	Moderate (10^4-10^5)	Large (10^5-10^6)	Very Large (>10^6)
Displacement rate	Low	Low	Low	Low	Low
	Moderate	Low	Moderate	Moderate	Moderate
	High	Low	Moderate	High	High
	Very high	Low	Moderate	High	Very high

Fig. 8 Criticality level based on the intersection between the GBInSAR-derived warning levels (rows; derived from Di Traglia et al. 2014, 2017) and the instability scenario based on the results of the slope stability analysis (columns; Schaefer et al. 2019)

Fig. 9 Location of the Xinmo landslide in the Ngawa Prefecture, Sichuan Province (China). Major historical earthquakes are reported on the map

methods to analyse multi-temporal stacks of SAR acquisitions. SqueeSAR algorithm is designed to identify a grid of measurement points for which it is possible to estimate the displacement time series and deformation rate (in mm/year) both along the satellite LOS. The purpose of the SqueeSAR analysis was twofold:

(i) detecting and recording any pre-event deformation in and around the village of Xinmo, leveraging on ground deformation maps;
(ii) identifying, within time series of displacement, potential pre-failure signals for the Xinmo landslide.

Results

SqueeSAR results (Fig. 10) highlight the presence of active movements in the upper sector of the slope above the Xinmo village, with deformation rates exceeding several millimetres per year. It is worth remarking that this specific sector of the slope was the origin of the sliding event. Close to the landslide scarp, velocity values range between −10 and −20 mm/year, with peaks of about −30 mm/year. Considering the acquisition geometry (descending), the westward orientation of the slope and the negative sign of the movement (i.e., away from sensor), the measured deformation is

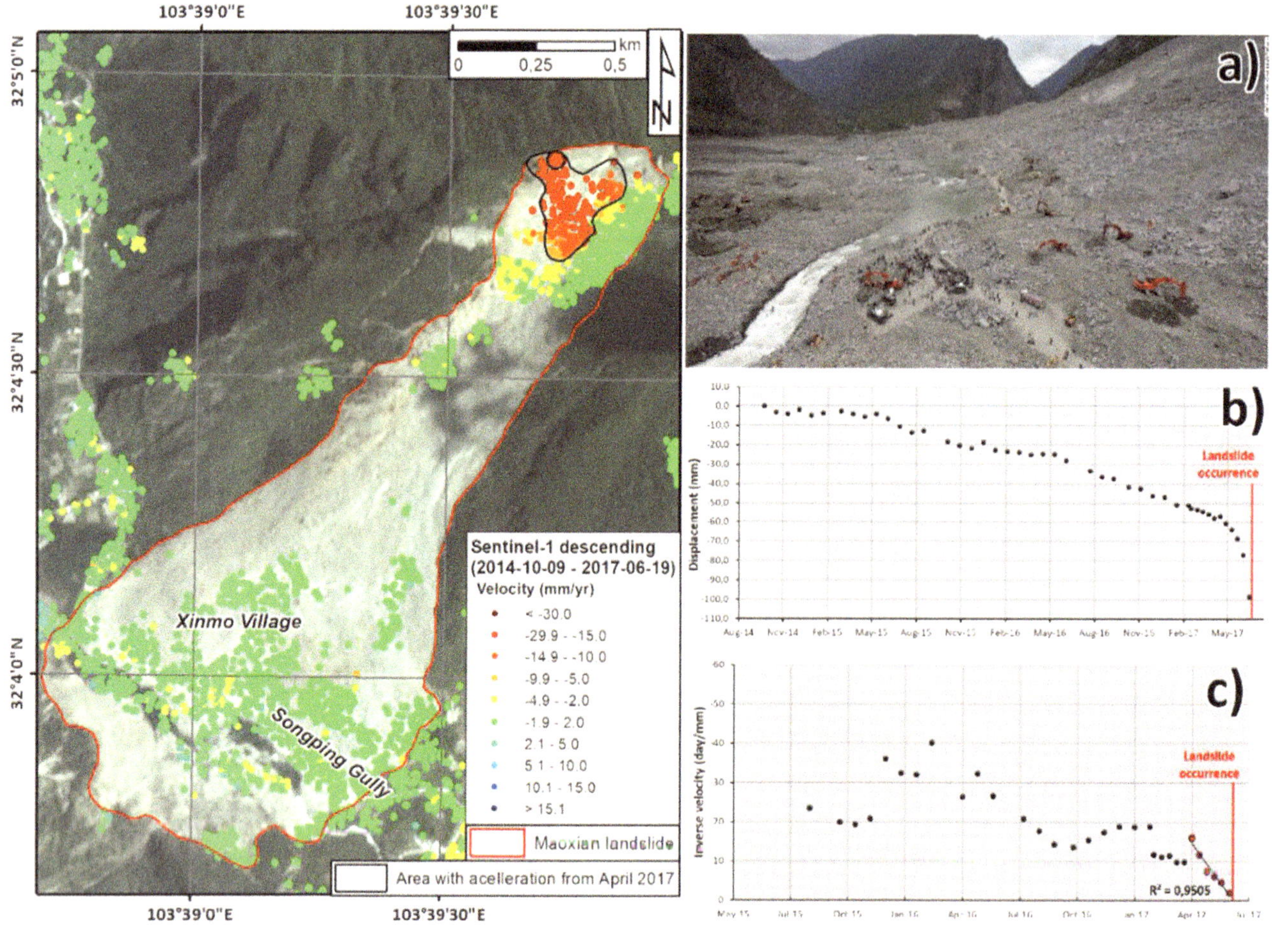

Fig. 10 Pre-failure sign of the Xinmo landslide as seen from Sentinel-1. Ground deformation maps are reported on the left side of the picture. The toe area along the Songping Gully (**a**), displacement time series (**b**) and respective inverse velocity (**c**) of a point taken from the accelerating area are included on the right. Modified from Intrieri et al. (2018b)

consistent with the occurrence of pre-existing slope instability in the area affected by the 24 June 2017 landslide.

An advanced and systematic analysis of the displacement time series allows for the identification of several points located in the NW sector of the source area exhibiting an acceleration starting from April 2017. Anomalous accelerations are the most important to detect for early warning purposes, as they can indicate that the landslide has entered the tertiary creep and it is approaching collapse (Saito 1969).

Specific forecasting methods exist to determine the probable time of failure (Intrieri et al. 2019). A retrospective forecast of the time of failure has been performed starting from the accelerating time series (Fig. 10b). The Fukuzono method (Fukuzono 1985) for forecasting the time of failure has been applied to the displacement data exhibiting progressive acceleration, pointing out that an accurate estimation of the failure time was already possible since the beginning of June 2017 (Fig. 10c).

Results obtained for the Xinmo landslide pointed out that the new generation of satellite radar sensors fosters a new paradigm in the ground deformation monitoring systems. The launch, in April 2014, of Sentinel-1 mission opened new opportunities for InSAR monitoring applications thanks to the increased acquisition frequency and the regularity of acquisitions.

Leveraging the enhanced imaging capabilities of Sentinel-1, advances of computing capacities and refinement of data screening tools, regional to national scale monitoring systems are now possible, supporting authorities with prioritization of the hazards deemed to be most urgent (Raspini et al. 2018).

The potential offered by Sentinel-1 constellation made InSAR suitable for landslide displacement monitoring, early identification of rainfall-triggered accelerations (Raspini et al. 2019) and—under certain circumstances and for some typologies of phenomena—for the prediction of catastrophic slope failures (Carlà et al. 2019a).

Monitoring and Early Warning of Rapid Landslides

Rockfall Monitoring

With the exception of open-pit mines and other civil engineering projects, the use of near-real-time slope monitoring for rockfall forecasting and risk management is still quite uncommon. Conventional strategies are in fact based on the design of physical barriers and slope reinforcement systems (e.g., embankments, ring nets, etc.) for the protection of areas at risk. With regards to monitoring, standard geotechnical tools (e.g. crackmeters, extensometers, survey prisms, etc.) are usually preferred, but point-wise information are often unable to give a full picture of the ongoing instabilities over large rock slopes. This is because it may be technically or economically impossible to place sensors in every potential rockfall source area. Other well-established methodologies include external trigger monitoring, the use of statistical approaches to assess the spatial or temporal distribution of rockfall risk, or the detection of pre-failure damage features through baseline monitoring techniques (Chau et al. 2003; Rosser et al. 2007; Stock et al. 2012; Macciotta et al. 2016; Kromer et al. 2017); however, these solutions may not be appropriate in cases where hourly or sub-hourly monitoring is required, or in areas characterized by a constant human presence.

The Italian National Road (NR) 36 is the main thoroughfare leading to the Chiavenna Valley (north of Lake Como, Italy). The Splügen Pass (2117 m a.s.l.) represents the endpoint of the road and marks the border between Italy and Switzerland. In its last stretch, the roadway climbs up the San Giacomo Valley (the northern branch of the Chiavenna Valley) with a quick succession of sharp hairpin bends that are constrained between towering rock walls and the left bank of the Liro Creek. The Gallivaggio sanctuary lies between km 126 of the NR 36 and a roughly 500 m-high, sub-vertical granitic slope (Fig. 11). In spite of the presence of a 4.5–9 m high protection embankment, a 3.5 m high catch fence on top of the embankment, and a 10 m wide catch ditch between the slope and the embankment, rockfalls have repeatedly reached and caused minor damages to both the NR 36 and the sanctuary in the recent past. Barriers were in fact not high enough to intercept every possible rockfall path, since blocks mostly bounce and fragment over the steep slope with almost no influence of the rolling component. Moreover, given the considerable fall height, no physical barrier can have sufficient energy absorption capacity to efficiently retain large failure volumes. Besides the obvious threat to people safety, rockfalls in this area can induce road closures lasting for extensive periods of time. Villages at higher elevation are thereby subject to losing the sole transportation corridor connecting them with the rest of Italy.

As crossing of the Splügen Pass is barred by snow during the winter, closures of the NR 36 in the Gallivaggio area can determine the isolation of almost 1500 permanent residents. In this circumstance, the exclusion zone can be bypassed only by walking a mountain trail running on the opposite side of the Liro Creek. A detailed description of the geological and geomorphological features of the study area can be found in Carlà et al. 2019b.

The Gallivaggio Landslide Monitoring System

In 2016, following up the results of several discontinuous campaigns performed in the period 2011–2015, a GBInSAR was permanently deployed on the left bank of the Liro Creek to survey the deformation of the overhanging slope in near-real-time. The discontinuous campaigns indicated the consistent creep of ~ 900 m^2 of rock face at the top of the cliff, as opposed to the substantial stability of the middle and lower sectors. The selection of this monitoring strategy was not made with the aim of detecting rockfalls of few m^3 in size, which may result invisible to the GBInSAR. It was trusted that these events would have been, for the most part, contained by the physical barriers already in place. The primary focus was the timely identification of instabilities with a volume in the order of 10^2–10^3 m^3 or more, against which the existing defense systems could not represent a significant hurdle. In the employed configuration, range and azimuth resolution of the pixels in the cliff sector of interest were approximately 75 cm and 180 cm, respectively. The frequency of image acquisition could be remotely set as low as 2 min (baseline frequency of 14 min), so as to avoid phase wrapping issues in the case of accelerating slope movements. The early-warning system was thus controlled

Fig. 11 View of the monitored sub-vertical granitic slope, as seen from the Gallivaggio sanctuary

by a sequence of velocity thresholds. The trigger action response plan was initially associated with different levels of surveillance activity, and finally combined with specific civil protection countermeasures.

Figure 12a supplies the reader with an image of the ~5000 m^3 instability that eventually failed on 29 May 2018 (approximate size 20–25 m in height, 21 m in width, and 8–11 m in thickness). Several large open fractures can be evidently observed across the entire rock face, as well as an indentation at the bottom of the picture that made the unstable volume jut out of the slope. This indentation likely originated from previous minor rockfalls, which critically removed support at the base of the overlying mass. The instability can be associated with a planar mechanism exploiting a primary basal failure surface (dipping at 55–60°) and a sub-vertical rear tension crack, opening up to as much as 2 m (Fig. 12b). At the top of the failed block, a chaotic pile of large boulders and crushed debris was formed by the stresses related to significant vertical movements of the ground.

10 highly coherent pixels were picked as representative surface points to monitor the evolution of the described instability. The respective time series of displacement were automatically updated on a dedicated web platform just minutes after each new image acquisition. Soon after the beginning of monitoring, all monitoring points started to show a steady-state creep behaviour occurring at rates of 1.5–2 cm/year. Potential triggering factors such as heavy rainfall or abrupt temperature changes did not seem to affect data trends in a noticeable fashion.

The Monitoring Campaign

At the end of 2017 and in the following months, velocities progressively increased to values of 3–5 cm/year, and then experienced an abrupt increment on 13 April 2018 (Fig. 13). This occurred concurrently with the fall of a few small blocks (each one less than 0.5 m^3) that reached the NR 36 and likely originated around the perimeter of the main instability.

The upper alarm code was consequently issued, and the response actions aimed at regulating human presence within the exclusion zone were undertaken. Velocities fluctuated around the value of 4 mm/day until the end of April, before decreasing temporarily to 1–2 mm/day for most of the month of May. Finally, displacements started to increase dramatically on the evening of 28 May 2018. The next morning, about 9 h before ultimate failure, the monitoring system captured the onset of a slope acceleration which

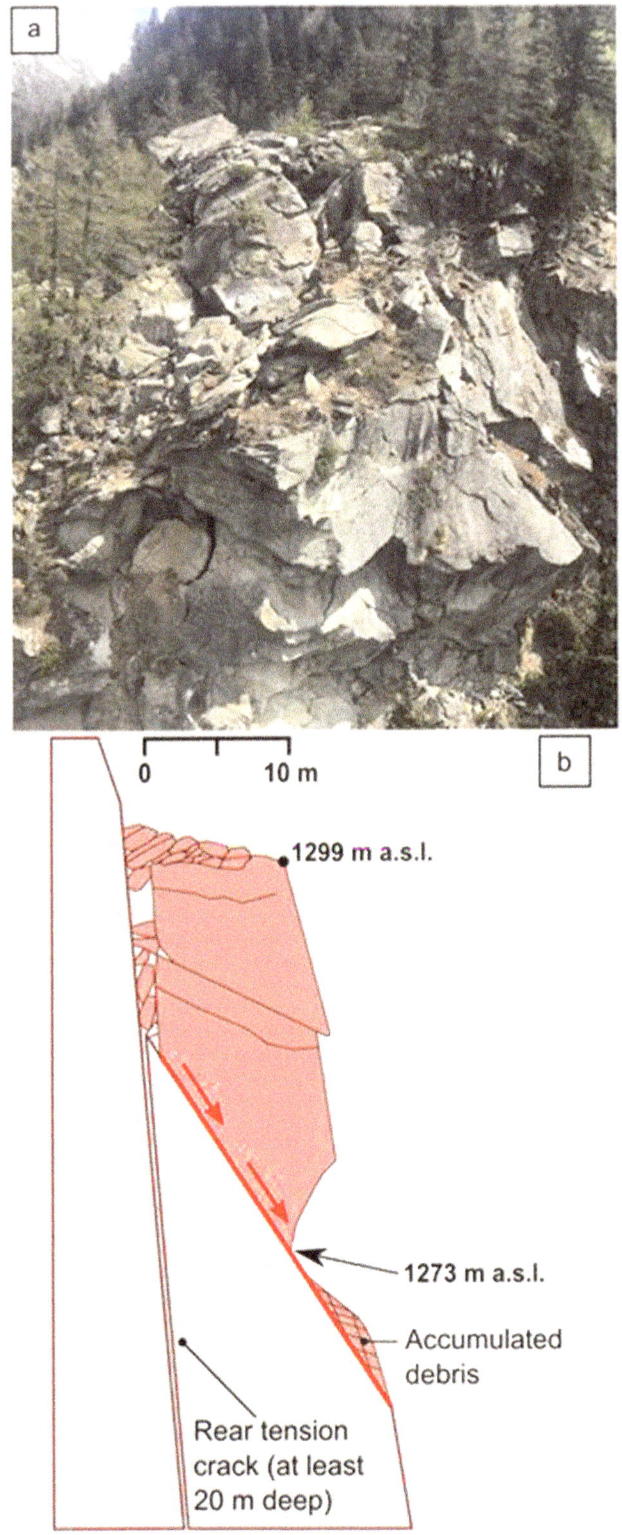

Fig. 12 **a** Frontal view and **b** schematic cross-section of the ~5000 m^3 instability (modified after Carlà et al. 2019b)

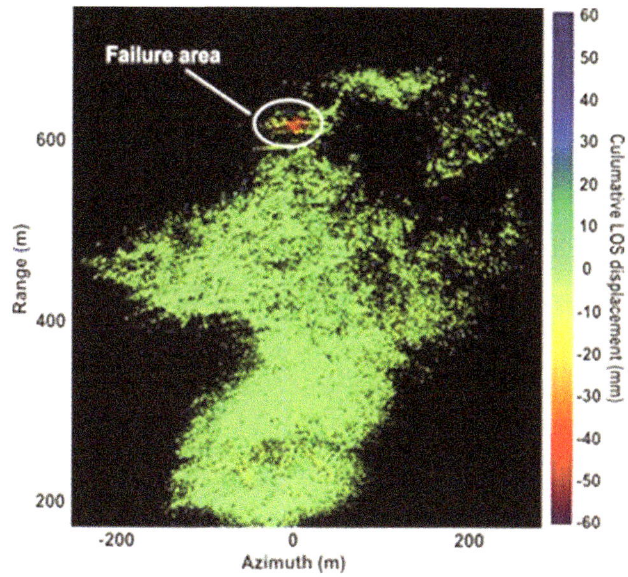

Fig. 13 Cumulative LOS displacements measured by the GBInSAR in the period 13–26 April 2018, in after a minor rockfall event (black points indicate no radar return signal)

closely resembled a classic tertiary creep curve (Saito 1969). Measured velocities widely exceeded the peak values detected before (up to 73.5 mm/h). The unstable mass failed at 4:32 pm local time and subsequently crumbled in a large number of smaller blocks and fragments, thereby forming a huge cloud of dust and debris which ran over the valley floor. An important fraction of the falling material reached the Gallivaggio sanctuary and the adjacent segment of the NR 36; nonetheless, the high degree of fragmentation meant that the exposed elements did not suffer from irreparable damages. No injuries or fatalities were counted since the area had been timely cleared.

In the hours leading to the event, failure-time predictions were continuously performed to assist the government officials in charge of the emergency and of the evacuation procedures. Predictions were updated each time a new displacement measurement became available. Figure 14 shows the inverse velocity analysis (Fukuzono 1985) for 6 control points, based on a 5-point moving average of the data: a clear linearity of all the time series is observed from 10:30 am (6 h before the failure). Afterwards, the extrapolated solutions consistently fell in the interval 4:25–4:45 pm. For instance, it is worth mentioning that the last prediction for Point_08 (made at 4:15 pm, before the appearance of phase wrapping issues in the interferograms) resulted in an expected failure at 4:28 pm, basically matching the actual failure-time (4:32 pm). The confidence about the reliability of the inverse velocity predictions was also strengthened by the analysis of the peak velocity and peak acceleration values of the rock mass as the tertiary creep stage progressed.

Federico et al. (2012) first demonstrated the existence of a broadly linear correlation between the logarithm of these two parameters in proximity of a large number of slope failures. The work was later developed by Carlà et al. (2017a, b, c), who collected GBInSAR data from nine open-pit mine instabilities in hard intrusive rocks; it was determined that peak velocity and peak acceleration were linearly correlated in non-logarithmic form too, and were substantially larger in those cases that actually evolved into failure (as opposed to those instabilities which did not fail despite an episode of notable acceleration). A graph of peak velocity vs. peak acceleration, based on the one presented in Carlà et al. (2017a, b, c), is shown in Fig. 15: peak velocity and peak acceleration registered by all the GBInSAR monitoring

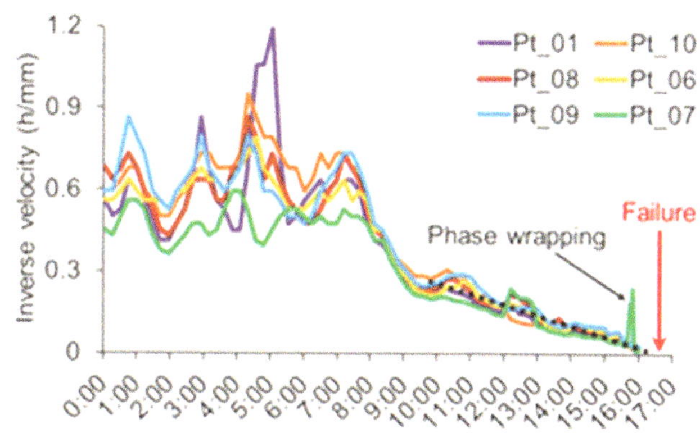

Fig. 14 Example of inverse velocity analysis on 6 GBInSAR monitoring points for the prediction of the 29 May 2018 failure

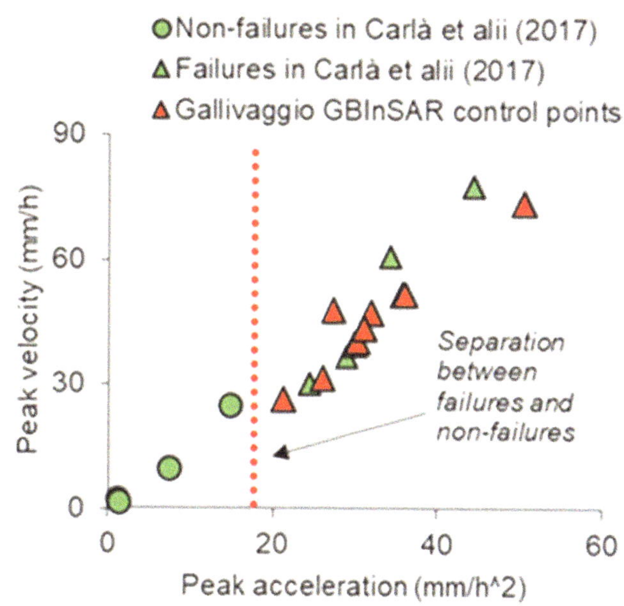

Fig. 15 Graph of peak velocity versus peak acceleration for a number of slope failure case studies (modified after Carlà et al. 2017a, b, c)

points were extremely consistent with the findings of the referenced study.

Moreover, it should be taken into account that the Gallivaggio data represent LOS measurements, and that these were also affected by phase wrapping ambiguity after 4:15 pm. All the collected evidences thus pointed towards a total and imminent failure of the ~ 5000 m^3 instability above the Gallivaggio sanctuary.

Discussion and Conclusions

Case studies analysis

Each of the cases presented above shows a peculiarity that can help us in the definition of the characteristics and potential of a modern landslide EWS.

Quincinetto case study is an example of a highly vulnerable and valuable element at risk—the highway connecting Italy to France and Switzerland—where the threat of even small rock fragments reaching the distal carriageway would cause the complete closure of the street and severe indirect economic losses. The high risk allows for a high-end EWS deploying a GBInSAR, periodical Sentinel-1 analyses and a WSN, all tailored within a communication and response system that enables to promptly carry out a partial or total closure or re-opening of the highway depending on the monitored data and visual verification of the operators.

Each of the three techniques employed fills a specific gap and has limitations that are compensated by the other two. GBInSAR is able to produce frequent SAR images (on the order of seconds to minutes), resulting in very high frequency slope maps and time series. With these fast repeat time intervals, GBInSAR has led InSAR from mapping to surveillance and early-warning applications. GBInSAR capability of producing 2D displacement maps is crucial to gain insight of the whole monitored area. On the other hand, it only records LOS movements, can be affected by atmospheric noise and by phase wrapping and ambiguity if high displacements occur.

The WSN equipped with tiltmeters and extensometers provides direct movement monitoring of critical unstable blocks. The measurements are accurate and easy to interpret, the main noise source being the thermic expansion of the metal wire of the extensometers. However, the information acquired is only point-wise, some parts of the instrumentation can be damaged by the activity of the landslide and maintenance requires access to the area.

Satellite InSAR monitoring delivers a different coverage and different LOS measurements with respect to GBInSAR. On the other hand, data require days to be acquired and elaborated and can only represent slow movements.

The full potential of satellites, specifically of Sentinel-1 constellation, is proved by the Xinmo case study. This experience clearly shows that a regular elaboration and interpretation of satellite data can effectively provide reliable early warning and ultimately save many lives. This application still shows apparent limitations: the landslide must exhibit a creep behaviour, a slow evolution (i.e. some weeks of acceleration), which is typically associated with very large landslides and must have a good data coverage (i.e. scarce vegetation, no snow, no intense human activity like mining, sufficient LOS alignment, no layover or shadowing due to steep slopes). Nevertheless, satellite data often represent the only existing monitoring information in an area, even remote and not accessible like the source area from where the Xinmo landslide originated. Recent experiences (Raspini et al. 2018) showed that a continuous satellite monitoring can be performed also for areas as large as Tuscany region in Italy (nearly 23,000 km^2, as opposed to the 460 km^2 area elaborated for the Xinmo landslide) implementing a semi-automatic detection of measurement points exhibiting anomalous displacements. In this case, the main difficulty in terms of early warning is represented by the large extension of the area and the consequent huge amount of data to be processed. Del Soldato et al. (2019) estimated in around one week the time required to run the entire monitoring chain at regional scale (covered areas of tens of thousands square km) from the acquisition of Sentinel-1 images to the interpretation of causes and persistence of anomalies of movement. The resulting time is still compatible with several cases of failure and would drop to around three days in case of areas of some hundreds of square km (this time would be mostly used for the elaboration of the raw data, while the automatic detection of anomalous displacements would only take a few minutes).

For the reasons explained above, GBInSAR can be considered a cutting-edge technology for the early warning of large landslides, but it stands out also for more peculiar applications, such as volcanoes and rockfall monitoring.

The Ku-band (17–17.1 mm wavelength) used by most GBInSAR apparatuses is able to maintain a good precision and resolution and at the same time can penetrate dust clouds that are abundant especially during collapse events and volcanic explosions and, contrarily to optical remote sensing techniques, enables to work in variable light and atmospheric conditions. At Stromboli, the GBInSAR systems have been used to monitor the slope instability in the northern sector of the SdF and also to detect the inflation and deflation of the N sector of the crater area, allowing for the definition of the unstable area within the SdF and also to outline the structural framework of the summit area. Therefore, GBInSAR devices resulted in an operational approach to mitigate landslide risk by defining a relationship between measured displacement rates and the associated instability and eruptive hazards.

Concerning rockfall hazard, this has always been a challenge in terms of early warning, owing to the sudden nature of collapses and the typically small dimensions of the detaching masses. Rose and Hungr (2007) even suggested that forecasting in brittle (rigid) rocks (where rockfall generally take place) was impossible, at least for the technical limitations at that time. However, Carlà et al. (2017a) demonstrated how the new generation of GBInSAR systems—with their increased acquisition frequency—were able to predict, a posteriori, the collapse of 500–1500 m^3 wedge and planar failures in high-grade metamorphic and volcanic rocks in an open-pit mine. The Gallivaggio landslide experience corroborates this result in the framework of a public safety application, where the prediction was made before failure rather than with hindsight.

Future Perspectives

Research in the landslide EWS field has been experiencing enticing advancements in the last decade, thanks to the continuing evolution of monitoring instrumentation and data analysis techniques.

One of the drawbacks of advanced, real-time monitoring tools (e.g. GBInSAR) is that they produce a huge amount of information that may cause logistic problems in data transfer and storage, that require specific procedures to be treated in a way that is compatible with a landslide EWS (Intrieri et al. 2017).

On the other hand, the advent of 5G wireless communications technologies could be a game-changer in a number of ways that is currently difficult to imagine.

Another issue with big data is that critical information is often amidst data having scarce or no use for early warning purposes. With this regard, Tordesillas et al. (2020), in this very volume, present an innovative method using knowledge of precursory failure dynamics of granular materials and advanced data analytics. They distinguish a high-risk small area from other areas exhibiting comparably higher than average velocity or even episodes of acceleration. What is most interesting is that this 0.3 km^2 area is detected among 130,000 Sentinel-1 measurement points in a 460 km^2 region around a year prior to the landslide.

Obviously, big data also represent a great potential albeit one that is rarely fully exploited. For example, GBInSAR data exhibit an unmatched spatial density of displacement measurement points but quantitative analyses are often made using the displacement time series of a few significant points. Advances in fields such as data analytics, material science and micromechanics show that, at laboratory sample scale, the location and geometry of failure are encoded in the grain motions and can be determined early in the precursory failure regime (Tordesillas et al. 2013). Recently, Singh and Tordesillas (2020) upscaled this problem from laboratory sample scale to landslide scale, where the pixels of a GBInSAR displacement map replace the grains of the soil sample. They managed to predict the geometry and location of the failure zone and time of failure before any tertiary creep-based forecasting method was applicable.

The increasing availability of monitoring techniques and the installation of redundant instruments also raises the need for integrated, user-friendly platforms for data visualization and management, to be used not only by scientists and specialized workers (e.g. operators working in open-pit mine safety) but also by technicians of local administrations or government institutions and civil protection agencies, which manage a relevant number of landslide EWSs (Pecoraro et al. 2019).

This also highlights that the connection with society is key to a successful landslide EWS, at least for those slope failures that represent a public threat. In fact, near real-time instruments, fast data transfers and elaboration algorithms make it possible to send reliable and timely warnings to the people exposed, the main problem being whether a correct communication and information campaign has been carried out beforehand to make this warning understandable and to ensure that the population takes the appropriate actions.

A connection with the society also means addressing the economic factor. A diffusion of low-cost monitoring instruments suitable for early warning would be beneficial to a larger diffusion of instrumented sites, which in turn would also increase the public awareness concerning landslide hazard. In this regard, there are several technologies that conceivably can play this role in the future, such as micro-electro-mechanical systems (MEMSs) (Cina et al. 2019), wireless sensor networks (WSN) (Mucchi et al. 2018; Muttillo et al. 2019), radio-frequency identification tags (RFID) (Le Breton et al. 2019), global navigation satellite system (GNSS) (Notti et al. 2020).

In conclusion, the recent and upcoming technological and scientific advancements are the premise of even more accurate and meaningful landslide EWSs. The real challenge will be played on the social side and it will concern the ability of scientists and social operators to engage the population, build capacity, increase risk awareness and make warnings understandable and credible. The increased landslide risk at global level caused by demographic growth, urbanization and climate changes might help to increase people awareness but will also make the need for effective responses even more pressing.

Acknowledgements The work on Gallivaggio and Stromboli case studies was financially supported by the "Presidenza del Consiglio dei Ministri – Dipartimento della Protezione Civile" (Presidency of the

Council of Ministers—Department of Civil Protection); this publication, however, does not reflect the position and official policies of the Department.

References

Adam N, Rodriguez-Gonzalez F, Parizzi A, Liebhart W (2011) Wide area persistent scatterer interferometry. In: 2011 IEEE International geoscience and remote sensing symposium (IGARSS), pp 1481–1484

Antonello G, Casagli N, Farina P, Leva D, Nico G, Sieber AJ, Tarchi D (2004) Groundbased SAR interferometry for monitoring mass movements. Landslides 1:21–28

Atzeni C, Barla M, Pieraccini M, Antolini F (2015) Early warning monitoring of natural and engineered slopes with ground-based synthetic-aperture radar. Rock Mech Rock Eng 48:235–246

Azimi C, Biarez J, Desvarreux P, Keime F (1989) Prévision d'éboulement en terrain gypseux. In: Bonnard C (ed) Proceedings of the 5th international symposium on landslides, Lausanne, vol 1. A. A. Balkema, Rotterdam, pp 531–536 (in French)

Bamler R, Hartl P (1998) Synthetic aperture radar interferometry. Inverse Prob 14:1–54

Bardi F, Frodella W, Ciampalini A, Bianchini S, Del Ventisette C, Gigli G, Fanti R, Moretti S, Basile G, Casagli N (2014) Integration between ground based and satellite SAR data in landslide mapping: the San Fratello case study. Geomorphology 223:45–60

Bardi F, Raspini F, Frodella W, Lombardi L, Nocentini M, Gigli G, Morelli S, Corsini A, Casagli N (2017) Monitoring the rapid-moving reactivation of earth flows by means of GB-InSAR: the April 2013 Capriglio Landslide (Northern Appennines, Italy). Remote Sens 9(2):165

Berardino P, Costantini M, Franceschetti G, Iodice A, Pietranera L, Rizzo V (2003) Use of differential SAR interferometry in monitoring and modelling large slope instability at Maratea (Basilicata, Italy). Eng Geol 68(1–2):31–51

Berti M, Corsini A, Franceschini S, Iannacone JP (2013) Automated classification of persistent scatterers interferometry time series. Nat Hazards Earth Syst Sci 13(8):1945–1958

Bianchini S, Pratesi F, Nolesini T, Casagli N (2015a) Building deformation assessment by means of persistent scatterer interferometry analysis on a landslide-affected area: the Volterra (Italy) case study. Remote Sens 7:4678–4701

Bianchini S, Ciampalini A, Raspini F, Bardi F, Di Traglia F, Moretti S, Casagli N (2015b) Multi-temporal evaluation of landslide movements and impacts on buildings in San Fratello (Italy) by means of C-band and X-band PSI data. Pure appl Geophys 172(11):3043–3065

Blikra LH (2012) The Åknes rockslide. In: Clague, Norway JJ, Stead D (eds) Landslides: types, mechanisms and modeling. Cambridge University Press, pp 323–335

Bossi G, Crema S, Frigerio S, Mantovani M, Marcato G, Pasuto A, Schenato L, Cavalli M (2015) The Rotolon catchment early-warning system. In: Engineering geology for society and territory, vol 3. Springer, Cham, pp 91–95

Brox D, Newcomen W (2003) Utilizing strain criteria to predict highwall stability performance. In: Proceedings of the 10th ISRM congress, Sandton, South Africa

Calvari S, Intrieri E, Di Traglia F, Bonaccorso A, Casagli N, Cristaldi A (2016) Monitoring crater-wall collapse at active volcanoes: a study of the 12 January 2013 event at Stromboli. Bull Volc 78(5):39

Canuti P, Casagli N, Catani F, Falorni G, Farina P (2007) Integration of remote sensing techniques in different stages of landslide response. In: Progress in landslide science. Springer, Berlin, Heidelberg, pp 251–260

Carlà T, Intrieri E, Di Traglia F, Nolesini T, Gigli G, Casagli N (2016) Guidelines on the use of inverse velocity method as a tool for setting alarm thresholds and forecasting landslides and structure collapses. Landslides 14:517–534

Carlà T, Farina P, Intrieri E, Botsialas K, Casagli N (2017a) On the monitoring and early-warning of brittle slope failures in hard rock masses: examples from an open-pit mine. Eng Geol 228:71–81

Carlà T, Intrieri E, Di Traglia F, Nolesini T, Gigli G, Casagli N (2017b) Guidelines on the use of inverse velocity method as a tool for setting alarm thresholds and forecasting landslides and structure collapses. Landslides 14(2):517–534

Carlà T, Intrieri E, Farina P, Casagli N (2017c) A new method to identify impending failure in rock slopes. Int J Rock Mech Min Sci 93:76–81

Carlà T, Farina P, Intrieri E, Ketizmen H, Casagli N (2018) Integration of ground-based radar and satellite InSAR data for the analysis of an unexpected slope failure in an open-pit mine. Eng Geol 235:39–52

Carlà T, Intrieri E, Raspini F, Bardi F, Farina P, Ferretti A, Colombo D, Novali F, Casagli N (2019a) Perspectives on the prediction of catastrophic slope failures from satellite InSAR. Sci Rep 9(1):1–9

Carlà T, Nolesini T, Solari L, Rivolta C, Dei Cas L, Casagli N (2019b) Rockfall forecasting and risk management along a major transportation corridor in the Alps through ground-based radar interferometry. Landslides. 16:1425–1435

Casagli N, Frodella W, Morelli S, Tofani V, Ciampalini A, Intrieri E, Raspini F, Rossi G, Tanteri L, Lu P (2017) Spaceborne, UAV and ground-based remote sensing techniques for landslide mapping, monitoring and early warning. Geoenviron Disasters 4(1):9

Casagli N, Morelli S, Frodella W, Intrieri E, Tofani V (2018) TXT-tool 2.039–3.2 Ground-based remote sensing techniques for landslides mapping, monitoring and early warning. In: Landslide dynamics: ISDR-ICL landslide interactive teaching tools. Springer, Cham, pp 255–274

Casagli N, Tibaldi A, Merri A, Del Ventisette C, Apuani T, Guerri L, Fortuny-Guasch J, Tarchi D (2009) Deformation of stromboli volcano (Italy) during the 2007 eruption revealed by radar interferometry, numerical modelling and structural geological field data. J Volcanol Geoth Res 182(3–4):182–200

Cerase A, Crescimbene M, La Longa F, Amato A (2019) Tsunami risk perception in southern Italy: first evidence from a sample survey. Nat Hazards Earth Syst Sci 19(12)

Chau KT, Wong RHC, Liu J, Lee CF (2003) Rockfall hazard analysis for Hong Kong based on rockfall inventory. Rock Mech Rock Eng 36:383–408. https://doi.org/10.1007/s00603-002-0035-z

Ciampalini A, Bardi F, Bianchini S, Frodella W, Del Ventisette C, Moretti S, Casagli N (2014) Analysis of building deformation in landslide area using multisensor PSInSAR™ technique. Int J Appl Earth Obs Geoinform 33:166–180

Ciampalini A, Raspini F, Bianchini S, Frodella W, Bardi F, Lagomarsino D, Di Traglia F, Moretti S, Proietti C, Pagliara P, Onori R, Corazza D, Duro A, Basile G, Casagli N (2015) Remote sensing as tool for development of landslide databases: the case of the Messina Province (Italy) geodatabase. Geomorphology 249:103–118

Ciampalini A, Raspini F, Frodella W, Bardi F, Bianchini S, Moretti S (2016a) The effectiveness of high-resolution LiDAR data combined with PSInSAR data in landslide study. Landslides 13(2):399–410

Ciampalini A, Raspini F, Lagomarsino D, Catani F, Casagli N (2016b) Landslide susceptibility map refinement using PSInSAR data. Remote Sens Environ 184:302–315

Cina A, Manzino AM, Bendea IH (2019) Improving GNSS landslide monitoring with the use of low-cost MEMS accelerometers. Appl Sci 9(23):5075

Confuorto P, Di Martire D, Centolanza G, Iglesias R, Mallorqui JJ, Novellino A, Plank S, Ramondini M, Kurosch T, Calcaterra D (2017) Post-failure evolution analysis of a rainfall-triggered landslide by multi-temporal interferometry SAR approaches integrated with geotechnical analysis. Remote Sens Environ 188:51–72

Crosetto M, Monserrat O, Cuevas-González M, Devanthéry N, Crippa B (2016) Persistent scatterer interferometry: a review. ISPRS J Photogramm Remote Sens 115:78–89

Crosta GB, Agliardi F (2003) Failure forecast for large rock slides by surface displacement measurements Can. Geotech J 40(1):176–191

Del Soldato M, Riquelme A, Bianchini S, Tomás R, Di Martire D, De Vita P, Moretti S, Calcaterra D (2018) Multisource data integration to investigate one century of evolution for the Agnone landslide (Molise, southern Italy). Landslides 15(11):2113–2128

Del Soldato M, Solari L, Raspini F, Bianchini S, Ciampalini A, Montalti R, Ferretti A, Pellegrineschi V, Casagli N (2019) Monitoring ground instabilities using SAR satellite data: a practical approach. ISPRS Int J Geo-Inf 8(7):307

Del Ventisette C, Intrieri E, Luzi G, Casagli N, Fanti R, Leva D (2011) Using ground based radar interferometry during emergency: the case of the A3 motorway (Calabria Region, Italy) threatened by a landslide. Natural Hazards Earth Syst Sci 11(9):2483–2495

Di Roberto A, Rosi M, Bertagnini A, Marani M P, Gamberi F (2010) Distal turbidites and tsunamigenic landslides of Stromboli volcano (Aeolian Islands, Italy). In: Submarine mass movements and their consequences. Springer, Dordrecht, pp 719–731

Di Traglia F, Nolesini T, Intrieri E, Mugnai F, Leva D, Rosi M, Casagli N (2014) Review of ten years of volcano deformations recorded by the ground-based InSAR monitoring system at Stromboli volcano: a tool to mitigate volcano flank dynamics and intense volcanic activity. Earth Sci Rev 139:317–335

Di Traglia F, Nolesini T, Casagli N (2017) Monitoring eruption-induced mass-wasting at active volcanoes: the Stromboli case. In: Workshop on world landslide forum. Springer, Cham, pp 669–676

Di Traglia F, Calvari S, D'Auria L, Nolesini T, Bonaccorso A, Fornaciai A, Esposito A, Cristaldi A, Favalli M, Casagli N (2018a) The 2014 effusive eruption at Stromboli: new insights from in situ and remote-sensing measurements. Remote Sens 10(12):2035

Di Traglia F, Nolesini T, Ciampalini A, Solari L, Frodella W, Bellotti F, Fumagalli A, De Rosa G, Casagli N (2018b) Tracking morphological changes and slope instability using spaceborne and ground-based SAR data. Geomorphology 300:95–112

Di Traglia F, Fornaciai A, Favalli M, Nolesini T, Casagli N (2020) Catching geomorphological response to volcanic activity on steep slope volcanoes using multi-platform remote sensing. Remote Sens 12(3):438

Dick GJ, Eberhardt E, Cabrejo-Liévano AG, Stead D, Rose ND (2015) Development of an early-warning time-of-failure analysis methodology for open-pit mine slopes utilizing ground-based slope stability radar monitoring data. Can Geotech J 52(4):515–529

Fan X, Xu Q, Scaringi G, Dai L, Li W, Dong X, Havenith HB (2017) Failure mechanism and kinematics of the deadly June 24th, 2017 Xinmo landslide, Maoxian, Sichuan, China. Landslides 14 (6):2129–2146

Farina P, Casagli N, Ferretti A (2008) Radar-interpretation of InSAR measurements for landslide investigations in civil protection practices. In: Proceedings of 1st North American landslide conference. Vail, Colorado, pp 272–283

Fathani TF, Karnawati D, Wilopo W, Crowley K (2016) An integrated methodology to develop a standard for landslide early warning systems. Nat Hazards Earth Syst Sci 16(9):2123–2135

Federico A, Popescu M, Elia G, Fidelibus C, Internò G, Murianni A (2012) Prediction of time to slope failure: a general framework. Environ Earth Sci 66(1):245–256

Fernández-Steeger T, Arnhardt C, Walter K, Haß SE, Niemeyer F, Nakaten B, Homfeld SD, Asch K, Azzam R, Bill R, Ritter H (2009) SLEWS—a prototype system for flexible real time monitoring of landslides using an open spatial data infrastructure and wireless sensor networks. Geotechnol Sci Rep 13:3–15

Ferretti A, Pratin C, Rocca F (2001) Permanent scatterers in SAR interferometry. IEEE Trans Geosci Remote Sens 39(1):8–20

Ferretti A, Fumagalli A, Novali F, Prati C, Rocca F, Rucci A (2011) A new algorithm for processing interferometric data-stacks: SqueeSAR. IEEE Trans Geosci Remote Sens 49(9):3460–3470

Ferrigno F, Gigli G, Fanti R, Intrieri E, Casagli N (2017) GB-InSAR monitoring and observational method for landslide emergency management: the Montaguto earthflow (AV, Italy). Nat Hazards Earth Syst Sci 17(6)

Fornaciai A, Favalli M, Nannipieri L (2019) Numerical simulation of the tsunamis generated by the Sciara del Fuoco landslides (Stromboli Island, Italy). Sci Rep 9(1):1–12

Frodella W, Ciampalini A, Gigli G, Lombardi L, Raspini F, Nocentini M, Scardigli C, Casagli N (2016) Synergic use of satellite and ground based remote sensing methods for monitoring the San Leo rock cliff (Northern Italy). Geomorphology 264:80–94

Frodella W, Salvatici T, Morelli S, Pazzi V, Fanti R (2017) GB-InSAR monitoring of slope deformations in a mountainous area affected by debris flow events. Nat Hazards Earth Syst Sci 17(10):1779

Frodella W, Ciampalini A, Bardi F, Salvatici T, Di Traglia F, Basile G, Casagli N (2018) A method for assessing and managing landslide residual hazard in urban areas. Landslides 15(2):183–197

Froude MJ, Petley D (2018) Global fatal landslide occurrence from 2004 to 2016. Nat Hazards Earth Syst Sci 18:2161–2181

Fruneau B, Achache J, Delacourt C (1996) Observation and modeling of the Saint-Etienne-de-Tine'e Landslide using SAR interferometry. Tectonophysics 265

Fukuzono T (1985) A method to predict the time of slope failure caused by rainfall using the inverse number of velocity of surface displacement. J Jpn Landslide Soc 22:8–13

Ghiglia DC, Romero LA (1994) Robust two-dimensional weighted and unweighted phase unwrapping that uses fast transforms and iterative methods. J Opt Soc Am 11(1):107–117

Gigli G, Fanti R, Canuti P, Casagli N (2011) Integration of advanced monitoring and numerical modeling techniques for the complete risk scenario analysis of rockslides: the case of Mt. Beni (Florence, Italy). Eng Geol 120(1–4):48–59

Guzzetti F, Mondini AC, Cardinali M, Fiorucci M, Santangelo M, Chang KT (2012) Landslide inventory maps: new tools for an old problem. Earth-Sci Rev 112:1–25

Guzzetti F, Gariano SL, Peruccacci S, Brunetti MT, Marchesini I, Rossi M, Melillo M (2020) Geographical landslide early warning systems. Earth Sci Rev 200:102973

Hao SW, Liu C, Lu CS, Elsworth D (2016) A relation to predict the failure of materials and potential application to volcanic eruptions and landslides. Sci Rep 6, Article 27877

Haque U, Blum P, Da Silva PF, Andersen P, Pilz J, Chalov SR, Malet J-P, Jemec Auflič M, Andres N, Poyiadji E, Lamas PC, Zhang W, Peshevski I, Pétursson HG, Kurt T, Dobrev N, García-Davalillo JC, Halkia M, Ferri S, Gaprindashvili G, Engström J, Keellings D (2016) Fatal landslides in Europe. Landslides 13(6):1545–1554

Hill CD, Sippel KD (2002) Modern deformation monitoring: a multi sensor approach. In: Proceedings of the FIG 22nd international conference, Washington, DC, USA, Apr. 2002, pp 1–12

Intrieri E, Gigli G (2016) Landslide forecasting and factors influencing predictability. Nat Hazards Earth Syst Sci 16:2501–2510

Intrieri E, Gigli G, Mugnai F, Fanti R, Casagli N (2012) Design and implementation of a landslide early warning system. Eng Geol 147:124–136

Intrieri E, Gigli G, Casagli N, Nadim F (2013) Brief communication. Landslide early warning system: toolbox and general concept. Nat Hazards Earth Syst Sci 13:85–90

Intrieri E, Gigli G, Nocentini M, Lombardi L, Mugnai F, Fidolini F, Casagli N (2015) Sinkhole monitoring and early warning: an experimental and successful GB-InSAR application. Geomorphology 241:304–314

Intrieri E, Bardi F, Fanti R, Gigli G, Fidolini F, Casagli N, Costanzo S, Raffo A, Di Massa G, Capparelli G, Versace P (2017) Big data managing in a landslide early warning system: experience from a ground-based interferometric radar application. Nat Hazards Earth Syst Sci 17:1713–1723

Intrieri E, Gigli G, Gracchi T, Nocentini M, Lombardi L, Mugnai F, Frodella W, Bertolini G, Carnevale E, Favalli M, Fornaciai A, Marturià Alavedra J, Mucchi L, Nannipieri L, Rodriguez-Lloveras X, Pizziolo M, Schina R, Trippi F, Casagli N (2018a) Application of an ultra-wide band sensor-free wireless network for ground monitoring. Eng Geol 238:1–14

Intrieri E, Raspini F, Fumagalli A, Lu P, Del Conte S, Farina P, Allievi J, Ferretti A, Casagli N (2018b) The Maoxian landslide as seen from space: detecting precursors of failure with Sentinel-1 data. Landslides 15(1):123–133

Intrieri E, Carlà T, Gigli G (2019) Forecasting the time of failure of landslides at slope-scale: a literature review. Earth Sci Rev 193:333–349

Intrieri E, Dotta G, Fontanelli K, Bianchini C, Bardi F, Campatelli F, Casagli N (2020) Operational framework for flood risk communication. Int J Disaster Risk Reduct 46:101510

Iovine G, Petrucci O, Rizzo V, Tansi C (2006) The March 7th 2005 Cavallerizzo (Cerzeto) landslide in Calabria—Southern Italy. In: Proceedings of the 10th IAEGCongress, Nottingham, Great Britain. Geological Society of London, 785, 12 pp

Kong VWW, Kwan JSH, Pun WK (2020) Hong Kong's landslip warning system—40 years of progress. Landslides 1–11

Kothari UC, Momayez M (2018) New approaches to monitoring, analyzing and predicting slope instabilities. J. Geol. Min. Res. 10 (1):1–14

Kromer R, Lato M, Hutchinson DJ, Gauthier D, Edwards T (2017) Managing rockfall risk through baseline monitoring of precursors using a terrestrial laser scanner. Can Geotech J 54:953–967. https://doi.org/10.1139/cgj-2016-0178

Le Breton M, Baillet L, Larose E, Rey E, Benech P, Jongmans D, Guyoton F, Jaboyedoff M (2019) Passive radio-frequency identification ranging, a dense and weather-robust technique for landslide displacement monitoring. Eng Geol 250:1–10

Li T (1983) A mathematical model for predicting the extent of a major rockfall. Zeitschrift für Geomorphologie 27(4):473–482

Lombardi L, Nocentini M, Frodella W, Nolesini T, Bardi F, Intrieri E, Carlà T, Solari L, Dotta G, Ferrigno F, Casagli N (2017) The Calatabiano landslide (Southern Italy): preliminary GB-InSAR monitoring data and remote 3D mapping. Landslides 14(2):685–696

Lu P, Casagli N, Catani F, Tofani V (2012) Persistent scatterers interferometry hotspot and cluster analysis (PSI-HCA) for detection of extremely slow-moving landslides. Int J Remote Sens 33(2):466–489. https://doi.org/10.1080/01431161.2010.536185

Luzi G, Pieraccini M, Mecatti D, Noferini L, Guidi G, Moia F, Atzeni C (2004) Ground-based radar interferometry for landslides monitoring: atmospheric and instrumental decorrelation sources on experimental data. IEEE Trans Geosci Remote Sens 42(11):2454–2466

Luzi G, Monserrat O, Crosetto M, Copons R, Altimir J (2010) Ground-based SAR interferometry applied to landslide monitoring in mountainous areas, 24–26. In: Mountain risks conference: bringing science to society, Firenze, Italy

Macciotta R, Hendry M, Roghani A (2016) Developing hazard management strategies based on tolerable risk to railway operations

Mak SH, Au Yeung YS, Chung PWK (2007) Public education and warnings in landslide risk reduction. A Commemorative Volume Published in Conjunction with the 40th Anniversary of the Southeast Asian Geotechnical Society, Kuala Lumpur, Malaysia, pp 367–375

Maramai A, Graziani L, Tinti S (2005) Tsunamis in the Aeolian Islands (southern Italy): a review. Mar Geol 215(1–2):11–21

Meisina C, Notti D, Zucca F, Ceriani M, Colombo A, Poggi F, Roccati A, Zaccone A (2013) The use of PSInSAR™and SqueeSAR™techniques for updating landslide inventories. In: Margottini C, Canuti P, Sassa K (eds) Landslide science and practice. Springer, Berlin, Heidelberg, pp 81–87

Meisina C, Zucca F, Macciotta R, Martin CD, Morgenstern NR, Cruden DM (2016) Quantitative risk assessment of slope hazards along a section of railway in the Canadian Cordillera—a methodology considering the uncertainty in the results. Landslides 13:115–127. https://doi.org/10.1007/s10346-014-0551-4

Monserrat O, Crosetto M, Luzi G (2014) A review of ground-based SAR interferometry for deformation measurement. ISPRS J Photogramm Remote Sens 93:40–48

Mucchi L, Jayousi S, Martinelli A, Caputo S, Intrieri E, Gigli G, Gracchi T, Mugnai F, Favalli M, Fornaciai A, Nannipieri L (2018) A flexible wireless sensor network based on ultra-wide band technology for ground instability monitoring. Sensors 18(9):2948

Mufundirwa A, Fujii Y, Kodama J (2010) A new practical method for prediction of geomechanical failure-time. Int J Rock Mech Min Sci 47(7):1079–1090

Muttillo M, Colagiovanni A, Pantoli L, Ferri G (2019) Landslides monitoring by means of low cost wired sensor networks. AISEM annual conference on sensors and microsystems. Springer, Cham, pp 143–147

Nadim F, Intrieri E (2011) Early warning systems for landslides: challenges and new monitoring technologies. In: 5th Canadian conference on geotechnique and natural hazards. Kelowna, BC, Canada, pp 15–17

Nolesini T, Frodella W, Bianchini S, Casagli N (2016) Detecting slope and urban potential unstable areas by means of multi-platform remote sensing techniques: the Volterra (Italy) case study. Remote Sens 8(9):746

Notti D, Cina A, Manzino A, Colombo A, Bendea IH, Mollo P, Giordan D (2020) Low-cost GNSS solution for continuous monitoring of slope instabilities applied to Madonna Del Sasso Sanctuary (NW Italy). Sensors 20(1):289

Pecoraro G, Calvello M, Piciullo L (2019) Monitoring strategies for local landslide early warning systems. Landslides 16(2):213–231

Petley D (2012) Global patterns of loss of life from landslides. Geology 40:927–930

Pieraccini M, Casagli N, Luzi G, Tarchi D, Mecatti D, Noferini L, Atzeni C (2002) Landslide monitoring by ground-based radar interferometry: a field test in Valdarno (Italy). Int J Remote Sens 24:1385–1391

Pieraccini M, Casagli N, Luzi G, Tarchi D, Mecatti D, Noferini L, Atzeni C (2003) Landslide monitoring by ground-based radar interferometry: a field test in Valdarno (Italy). Int J Remote Sens 24 (6):1385–1391

Pinggen Z (2004) Indicator system and techniques of landslide monitoring. J Geomech 1:19–26

Pratesi F, Nolesini T, Bianchini S, Leva D, Lombardi L, Fanti R, Casagli N (2015) Early warning GBInSAR-based method for monitoring Volterra (Tuscany, Italy) city walls. IEEE J Sel Top Appl Earth Obs Remote Sens 8:1753–1762

Ramesh MV, Kumar S, Rangan PV (2009) Wireless sensor network for landslide detection. Proc ICWN. 2009:89–95

Raspini F, Ciampalini A, Del Conte S, Lombardi L, Nocentini M, Gigli G, Ferretti A, Casagli N (2015) Exploitation of amplitude and phase of satellite SAR images for landslide mapping: the case of Montescaglioso (South Italy). Remote Sens 7(11):14576–14596

Raspini F, Bianchini S, Ciampalini A, Del Soldato M, Solari L, Novali F, Del Conte S, Rucci A, Ferretti A, Casagli N (2018) Continuous, semi-automatic monitoring of ground deformation using Sentinel-1 satellites. Sci Rep 8(1):1–11

Raspini F, Bianchini S, Ciampalini A, Del Soldato M, Montalti R, Solari L, Tofani V, Casagli N (2019) Persistent Scatterers continuous streaming for landslide monitoring and mapping: the case of the Tuscany region (Italy). Landslides 16(10):2033–2044

Rickenmann D (2005) Runout prediction methods. In: Jakob M, Hungr O (eds) Debris-flow hazards and related phenomena. Praxis, Chichester, pp 305–324

Rose ND, Hungr O (2007) Forecasting potential rock slope failure in open pit mines using the inverse velocity method. Int J Rock Mech Min Sci 44(2):308–320

Rosi A, Berti M, Bicocchi N, Castelli G, Corsini A, Mamei M, Zambonelli F (2011) Landslide monitoring with sensor networks: experiences and lessons learnt from a real-world deployment. Int J Sens Netw 10(3):111–122

Rosi M, Levi ST, Pistolesi M, Bertagnini A, Brunelli D, Cannavò V, Di Renzoni A, Ferranti F, Renzulli A, Yoon D (2019) Geoarchaeological evidence of middle-age tsunamis at Stromboli and consequences for the tsunami hazard in the Southern Tyrrhenian Sea. Sci Reports 9(1):1–10

Rosser N, Lim M, Petley D, Dunning S, Allison R (2007) Patterns of precursory rockfall prior to slope failure. J Geophys Res 112: F04014. https://doi.org/10.1029/2006JF000642

Saito M (1969) Forecasting time of slope failure by tertiary creep. In: Proceedings of the 7th international conference on soil mechanics and foundation engineering, Mexico City, vol 2, pp 677–683

Sassa K (2015) ISDR-ICL Sendai Partnerships 2015–2025 for global promotion of understanding and reducing landslide disaster risk. Landslides 12(4):631–640

Sassa K (2019) The fifth world landslide forum and the final draft of the Kyoto 2020 commitment. Landslides 16(2):201–211

Scaringi G, Fan X, Xu Q, Liu C, Ouyang C, Domènech G, Dai L (2018) Some considerations on the use of numerical methods to simulate past landslides and possible new failures: the case of the recent Xinmo landslide (Sichuan, China). Landslides 15(7):1359–1375

Schaefer LN, Di Traglia F, Chaussard E, Lu Z, Nolesini T, Casagli N (2019) Monitoring volcano slope instability with synthetic aperture radar: a review and new data from Pacaya (Guatemala) and Stromboli (Italy) volcanoes. Earth Sci Rev 192:236–257

Singh K, Tordesillas A (2020) Spatiotemporal evolution of a landslide: a transition to explosive percolation. Entropy 22:67

Solari L, Raspini F, Del Soldato M, Bianchini S, Ciampalini A, Ferrigno F, Casagli N (2018) Satellite radar data for back-analyzing a landslide event: the Ponzano (Central Italy) case study. Landslides 15(4):773–782

Solari L, Del Soldato M, Montalti R, Bianchini S, Raspini F, Thuegaz P, Casagli N (2019) A Sentinel-1 based hot-spot analysis: landslide mapping in north-western Italy. Int J Remote Sens 40(20):7898–7921

Stock GM, Martel SJ, Collins BD, Harp EL (2012) Progressive failure of sheeted rock slopes: the 2009–2010 Rhombus Wall rock falls in Yosemite Valley, California, USA. Earth Surf Process Landforms 37:546–561. https://doi.org/10.1002/esp.3192

Tarchi D, Ohlmer E, Sieber AJ (1997) Monitoring of structural changes by radar interferometry. Res Nondestr Eval 9:213–225

Tarchi D, Casagli N, Fanti R, Leva D, Luzi G, Pasuto A, Pieraccini M, Silvano S (2003) Landslide monitoring by using ground-based SAR interferometry: an example of application to the Tessina landslide in Italy. Eng Geol 1(68):15–30

Terzis A, Anandarajah A, Moore K, Wang I-J (2006) Slip surface localization in wireless sensor networks for landslide prediction. In: Proceedings of the international conference on sensor network, pp 109–116

Thiebes B, Glade T (2016) Landslide early warning systems–fundamental concepts and innovative applications. In: Aversa S, Cascini L, Picarelli L, Scavia C (eds) Landslides and engineered slopes: experience, theory and practice. Proceedings of the 12th international symposium on landslides, Napoli, Italy, pp 12–19

Tofani V, Raspini F, Catani F, Casagli N (2013) Persistent Scatterer Interferometry (PSI) technique for landslide characterization and monitoring. Remote Sens 5(3):1045–1065

Tordesillas A, Walker DM, Andò E, Viggiani G (2013) Revisiting localized deformation in sand with complex systems. Proc R Soc A Math Phys Eng Sci 469:20120606

Tordesillas A, Zhou S, Di Traglia F, Intrieri E (2020) New insights into the spatiotemporal precursory failure dynamics of the 2017 Xinmo landslide and its surrounds. WLF Kyoto, Japan (this volume)

Turchi A, Di Traglia F, Luti T, Olori D, Zetti I, Fanti R (2020) Environmental aftermath of the 2019 Stromboli eruption. Remote Sens 12(6):994

UNISDR (2006) Developing an early warning system: a checklist. The Third International Conference on Early Warning (EWC III). https://www.unisdr.org/2006/ppew/info-resources/ewc3/checklist/English.pdf. Last accessed March 2016, 2006

UNISDR (2009) Terminology on disaster risk reduction. https://www.undrr.org/publication/2009-unisdr-terminology-disaster-risk-reduction. Last accessed March 2020

Varnes DJ, IAEG Commission on Landslides (1984) Landslide hazard zonation: a review of principles and practice, vol 3. UNESCO, Paris, France, 63 pp

Voight B (1989) Materials science law applies to time forecasts of slope failure. Landslide News 3:8–11

Xu Q, Yuan Y, Zeng YP, Hack R (2011) Some new pre-warning criteria for creep slope failure. Sci China Tech Sci 54(1):210–220

Zavodni ZM, Broadbent CD (1978) Slope failure kinematics. Bull Can Inst Min 73:69–74

Part I

Monitoring and Remote Sensing for Landslide Risk Mitigation

Defining Kinematic and Evolutive Features of Earth Flows Using Integrated Monitoring and Low-Cost Sensors

Paola Revellino, Luigi Guerriero, Giuseppe Ruzza, and Francesco M. Guadagno

Abstract

Mid to long-term monitoring of earth flow displacements is essential for the understanding of their kinematic features, process dynamic and evolution, and designing of mitigation measures. This paper summarizes methods, results, and interpretations of monitoring activities carried out between 2006 and 2020 at three earth flow sites in southern Italy characterised by structurally and lithologically complex slopes: (1) the Montaguto, (2) the Mount Pizzuto, and (3) the Pietrafitta landslides. By integrating traditional monitoring techniques and specifically developed low-cost sensors, kinematic and evolutive features of the three earth flows were analyses allowing detailed reconstruction of the relationship among basal-slip surface geometry, deformation styles and pattern, geomorphic structures, movement velocity and sediment discharge during ordinary and extraordinary movements. Final results highlight that earth flows are composed of distinct kinematic zones with characteristic longitudinal velocity profiles. Velocity variation along a kinematic zone, which is controlled by the basal and lateral geometry of the slip surface, is consistent with the distribution of structures on the ground surface of the flows, reflecting stretching and shortening of material during movement. Seasonal movements characterized by alternation between relatively slow persistent movement and acceleration are induced by material recharge passing through each kinematic zone and depends from this amount. Finally, it is empathised the use of low-cost sensors for displacement monitoring associated with traditional instrumentations, which give the advantage to obtain multiple stations distributed over large areas and reduce the cost of expensive monitoring campaigns.

Keywords

Earth flow • Kinematic • Displacement • Evolution • Monitoring • Low-cost sensors

Introduction

Earth flows are among the most common mass movement in nature, and are pervasive in many rapidly eroding landscapes (Mackey et al. 2009; Revellino et al. 2010). Their architecture derives from a complex evolution regulated by climatic aspects and the availability of material for sediment-pulse formation (Guerriero et al. 2014, 2015a, b).

Earth flow activity alternates between long periods of slow and/or localized movements and surging events (e.g. Guerriero et al. 2015a).

Persistent-slow movement of earth flows creates deformational structures at their surface (i.e. faults and folds; Guerriero et al. 2013a, b). Observation of mesoscopic structures forming their surface (e.g. Guerriero et al. 2014) shows that earth flow material is affected by both longitudinal extension and shortening, controlled by the geometry of the basal slip surface (Guerriero et al. 2014). In structurally and lithologically controlled earth flows (e.g. Pinto et al. 2016), the basal slip surface can be a series of alternating steeply and gently sloping surfaces (e.g. Guerriero et al. 2014) along the earth flow profile, which confine distinct kinematic zones operating in unison to transmit

P. Revellino (✉) · G. Ruzza · F. M. Guadagno
Department of Sciences and Technologies, University of Sannio, via de' Sanctis, 82100 Benevento, Italy
e-mail: paola.revellino@unisannio.it

G. Ruzza
e-mail: gruzza@unisannio.it

F. M. Guadagno
e-mail: guadagno@unisannio.it

L. Guerriero
Department of Earth, Environment and Resources Sciences, Federico II University of Naples, Naples, Italy
e-mail: luigi.guerriero2@unina.it

© Springer Nature Switzerland AG 2021
N. Casagli et al. (eds.), *Understanding and Reducing Landslide Disaster Risk*, ICL Contribution to Landslide Disaster Risk Reduction,
https://doi.org/10.1007/978-3-030-60311-3_2

sediment pulses along the length of the flow (Guerriero et al. 2016a, b).

While slow movement can persist for days, months, or years (Coe et al. 2003), surges in earth-flow movement are less common. Velocities of earth flows during slow, persistent, movement range from less than 1 mm/d to several meters per day (Keefer and Johnson 1983). However, velocity of several meters per day have been observed also during earth flow surge (Guerriero et al. 2013a, b). The highest localized earth-flow surge speed documented in the literature is 0.13 m/s (Hutchinson et al. 1974).

Movement velocity is controlled by hydrologic forcing, and seasonal acceleration and deceleration are induced by variation of the pore-water pressure (e.g. Grelle et al. 2014). Thus, most earth flows move faster during periods of high precipitation or snowmelt than during drier periods, and the correlation between precipitation and velocity is normally complex (Schulz et al. 2009).

Earth flow surges can occur when prolonged rainfalls are associated with the loss of efficient drainage pathways (Handwerger et al. 2013) and new sediment becomes available in the source area through retrogression of the upper boundary (e.g., Guerriero et al. 2014). In these conditions, the earth flow material can fluidize and fail catastrophically.

Each different kinematic behaviour materializes a specific hazard level that needs to be quantified on the basis of monitoring data. An accurate identification of hazard also includes the understanding of factors controlling earth flow movement (Schulz et al. 2009). In this way, a continuous monitoring record of earth flow displacement and its environmental drivers is essential in defining the dynamic of the process. Additionally, for earth flow involving human infrastructures (e.g. roads and railroads) displacement monitoring is crucial for understanding the ongoing evolution and designing mitigation measures.

Based on the above consideration, this paper summarizes methods, results and interpretation of monitoring activities carried out between 2006 and 2020 at three earth flow sites in southern Italy located on structurally complex slopes: (1) the Montaguto landslide, (2) the Mount Pizzuto landslide and (3) the Pietrafitta landslide. In particular, analyses and monitoring investigations helped to define (i) relation between structures, velocity distributions and basal slip-surface geometry within individual kinematic domains of earth flows; (ii) earth flow movement and related sediment transport during both ordinary and extraordinary movements, understanding relation and connectivity between successive kinematic zones; (iii) monitoring earth flow displacement from specifically developed low-cost sensors.

The Montaguto Earth Flow

Landslide Description

The Montaguto earth flow (Fig. 1), in southern Italy, is one of the largest active earth flows in Europe. It is located along the northern side of the Cervaro River valley.

The landslide is approximately 3 km long and involves 4–6 Mm (Giordan et al. 2013, Guerriero et al. 2013a, b). Its width ranges from 75 m at the earthflow neck to 450 m in the upper part of the earth-flow source area. The earth flow affects an area of about 67 ha and its thickness ranges from a few meters ($\sim$4) to more than 20 m near the toe. The average slope angle, excluding the headscarp, is approximately 7.2°.

Early in 2006, the earth flow remobilized and covered the east–west-trending Italian National Road SS90, damaging several farmhouses. This reactivation marked the start of the most intense period of activity because of the large volume of mobilized material and the impacts to linear infrastructure. This period culminated on March 2010, when the earth flow destroyed a segment of the national railway.

Historical records indicate that the Montaguto earth flow has about a 100-year history of periodic activity characterized by rapid surges in 1958 and 2006, alternating with longer periods of quiescence with small and localized displacements (Guerriero et al. 2013a, 2015a, b). Additional surges involving only part of the earth flow occurred in 2009 and 2010.

Starting from April 2010, current activity is controlled by mitigation measures consisting of deep and surficial drainage and a retaining wall at the toe.

The earth-flow path is complex as it is controlled by major geologic structures associated (Guerriero et al. 2014, 2019b). The slope is constituted by Miocene and Pliocene flysch formations that are lithologically complex, containing a wide variety of clay beds, marls, sandstones, and conglomerates 11. This geological complexity influences groundwater flow and many springs are present from 600 m above sea level to the top of the mountain (Diodato et al. 2014). Several groups of springs are located along the western flank of the earth-flow and at/near the head of earth-flow (Fig. 2).

Deformational Structures and Segmentation

Deformational structures and hydrologic features of the Montaguto earth flow were mapped and monitored starting form May 2010 (Guerriero et al. 2013a, b) using real-time kinematic (RTK) Global Positioning System (GPS) technique, with a dual-frequency GPS receiver.

Fig. 1 The Montaguto earth flow on 27 April 2006. Photo taken from a helicopter looking south toward the earth-flow toe

The mapped distribution of the structures was used to identify kinematic zones formed by major paired driving and resisting earth-flow elements.

Deformational structures comprise normal faults indicating earth-flow stretching (i.e. driving element), thrust faults indicating earth-flow shortening (i.e. resisting element), back-tilted surfaces indicating backward rotation, and strike-slip faults bounding the earth-flow moving core. This mapped distribution of the structures was used to identify kinematic zones formed by major paired driving and resisting elements.

Each kinematic zone had an area of stretching with one or more normal faults at its head and an area of shortening with one or more back-tilted surfaces or thrust faults at its toe (Fig. 3). In some cases, kinematic zones had a central area of no stretching or shortening where earth-flow movement occurred largely by translation along discrete basal and lateral-slip surfaces. Strike-slip faults represented the surface expression of lateral-slip surfaces.

Figure 2 shows the configuration of the Montaguto earth flow in 2010 and the deformational structures. Five active kinematic zones containing structures indicating both stretching and shortening were recognized along the earth flow: the Head, the Hopper, the Neck, the Body and the Active Toe (see Guerriero et al. 2013a, 2014, b for details). Two kinematic zones were in the earth-flow source area.

Monitoring of the Movement Velocity

In order to monitoring velocity and structures within the Neck of the Montaguto earth flow, the displacement of 25 natural and artificial objects (i.e., large rock fragments, drain pipes, etc.) on the surface of the Neck kinematic zone of the Montaguto earth flow, visible in successive sets of Eros-B satellite orthoimages (25/05/2010 and 25/08/2010) were measured. The EROS-B satellite acquires panchromatic images with a nadiral Ground Sampling Distance (GSD; i.e., spatial resolution) of 0.7 m (single sided pixel dimension) and the radiometric resolution is within a spectral range of 0.5 to 0.9 μm. Objects consisting of groups of pixels were recognized on the basis of their geometry and colour (i.e., Digital Number) distribution. Corners of object were visually picked from a computer display, and displacement was manually measured in a GIS. The irregularly distributed displacement values were interpolated using the Inverse power to a distance method to produce a displacement map; deformational structures were used as breaklines. The error in displacement was assigned on the basis of the computed east–west and north–south root mean square errors in the position of 16 stable ground control points. In the Easting direction, the 2xRMS values ranged from 0.04 to 0.12 m, but averaged about 0.08 m. In the Northing direction, the 2xRMS values ranged from 0.02 to 0.18 m, but averaged

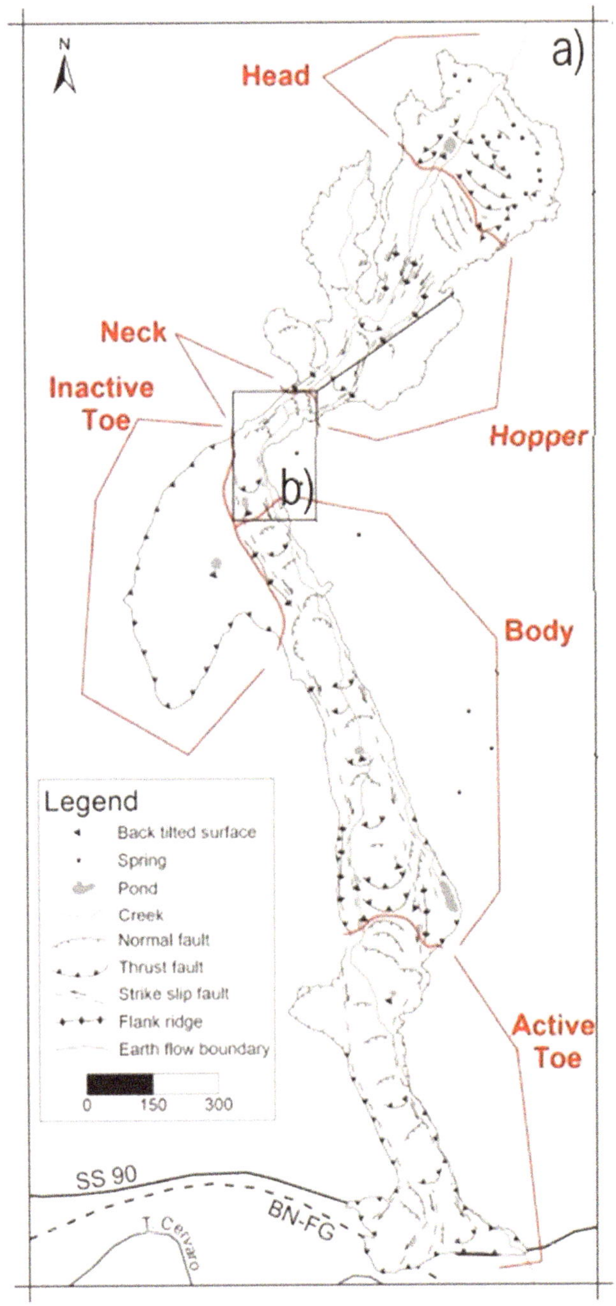

Fig. 2 Structures and kinematic zones of the Montaguto earth flow. Box **b** correspond to the sector of Fig. 4

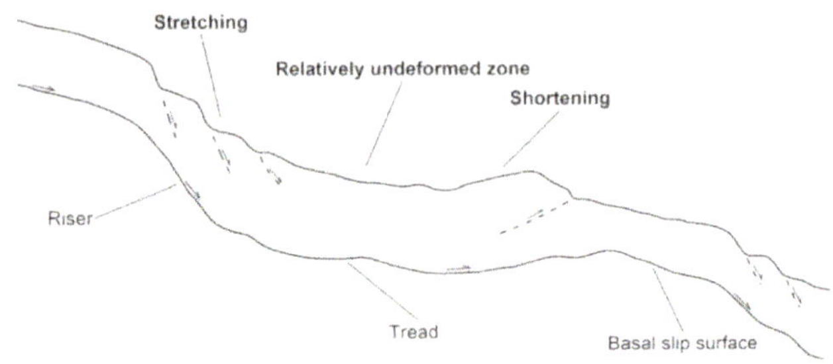

Fig. 3 Schematic longitudinal profile of a kinematic zone within the earth flow

about 0.06 m. Considering that Eastting and Northing RMS values are not equal, the error in displacement depends on the azimuth of the displacement vector, with the Easting and Northing RMS values being the maximum and the minimum errors, respectively.

Results form displacement monitoring showed that the Neck was actively moving between May and August 2010 (90 days). The average earthflow velocity over the entire monitoring period ranges from 0.016 m/d (1.4 m/90 ds) in the upper and lower parts of the kinematic zone, to 0.066 m/d (6 m/90 ds) in the middle part (Fig. 4). Variation of earth-flow velocity along the longitudinal axis of the kinematic zone is shown in the velocity profile in Fig. 4b. The distribution of displacement and average velocity (Fig. 4a) corresponds with three structural sectors characterized by (i) normal faults and a linear increase of flow velocity; (ii) a peak in velocity where the flow moving core is bounded by strike-slip faults; (iii) a decrease of earth-flow velocity associated with the existence of thrust faults and a rapid narrowing of the active flow. Displacement monitoring, and data from Guerriero et al. (2014, 2016a), indicates that the ground surface of the Montaguto earth flow consist of structures moving with different styles of deformation. This structure distribution can be used to infer the geometry of the basal-slip surface, because of relations between extensional structures and risers in the basal slip surface and compressional structures and treads in the slip surface. Such structures accommodate deformation caused variation in earth-flow velocity (i.e., acceleration and deceleration). Moreover, comparison with similar analysis in other earth flows (e.g. Guerriero et al. 2016c, d) showed that the density of structures appears to be independent of the magnitude of velocity change. From this observation, it appears that the slope angle of the basal-slip surface might control the density of deformational structures forming during flow movement.

The Mount Pizzuto Earth Flow

Landslide Description

The Mount Pizzuto earth flow (Guerriero et al. 2016a) is among the most active landslide of the Benevento province (Revellino et al. 2010; Grelle et al. 2011). It affects the northeastern side of Mount Pizzuto from about 720 m above sea level (a.s.l.) to about 550 m (a.s.l.), and involves an estimated volume of 300,000 m^3 of fine-grained flyschoid material.

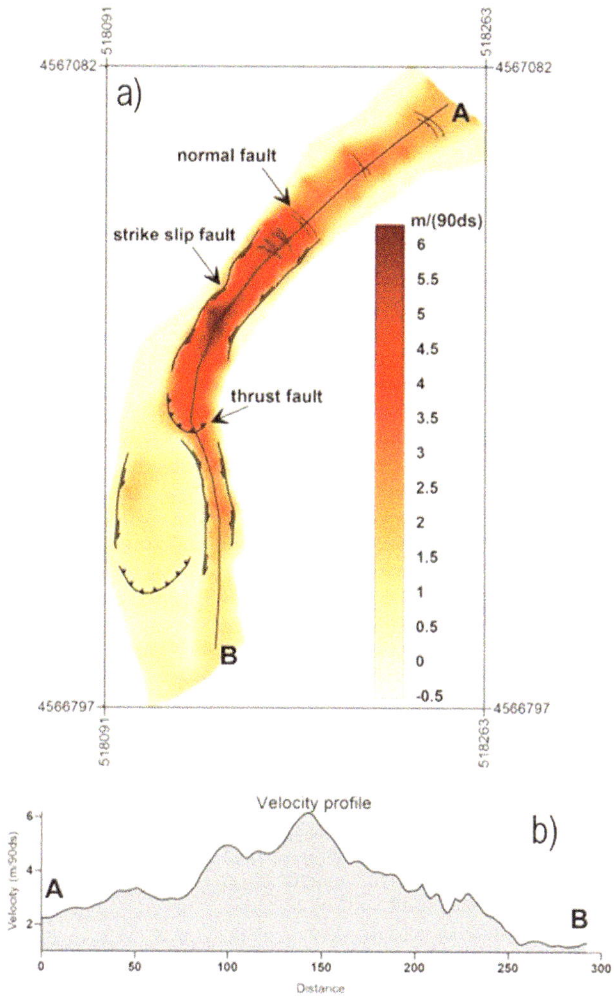

Fig. 4 **a** Structures and velocity within the earth flow Neck, and **b** velocity profile of sector in **a**. See Fig. 2b for location

From a geological viewpoint, the landslide is located at an overthrust fault between two different flysch formation. The tectonic contact between these formations materializes a WNW-ESE trending thrust fault that constitutes a weak zone where several landslide source-areas are located. It has a complex source area with two branches, a 500 m long transport zone and a fan-shaped bulging toe.

The landslide has been periodically active in the last decades and, as described by local people, early in 2006, it surged damming the Ginestra Torrent at its toe.

The earth-flow dam induced episodic floods that periodically damaged a segment of a local road and power and telephone service lines. In 2008, few mitigation structures were arranged at the foot zone consisting in a man-made ditch excavated along the torrent course and in the installation of a large diameter drain. These structures were destroyed in 2011 by a flood together with the local road and service lines. In October 2015, two new flooding events generated by large rainfall affected the landslide area considerably enlarging the bed of the Ginestra torrent.

Deformational structures (e.g. normal faults, thrusts, etc.) are disseminated along the earth flow surface. Guerriero et al. (2016a, b) mapped deformational structures and hydrologic features of the earth flow in September 2014 using real-time kinematic (RTK) Global Positioning System (GPS). Structures were identified, mapped, and processed in the same way as at the Montaguto earth flow. Based on the distribution of these structures, five kinematic zones were recognised along the earth flow: the head, the hopper, the neck, the body, and the toe (Fig. 5).

Installation, Distribution and Monitoring of GPS Points

In April 2014, 35 monitoring points were installed within the five kinematic zones (from 4 to 9 points for each zone) of the earth flow (red points 1–35, Fig. 5). Each point consists of a 70 cm long wooden-rebar quipped on its head with a red hat.

Monitoring points were placed approximately along the longitudinal axis of the earth flow or following two approximately parallel lines (i.e. along the hopper and at the toe). The monitoring points placed with a clear view to the sky were surveyed using Real Time Kinematic GPS techniques (RTK-GPS) during 17 GPS campaigns of 1 day over a period of 694 days (Table 1). In should be noted that points 16 and 24 were lost within the earth flow material after some surveys and replaced with new points on similar locations, thus compromising the recording continuity for these points.

We used a Leica Viva-Net rover equipped with a Leica GS08 dual-frequency antenna for survey 1 and surveys from 12 to 17, and a Geomax Zenith 10/20 rover for surveys from 2 to 11. Real-time correction for high-accuracy measurements was obtained through the "Regione Campania" network. Minimum and maximum horizontal errors for the Leica device are 7 and 19 mm, respectively; while for the Geomax device they are 20 and 130 mm. Earth flow displacement vectors were plotted in a GIS environment and used to describe earth flow kinematics Guerriero et al. (2019a).

Earth Flow Movement and Sediment Discharge

The sediment discharge along the earth flow transport zone was estimated using displacement/velocity data from successive sets of GPS surveys and the cross-sectional geometry of the earth flow from seismic refraction profiles (Fig. 5, green lines S1–S4). Sediment discharge was calculated considering average velocity of monitoring points moving

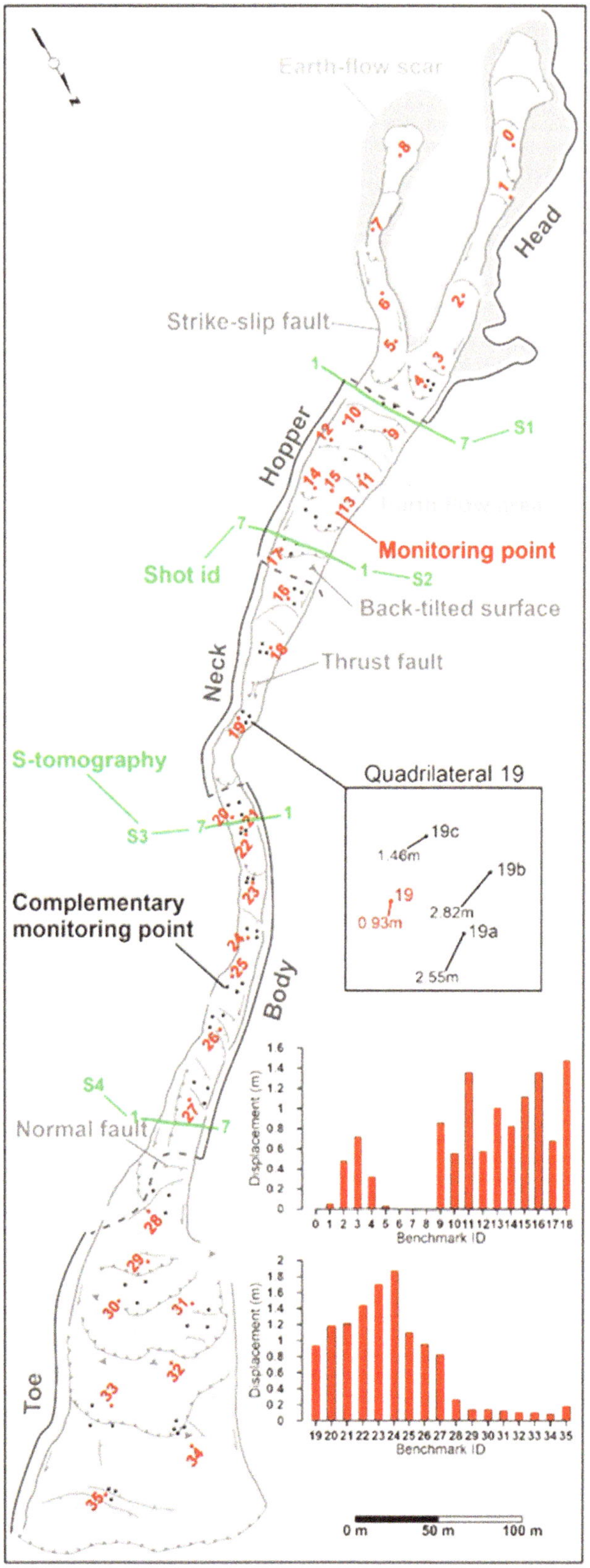

Fig. 5 Major structures, kinematic zones, monitoring points, boreholes, seismic tomographic profiles and displacement magnitude for the period May 8-Nov. 28, 2015 (bar graphs)

through the cross sections during different reference periods (Guerriero et al. 2017a, b).

The total movement of all points was largely dominated by the horizontal component. Figure 6 displays the horizontal displacement of the monitoring points installed within the different kinematic zones of the landslide. As shown, the highest cumulative horizontal displacement of the points over the entire monitoring period were recorded within the neck.

The average daily velocity ranges between 14 cm/day at point 20–0.1 cm/day at point 33 for the all period.

Most of the earth flow movement occurred during two main episodes (Fig. 6): a first surging event between April and August, 2014, and a second event between December 2014 and May/June 2015. During these episodes, a maximum velocity of 250 cm/day was recorded in 2014 and the earth flow movement assumed the characteristics of a surge.

To calculate sediment discharge, the earth flow transport zone comprises from sections S1 to S4 (Fig. 5) were considered. Estimation of sediment discharge passing through sections S1–S4 is displayed in Fig. 7, were bar graphs and line graphs show sediment discharge calculated over shorter and larger periods, respectively.

During the surging episode of the spring–summer 2014, the sediment discharge averaged over almost 5 months of activity was approximately constant through the transport zone.

Therefore, the earth flow was active during the entire monitoring period with both ordinary and extraordinary movements and a major phase of acceleration (i.e. earth flow surge) during the spring 2014, also in response to the intense rainfall occurred between end of winter and beginning of spring (e.g. Guerriero et al. 2015a, b).

Field observations and displacement data suggest that the surging reactivation/phase started from the upper part of the source area and propagated downslope zones. During such an episode, the kinematic zones were completely active and operated in unison to transmit sediment along the transport zone with a constant sediment discharge. In this period, sediment discharge calculated at transition sections between the head and hopper, and the hopper and neck decreased in time indicating a progressive depletion.

Conversely, in the transition sections between the neck and body, and between the body and toe, sediment discharge increased in a first period and then successively decreased (Fig. 7). These observations seem to suggest that earth flow movement propagated from the source area towards the toe.

The second phase of acceleration occurs in 2015. The displacement/velocity data (Fig. 6) seem to indicate that this seasonal acceleration initiated almost simultaneously within all kinematic zones. It started at the upper end of each kinematic zone and propagated downslope (i.e. through the driving element) while the lower part of the zone was not

Table 1 Survey campaigns at the Mount Pizzuto earth flow

GPS surveys	Date	GPS surveys	Date
1	April 8, 2014	10	December 19, 2014
2	May 9, 2014	11	February 20, 2015
3	May 21, 2014	12	May 8, 2015
4	June 5, 2014	13	July 11, 2015
5	June 17, 2014	14	September 6, 2015
6	July 17, 2014	15	November 28, 2015
7	July 29, 2014	16	January 29, 2016
8	August 30, 2014	17	March 2, 2016
9	October 25, 2014		

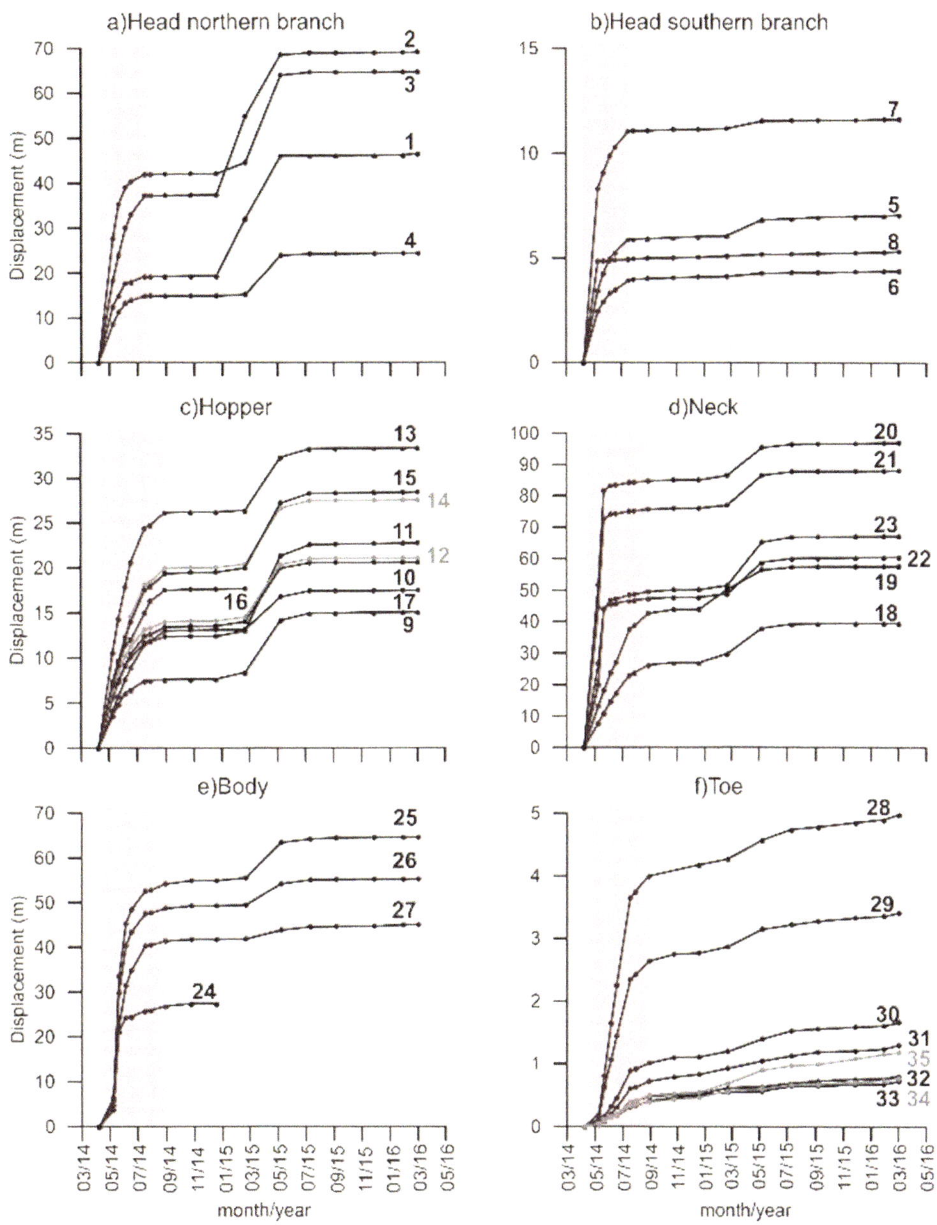

Fig. 6 Incremental horizontal displacement of monitoring points installed within the different kinematic zones of the earthflow. Pink and grey rectangles indicate periods of acceleration in 2014 and 2015, respectively

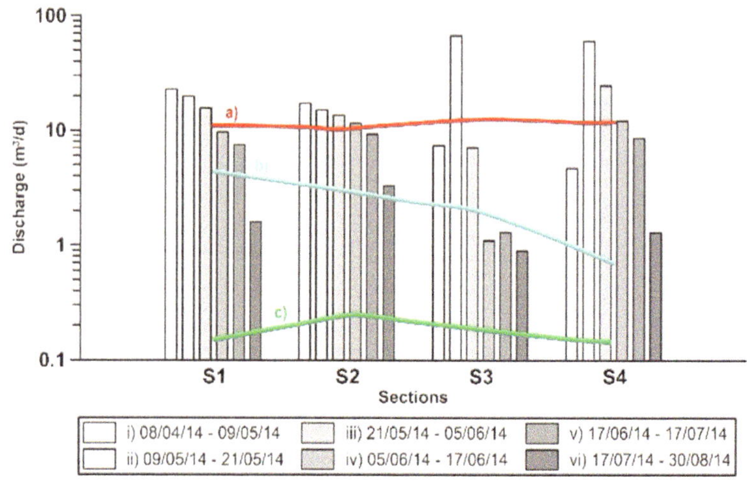

Fig. 7 Computed sediment discharge at cross sections S1, S2, S3, S4 (Fig. 5). **a** 08.04.14–30.08.14; **b** 30.08.14–08.05.15; **c** 05.08.15–03.02.15. See Table 1 for periods

moving or was moving with a low velocity. After this acceleration a period of ordinary activity with local slow movement interrupted by short periods of dormancy characterizes the earth flow underlying the intermittent nature of these phenomena. In this period, the sediment discharge was not constant along the flow.

Displacement and Velocity Profiles

Monitoring points shown in Fig. 5 were used for analysing displacement and velocity of movement within the different sector of the earth flow. Considering only position data derived by the first and the twelfth survey, earth-flow displacement and direction were determined. All of the monitoring points moved between the monitoring period and total movement was largely dominated by the horizontal component.

The blue line of Fig. 8 shows how displacement varies along the length of the flow. Similar variation was also observed within the neck of the Montaguto earth flow. Each kinematic zone is characterized by an upslope area of acceleration and a downslope area of deceleration with a peak of displacement in the middle of each zone. An exception to this statement is the earth flow Toe where the displacement vectors linearly decrease toward the toe of the flow.

Excluding point 18, displacement increases linearly from the upper end of the Hopper (i.e., point 9) to the earth-flow neck (i.e., point 21). Considering that this part of the flow is characterized by a consistent reduction in the width of the earth-flow, and that field observations indicate constant sediment supply, we infer that displacement in this part of the flow is controlled by both lateral and basal geometry of the basal-slip surface. The presence of a tread causes deceleration of the flow around point 18 while the presence of risers induces acceleration of the flow.

In this context, the velocity profile along kinematic zones appear to have a consistent pattern. These observations suggest that a relation between the velocity profile of an active earth flow and its slip surface geometry does exist and that local variation of flow velocity are controlled by variations in the slope angle of the slip surface. In other words, the lateral and basal geometry of the basal slip surface controls both the kinematic segmentation of an earth flow and its velocity profile.

Deformational Pattern and Strain Computation

On April 2015, associated to the 35 monitoring points inside the Mount Pizzuto earth flow, 64 complementary monitoring points were positioned to form quadrilaterals with the existing monitoring points (Fig. 5). In each kinematic zone, we installed from 12 to 28 points and from 1 to 7 quadrilaterals. In surveying, a quadrilateral is a foursided plane polygon defined by the survey stations at its four corners that permit an estimation of the earth-flow's strain (Baum and Fleming 1991). In order to determine the size and shape of the quadrilaterals, the mapped distribution of structures of Guerriero et al. (2016c, d) were used. Especially, survey stations were positioned on the ground encompassing an area characterized by structures indicating a homogeneous style of deformation.

The points were surveyed using Real Time Kinematic GPS and real-time correction. For the analysis, all of the points were surveyed on May 08, 2015 and on November 28, 2015 (a time span of 204 days).

Earth-flow strain was computed using the program SSPX (Cardozo and Allmendinger 2009). SSPX calculates

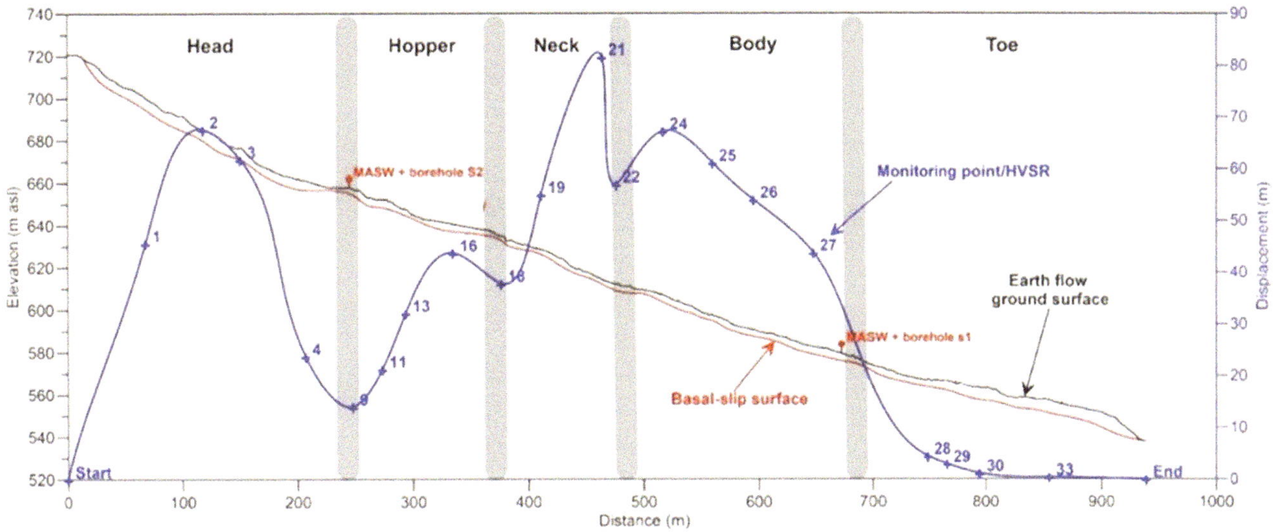

Fig. 8 Longitudinal profile of the Mount Pizzuto earth flow showing the geometry of the basal-slip surface (red line) and the displacement profile (blue line). Start and end points of the displacement profile are arbitrary taken at the beginning and at the end of the topographic profile

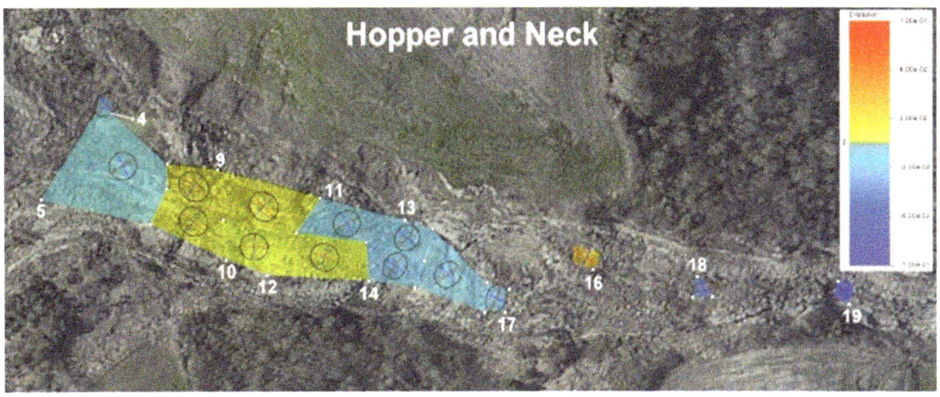

Fig. 9 Some results from strain computation. Polygons are coloured by dilatation. Strain ellipses, and principal extension (red) and shortening (blue) axes are also included. Principal strain axes are exaggerated 10 times

best-fitting strain tensors given displacement vectors at a minimum of three points in 2D. The program offers several options to compute strain. For our analysis, we used a modified version of the Delaunay routine (implemented in SSPX v. 5.8) to compute strain from the displacement of groups (or polygons) of 4 to 6 monitoring points (or vertices). Since there are more than 3 points in each polygon, 2D strain and its associated error can be calculated. The results comprise earth-flow 2D finite strain within the polygons defined by the monitoring points, magnitude and orientation of principal extension and shortening axes.

An example of the spatial distribution of analysed groups of monitoring points is shown in Fig. 9, where only the monitoring points installed in April 2014 and used for strain computation are labelled. Polygons are coloured based on the amount of dilatation, which in this 2D case gives an idea of increase or decrease of area (positive and negative dilatation, respectively).

Since the earth flow is longitudinal, dilatation also gives an indication of longitudinal stretching (positive dilatation) and longitudinal shortening (negative dilatation). Within the Hopper and Neck (from point 5 to 19), 15 polygons (13 of which are quadrilaterals) are characterized by both longitudinal stretching and shortening (positive and negative dilatation, respectively). Polygons containing point 4, 5, 11, 13, 14 and 17 are characterized by negative dilatation, while polygons containing points 10, 11, 12 and 14 are characterized by positive dilatation. In all of these cases principal strain axes orientations (and strain ellipses) are consistent with the structures orientation and deformation style in Fig. 5.

The Pietrafitta Earth Flow

Landslide Description

The Pietrafitta earth flow is an active landslide which involves the national road SS 87 connecting the cities of Benevento and Campobasso in Southern Italy. Mobilizing flysch and clay sequences, the earth flow has a length of 250 m from the source area to the toe and a width of about 100 m on the accumulation area.

The landslide affects the left side of the Reventa torrent valley from 200 to 250 m above sea level. The landslide had a complex source area with a, southern, flow-like source zone and a larger, northern, rotational-slide zone, a short transport zone, and a 40 m wide active bulging toe. Next to the southern flow-like source zone, landslide material was disrupted by multiple sets of tension cracks and isolated normal faults while, within the northern rotational-slide-like source, normal faults indicated extending rotation toward the transport zone.

This deformational pattern had some similarities with those characterizing the Montaguto and the Mount Pizzuto earth flows and was related to the longitudinal geometry of the basal slip surface controlled by the geologic complexity of the slope. In fact, the Pietraffitta landslide is an example of earthflow affecting structurally complex formations and composed of multiple kinematic zones.

Integrated Monitoring System

Due to the landslide, since 2014 one lane only of the SS87 allows the passage of the motor vehicles in an alternating way; the other lane is occupied by part of the landslide toe. In order to mitigate the risk of sudden invasions of landslide material on the transit lane and to design mitigation measures, an integrated monitoring system of the earth flow displacement was installed in March 2016 consisting of: (1) a Ground-based synthetic aperture radar interferometry (GBInSAR, Fig. 10), located in front of the landslide on the opposite slope and with an accuracy of less than 1 mm and an acquisition rate of 4 min; (2) an automatic Total Station (Robotic station), located near the GBInSAR, looking at 23 reflectors with acquisition rate of 2–6 h; (3) multi-temporal scans with terrestrial laser scanner (TLS); (4) a video surveillance system, installed at the toe and h24 working; and (5) an experimental low-cost Arduino®-based wire extensometer (Guerriero et al. 2017a, b) placed along the left flank of the earth flow toe discussed in the following paragraph.

The joint use of different accurate monitoring techniques allowed to detect not only real-time displacement of the landslide body but also critical conditions of movement acceleration or material invasion on the road. The integration and comparison among different displacement information from the simultaneous combination of different monitoring techniques adequately allowed to undertake safety countermeasure and alert procedures in order to stop the traffic at the right time on the road.

Starting from 2018, due to some mitigation works, landslide activities reduced at the landslide toe and a new configuration of the monitoring system was progressively undertaken. The monitoring system consist of (Fig. 11): (1) 10 GPS targets (Fig. 12); (2) an experimental continuous low-cost GPS; (3) an experimental low-cost multi-sensor station for measuring water discharge from the drainage trench, consisting of rain gauge, weir and sensor for water level measurement; (4) an experimental alarmed network for load-based early warning system; (5) 2 video surveillance systems, installed at the toe and along the channel and h24 working; (6) a wire extensometer at the head scarp; (7) an experimental ultrasonic barrier at the landslide toe; (8) multi-temporal scans with terrestrial laser scanner (TLS) and UAV.

In particular, as for the experimental instrumentations, with the continuous low-cost GPS (see Fig. 11 for location) it has been experimented the potential of using a Master-Rover RTK Ublox modules to derive high precision landslide displacements. The Master station is installed outside from the landslide broadcasting corrections to the Rover with a 1hz frequency. In case of displacement higher than 40 cm, the system managed through an Arduino UNO board, is programmed to send alert SMSs to a list of numbers. This system was tested for the first time in a high risk condition demonstrating its suitability.

Moreover, for monitoring water drained by a drainage trench installed within the landslide, a multi-sensor station was assembled (picture and location in Fig. 11). More specifically, a squared concrete box was modified in order to allow the anchorage of an aluminium weir. It was triangularly shaped with an apex angle of 30° allowing a measuring range of between 1 and 100 l/min. Discharge estimation is completed measuring the level within the box through an ultrasonic sensor characterized by a resolution of 1 mm and a range of 5 m.

The monitoring system is equipped with a rain gage of 0.25 mm of resolution. Sensed data are register every 5 min. Figure 13 shows flow discharge and rainfall recorded by the multi-sensor station for a four-day period.

Due to the clay nature of the involved material and the flow mechanic, the Pietrafitta landslide alternates between long term slow movement and rapid acceleration that in some cases mimics a surging event. Figure 14 shows a recent surging event, materialised by a difference of DEM,

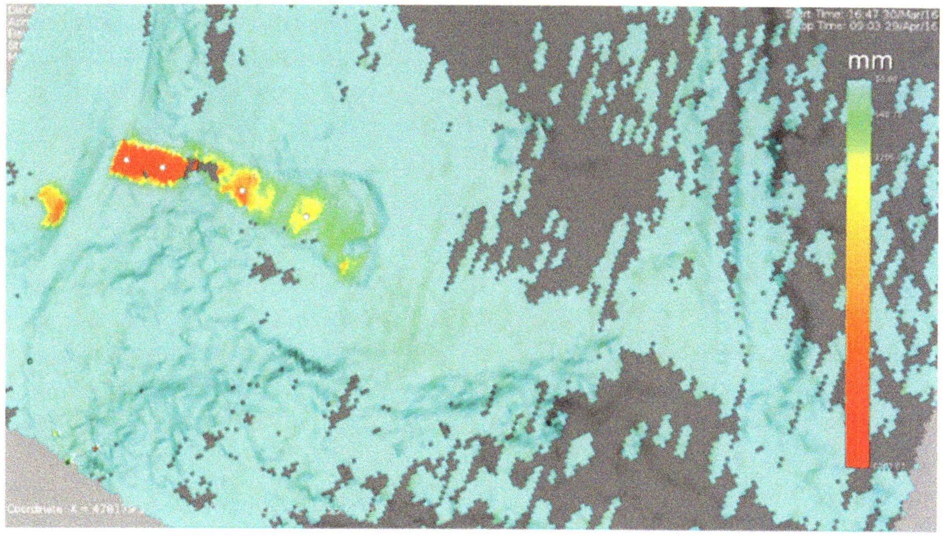

Fig. 10 Example of displacements measured by the GB-InSAR from March, 30th and April, 29th 2016

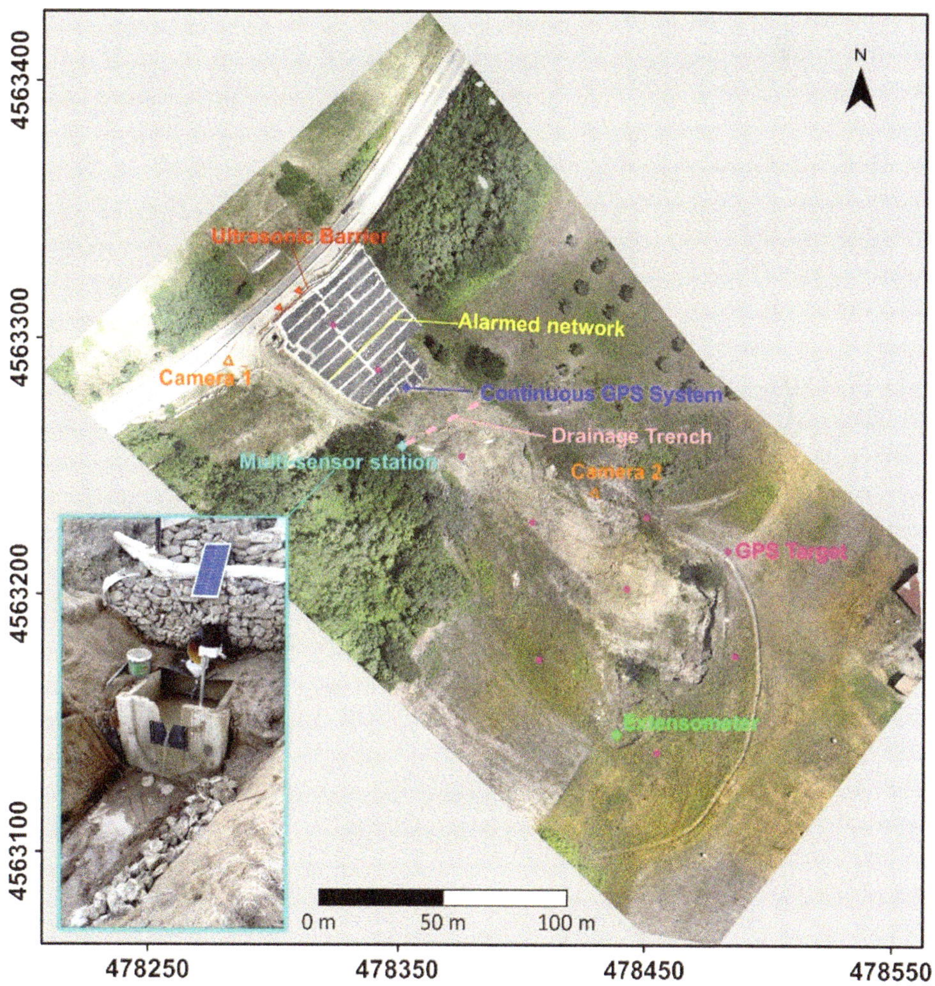

Fig. 11 New configuration of the monitoring system of the Pietrafitta landslide between early 2018 and late 2019. Image of 27.07.2018

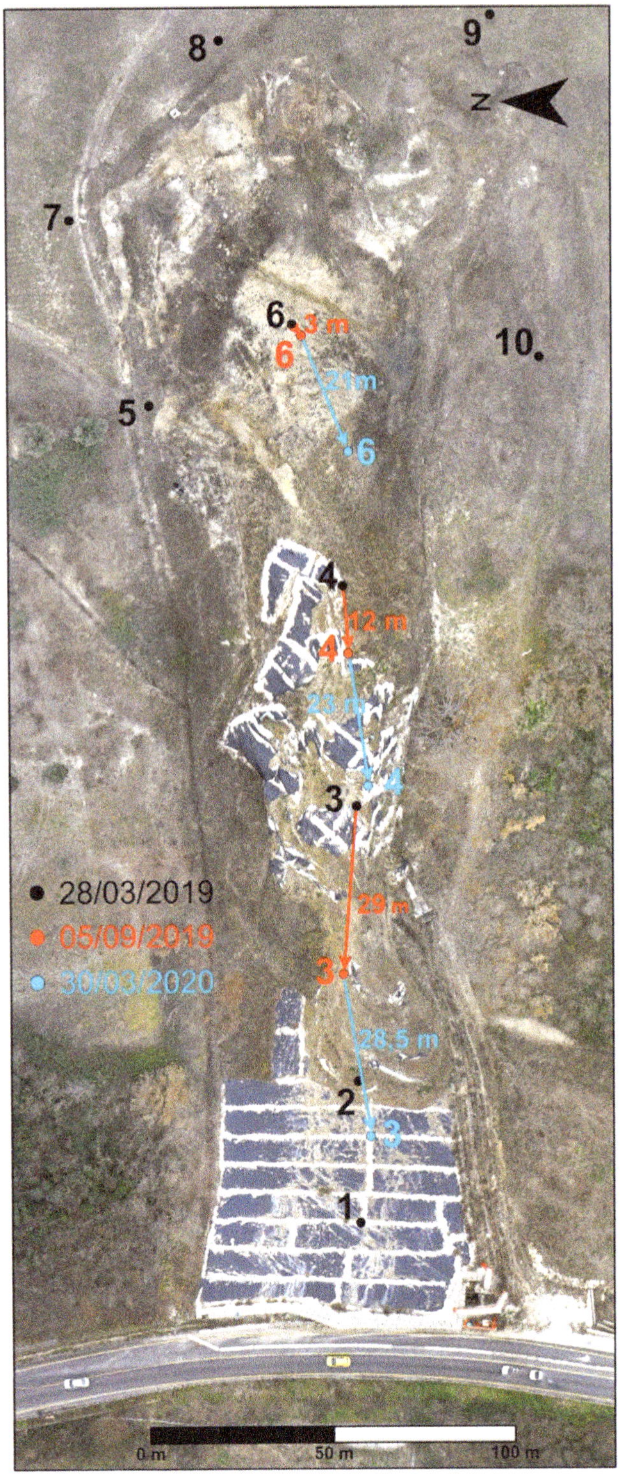

Fig. 12 Example of vectors of horizontal displacements obtained from GPS data for the range periods 28.03.2019–05.09.2019 (red vectors) and 05.09.2019–30.03.2020 (light blue vectors)

that is superimposing on the landslide toe and approaching to the road. On this basis and in order to manage landslide risk associate to the presence of the National road, an early warning system based on the use of an instrumented steel network barrier was developed and installed. Especially, using steel rebars a steel network was installed normal to the direction of the movement. For early warning purpose, a load cell was installed at the network and connected with a stable point through an auxiliary cable. The system was configured considering a threshold of force that, once exceeded, trigger an alert email message. For this purpose a MRK 1200 Arduino was used. This board is equipped with a SIGFox communication shield that allow for communicating also in absence of a GPRS signal.

Finally, for controlling landslide toe advancement, an array of ultrasonic sensors was installed on the Jersey bars protecting the road from the landslide in case of sudden advancement. Especially, since the National road operates only in one direction and bidirectional vehicular traffic was managed by alternating flux, two sets of Jersey bars were present.

For monitoring purpose, sensors were installed on a set of bars delimiting the left side of the closed roadway and used the other set as target. The distance between the two set is continuously measured by ultrasonic sensors and an Arduino UNO board registers data every 20 min. Monitoring data are used for early warning purpose by selecting a displacement threshold. Especially, since the roadway is approximately 4 m wide, we selected an alert threshold of 2 m and an alarm threshold of 3 m.

The choice to joint low-cost sensors to traditional instrumentations came from the need to find practical solutions and spatial distributed monitored points. In fact, physical monitoring can be completed with a variety of instrumentations, but most of them (i) do not allow nearly continuous monitoring, (ii) imply time-consuming and expensive monitoring campaigns and/or (iii) cannot be integrated with additional sensors.

A Low-Cost Arduino®-Based Wire Extensometer for the Earth Flow Monitoring

Wire extensometers are particularly suitable for continuous monitoring, especially when it is concentrated along well-defined shear surfaces, and can be easily integrated in multisensor monitoring systems. Major disadvantages of extensometers are their cost (a single-point

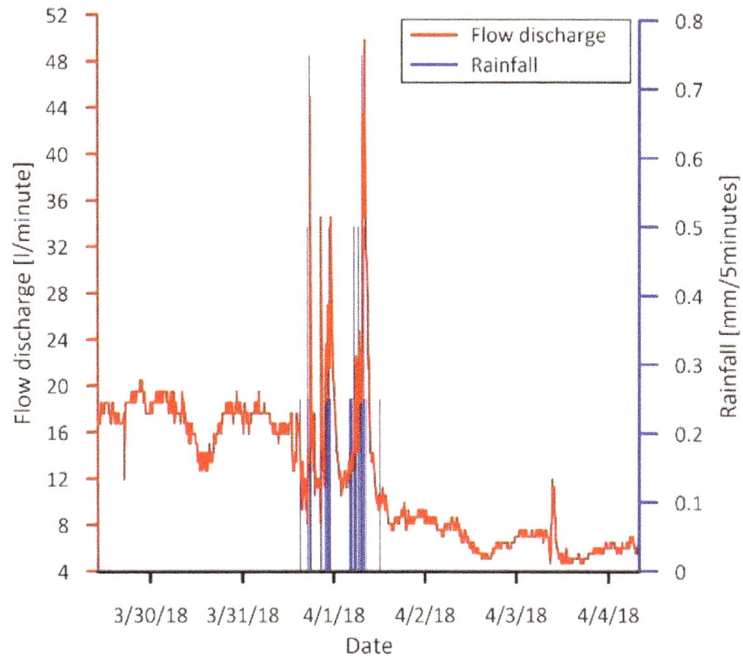

Fig. 13 Flow discharge and rainfall recorded by the multi-sensor station for a four-day period

high-performance sensor is sold at (EUR 1000) and their sensitivity to temperature is also a function of the characteristics of cabling systems.

For the Pietrafitta earth flow, we assembled a new Arduino®-based wire-rail extensometer specifically developed for monitoring earth flow movement (Guerriero et al. 2017a, b). We chose the Arduino board because it has been successfully used for the development of monitoring systems for different applications (e.g. Bitella et al. 2014). The system integrates a power unit, a data logger and an operating temperature sensor, has a very low cost (EUR 200), is configurable with different measurement ranges and accuracy and has the potential to work with additional sensors. We tested extensometer performance at the Pietrafitta earth flow and compare its measurements with those derived by successive GPS surveys and discrete rTPS measurements (Fig. 15).

The low-cost Arduino-based extensometer was installed along the left flank of the earth flow toe (Fig. 15a). The installation was completed using a 2.5 m long wire supported by several rebars, which forms a rail parallel to the strike-slip fault, materializing the left flank of the flow. In this way, the extensometer is dragged/moves along the flank registering the cumulative displacement (scheme is shown in Fig. 15b) every 30 s.

We used available displacement data to make a comparative analysis of the monitored displacement. To compare displacement measured with different systems, we installed a GPS antenna screw mounting and a rTPS target on the wire extensometer (Fig. 15b).

In particular, the earth flow toe moved approximately 1 m in 6 days and 6 h. The average velocity calculated on the basis of these data was about 6.6 mm h-1. In this part of the flow, the movement was largely dominated by the horizontal component. This makes it possible to compare the displacement measured by the extensometer and the horizontal component of the displacement vectors reconstructed with both the GPS surveys and the rTPS.

The comparison of the results indicates that the total displacement measured by the extensometer was approximately the same as that measured by combining GPS and rTPS surveys, whit a difference of about 1.5 cm (1.5%). Additionally, the displacement time series reconstructed using rTPS data perfectly fits the extensometer time series for the first 4 days of rTPS monitoring, despite a slight thermal drift of the rTPS (see red curve of Fig. 15c). In the successive 2 days the degree of fit seems to decrease and affects the measured total displacement. This was probably also caused by the deformation of the rail induced by the tilting of the ground surface around the extensometer.

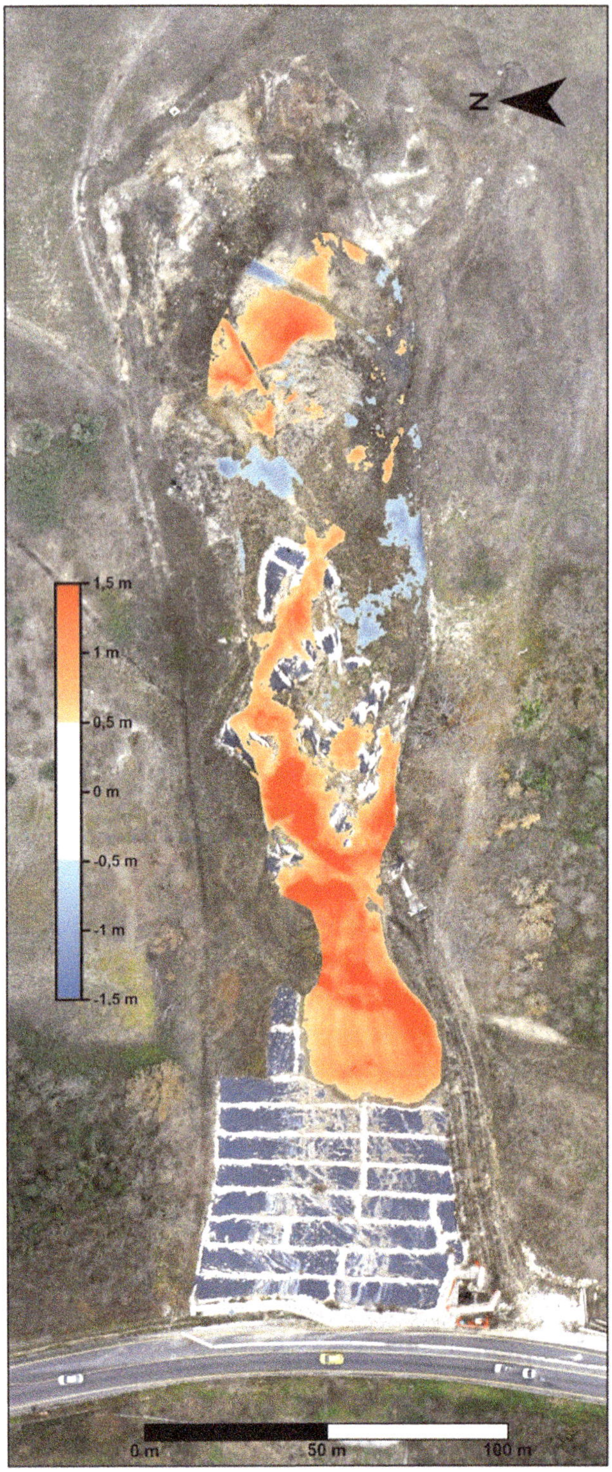

Fig. 14 DoD (Difference of DEM) from orthorectified images of 12.11.19 and 21.01.20 at the Pietrafitta landslide. Coloured scale shows the thickness of the material moved

Despite this drawback, the system exhibits a very high monitoring stability without visible thermal drift, also at operating temperatures higher than 35 °C.

Conclusion

Mid to long-term monitoring of earth flow displacements is essential for the understanding of their kinematic features, process dynamic and evolution, and designing of mitigation measures.

Monitoring activities here presented allowed to explain important aspects related to earth flow activity that can be summarised below:

(1) Earth flows are composed of distinct kinematic zones with characteristic longitudinal velocity profiles. Velocity variation along a zone, being consistent with the distribution of structures on the ground surface of the flows, is controlled by geometry of the slip surface and in particular by local variation of the slope angle;
(2) Zones of longitudinal stretching and shortening exist within earth flows. Such zones correspond to zones of acceleration and deceleration of the flow, respectively. Strain analysis indicates that the structures forming at the surface of the earth flow are consistent with dilatation computed at polygons defined by GPS monitored points.
(3) Earth flow exhibits a highly seasonal movement characterized by an alternation between acceleration surges in movement and localized and/or relatively slow persistent movement. This induced an intermittent cascade effect between kinematic zones controlled by the amount of material passing through each zone.

Finally, for landslides which need extensive and distributed monitoring points, monitoring instruments based on low-cost technology may provide several advantages including (i) the very low cost, (ii) the potential to integrate it with additional sensors, (iii) the possibility for use with different types of landslides. Therefore, the use of low-cost sensors for displacement monitoring associated with traditional instrumentations gives the advantage to obtain multiple stations distributed over large areas and reduce the cost of expensive monitoring campaigns.

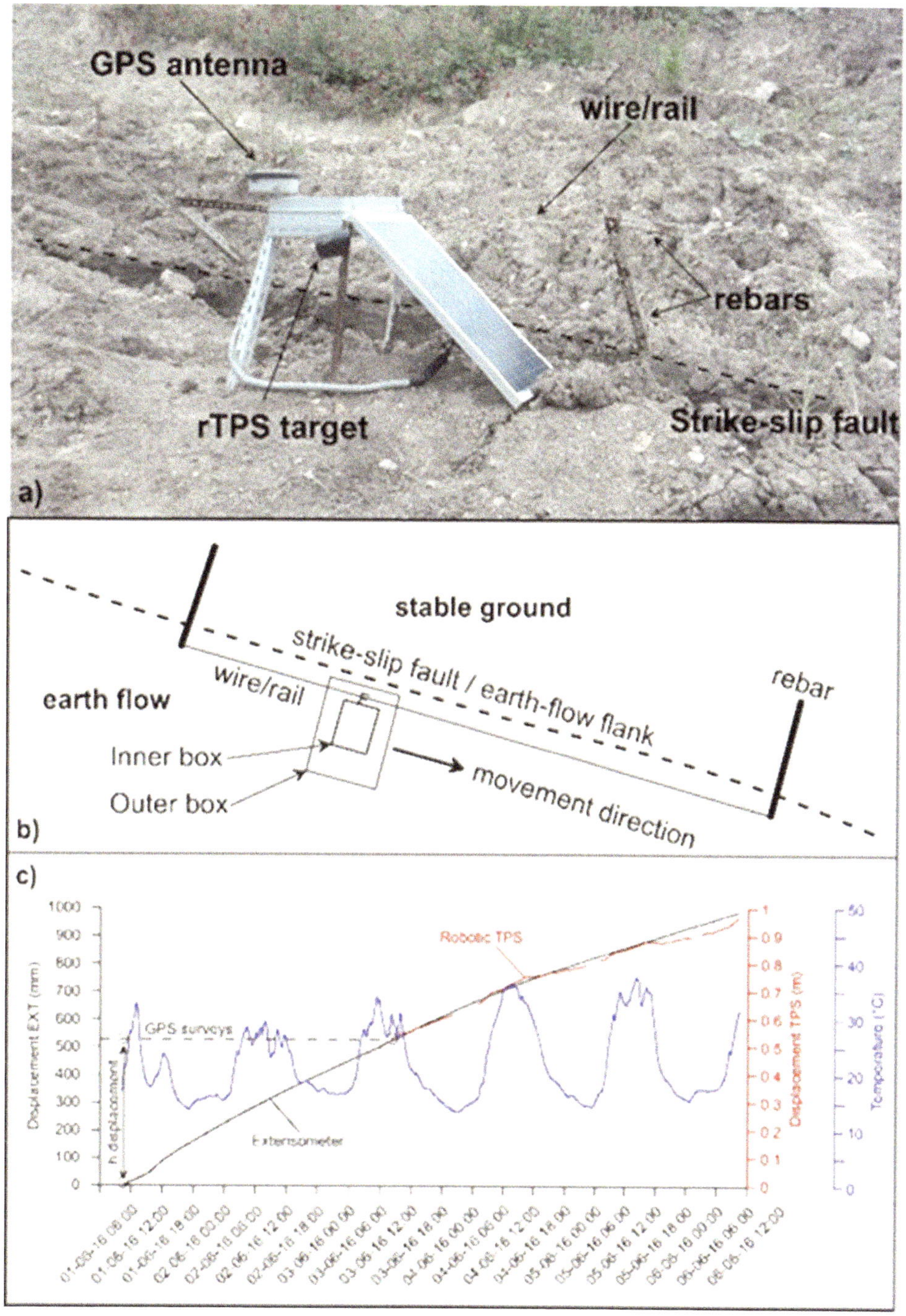

Fig. 15 **a** Installation configuration and monitoring equipment of the extensometer. **b** Installation scheme. **c** Results of displacement monitoring with the extensometer and comparison with GPS derived and rTPS results. Black circles indicate error associated with GPS surveys. Operating temperature measured by the extensometer is also shown

References

Baum RL, Fleming RW (1991) Use of longitudinal strain in identifying driving and resisting elements of landslides. Geol Soc Am Bull 103:1121–1132

Bitella G, Rossi R, Bochicchio R, Perniola M, Amato M (2014) A novel low-cost open-hardware platform for monitoring soil water content and multiple soil-air-vegetation parameter. Sensors 14:19639–19659

Cardozo N, Allmendinger RW (2009) SSPX: a program to compute strain from displacement/velocity data. Comput Geosci 35:1343–1357

Coe JA, Ellis WL, Godt JW, Savage WZ, Savage JE, Michael JA, Kibler JD, Powers PS, Lidke DJ, Debray S (2003) Seasonal movement of the Slumgullion landslide determined from Global Positioning System survey and field instrumentation. Eng Geol 68:67–101

Diodato N, Guerriero L, Fiorillo F, Esposito L, Revellino P, Grelle G, Guadagno FM (2014) Predicting monthly spring discharge using a simple statistical model. Water Resource Manage 28:969–978

Giordan D, Allasia P, Manconi A, Baldo M, Santangelo M, Cardinali M, Corazza A, Albanese V, Lollino G, Guzzetti F (2013) Morphological and kinematic evolution of a large earth flow. The Montaguto landslide, southern Italy. Geomorphology 187:61–79

Grelle G, Revellino P, Donnarumma A, Guadagno FM (2011) Bedding control on landslides: a methodological approach for computer-aided mapping analysis. Nat Hazards Earth Syst Sci 11:1395–1409

Grelle G, Soriano M, Revellino P, Guerriero L, Anderson MG, Diambra A, Fiorillo F, Esposito L, Diodato NG, Guadagno FM (2014) Space–time prediction of rainfall-induced shallow landslides through a combined probabilistic/deterministic approach, optimized for initial water table conditions. Bull Eng Geol Environ 73:877–890

Guerriero L, Revellino P, Coe JA, Focareta M, Grelle G, Albanese V, Corazza A, Guadagno FM (2013a) Multi-temporal maps of the Montaguto earth flow in southern Italy from 1954 to 2010. J Maps 9:135–145

Guerriero L, Revellino P, Grelle G, Fiorillo F, Guadagno FM (2013b) Landslides and infrastructures: the case of the Montaguto earth flow in southern Italy. Italian J Eng Geol Environ, Book Series 6:447–454

Guerriero L, Coe JA, Revellino P, Grelle G, Pinto F, Guadagno FM (2014) Influence of slip-surface geometry on earth flow deformation, Montaguto earth flow, southern Italy. Geomorphology 219:285–305

Guerriero L, Diodato N, Fiorillo F, Revellino P, Grelle G, Guadagno FM (2015a) Reconstruction of long-term earth-flow activity using a hydro-climatological model. Nat Hazards 77:1–15

Guerriero L, Revellino P, Mottola A, Grelle G, Sappa G, Guadagno FM (2015b) Multi-temporal mapping of the Caforchio earth flow, southern Italy. Rendiconti Online Della Società Geologica Italiana 35:166–169

Guerriero L, Revellino P, Luongo A, Focareta M, Grelle G, Guadagno FM (2016a) The Mount Pizzuto earth flow: deformational pattern and recent thrusting evolution. J Maps 12:1187–1194

Guerriero L, Revellino P, Bertello L, Grelle G, Berti M, Guadagno FM (2016b) Kinermatic segmentation and velocity in earth flows: a consequence of complex basal-slip surfaces. Procedia Earth Planet Sci 16:146–155

Guerriero L, Mascellaro N, Grelle G, Guadagno FM (2016c). Combined monitoring of earth flow displacement and its environmental drivers. Rendiconti Online della Società Geologica Italiana 41:167–170

Guerriero L, Cardozo N, Revellino P (2016d) Earth-flow deformation from GPS surveys, Mount Pizzuto earth flow, southern Italy. Rendiconti Online Societa Geologica Italiana 41:163–166

Guerriero L, Guerriero G, Grelle G, Guadagno FM, Revellino P (2017a) Brief communication: a low-cost Arduino®-based wire extensometer for earth flow monitoring. Nat Hazards Earth Syst Sci 17:881–885

Guerriero L, Bertello L, Cardozo N, Berti M, Grelle G, Revellino P (2017b) Unsteady sediment discharge in earth flows: a case study from the Mount Pizzuto earth flow, southern Italy. Geomorphology 295:260–284

Guerriero L, Guadagno FM, Revellino P (2019a) Estimation of earth-slide displacement from GPS-based surface-structure geometry reconstruction. Landslides 16(2):425–430

Guerriero L, Ruzza G, Cusano A, Focareta M, Revellino P, Guadagno FM (2019b) Landslide change detection and displacement tracking using nanosatellite imagery: La Montagna landslide, southern Italy. Italian J Eng Geol Environ (special issue 1):53–58

Handwerger AL, Roering JJ, Schmidt DA (2013) Controls on the seasonal deformation of slow-moving landslides. Earth Planet Sci Lett 377–378:239–247

Hutchinson JN, Prior DB, Stephens N (1974) Potentially dangerous surges in an Antrim [Ireland] mudslide. Quater J Eng Geol 7:363–376

Keefer DK, Johnson AM (1983) Earthflows: morphology, mobilization and movement. U.S. G. S. Prof. Paper 1264, p 56. Korup O, Densmore AL, Schlunegger F (2010) The role of landslides in mountain range evolution. Geomorphology 120(1–2):77–90

Mackey BH, Roering JJ, McKean JA (2009) Long-term kinematics and sediment flux of an active earthflow, Eel River California. Geology 37(9):803–806

Pinto F, Guerriero L, Revellino P, Grelle G, Senatore MR, Guadagno FM (2016) Structural and lithostratigraphic controls of earth-flow evolution, Montaguto earth flow, southern Italy. J Geologi Soc 173(4):649–665

Revellino P, Grelle G, Donnarumma A, Guadagno FM (2010) Structurally-controlled earth flows of the Benevento Province (southern Italy). Bull Eng Geol Environ (69):487–500

Schulz WH, Mackenna JP, Kibler JD, Biavati G (2009) Relations between hydrology and velocity of a continuously moving landslide—evidence of pore pressure feedback regulating landslide motion? Landslides 6:181–190

Monitoring of Thermoelastic Wave Within a Rock Mass Coupling Information from IR Camera and Crack Meters: A 24-Hour Experiment on "Branická Skála" Rock in Prague, Czechia

Ondřej Racek, Jan Blahůt, and Filip Hartvich

Abstract

Results from a 24-hour time-lapse IR camera monitoring experimental study performed on a rock mass in the city of Prague are presented. The thermal images were processed and analysed. Acquired temperatures were coupled with information from crack meters monitoring of an unstable block. It has been shown that it is feasible to directly observe the thermoelastic wave on the monitored block. Correlations of movements with monitored temperatures showed that the rock surface temperature is not the only variable that influences thermally-induced movements. The movements are probably controlled more by the overall air/rock mass temperature rather than maximum and minimum peaks. However, for better understanding of this phenomena, we suggest monitoring for a longer period including measurement of temperatures inside the rock mass.

Keywords

Thermoelastic wave • Rock mass monitoring • IR camera • Crack meters • Branická skála • Czechia

O. Racek · J. Blahůt (✉) · F. Hartvich
The Czech Academy of Sciences, Institute of Rock Structure and Mechanics, V Holešovičkách 94/41, 18209 Prague, Czechia
e-mail: blahut@irsm.cas.cz

O. Racek
e-mail: racek@irsm.cas.cz

F. Hartvich
e-mail: hartvich@irsm.cas.cz

O. Racek
Charles University in Prague, Faculty of Science, Department of Physical Geography and Geoecology, Albertov 6, 12843 Prague, Czechia

Introduction

Influence of the temperature changes on rock mass stability trough thermoelastic dilation of rock blocks has been thoroughly discussed in recent literature (Gunzburger et al. 2005; Gischig et al. 2011; Bakun Mazor et al. 2013; Collins and Stock 2016). In the long term, thermal dilatation can lead to irreversible movements and block destabilization (Bakun Mazor et al. 2013; Collins and Stock 2016). The temperature effect is usually measured directly using temperature sensors placed on the rock surface, or meteorological stations placed near the rock slope, whether dilatations of blocks are measured on discontinuities that limit the particular blocks (Boyd et al. 1973; Fantini et al. 2016; Janeras et al. 2017). However, this approach is limited to single-point measurement of rock temperature and there is no information on its spatial distribution over the whole rock mass. As a consequence, the temperature of the rock mass is known only in few points or only air temperature from the weather station is known, which can differ from the rock face significantly (Gruber et al. 2003).

This limitation can be overcome by using infra-red (IR) cameras, which enable to measure temperature changes over the whole rock face (Teza et al. 2012). Today the IR cameras are becoming affordable and are often used in geological or engineering research (Sobrino et al. 2016). Using an IR camera, it is possible to measure the spatial and temporal distribution of temperature values over the visible part of a rock mass. From the acquired thermal images or videos it is possible to visualize heating/cooling trends using appropriate software (Pappalardo et al. 2016). Thermal images are also suitable for discontinuity detection (Barla et al. 2016; Seo et al. 2017), porosity measurements (Mineo and Pappalardo 2016) heat propagation (Fiorucci et al. 2018a), or shading effect (Pappalardo and D'Olivo 2019). By further examination of thermal images, it is possible to detect sites with the highest temperature-induced stress caused by temperature changes (Fiorucci et al. 2018b).

© Springer Nature Switzerland AG 2021

N. Casagli et al. (eds.), *Understanding and Reducing Landslide Disaster Risk*,
ICL Contribution to Landslide Disaster Risk Reduction,
https://doi.org/10.1007/978-3-030-60311-3_3

In the case of rock mass monitoring, IR cameras can be used for several applications. Guerin et al. (2019) proposed a methodology which uses IR images to detect rock bridges, that are important in case of detached block stability. IR camera can also provide information about temperature anomalies. These anomalies can be caused by different lithology/mineralogic content (Yousefi et al. 2016; Liu et al. 2016), by shading effect (Pappalardo and D'Olivo 2019) or moisture-seepage (Frodella et al. 2017). During the cold season, wide continuous cracks can be mapped as they appear warmer due to duction of warmer air from inside the mass (Baroň et al. 2014). IR camera can be also used to detect rock flakes (Guerin et al. 2016) that are parts of the rock mass, which are detached from the mass itself and highly susceptible to form rockfall event (Matasci et al. 2011).

In this study, we used the IR camera as a supplement to long term monitoring performed on a studied rock face in Braník, Czechia. Experimental monitoring of a part of rock mass was performed for a 24 h period during a warm, sunny July day. The behaviour of minimal, average and maximal rock mass temperatures during the monitored period was analysed and this information was coupled with direct measurements of crack dilation on site.

Experiment Settings

The studied rock slope is called "Branická skála" and forms part of an abandoned limestone quarry situated on the right bank of Vltava River in Prague. The rock face has prevailing SW orientation, with the general aspect of 240°, dipping at 70°. The rock mass is formed by biomicrite and micrite of light grey to a grey colour layers (Czech Geological Survey 1998) with dip direction and dip of 325/50. The rock slope reaches 39 m at its highest point and it is highly dissected by numerous joints and cracks (Fig. 1). It includes frequent overhangs and loose blocks with sizes ranging from 0.2 to 3 m. Three main joint sets (325/50, 197/89, 085/62) define the rockfall detachment planes. The strength of the particular limestone layers is very variable from rather fresh rock, reaching UCS of 240 MPa to weathered rock with UCS of 90 MPa (Petružálek 2019). The slope stability of this rock mass has been assessed using different methodologies (RQD, RSR, Qslope, SMR, GSI and QTS). Despite rather variable results of the individual methods, the rock mass can be classified as conditionally unstable (Racek 2020).

An automated meteorological station is installed directly on the rock mass providing readings of air temperature, air humidity, atmospheric pressure and precipitation every 10 min. An unstable block is being monitored using three Gefran PZ-67-A-200 displacement transducers (potentiometer) with a resolution of 0.05 mm coupled with Tertium Beacon datalogger recording also the air temperature (Fig. 2). Readings are normally taken once an hour, for the sake of this experiment, the sampling time has been increased to every 10 min to match the frequency of the meteodata.

Spatial distribution of rock surface temperatures has been monitored using an infrared camera FLIR E95 with a 464 × 348 resolution @30 Hz and using 42° lens. The camera has been placed on a rock outcrop facing the upper part of the rock mass at a distance of approximately 90 m. Unfortunately, the monitored unstable block was not directly visible from this site due to dense trees, but a representative area nearby was chosen to be monitored. The experiment took accidentally place during the hottest day recorded in Prague in summer 2019. The IR monitoring started at 7:30 UTC on 25 July and ended at 7:20 UTC on 26 July. The IR images have been taken from a fixed position every 10 min. Every IR image recorded has been calibrated afterwards using air temperature and humidity records from the metro station. For this calibration, the emissivity of the rock surface has been fixed on 0.95 (Rubio and Caselles 1997; Rubio et al. 2003; Danov et al. 2007).

Results and Discussion

The temperatures from the three areas of calibrated IR thermal images have been analysed: the whole image area and on two sub-regions—rock slabs above and near the installed crack meters. 10-minute evolution of air and rock temperatures is shown in Fig. 3. It can be seen, that air temperature gradually increased and reached its peak at 14:20 UTC with 38.1 °C. After 16:40 the temperature began to drop reaching its minimum of 17.9 °C at 3:20.

The differences in temperatures of the rock surface are controlled by shadowing effect of the overhangs, orientation of the rock surface (aspect and dip) and by the elevation of the Sun above the horizon. Heating of the rock mass started at 7:40 from the southeast part of the rock mass in its upper parts that are not shadowed by trees and are close to the vertical. After these upper parts of the rock mass were heated, direct insolation started to influence rest of the mass, starting from southwest oriented slabs (left bottom corner on Fig. 4. This process started to be more obvious at 9:10. In the initial state of heating, the temperature differences were relatively high even on small distances due to the shadowing effect. The central part of the rock face with steeper inclination remained relatively cooler until the afternoon when became affected by direct sunlight. This effect was most obvious after 14:00 when the highest temperature of 73.3 °C as well the highest temperature difference of 40 °C on the rock mass surface were observed. Heating of the rock continued until 15:20 when the maximum average temperature of 47.1 °C was reached. After that, the whole rock started to cool down and the spatial distribution of temperatures

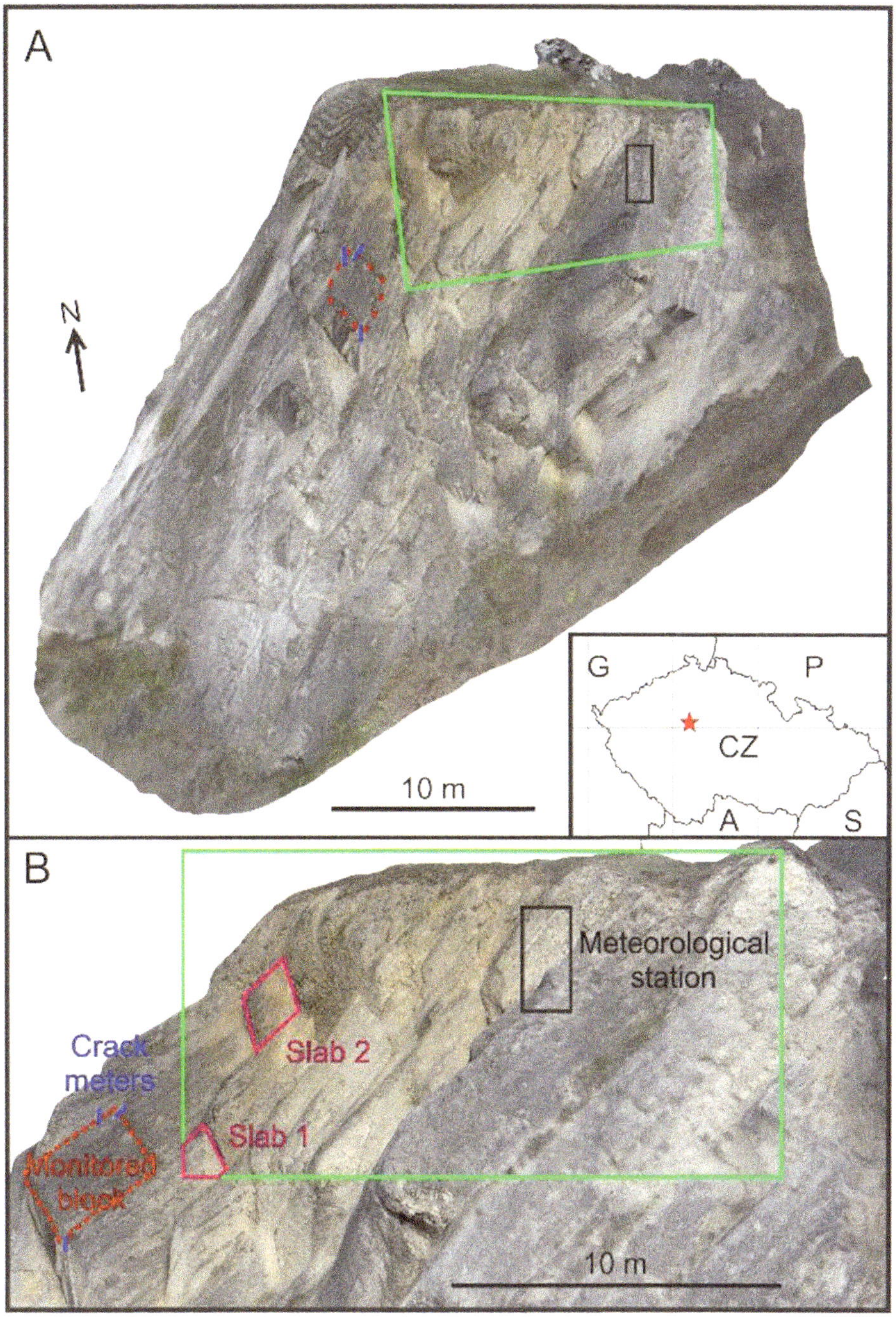

Fig. 1 Experiment location on UAV-derived model of the rock face. **a** Overview of the "Branická skála" rock mass; inset figure shows the location with a red star (CZ: Czechia, G: Germany, P: Poland, A: Austria, S: Slovakia). **b** Detail of the monitored site (green rectangle) with the location of the separately analysed slabs

became more even. From 17:30, slabs which showed high temperatures during the heating phase started to cool at a rather fast pace. However, a major part of the rock remained warm with an average of almost 30 °C at 21:30. The lowest average temperature was reached at 5:30 with a value of 21.7 °C. The lowest measured temperature (16.9 °C) was recorded at 5:00. At 7:20 we measured the lowest difference between maximal and minimal temperature: only 7.4 °C.

The parts of the rock mass warming fastest also showed fastest cooling. These parts also showed highest temperature range. Rock slab 1 is a good example of such behaviour. This slab is southwest oriented and is built by a relatively light-coloured limestone. According to the record, it is situated in one of the rock face parts with the highest daily temperature amplitude. At the beginning of the experiment, the average temperature of this slab was 22.9 °C, which is about a degree lower than the whole rock average temperature. Until 9:30, when direct sunlight reached it, the slab 1, was warming up relatively slowly about 1 °C per hour. Since 9:30, however, the heating became rather fast with a 1 °C increase every 10 min. This went on until 12:40 when the whole slab was heated up to 52.5 °C, i.e. about 10 °C warmer

Fig. 2 Crack meters on the monitored block

than the whole rock average at that time. Afterwards, the slab temperature was still increasing, but at a slower pace until reaching the maximum average temperature (54.9 °C) as well as the highest maximal temperature (64.2 °C) at 14:00. From that time, the rock slab average temperature fluctuated around 55 °C until 15:20 when direct sunlight disappeared and the cooling of the slab started. At 18:20 the slab 1 average temperature was already lower than that of the whole rock face and relatively rapid cooling continued. Slab 1 continued to cool down faster than the rest of the rock mass, the maximum difference between slab 1 and whole rock face temperature occurred at 3:30. From this point, cooling of the slab 1 slows down, however at the end of measuring it was still about 1 °C cooler than whole rock face.

The rock slab 2 is located slightly lower and more to the west with respect to the slab 1, however, their geometric orientation is similar. Also, this part of the rock is formed by grey limestone with higher roughness than slab 1. Thus it is more representative compared to conditions of the whole rock face. This is obvious from the start of measuring when the average temperature of the rock face and the slab 2 differed only by 0.3 °C. The temperature started to rise rapidly after 9:10 under direct sunlight. Peak average temperature (47.8 °C) was reached at 13:50 and it is about 0.6 °C higher and 80 min earlier than for the whole rock. Slab two is warmer than the whole rock until 15:10 when it lost the direct sunlight and began to cool down about 3 h earlier than slab 1. The biggest temperature difference between a whole rock and slab 2 was measured at 17:40 when slab 2 was 2.8 °C cooler than the whole rock. From that point, the temperature of slab 2 and the whole rock started to equalize till 6:30 with a difference of only 0.2 °C.

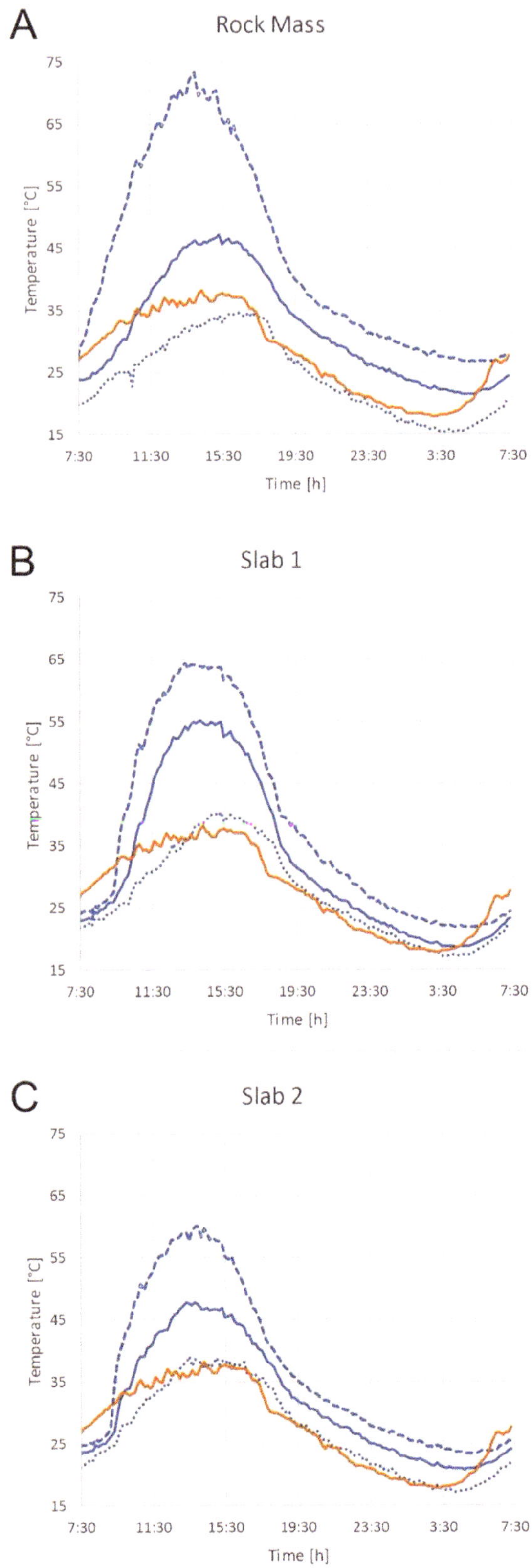

Fig. 3 Temperature graphs. Orange: Air temperature; Blue: rock surface temperatures derived from IR camera; dashed line: maximum temperatures; bold line: average temperatures; dotted line minimum temperatures. **a** whole observed area; **b** rock slab 1; **c** rock slab 2

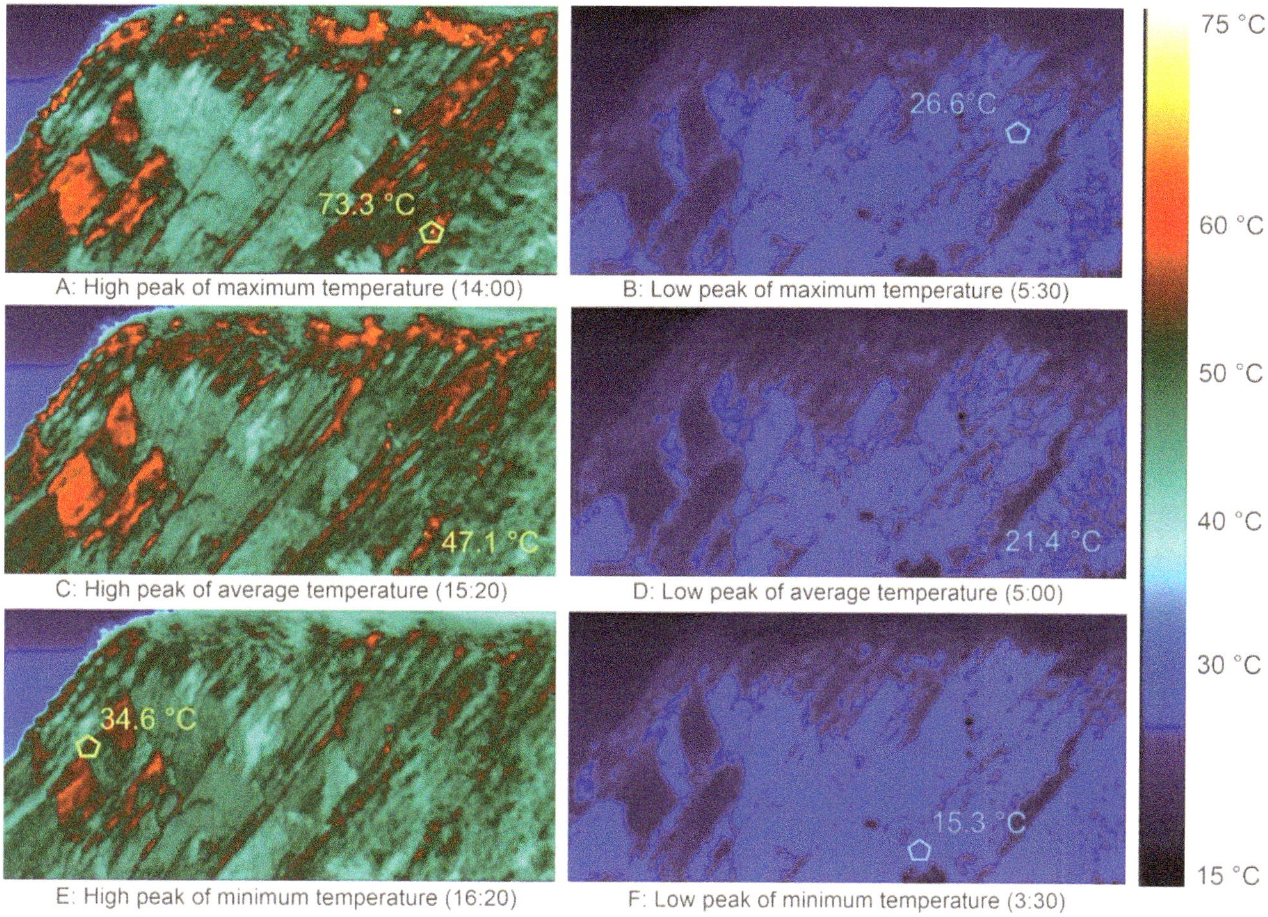

Fig. 4 IR images of the high and low peaks of maximal, average and minimal rock mass temperatures

Three crack meters are installed at "Branická skála". They are codenamed 102D_1; 102D_2 and 0CB7. The crack meters are placed on a rock block in the vicinity of rock slab 2 (Figs. 1 and 2). Unfortunately, it was not possible to monitor this particular rock block with the IR camera as the view was obscured by trees. Two of the crack meters showed similar behaviour of rock dilation when analysed using 1-h moving average (Fig. 5). A 1-h moving average was calculated because it better captures the general crack behaviour in time than direct results that are influenced by the resolution of the crack meter (0.05 mm).

All crack meters responded to increased rock surface temperature by the closing of the monitored cracks. The maximum closing time is delayed by several tens of minutes to hours after the maximum air and rock surface temperatures. Crack meter 102D_2 (Fig. 5b) did not record any obvious response to temperature-induced rock dilation. However, closing and consequent opening can still be observed during the 24-hour experiment. This less direct response of this crack meter to temperature changes might be explained by the position of the crack meter, which is not perpendicular to the monitored crack (Fig. 2) and by the resolution of the readings.

The maximum daily amplitude of movements was recorded on crack meter 102D_1 (0.35 mm). On the other crack meters, the amplitude reached 0.1–0.15 mm. This is far below the long-term amplitude, which reaches 1.40 mm in case of crack meter 102D_1.

A correlation analyses between the monitored temperatures and crack meter movements was made using Spearman correlation analyses (Spearman 1904). The best correlation was observed in the case of 102D_1 crack meter and the average temperature of the whole rock mass (Table 1). This crack meter also shows higher correlation coefficients with temperatures than the two others. In case of crack meter 102D_2 no significant correlation was found, probably due to small measured movements and position of the crack meter, which is not perpendicular to the monitored crack. However, in Fig. 5 it is still visible relation between the movement and the temperature. Lower correlation with temperature in case of crack meter OCB7 can be caused by the location of the device under the

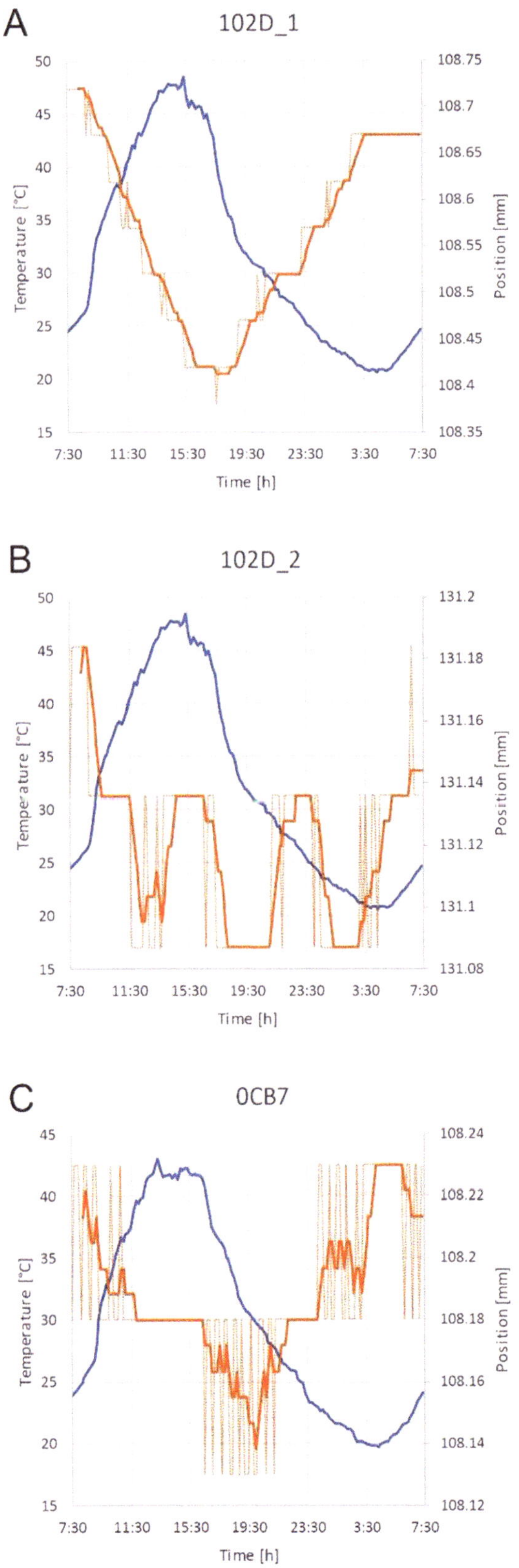

Fig. 5 Graphs of the movements recorded on the crack meters. Blue: air temperature at the datalogger; dotted orange: measured movements; bold orange: 1-h moving average

monitored block. At this place, three cracks intersect, so the measured dilatation is more complicated than the movement measured by crack meter 102D_1.

Conclusions

This experiment has shown that the rock surface temperature can significantly differ (by tens of °C) at very short distances and within a short time interval. This leads to rapid changes in rock stress at short distances, followed by reversible closing and opening of the cracks. Gradually, this frequently repeated thermal stress can lead to weathering inside the rock mass, decrease of the stability and to a rockfall (Collins and Stock 2016). It has been shown that for proper monitoring of crack movements and consequent rockfall hazard assessment the sub-daily changes are important. When using lower monitoring frequency, these changes might be recorded unconvincingly or not at all. Another issue arose from the resolution of the crack meters, as the smallest movements might not be recorded with devices with lower resolution than 0.05 mm.

The measured data showed that the thermally-induced crack movements can be monitored in short time-steps. Correlations of movements with monitored temperatures indicate that surface temperature is not the only variable that influences thermally-induced movements. The movements are probably controlled more by overall air/rock mass temperature rather than maximum and minimum peaks. Long-term heat transition inside the rock mass possibly also plays an important role. To proof this assumption, it is necessary to perform more detailed research combined with subsurface temperature monitoring and modelling.

We have shown that the time-lapse IR imaging brings new spatial information about the thermal behaviour of the rock surface and helps with the interpretation of movements on cracks. Thus the use of IR imaging contributes to a better hazard assessment of rock mass stability. For future applications, a longer (several days with varying insolation and temperature conditions) monitoring should be tested together with more in-depth analysis of induced rock stresses. This might lead to identification if the areas of high thermally-induced rock stresses lead to destabilisation of the rock mass. Yet another aspect of the monitoring should be the presence and content of water, namely during the winter as the freezing water is also an important factor in rockfall initiation. Therefore, the combined monitoring performed during winter months appears to be essential as, in some countries including Czechia, most of the recorded rockfalls occur during winter and spring months (December to March/April; Adamovič 2020) when temperatures likely oscillate around 0 °C causing repeated freezing and thawing of the water present in the rock mass.

Table 1 Correlation coefficients (95% confidence interval) between recorded temperatures and monitored crack movements

Temperature	102D_1	102D_2	0CB7
Air	−0.699	0.004	−0.545
Mass MAX	−0.638	−0.045	−0.532
Mass AVG	−0.806	−0.101	−0.617
Mass MIN	−0.756	−0.025	−0.587
Slab 1 MAX	−0.732	−0.066	−0.585
Slab 1 AVG	−0.728	−0.028	−0.577
Slab 1 MIN	−0.773	−0.025	−0.596
Slab 2 MAX	−0.702	−0.103	−0.575
Slab 2 AVG	−0.719	−0.053	−0.451
Slab 2 MIN	−0.723	−0.010	−0.569

Acknowledgements We would like to thank the OP VVV project RINGEN + (CZ.02.1.01/0.0/0.0/16_013/0001792) and Strategy AV21 for financial support of this research.

References

Adamovič J (2020) Records and classification of rockfall phenomena in sandstone areas of the Czech Republic. http://rockfall.gli.cas.cz. Last Accessed 12 March 2020

Bakun Mazor D, Hatzor YH, Glaser SD, Santamarina JC (2013) Thermally vs. seismically induced block displacements in Masada rock slopes. Int J Rock Mech Min Sci (1997) 61:196–211

Barla G, Antolini F, Gigli G (2016) 3D Laser scanner and thermography for tunnel discontinuity mapping. Geomech Tunnel 9(1):29–36

Baroň I, Beckovský D, Míč L (2014) Application of infrared thermography for mapping open fractures in deep-seated rockslides and unstable cliffs. Landslides 11:15–27. https://doi.org/10.1007/s10346-012-0367-z

Boyd JM, Hinds DV, Moy D, Rogers C (1973) Two simple devices for monitoring movements in rock slopes. Quart J Eng Geol Hydrogeol 6(3–4):295–302

Collins B, Stock GM (2016) Rockfall triggering by cyclic thermal stressing of exfoliation fractures. Nat Geosci 9:395–400

Czech Geological Survey (1998) Database of geologically important places: 232. http://lokality.geology.cz/d.pl?item=3&id=232&vyb=1&odlok=7 [Online; 30-December-2019, in Czech]

Danov M, Petkov D, Tsanev V (2007) Investigation of thermal infrared emissivity spectra of mineral and rock samples. In: Bochenek Z (ed) New developments and challenges in remote sensing. Millpress, Rotterdam, pp 145–152

Fantini A, Fiorucci M, Martino S, Marino L, Napoli G, Prestininzi A, Salvetti O, Sarandrea P, Stedile L (2016) Multi-sensor system designed for monitoring rock falls: the experimental test-site of Acuto (Italy). Rendiconti Online Societa Geologica Italiana 41:147–150

Fiorucci M, Marmoni GM, Martino S, Mazzanti P (2018a) Thermal response of jointed rock masses inferred from infrared thermographic surveying (Acuto Test-Site, Italy). Sensors 18:2221

Fiorucci M, Marmoni GM, Martino S, Paciello A (2018b) Experimental evidences of thermo-mechanical induced effects on jointed rock masses through infrared thermography and stress-strain monitoring. In: Litvinenko V (ed) Geomechanics and geodynamics of rock masses. Taylor & Francis Group, London, pp 263–268

Frodella W, Gigli G, Morelli S, Lombardi L, Casagli N (2017) Landslide mapping and characterization through infrared thermography (IRT): suggestions for a methodological approach from some case studies. Remote Sens 9(12):1281

Gischig V, Moore, JR, Evans KF, Amann F, Loew S (2011) Thermomechanical forcing of deep rock slope deformation: 1. Conceptual study of a simplified slope. J Geophys Res 116

Gruber S, Peter M, Hoelzle M, Woodhatch I, Haeberli W (2003) Surface temperatures in steep alpine rock faces—a strategy for regional-scale measurement and modelling. In: Proceedings of the 8th international conference on Permafrost, Swets & Zeitlinger Lisse, vol 1, pp 325–330

Guerin A, Derron MH, Jaboyedoff M, Abellán A, Dubas O, Collins BD, Stock GM (2016) Intraday monitoring of granitic exfoliation sheets with LiDAR and thermal imaging (Yosemite Valley, California, USA). Geophysical Research Abstracts 18: EGU2016-12855-2

Guerin A, Jaboyedoff M, Collins BD, Derron M-H, Stock GM, Matasci B, Boesiger M, Lefeuvre C, Podladchikov YY (2019) Detection of rock bridges by infrared thermal imaging and modeling. Sci Rep 9:13138

Gunzburger Y, Merrien-Soukatchoff V, Guglielmi Y (2005) Influence of daily surface temperature fluctuations on rock slope stability: case study of the Rochers de Valabres slope (France). Int J Rock Mech Min Sci (1997) 42:331–349

Janeras M, Jara J-A, Royan MJ, Vilaplana J-M, Aguasca A, Fabregas X, Gili JA, Buxo P (2017) Multi-technique approach to rockfall monitoring in the Montserrat massif (Catalonia, NE Spain). Eng Geol 219:4–20

Liu W, Apel DB, Bindiganavile VS (2016) Cylindrical models of heat flow and thermo-elastic stresses in underground tunnels. Int J Numer Methods Heat Fluid Flow 26:2139–2159. https://doi.org/10.1108/HFF-10-2014-0331

Matasci B, Carrea D, Jaboyedoff M, Pedrazzini A, Stock GM, Oppikofer T (2011) Structural characterization of rockfall sources in Yosemite Valley from remote sensing data: towards more accurate susceptibility assessment. In: Proceedings of the 14th ISSMGE Pan-American conference, 8 pp

Mineo S, Pappalardo G (2016) The use of infrared thermography for porosity assessment of intact rock. Rock Mech Rock Eng 49:3027–3039

Pappalardo M, D'Olivo M (2019) Testing a methodology to assess fluctuations of coastal rocks surface temperature. J Mar Sci Eng 7:315

Pappalardo G, Mineo S, Zampelli SP, Cubito A, Calcaterra D (2016) InfraRed thermography proposed for the estimation of the cooling rate index in the remote survey of rock masses. Int J Rock Mech Min Sci 83:182–196. https://doi.org/10.1016/j.ijrmms.2016.01.010

Petružálek M (2019) Description and mechanical parameters of limestones from Prague—Branik. Final unpublished technical report. Geological Institute of the Czech Academy of Sciences [in Czech]

Racek J (2020) Use of rock mass classifications for rock fall susceptibility analysis in the conditions of the Bohemian Massif. Bachelor Thesis, Charles University in Prague, Faculty of Sciences, Geotechnology, Prague, 63 pp

Rubio E, Caselles V (1997) Emissivity measurements of several soils and vegetation types in the 8–14 µm wave band: analysis of two field methods. Remote Sens Environ 59(3):490–521

Rubio E, Caselles V, Coll C, Valour E, Sospedra F (2003) Thermal-infrared emissivities of natural surfaces: improvements on the experimental set-up and new measurements. Int J Remote Sens 24(24):5379–5390

Seo H, Choi H, Park J, Lee IM (2017) Crack detection in pillars using infrared thermographic imaging. Geotech Test J 40:20150245. https://doi.org/10.1520/GTJ20150245

Sobrino J, Del Frate F, Drusch M, Jimenez-Munoz JC, Manunta P, Regan A (2016) Review of thermal infrared applications and requirements for future high-resolution sensors. IEEE Trans Geosci Remote Sens 54:2963–2972

Spearman C (1904) The proof and measurement of association between two things. Am J Psychol 15:72–101

Teza G, Marcato G, Castelli E, Galgaro A (2012) IRTROCK: A MATLAB toolbox for contactless recognition of surface and shallow weakness of a rock cliff by infrared thermography. Comput Geosci 45:109–118

Yousefi B, Sojasi S, Castanedo CI, Beaudoin G, Huot F, Maldague XP, Chamberland M, Lalonde E (2016) Mineral identification in hyperspectral imaging using Sparse-PCA. Thermosense: Therm Infrared Appl XXXVIII 9861:986118

The Role of Measure of Deep-Seated Displacements in the Monitoring Networks on Large-Scale Landslide

Paolo Allasia, Marco Baldo, Francesco Faccini, Danilo Godone, Davide Notti, and Flavio Poggi

Abstract

The aim of this research is to obtain a hydro-geomorphological and geotechnical model of the Arzeno and Prato di Reppia large scale landslide for the geo-hydrological risk management. Arzeno and Reppia are typical rural villages built in historical times at about 600 m a.s.l. near the main watershed, in the Entella river catchment (Ligurian Apennine): this is a very common situation in the Ligurian hinterland, where large landslides represent about 15% of the territory. For decades, there have been reports of slope instability phenomena in the villages of Arzeno and Prato di Reppia, which are outlined by several indirect kinematic indicators (mainly damage on buildings and infrastructures). In order to improve the comprehension of the landslide kinematic and the consequences on buildings and infrastructures, in 2017, a drill survey was carried out: one inclinometer case with robotized system and one piezometer with pressure transducer for continuous measurement have been installed in Arzeno. One inclinometer case and one piezometer case have been installed in Prato di Reppia village. The monitoring activities show a complex situation for the studied large-scale landslide: an evolution of the slope instabilities in several compartments with different kinematics and hydrogeological set-up is recognized, as well as a good convergence between satellite, GPS and inclinometric monitoring data.

Keywords

Deep-seated ground deformations • Monitoring network • InSAR

Introduction

Landslide hazard in mountain environment is very frequent, representing one of the main issues in natural hazard control. Among natural causal factors, we can recognize the ground conditions (lithological and structural features), the geomorphological processes (tectonic uplift and fluvial erosion of the slope toe) and physical processes (heavy and short rainfalls and prolonged high precipitation). Nevertheless, landslide hazard is also related to man-made processes, eg. to the insufficient maintenance of the drainage systems or land use changes. Climate change in Mediterranean environment accelerate landslide dynamic, and this often causes damages and human losses. A policy based on risk prediction and prevention has been adopted in Italy, and now there is more interest about the understanding of the landslides causal factors and their mitigation. A detailed knowledge of landslide failure mechanisms becomes a main aspect for the risk reduction management.

In Italy almost all the settlements on slopes were built in historical times above large scale landslide: these portions of land were preferred because of their low acclivity, groundwater and stratigraphy suitable for agriculture. When studying the features of these landslides, a different morphoclimatic context from the current one (paleolandslides) must be taken into account; the current movement is mostly due to a residual-state landslide, with very slow kinematics.

P. Allasia (✉) · M. Baldo · D. Godone · D. Notti
IRPI-CNR, Strada delle Cacce 73, 10135 Torino, Italy
e-mail: paolo.allasia@irpi.cnr.it

M. Baldo
e-mail: marco.baldo@irpi.cnr.it

D. Godone
e-mail: danilo.godone@irpi.cnr.it

D. Notti
e-mail: davide.notti@irpi.cnr.it

F. Faccini
University of Genova DISTAV, 16132 Genova, Italy
e-mail: faccini@unige.it

F. Poggi
Liguria Region Department, 16121 Genova, Italy
e-mail: flavio.poggi@regione.liguria.it

© Springer Nature Switzerland AG 2021

N. Casagli et al. (eds.), *Understanding and Reducing Landslide Disaster Risk*,
ICL Contribution to Landslide Disaster Risk Reduction,
https://doi.org/10.1007/978-3-030-60311-3_4

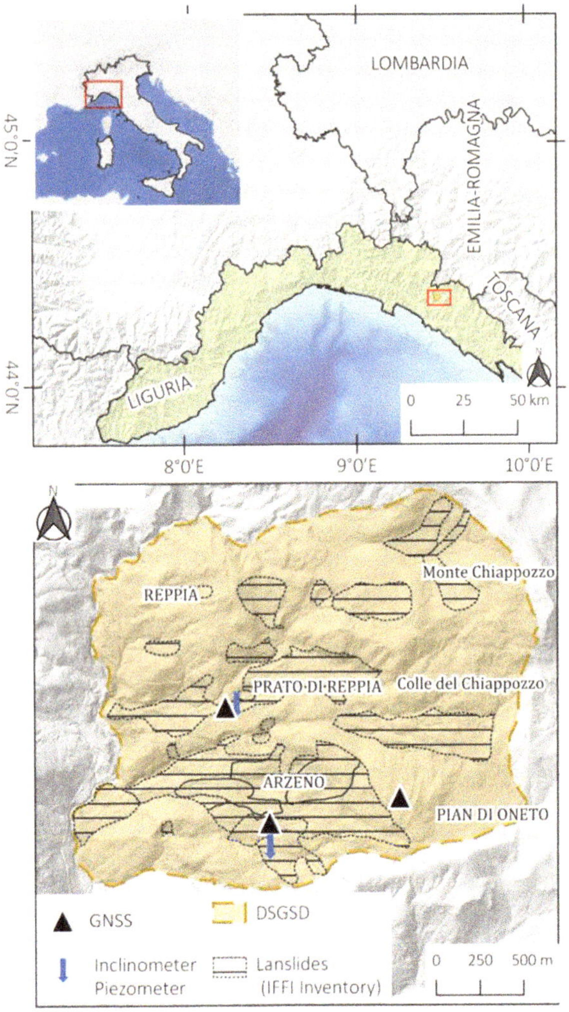

Fig. 1 Upper panel, location of the study area. Lower panel, location of the installed instruments over landslides: GNSS, Piezometers and inclinometers (both manual and AIS)

The aim of this research is to obtain an hydro-geomorphological and geotechnical model of the Arzeno and Prato di Reppia large scale landslide for the geo-hydrological risk management. Arzeno and Reppia are typical rural villages built in historical times at about 600 m a.s.l. near the main watershed, in the Entella river catchment (Ligurian Apennine): this is a very common situation in the Ligurian hinterland, where large landslides represent about 15% of the territory.

Both the IFFI project (Inventory of Italian landslide phenomena) and the Entella River Master plan specifically mention it as a very high geomorphological hazard area (Fig. 1). For decades, there have been reports of slope instability phenomena in the villages of Arzeno and Prato di Reppia, which are outlined by several indirect kinematic indicators (mainly damage on buildings and infrastructures). Although it is a low velocity landslide, geomorphological risk mitigation strategies in this area representing a priority for public administration: since the third millennium the municipality of Ne and Liguria Region has been paying great attention to a better knowledge of the phenomenon and to activate landslide monitoring activities. The geological structure of this catchment is internationally well known by earth science researchers for the ophiolitic sequence with sedimentary covers.

A careful field survey has made it possible to identify the most important geological and geomorphological features: several rock masses with different strength and deformability behaviour have been recognized: shales with limestone interlayers, limestones, cherts, ophiolitic breccias, basalts, serpentinites.

The whole upper catchment examined, on which the villages of Prato di Reppia and Arzeno are located, shows evident signs of neotectonic influence: numerous morphological indicators (wetlands in flat areas, trench filled by swampy deposits, closed depression, counterslope) suggest, at the base of the large scale landslides, the presence of a Deep Seated Gravitational Slope Deformation that affects the whole ridge, slope and valley bottom system (Fig. 1).

Material and Methods

In recent years, the study area has received a lot of attention from the regional management offices and the scientific community (Brandolini et al. 2007). Contemporarily, a small but constant buildings damages and road deformations has been observed (Faccini et al. 2019). These issues did not create substantial problems for the inhabitants life, but pushed authorities and scientists to an accurate analysis of the kinematic of the landslide. To increase the knowledge of the phenomenon, starting from 2018, a terrestrial monitoring network has been implemented mainly aimed to measuring subsoil deformations and the groundwater level (Table 1). The goals were (i) describe the behaviour on the subsoil (ii) identify kinematic domains (iii) evaluate (if any) seasonal deformation accelerations (iv) describe the behaviour of the groundwater level in a pseudo-continuous way (Herrera et al. 2017). To achieve these goals, a geo-gnostic activity was carried out and 2 inclinometers and 2 piezometers (coupled) have been installed. In addition to the deep-seated measurements, to make a robust instrumental crosscheck, a periodic GNSS network was organized with 3 benchmarks (Fig. 1). At the same time, the analysis of the InSAR satellite data continued by the Liguria Region, acquiring dataset from Sentinel and Cosmo-SkyMed platforms (2014–2019). These data will be available in the near future.

Table 1 Main characteristics of the used instrumentations

Type of measure	Spatial resolution	Temporal resolution	Notes
Robotized inclinometer system (AIS)	Single point. Vertical step 500 mm	1/2 per days	Tube length 63 m. Remote controlled
Manual inclinometer	Single point. Step 500 mm	Every 3/4 months	Tube length 63 m
Automatic piezometer	Single point	Every 3 h	Tube length 21 m. Remote controlled
GNSS benchmark	Single point	Every 3/4 months	

Ground Deformations (On the Surface)

The development of the last twenty years of remote sensing techniques allows often and in a quite simple way, to define a large-scale surface deformations. The availability of previous data (up to 1992) is very useful to investigate the history of the landslide with some constraints: i) a sufficient number of persistent scatterers ii) low landslide velocity. Persistent scatterer interferometry is one of the most used techniques for very-slow landsides monitoring (Béjar-Pizarro et al. 2017).We used Persistent Scatterers (PS) InSAR data available from extraordinary plan of remote sensing (Piano Straordinario di Telerilevamento-PST) of Italian environment minister (https://www.pcn.minambiente.it/mattm/en/project-pst-interferometric-products/). The data cover the period 1992–2014 using ERS and ENVISAT satellites of European Space Agency ESA, Cosmo-SkyMed (CSKM) satellites of Italian Space Agency (ASI). Data from Sentinel-1 satellites are actually under interpretation and will be used in further studies.

In parallel to the use of the InSAR data, a small GNSS network has been installed aimed to: (i) validation of the inclinometric data (cumulated displacements on the top of the borehole) (ii) assessing the consistency of deformation measured by the GNSS system and by the InSAR data recent (when available).

Deep-Seated Ground Deformations and Water Table

In order to investigate the subsoil kinematic, two pairs of inclinometers and piezometers were installed in the areas of Arzeno and Prato di Reppia respectively. Based on geological data, the inclinometer tubes were drilled up to about 63 m deep while the piezometric tubes up to about 20 m. The borehole located in Arzeno has been equipped with a robotic inclinometer system developed by the CNR IRPI (Fig. 2).

The robotized inclinometer system was designed with the goal to automatize the measurement carried out, conventionally, by an operator (Allasia et al. 2018). The system is composed by two main parts: the Ground Control Unit (GrCU) and the Inclinometer Control Unit (InclCU). The first one has the task of probe up-lift/down-lift into the inclinometer tube while the second, together with the probe, is the measuring element. These two elements are connected only by a thin synthetic fiber cable (Dyneema®, 2 mm diameter) finalized to the mechanical support (and moving) of the probe into the inclinometer tube. The communications between GrCU and InclCU take place, via radio, exclusively during the idle phase. During the down-lift and up-lifts operations, there are no communications between the systems and the InclCU is fully autonomous from power supply and signal analysis point of view. Regarding the second borehole, located in Prato di Reppia, manual measures with portable probe have been carried out every 3 months. At the beginning of 2020 after about 2 years of measures, the robotic system was moved from Arzeno to Prato di Reppia (Figs. 2 and 3). This change was necessary because the periodic measures were unable to describe correctly the complex evolution of the Prato di Reppia area. The groundwater measurements were carried out using wireless and remote controlled electric pressure transducers.

Results and Discussion

The use of a terrestrial monitoring network (subsoil and surface measurements) allowed deepening what has already been observed by the InSAR data. An important element to keep in mind is that the deformations measured on the surface are a sum of what happens in the subsoil. Without deep-seated deformation measurements, the surface measurement requires a series of geological/geomorphological considerations and hypotheses with, sometimes, a strong subjective component. In the illustrated case, a first analysis based on geology analysis and InSAR data led to the

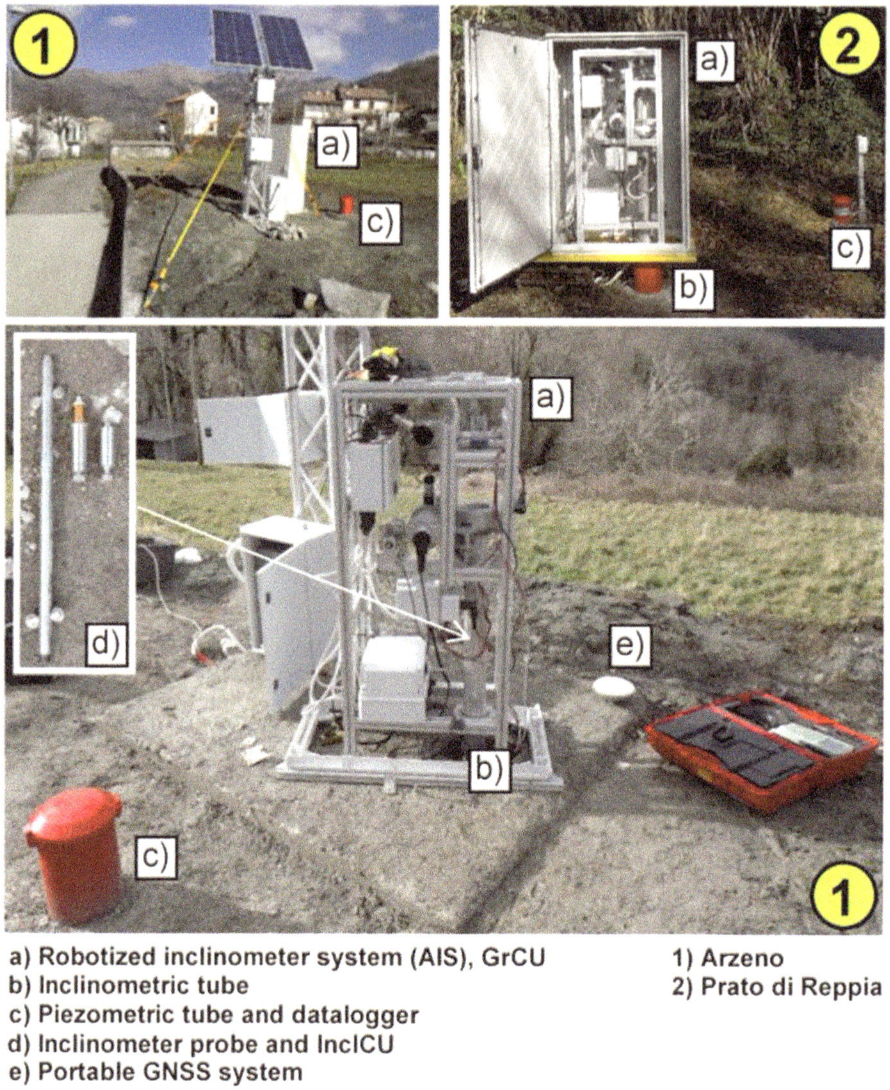

Fig. 2 Robotized inclinometer system, piezometer and GNSS receiver

identification of large areas with very similar characteristics (from geological point of view and average annual speed).

The obtained data from terrestrial monitoring have made it possible to improve this interpretation by identifying two macro sectors (Arzeno and Prato di Reppia) which on the surface have comparable deformation rates, but very different behaviors and characteristics from the subsoil point of view. Other important differences have been observed on the groundwater levels both in terms of maximum excursions and charging/discharging times and velocity.

Ground Deformations (On the Surface)

Concerning the InSAR dataset, both ascending and descending geometries are available on this area and this allowed us to estimate vertical and E-W component of displacement. In this work we used the descending one as it is more suitable for the slope facing West. Deformation time series of InSAR and GNSS were summarized in order to compute the displacement cumulated in the last 28 year of measurements. In this study the InSAR data were used to understand the long-term kinematic of landslides and to compare them with in-situ monitoring. From the Fig. 4 it is possible to see that the pattern and the rate of displacement is almost the same for the 3 time interval considered, with slight low velocity during ENVISAT (2003–2009) monitoring. The difference of PS density is mainly related to the type of processing and to the spatial resolution of satellite (very high for CSKM). The rate of movement show a peak in the eastern part of Arzeno village (>30 mm/yr), and an average velocity of 20 mm/yr for both Arzeno and Prato di Reppia. Here most of PS are represented by buildings or structure. A second area with lower displacement rate

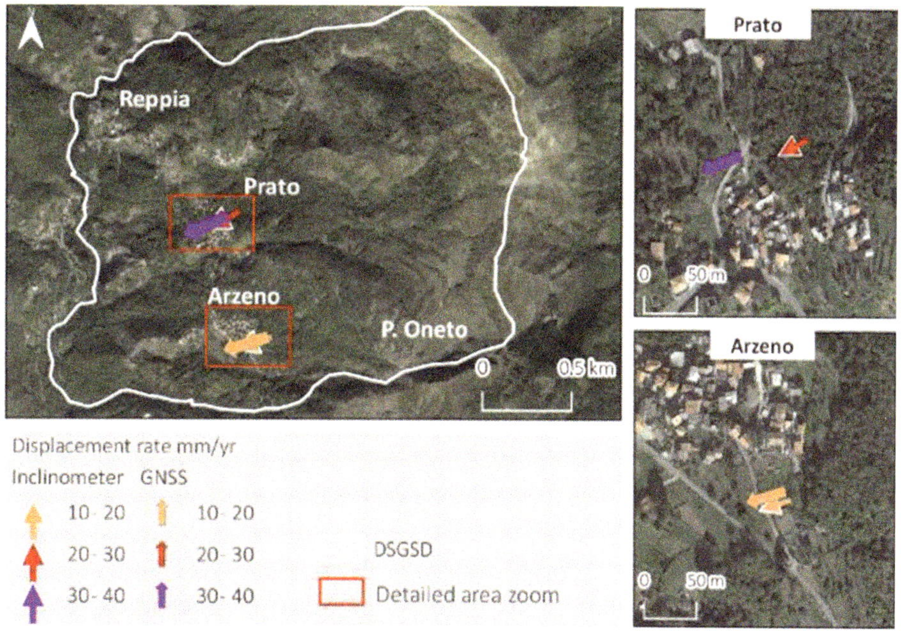

Fig. 3 Map with displacements vector recorded by GNNS an inclinometer (manual and robotized)

(5–10 mm/yr) is located on the outcrop of disaggregated limestone ('Calcari a Calpionelle' formation) between Arzeno village and the pseudo-karst depression of Piani di Oneto. In Fig. 5 are shown the time series of displacement for InSAR and GNSS data for Arzeno (Fig. 5a) and Prato di Reppia (Fig. 5b). It is possible to note that for the case of Arzeno the rate of movement is almost constant along time, while for Prato di Reppia the time series show variability. A hypothesis on the different rate of displacements in the two locations, is tied to the stratigraphic and geomechanical differences. These aspects coupled to the different groundwater patterns, can be induce different geometries and dynamics of the slip surfaces. In particular, if we focus on the most recent period covered by GNSS data it is possible to see an acceleration of the movement in the last months of 2019 in correspondence of extreme and persistent rainfalls event. Preliminary results of Sentinel-1 satellites (2014–2019) show a great correspondence both in displacement rate and time series with in-situ installed monitoring system.

Deep-Seated Ground Deformations and Water Table

The analysis of inclinometer data led to the identification of different behaviors between the Prato di Reppia and Arzeno Areas. The Prato di Reppia inclinometer identifies two active sectors with very different displacement rates. The first one, the most important, is observed at 49 m depth in a sector characterised by fractured ophiolitic breccia in fine sand matrix. The second sector, located at a lower despth, provided considerable deformations only after about a year of measurements and are approximately a quarter of what was measured in the deepest sector. Also in this case, the shear band developed in a ophiolitic breccia. Overall, the cumulated displacement rate has grown in the last two years from around 20 mm/yr to 28 mm/yr. This increase is mainly due to the velocity increase in the deepest sector (Fig. 6). As often happens for periodic measurements, also in this case the temporal trend was estimated by interpolation between subsequent measurements (Fig. 7). The observed increase, was also confirmed by the GNSS surveys and suggested moving the robotic system on this tube in order to follow accurately the temporal evolution of the phenomenon. Thanks to the automation of the system and the ability to customize the starting point of the robotic measurements, it was possible to easily connect the manual measurements with the robotic ones (Fig. 7). The first robotic measurements (about 45 measurements in 40 days) confirmed the noteworthy deformation trend started following the persistent rainfalls that affected the whole of Liguria region in November 2019.

Unlike Prato di Reppia, the Arzeno borehole was immediately equipped with the robotic measuring system. The obtained results highlighted a single active sector at about 17 m depth at the contact between Loamy and sandy silt with various sized clasts and Gabbro. Overall, at the cumulative level, the displacement rate remained fairly constant over the two years and equal to about 18 mm/yr. Also in this case, the overall displacement was confirmed by GNSS benchmark located at about 1 m from the inclinometer casing (Figs. 2 and 3). The availability of the measured data in almost continuous for the entire length of the tube, has also allowed observing the real trend of deformation over time

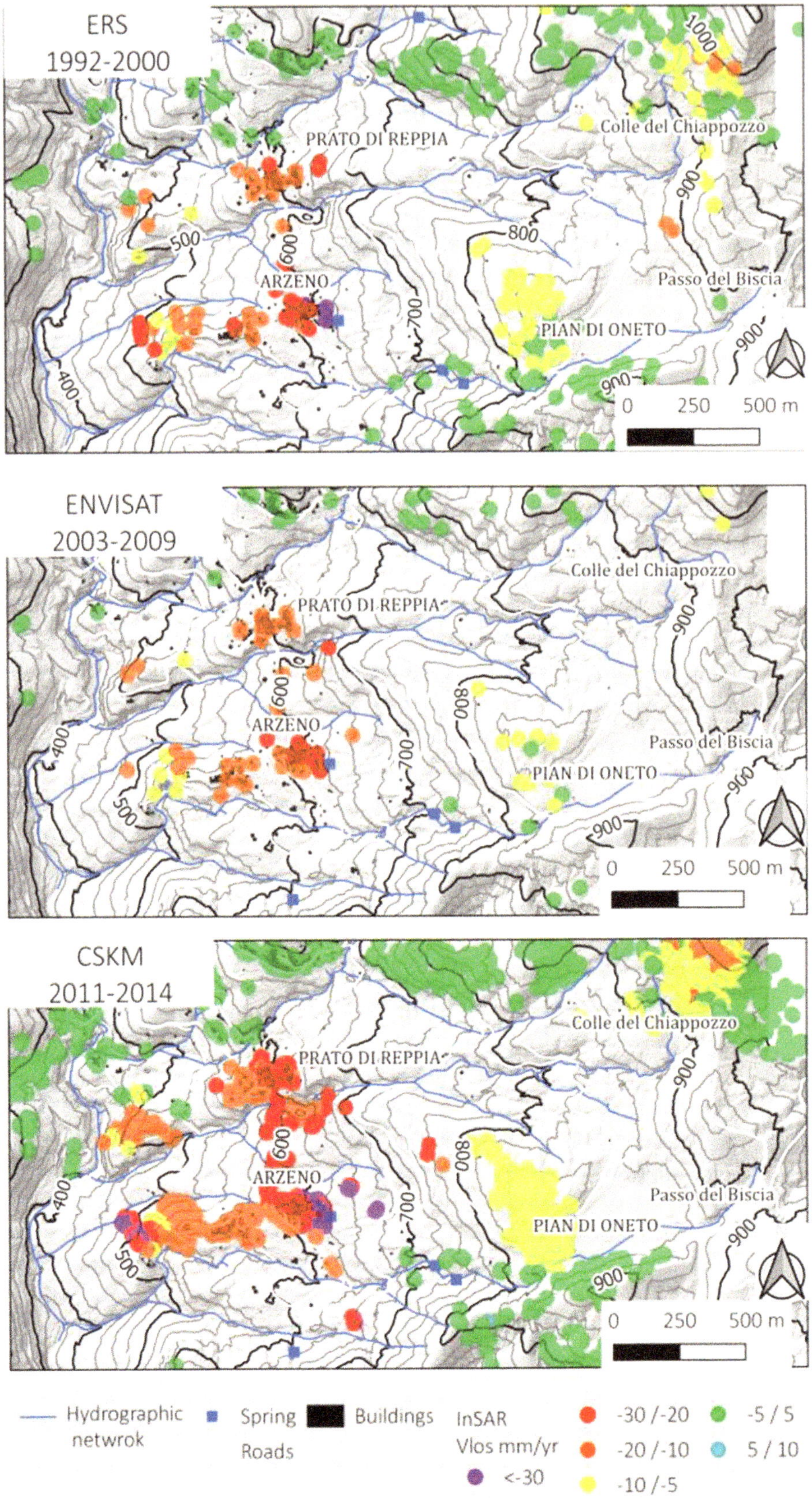

Fig. 4 LOS velocity derived from InSAR data from descending geometry dataset of ERS, ENVISAT and CosmoSky-Med satellites

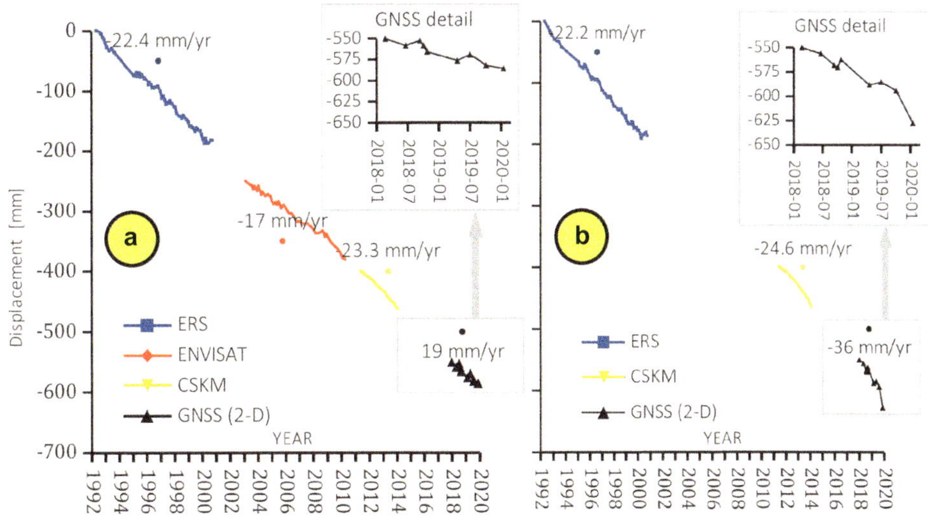

Fig. 5 Displacement time series from 1992 to 2020 based on InSAR and GNSS data for Arzeno (**a**) and Prato di Reppia (**b**). In the case **b** there are no close data for ENVISAT

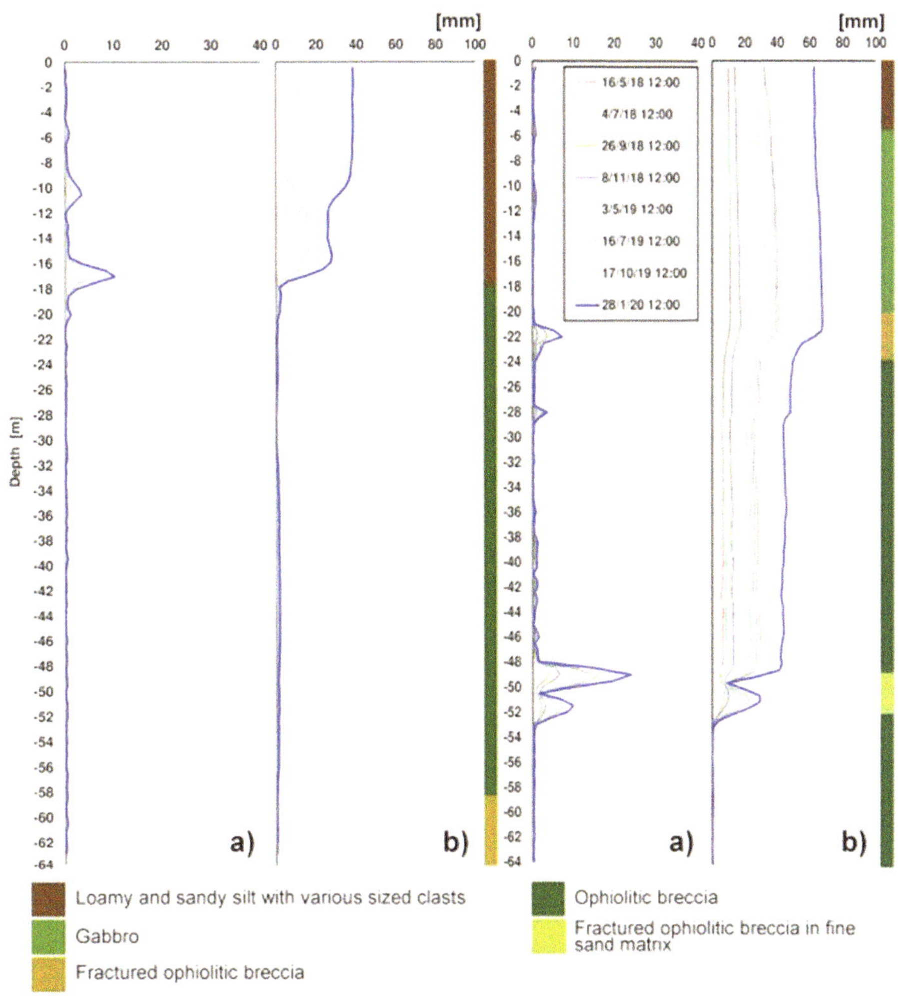

Fig. 6 Local (**a**) and cumulated (**b**) displacements for the Arzeno (left) and Prato (right) boreholes. For the Arzeno (plotted only 30 over more than 1000 measure) and for Prato First measure 7/3/2018—Last measure 23/1/2020

(Figs. 7 and 8). At the maximum deformation depth (17 m), the local displacement rate is fairly constant (5.2 mm/yr) and moderate accelerations are observed at the end of the winter period (up to 6.6 mm/yr). As regards the groundwater level, the trend between the two sites is generally very different in terms of maximum excursions, charge and discharge velocity. The levels observed in the Prato di Reppia piezometer are characterized by fast increases (even just a few hours after the

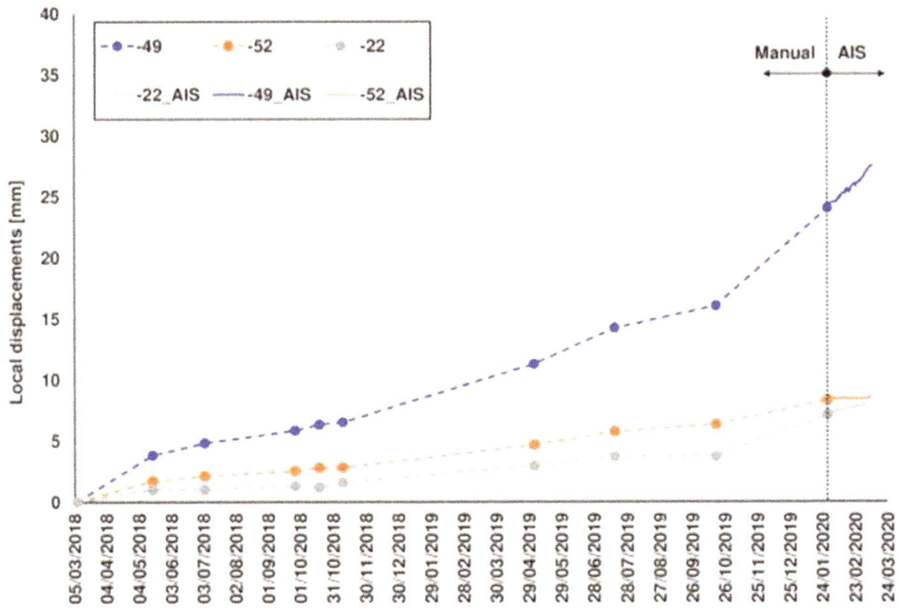

Fig. 7 Time series of selected depths for Prato borehole (local displacements). Starting of end January 2019, the inclinometer measurements was acquired automatically using the AIS

rainfalls start) and by equally rapid decreases. Since November 2019, some increases have been observed that have reached and exceeded the ground level (Fig. 8). Ongoing studies seem to attribute this behaviour to the presence of complex karst circuits with remarkable water flows. Concerning the Arzeno piezometer, the response of rainfalls is slightly slower and the excursions are not comparable with those observed in Prato di Reppia. At the moment, for both cases there are no direct and immediate relationships with the water level changes but the study is still ongoing (Lollino et al. 2006). However, as a preliminary point, for the Prato di Reppia area it is possible to observe a qualitative relationship between increased of displacements rate and the high groundwater levels recorded between November and January (Figs. 7 and 8).

Conclusions

With this preliminary research we have been trying to highlight the importance of the role of measure of deep-seated displacements in the monitoring networks on large-scale landslide for the geo-hydrological risk reduction and management.

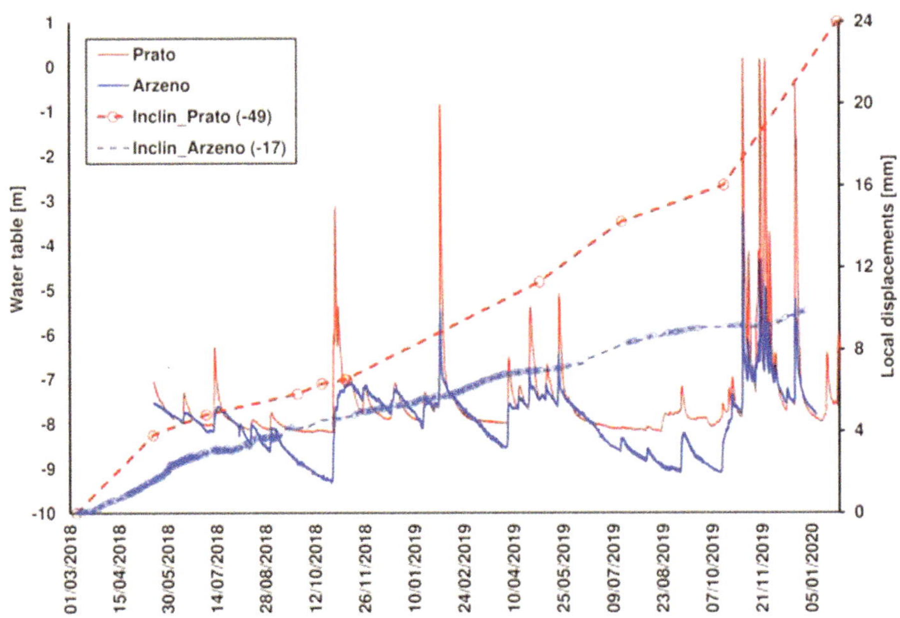

Fig. 8 Water levels versus local displacements in Prato and Arzeno areas. The inclinometer time series was chosen using the depth with maximum displacements values

The latest monitoring experiences with Remote Sensing techniques allow to obtain accurate dataset in space and time but related only to the surface dynamics. The detected surface displacements in any case must be coupled with deep-seated on-site measurements in order to identify the evolution mechanisms of the slip surface.

To this end, continuous real time on-site measurements are mandatory to the aim of defining the geological kinematic of the landslide and to follow his evolution. Moreover, this approach is crucial for the coexistence with the phenomenon in safety conditions (also using early warning systems) or to define more efficient remedial works for risk reduction.

Acknowledgements The authors gratefully acknowledge the support provided by Environment Department of Liguria Region e and Municipality of Ne (Genova).

References

Allasia P, Lollino G, Godone D, Giordan D (2018) Deep displacements measured with a robotized inclinometer system. In: Proceedings of the 10th international symposium on field measurements in GeoMechanics, Rio de Janeiro, Brasil, 16–20 July 2018

Béjar-Pizarro M, Notti D, Mateos RM, Ezquerro P, Centolanza G, Herrera G, Bru G, Sanabria M, Solari L, Duro J, Fernández J (2017) Mapping vulnerable urban areas affected by slow-moving landslides using Sentinel-1 InSAR data. Remote Sens 9:876

Brandolini P, Canepa G, Faccini F, Robbiano A, Terranova R (2007) Geomorphological and geo-environmental features of the Graveglia Valley (Ligurian Apennines, Italy). Geografia Fisica E Dinamica Quaternaria 30:99–116

Faccini F, Elter FM, Allasia P, Berruti L, Godone D, Notti D, Poggi F (2019) Geological-geomorphological characterisation and monitoring activities of a large slope instabilities in Upper Graveglia Valley (Ligurian Apennine, Italy). Geophysical research abstracts, vol 21, EGU2019–18661–1. EGU General Assembly 2019

Herrera G, García López-Davalillo JC, Fernández-Merodo JA, Béjar-Pizarro M, Allasia P, Lollino P, Lollino G, Guzzetti F, Fernández I, Manconi A, Duro J, Sánchez C, Iglesias R (2017) The differential slow moving dynamic of a complex landslide: multi-sensor monitoring. In: 4th world landslide forum, Ljubljana, Slovenia, 29 May–2 June 2017

Lollino G, Arattano M, Allasia P, Giordan D (2006) Time response of a landslide to meteorological events. Nat Hazards Earth Syst Sci 179–184

Not ordinary plan of remote sensing. Interferometric. https://www.pcn.minambiente.it/mattm/en/project-pst-interferometric-products/. Accessed 09 Jan 2020

Flow Slides in Uzbekistan: Overview and Case Studies

2020 Kyoto Japan

Rustam Niyazov, Bakhtiar Nurtaev, Gani Bimurzaev, and Mansur Tashpulatov

Abstract

Paper describes in brief spatial distribution of the landslide-prone areas in Uzbekistan, temporal evolution of these phenomena during last 60 years and organization of the landslide monitoring in in the country. Special emphasis is given to flow slides in loess and clayey often triggered by the prolonged low-frequency seismic vibrations of the distant deep Hindu Kush earthquakes. Three typical case studies of such flow slides that occurred in the recent years are presented and their evolution and motion characteristics are described. Seismometric measurements performed at the source zone of the Achiyak landslide prove that the vibration frequency of the deep Hindu Kush earthquakes coincides with natural frequency of the loess blanketing the slopes in the foothill areas of Uzbekistan that cause resonance effects. Multi-stage evolution of large flow slides some of which transform into mud flows that can last from several days to several years is described by examples of the Khandiza and the Otbokarsai flow slides.

Keywords

Loess • Monitoring • Flow slide • Mudflow • Blockage

R. Niyazov · G. Bimurzaev
Hydroengeo Institute, Tashkent, 100041, Uzbekistan
e-mail: gany82@mail.ru

B. Nurtaev (✉)
Institute of Geology and Geophysics, Tashkent, 100041, Uzbekistan
e-mail: nurtaevb@gmail.com

M. Tashpulatov
State Hazardous Geological Processes Monitoring Survey, Tashkent, 100041, Uzbekistan
e-mail: mansur.uzbek@mail.ru

Introduction

A systematic study of landslides and their monitoring in Uzbekistan began in 1958. It was performed, first, by the special Landslide engineering-geological party, and, since 1994, by the State Hazardous Geological Processes Monitoring Survey (hereafter named Survey) that consists of 7 regional stations located in the foothill areas of Uzbekistan.

The State monitoring system is divided into 2 parts—the general regular regional monitoring and the long-term comprehensive stationary observations at reference sites. The main task of the Survey is to provide information, alert and warn public and authorities about the possible activation of hazardous geological processes close to settlements and economical facilities in order to arrange security measures timely.

Regional monitoring is carried out in the spring seasons (from mid-February till end of May) in high alert mode. Observations are conducted at 500–570 sites located in the landslide-prone areas in the Eastern part of Uzbekistan with mountainous or hilly relief (Fig. 1). Each year, before the start of the spring, the Survey issues warning information about possible manifestations of dangerous exogenous processes in the territory. If signs of slope processes activation are detected, the Survey issues an order to the local administration to start on-site monitoring.

To protect people during the landslide-prone period temporary evacuation to a safe place is recommended, as a priority measure. During this period, particular attention is paid to monitor the climatic conditions to predict large-scale slope failure. Manifestations of all types of hazardous geological processes are recorded in the daily regime. In addition, according to the results of the on-site monitoring, special 1:25,000–1:10,000 warning maps for linear structures, for recreation areas and for endangered settlements are compiled and updated annually.

© Springer Nature Switzerland AG 2021
N. Casagli et al. (eds.), *Understanding and Reducing Landslide Disaster Risk*, ICL Contribution to Landslide Disaster Risk Reduction, https://doi.org/10.1007/978-3-030-60311-3_5

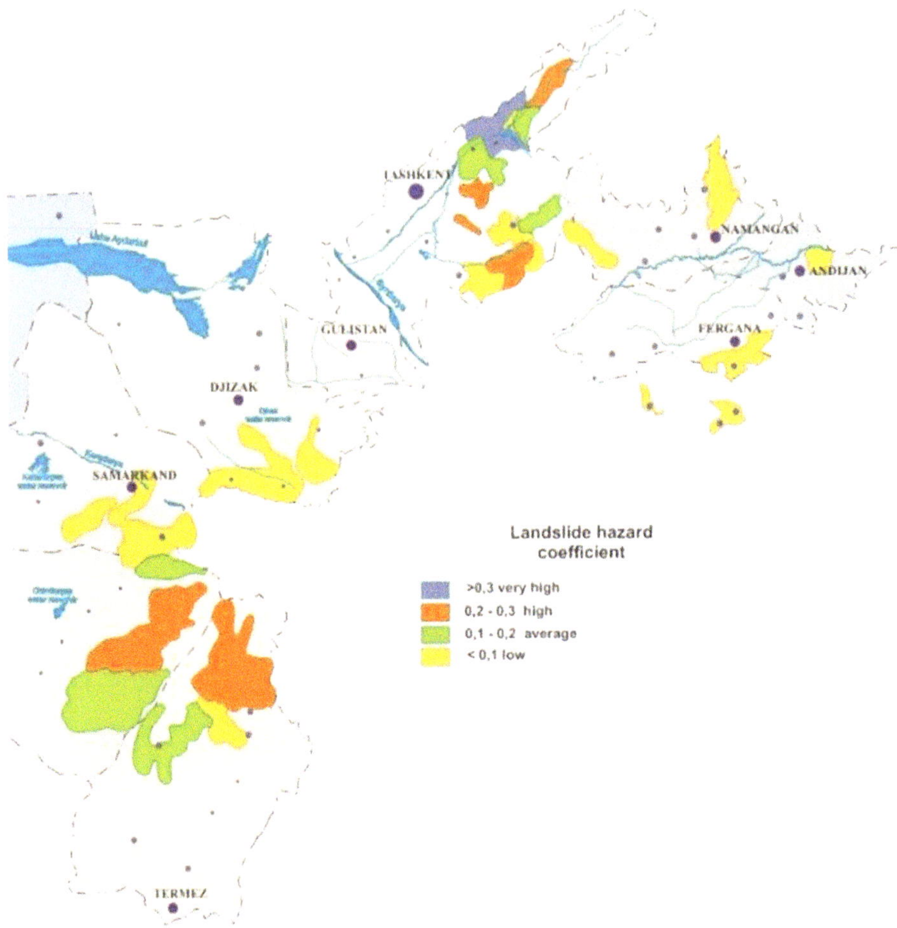

Fig. 1 Landslide hazard zoning map of Eastern Uzbekistan. Notice that such zoning has been performed not over the entire territory but at some particular areas

General Landslides Statistics

The total number of landslides with volume more than 1 thousand m^3 that were formed or reactivated in Uzbekistan during a 60-years long period is about 3300–3500. If we consider smaller events there have been more than 12,000 cases. Over decades, the largest number of landslides occurred in 1958–1970 (991 cases), and in 1991–2000 (934 cases), much more than in 1971–1980 (238 cases), in 1981–1990 (245 cases), and in 2001–2019 (from 340 to 545 cases). The largest yearly number of landslides was recorded in 1969 (721 cases), in 1987 (191); in 1993(350), in 1994 (135), in 1998 (142), in 2005 (185) and in 2012 (72) (Fig. 2). 340–350 sites have been affected by large landslides exceeding 10^5 m^3 in volume and 120–130 events exceeded 1 million m^3. The largest historical event in Uzbekistan is the Atcha block slide (41.012° N, 70.184° E) about 800 million m3 in volume (Niyazov 2009).

No clear and justified tendency of the increase or the decrease of the mean annual number of dangerous exogenous processes have been identified. Many modern landslides are just the reactivation of the older (prehistoric or ancient) and often larger landslides. These secondary landslides are of various scales and types. Such interrelations complicate the assessment of the landslides' frequency at a large extent (Niyazov and Nurtaev 2014).

1270–1400 householders appeared to be in the landslide prone zones in different periods of time. By 2015–2019, their number decreased to 40–80. During last 30 years new villages have been built, and about 2,000 families have been relocated there, and, thus, the number of householders that require resettlement decreases every year.

Landslides caused by anthropogenic impact in different years ranged from 40 to 60% of their total amount. The flow slides, which number exceeds 250, though being relatively rare, in comparison with landslides of other types, represent the most dangerous type of slope processes due to high speed of their motion combined with rather large size. It was found (Niyazov 2009; Niyazov and Nurtaev 2014) that many of them were associated with Hindu Kush earthquakes. Study of the effect of resonance caused by the long-duration low-frequency oscillations produced by such earthquakes and combined with the influence of precipitation

Fig. 2 Annual occurrence of landslides in Uzbekistan since 1958

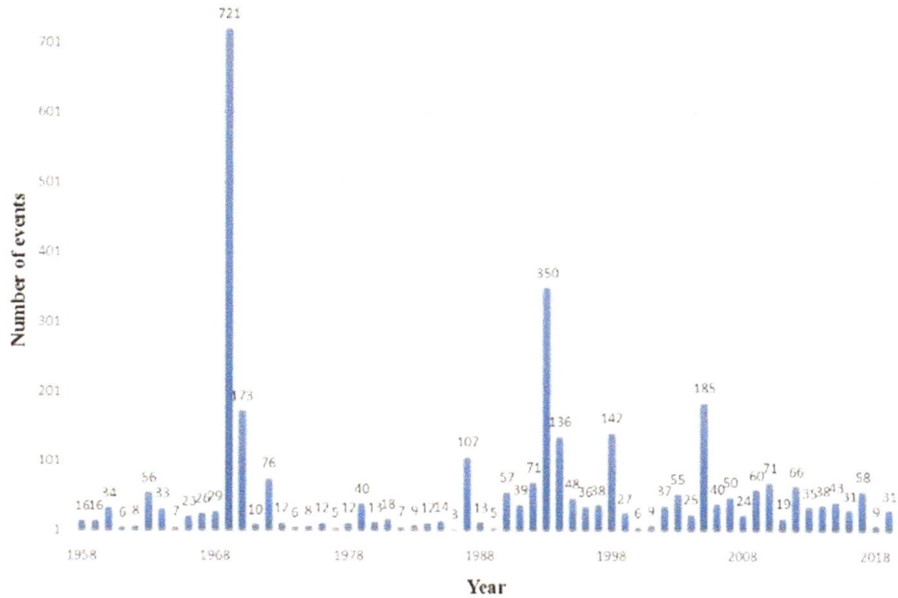

on slopes stability was performed to fulfill the decisions of the Sendai Framework Program (Sassa 2017).

Self-excited Flow Slides in Loess

Loess deposits, widely developed in Uzbekistan, in the foothills in particular, are affected by numerous landslides. Many of them are the flow slides that demonstrate simultaneous crushing of soil over the entire landslide area. It results from the combination of several factors. It was found, in particular, that the frequency of seismic vibrations caused by the distant deep Hindu Kush earthquakes coincides with the natural frequencies on the slope causing resonance phenomena and the formation of landslides that we called "the self-excited".

The simultaneous destabilization of loess deposits over the entire flow slides source area differs from the formation mechanism of landslides of other types. Energy provided by seismic shaking leads to very fast propagation of the destruction of loess within the source area but has very limited effect on its further motion, in other words on its runout. Such landslides are characterized by the distinct sharp headscarp boundaries, where liquefaction occurs within the interbeds near the sliding surface, and sliding concentrates along a thin clayey layer. Soil crushing is caused by the compression-tension deformations mainly rather than by shearing, and develops simultaneously all over the affected slope from its top to base.

Several typical examples of such flow slides that occurred during last years, coinciding with deep Hindu Kush earthquakes are described hereafter.

The Achiyak Landslide

The Achiyak flow slide (41.6383° N, 69.7813° E) was formed on March 25, 2018, almost simultaneously with the M 5.1 Hindu Kush earthquake recorded on March 25, at 3 h 17 min (local time), with a focal depth of 297 km. Duration of oscillations was 90–95 s, dominant frequency 0.8–2.1 Hz. A landslide was formed in loess with a thickness of 15–20 m, lying on the water-encroached sandy-clayey rocks. The headscarp shape is rectangular, 80 m wide, 60 m long and 17–18 m deep. The estimated failure volume of 86.4 thousand m^3 (Fig. 3). The landslide mass broken in separate blocks moved into the riverbed, forming a blockage up to 100 m wide, 15–25 m long and 5–8 m high. The headscarp wall is almost vertical and is rather straight in plan view. It can be assumed that the formation of a landslide was associated first by tension with a successive sliding.

Fig. 3 The Achiyak landslide and location of seismometers (1–7)

The sliding surface zone of the landslide is gently dipping with an angle of 4–6°. The seismometric measurements were carried out at 7 points located around the headscarp (see Fig. 3). Recorded resonance frequencies at measurement points 1, 3, 4, 6—were 1, 2; 2.2; 2.5 Hz, while at the lower marginal zones at points 5 and 7, they vary from 3.0 to 3.5 Hz. The horizontal to vertical ratios of the spectra (HVSR) vary from 2.6 to 3.4. The coefficient of seismic liquefaction of soils ranges from 3 to 9.6, which characterizes fairly dense rocks. It can be assumed that loess mass excitation most likely started at the upper and central part of the landslide where frequency of earthquake vibrations was close to natural vibration frequency of 2.2–2.5 Hz.

The Khandiza Flow Slide

The Khandiza flow slide (38.587° N, 67.5727° E) 1.5 million m^3 in volume was formed on April 6, 2015 (Fig. 4). It occurred in loess with a thickness of 12–15 m to 30 m, lying on the Cretaceous clays with interbeds of sandstones. It was preceded by the intensive (43 cm in one day) snow fall on February 24, 2015, and its fast melting with a rate of 2–8 cm per day during the last days of March that caused significant watering of subsurface soils. In addition, according to local residents, from April 2 to 5 there was watering the garden located on the landslide body. Landslide was triggered, likely, by the M 4.2 Hindu Kush earthquake that occurred on April 5, 2015 at 5:22 (local time) at a focal depth of 128 km.

In 23 days, the total horizontal movement of the landslide front was 960 m, while during first four days it moved for 570 m. After that the front displacement rate started decreasing. On the 12th day, when the front displacement was 954 m, the height of the tongue part began to grow from 10 to 25–30 m and the front width increased from 20–30 m to 40–75 m. The proximal part of the landslide displaced by 250 m in 23 days, and its displacement rate was relatively uniform—20 m/day on average. Here the stabilization process also began on April 12, when speed started decreasing up to 7 m/day and, later to 0.1 m/day. In the middle part of the landslide total horizontal displacement was 317 m. This is three times less than the movement of the frontal part. Moreover, in the first four days this part moved for 25 m only, i.e. the rate was the lowest, and the highest occurred from April 11 to 17 at a speed of 40–70 m/day. During this period different parts of the landslide moved uniformly. The stabilization process began on day 16, April 21, when speeds fell to 2–0.2 m/day. In the process of stabilization of the landslide, the most mobile was the middle zone, i.e. the flow slide moved in pulses.

Fig. 4 The Khandiza flow slide after its reactivation in March 2017

Next activation occurred in the spring of 2016 when, due to rainfall, surface displacements were observed in the tongue part of the landslide. Mudflows formed a furrow along the right side of the flow slide up to 800 m long, 5–6 m deep, and 8–10 m wide, from which about 20 thousand m^3 of landslide masses were eroded.

Two years later, on March 31, 2017, at 10 a.m. landslide reactivated again. It can be assumed that this activation (Fig. 5) was predispose by the formation of this erosion feature.

About 150 thousand m^3 of heavily watered loess up to 100 m wide and 5–10 m thick from the upper part of the slope, along with up to 250 thousand m^3 of loam located downslope that had been displaced in 2015, moved along a slope of 10–12° and created the flow slide up to ca. 1200 m long and from 30 to 100 m wide. Experts of the Surkhandarya monitoring station organized regular monitoring of its movement. It was found that the head and middle part of the landslide on the first day moves at a speed of 3 m/h, tongue 4–5 m/h, and the horizontal displacement was 70 m. As a result in the second day (April 1), the accumulation zone of the 2015 landslide was severely deformed by a series of arcuate cracks up to 30–40 m long. The bulging bars up to 2.0 m high divided with furrows up to 7.0 m deep and the activated landslide deposits moved into the old creek channel at a speed of 2.5–3.0 m/h. The flow width increased sharply from 1.5 to 60–100 m, the flow rate was 3.0–4.0 m/day and increased up to 50 to 80 m/day, being 60–80 m wide and 5–10 m high.

On the fourth day (April 3), the total displacement of the 80 m wide and 10 m thick flow slide reached 230 m at a speed of 3–4 m/h. The landslide reached the school building and began destroying it (see photo at Fig. 5).

In the next two days (April 4–5), the displacement of the landslide masses was divided into two directions: one went along the Kharkushsai channel in the form of a mud stream at a speed of 20–25 m/h; another in the form of a slide towards the school at a speed of 4.0 m/h. As a result, on the territory of the destroyed school, the thickness of landslide

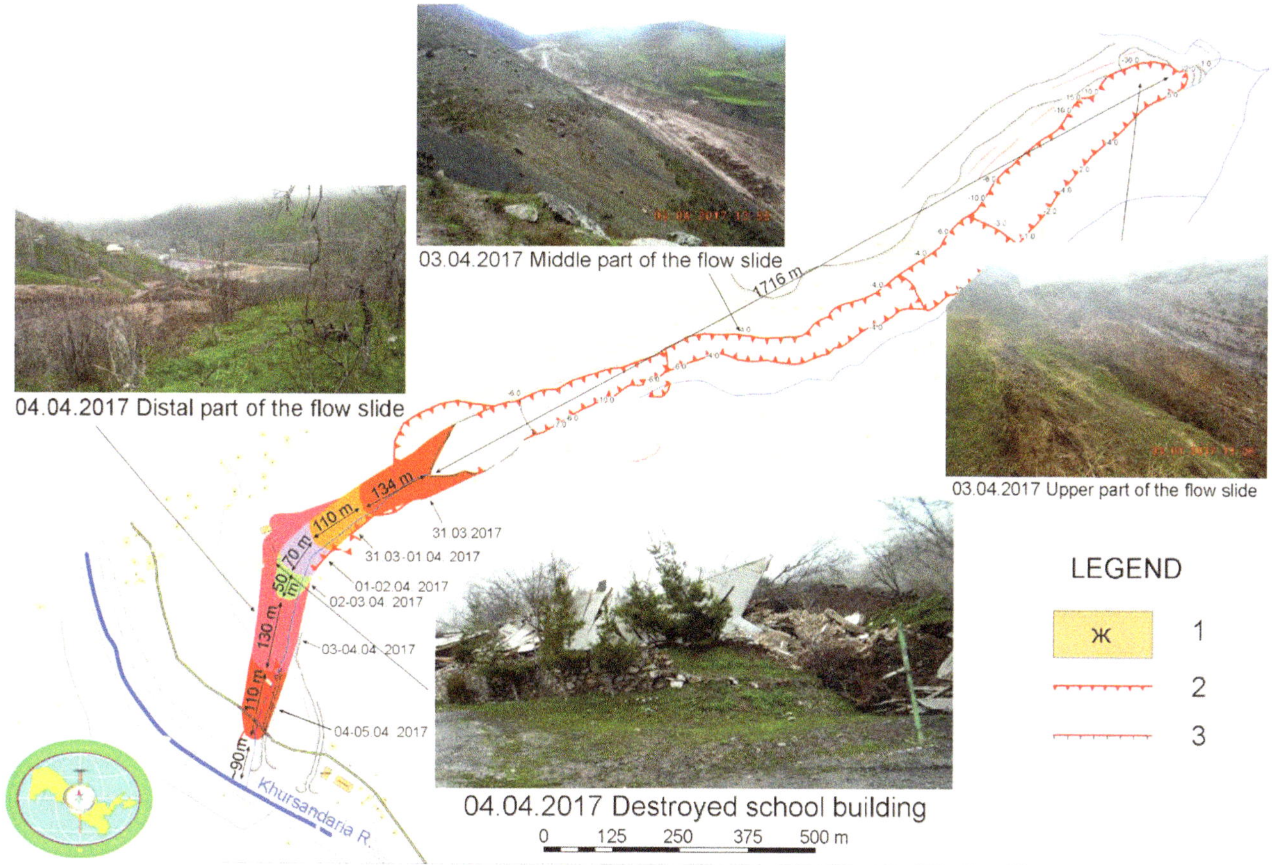

Fig. 5 Schematic map of the Khandiza flow slide activation in March–April 2017 compiled by the State Hazardous Geological Processes Monitoring Survey. Legend: 1—buildings, 2—main headscarp, 3—additional scarps and fissures. Different colors mark parts of the flow slide where it moved from March 31 till April 5.

masses increased up to 18–20 m, the width of the landslide reached 150 m, and the total amount of landslide displacement was 460 m. This displacement of the landslide provoked the movement of soil downstream.

On April 6, 80–100 thousand m^3 of debris moved further downstream the gulley at a speed of 5–6 m/h, reached the Khursandarya River and partially blocked its riverbed. The width of this partial blockage was 70 m, it was up to 3.0 m high and the volume of the dam was estimated as about 50.0 thousand m^3. During the following days the liquefied flow slide mass continued moving into the river and was gradually eroded by water flow.

Fig. 6 The Otbokarsai flow slide source zone

The Otbokarsai Flow Slide

On April 23–28, 2019 a landslide of 1.5 million m^3 in volume originated on the upper part of the left-bank slope of the Otbokarsai River—the tributary of the Djinnidarya River (39.2048° N, 67.3735° E) (Fig. 6) and converted into highly mobile flow slide.

Its source zone was composed of loess and of the underlying Cretaceous sandy-clayey red beds. The landslide formation took place in three nearby circus-like headscarps. The first—the central one—was located at the transition from the slope to the watershed surface, the second one occurred simultaneously to the right from the firs. In plan view the landslide had a conical shape with a very narrow (12–15 m) exit at the headscarp base trough which landslide had to pass moving downslope. The steep backwall was up to 300 m long and 35–40 m high.

The third, the largest circus-like headscarp, originated on the watershed surface. According to the local shepherd who attended the event, it occurred at 6–7 p.m. (local time), i.e. 11–12 h after the first slope failure. The 270–290 m long and up to 31 m high backwall crossed the watershed surface. The initial block slide was 300 m wide, and 120–140 m long and up to 20 m deep. The entire landslide mass rapidly converted into fragmented and liquefied state. As a result of the simultaneous movement, there was local blocking of this flow-like motion. On the left side there are traces of splashes of liquefied mass, 15–20 m high that left patches of debris 0.2–0.3 m thick. As a result, the flow slide passed for about 240 m and stopped in the upper zone, forming a blockage. The trigger for the start of the landslide was probably the Hindu Kush M 4.2 earthquake that occurred, according to the catalog, on April 23, 2019 at 6 h 31 min (local time) at a depth of 198 km, i.e. at the same time when landslide had started forming. And intense rainstorm on April 23 with cumulative precipitation 35.2 mm followed next day by even stronger (43.6 mm) rainstorm, provoked watering of the liquefied soil and the formation of a rapid flow slide.

The first time, on April 25, 2019 morning, this flow slide passed along the gulley and blocked the Otbokarsai Creek. The natural dam was 5–8 m high only and 30–40 m long. The discharge of this creek was up to 0.5–0.7 l/s. Water accumulated until April 26, when the blockage was breached producing the mud flow that passed along the Otbokarsai channel at a distance of 810 m, being 20–40 m wide. It reached the larger Djinnidarya River and partially blocked it.

The second time a mud flow occurred on April 27 when the flow 8.0–10.0 m thick passed along the creek. This mudflow also passed 810 m and blocked the Djinnidarya River again. This time the blockage was 270 m long, 80–110 m wide and from 8 to 15.0 m high. Volume of the blockage was 350 thousand m^3. The discharge of this river was 5–6 m^3/s. Within 3 h a dammed lake 200 m long, up to 110 m wide and up to 8–10 m deep was formed. After that its erosion started at the right bank river side.

The third mud flow originated on April 28 and blocked the Djinnidarya River channel again forming the dam 4–6 m higher than at the second time, so that the highway was blocked. The more than 200 m long dam was eroded in the zone of the old channel. The erosion channel was 7–10 m deep and 8–12 m wide.

Photo of the Otbokarsai Creek made on May 16, 2019 (Fig. 7) show that mud flows were 10–12 m thick, with a flow width of 35–40 m, a gradient of 6–8°, and that practically no debris accumulated in the channel.

Fig. 7 The Otbokarsai channel after the mud flow

Conclusions

Uzbekistan is a very landslide-prone country where more than 3500 landslides occurred during the last 60 years. The flow slides in loess are the most hazardous and unpredictable and pose especial threat. Analysis of the disasters associated with such events show that in many cases it very difficult to foresee their runout, location of the sites where the liquefied loess could be ejected from the channel on the opposite slope and height of such ejection. Timely prediction of such events is even more complicated due to their association with seismic shaking.

Such self-excited flow slides are triggered by low-frequency (0.5—3.5 Hz) prolonged (90–140 s.) vibrations produced by P-waves of very distant (400–700 km) deep (180–270 km) Hindukush earthquakes that cause simultaneous liquefaction of subsurface saturated sediments and tension in the surficial layers.

The characteristic feature of such flow slides is their recurrent simultaneous activation. Increase of the amount of the affected soil does not increase the velocity of motion, but enlarges the runout. Thickness of moving flow and its mechanical properties are often almost the same along the entire length. Loess flow slide are usually 1.5–3 m thick, while clayey flow slides can be 8–10 m thick.

Study of the eroded landslide dams produced by the flow slides show that their volume can comprise up to 30% of the entire flow slide volume if it crosses the dammed valley, and up to 80% if the flow slide moves along the valley. In both cases dams' height rarely exceeds 2.5–4.0 m.

Acknowledgements Authors want to express their gratitude to the anonymous reviewer for valuable comments and to Alexander Strom for useful discussions and support.

References

Niyazov RA (2009) Landslides in Uzbekistan. FAN, Tashkent, 207 pp (in Russian)

Niyazov RA, Nurtaev BS (2014) Landslides of liquefaction caused by single source of impact Pamir-Hindu-Kush Earthquakes in Central Asia. In: Sassa et al. (eds) Landslide Science for a safer geoinvironment, vol. 3. Springer, Switzerland, pp 225–232

Sassa K (2017) The ISDR-ICL Sendai partnerships 2015–2025: background and content. In: Sassa K, Mikoš M, Yin Y (eds) Advancing culture of living with landslides. Springer, pp 3–21

Long-Term Geophysical Monitoring of Moisture Driven Landslide Processes

Jonathan Chambers, Philip Meldrum, Paul Wilkinson, Jessica Holmes, David Huntley, Peter T. Bobrowsky, David Gunn, Sebastian Uhlemann, and Nick Slater

Abstract

Here we describe the development of a novel characterization and monitoring technology for unstable natural and engineered slopes. The system is based on time-lapse electrical resistivity tomography (ERT), which is a geophysical technique used to non-invasively image subsurface resistivity to depths of tens of meters. Resistivity is a useful property because it is sensitive to compositional variations, changes in moisture content, and ground movement. We have developed a low-cost system designed for remote operation, allowing resistivity images to be captured automatically and streamed via a web interface. It comprises four key elements: (1) low-power field instrumentation; (2) data telemetry and storage; (3) automated data processing; (4) and web dashboard information delivery. These elements form the basis of slope condition monitoring approach that provides near-real-time spatial information on both subsurface processes and surface responses. The use of this approach is illustrated with reference to the Ripley Landslide, a case study that demonstrates this approach as a means of spatially tracking complex subsurface moisture driven processes that would be very difficult to characterize using other approaches (e.g. surface observations or intrusive sampling). We propose that this approach could provide sub-surface information in the context of slope-scale landslide early warning systems.

Keywords

Electrical resistivity tomography • Monitoring • Geophysics

J. Chambers (✉) · P. Meldrum · P. Wilkinson · J. Holmes · D. Gunn
British Geological Survey, Nottingham, UK
e-mail: jecha@bgs.ac.uk

P. Meldrum
e-mail: pime@bgs.ac.uk

P. Wilkinson
e-mail: pbw@bgs.ac.uk

J. Holmes
e-mail: jeho@bgs.ac.uk

D. Gunn
e-mail: dgu@bgs.ac.uk

D. Huntley
Geological Survey of Canada, Vancouver, British Columbia, Canada
e-mail: dhuntley@tru.ca

P. T. Bobrowsky
Geological Survey of Canada, Sidney, British Columbia, Canada
e-mail: peter.bobrowsky@canada.ca

S. Uhlemann
Lawrence Berkeley National Laboratory, California, USA
e-mail: suhlemann@lbl.gov

N. Slater
Socotec Monitoring, Uckfield, UK
e-mail: nick.slater@socotec.com

Introduction

Context and Rationale

Assessment of slope condition and subsurface moisture movement is essential for providing early warning of hazardous failure events. Conventional approaches to slope monitoring are often inadequate for predicting failure events. They are heavily dependent on walk-over surveys, intrusive investigations, or remotely sensed data such as aerial photography or LiDAR. However, surface observations (from walkover or remote sensing) cannot detect the subsurface precursors to failure; instead they identify failure once it is expressed at the ground surface, which is often at a late stage when there is insufficient time to implement low cost re-medial solutions. Furthermore, they generally provide relatively low temporal resolution (e.g. weeks to years) due to the cost of manual site visits or flights. For intrusive sampling, even with significant numbers of boreholes, it is

© Springer Nature Switzerland AG 2021
N. Casagli et al. (eds.), *Understanding and Reducing Landslide Disaster Risk*, ICL Contribution to Landslide Disaster Risk Reduction, https://doi.org/10.1007/978-3-030-60311-3_6

only ever possible to sample a tiny proportion of the overall volume of the slope, which means that in heterogeneous ground conditions small-scale processes (e.g. seepages or piping) can be extremely difficult to detect and characterize. Consequently, these conventional approaches are often inadequate for providing early warning of deteriorating condition or failure.

Here we describe the development and application of a low-cost low-power geophysical ground imaging system, for fully automated remote monitoring of unstable slopes. The purpose of the system is to provide improved decision support and early warning of deteriorating condition. The system is called PRIME—'Proactive Infrastructure Monitoring and Evaluation', and is designed to non-invasively visualize the interior of both natural and engineered slopes. It has been applied to a range of moisture driven slope stability applications (e.g. Huntley et al. 2019; Holmes et al. 2020), indicating its suitability for volumetric tracking of moisture content changes and ground movement to identify problems at an early stage. We propose that this system can support the development of landslide early warning systems (Guzzetti et al. 2020).

Development of Monitoring Approach

PRIME System Overview

The PRIME system is based on time-lapse electrical resistivity tomography (ERT), which is a geophysical technique used to generate resistivity images of the subsurface. Resistivity data are useful measurements to make as they are sensitive to compositional variations, changes in moisture content, and also ground movement. PRIME is designed for remote operation using telemetry, so that ERT images can be captured automatically and streamed in near-real-time via a web interface—thereby providing a remote condition monitoring system to reveal sub-surface changes within earth structures.

The system comprises four key elements (Fig. 1): (1) Field measurement instrumentation comprising an array of electrode sensors, the PRIME instrument, and communications hard-ware; (2) Data telemetry (via wireless transfer to the office), storage and indexing; (3) Automated data processing, incorporating quality assessment, filtering, extraction of movement information, generation of resistivity images and translation to distributions of moisture content; and (4) Information delivery via a web dashboard for analysis, interpretation and early warning alerts.

Measurement Instrumentation

The PRIME system (Fig. 2), comprising the low power (10 W) PRIME ERT instrument (standard 19-inch module) connected to the sensor array cables and communications hard-ware, is housed in a small equipment enclosure and powered from 12 V batteries that are charged using a small solar array. The low power consumption of the instrument is essential to maintain operation in remote areas with no access to mains electricity.

Measurement data is collected from sensors (metal electrodes) connected to the system by cables, which are typically deployed at the ground surface in lines (for 2D images) or grids (for 3D images). The depth of investigation of the technique is determined by the electrode spacing and spread, not by the length of the sensor (which is typically <10 cm). Arrays of surface sensors can give information non-invasively to tens of metres below the surface. The resolution of resistivity imaging is limited by the electrode spacing, which is typically between 0.5 and 5 m. Despite this constraint however, even small features (whose size is below the image pixel size, as small as 25 × 25 cm for a 0.5 m electrode spacing) often remain detectable provided that property contrasts are sufficiently high and the sensor array configuration is sufficiently sensitive to the region in which the defect occurs.

The system has 7 measurement channels and can address up to 256 electrodes (with the option to expand to >1000 electrodes). Tests have shown that the measurement quality is similar to existing resistivity imaging systems when using arrays covering 200 m with electrode spacings of 1–2 m. An SDI-12 interface has been incorporated into the system, meaning that geotechnical and environmental sensors (e.g. pore pressure, temperature, rainfall) can be attached to the system, and data from this interface is incorporated in the system telemetry. These data assist with the calibration and interpretation of the geophysical monitoring results.

Data Telemetry and Control

The PRIME system has been designed to be fully automated/autonomous with measurements scheduled to run at given times during the day. Measurement data is stored locally then relayed using wireless telemetry, such as a GSM/Mobile network (or even a satellite) link. Along with the measurement data, health logs of system and sensor performance are also transmitted (Fig. 3).

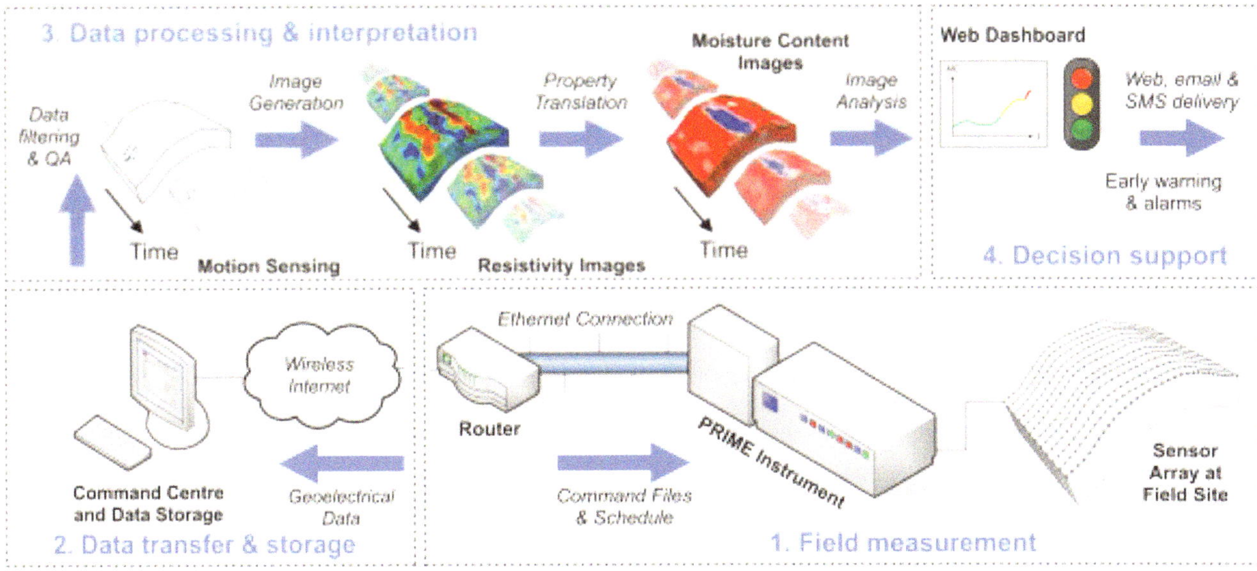

Fig. 1 PRIME system—geoelectrical monitoring and early warning workflow

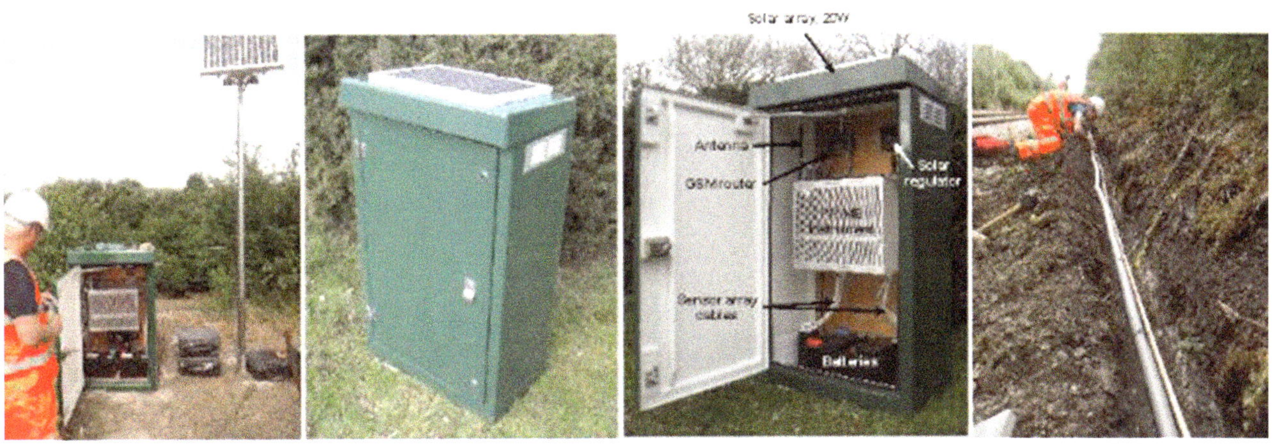

Fig. 2 PRIME system field deployments showing key components and sensor array installation

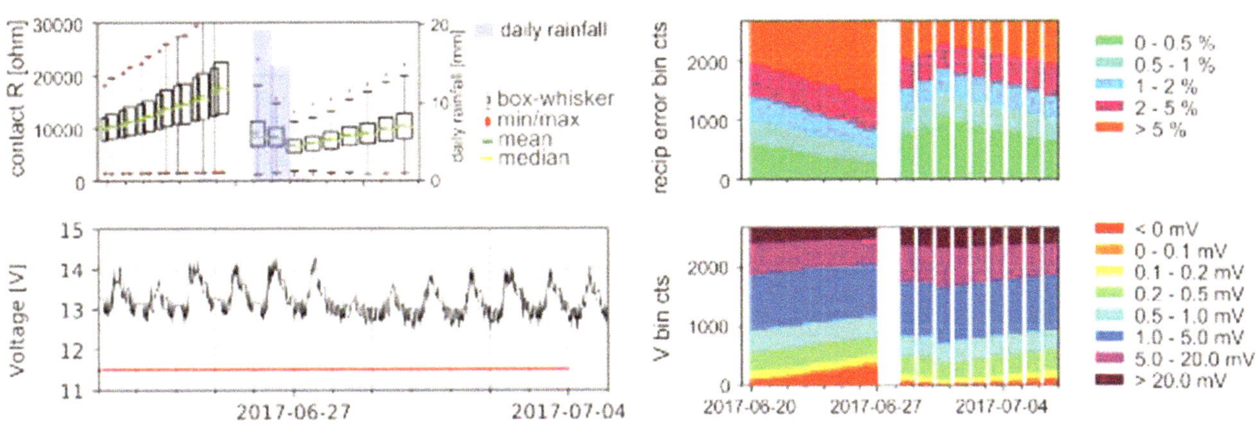

Fig. 3 PRIME system health logs (contact resistance; reciprocal error, battery voltage, measured voltages)

Automated Data Processing

For large ERT time-series data sets from multiple sites, manual processing and interpretation can be time-consuming and impractical. Consequently, an automated data processing work-flow (Fig. 1—Sect. 3) that comprises several stages is being implemented.

Stage 1: electrode displacement information is extracted from the measured data using the method described by Wilkinson et al. (2016). This information is used as a means of detecting ground motion, and to provide updated electrode positions.

Stage 2: the data are then inverted to produce 2D or 3D time-lapse images of subsurface resistivity distribution, to which a temperature correction is applied to normalise the resistivity images to mean air temperature (e.g. Chambers et al. 2014; Uhlemann et al. 2017).

Stage 3: if suitable geotechnical—geophysical property relationship information is available (e.g. Gunn et al. 2014), the resistivity images will be converted into images of moisture content.

Stage 4: we are developing automated approaches to image analysis drawing upon pattern recognition and change detection algorithms to assist with interpretation of results (e.g. Chambers et al. 2015).

Decision Support

The timely and intelligible communication of monitoring results to stakeholders is essential for an effective decision support in the context of condition assessment and early warning. The PRIME system delivers near-real-time information in a number of ways. Firstly, the system has an autonomous monitoring capability, in that it can respond to environmental triggers (Fig. 4); it uses environmental and geotechnical sensors (e.g. a rain gauge or pore pressure sensor) that are integrated into the system to trigger high intensity monitoring and SMS and email-based alerts to stakeholders.

Also, information from the system can be delivered through a web-dashboard (Fig. 5), where 4D monitoring results can be interrogated (Fig. 5, top), alarm thresholds can be set, and the monitoring results can be analysed on a cell by cell basis (Fig. 5, bottom) and compared with other environmental and geotechnical monitoring data to facilitate decision support.

Case History—Ripley Landslide, Canada.

The Ripley Landslide is a small (0.04 km^2), slow-moving (355 mm/year), translational landslide on a natural slope in the Thompson River Valley, and is one of 14 active landslides along a 10 km stretch of this vital transport corridor (Fig. 6). Slope failures in this area are having negative impacts on railway infrastructure, terrestrial and aquatic ecosystems, public safety, communities, local heritage, and the economy, and the Ripley Landslide threatens the serviceability of two national railway lines (Canadian National (CN) and Canadian Pacific Railway (CPR)), which drives the need for long-term monitoring at the site (Huntley and Bobrowsky 2014).

A PRIME system was installed on the Ripley Landslide in November 2017 and has been actively collecting data for over 2 years. The installation consists of two sensor arrays: a 91 m long array with 45 evenly spaced, buried rod electrodes, oriented NE-SW across the slope, and crossing the head scarp at the southern end; and a 54 m long array with 27 evenly-spaced, buried rod electrodes, oriented SE-NW downslope, and spanning the eastern extent of the head scarp. Measurements are taken every 12 h at the Ripley Landslide, providing high temporal resolution information on moisture driven changes to subsurface resistivity. The data from PRIME were inverted using an iteratively reweighted Gauss–Newton least-squares method (Loke and Barker1996) with an L1 norm on the data misfit, an L1 spatial smoothness constraint and an L2 temporal smoothness constraint, and data were inverted in 3D.

The baseline ERT image of the Ripley Landslide from December 2017 (Fig. 7) shows the resistivity profile of the site, and the dominant lithological units (high resistivity, coarse alluvial sediments overlying lower resistivity, high plasticity glaciolacustrine sediments and Fraser Glaciation diamicton). Key features of the landslide, including tension cracks, which are important in providing pathways for surface water to infiltrate into the body of the landslide, are also illustrated. An inferred shear surface is presented, although the ERT has been unable to resolve the exact location of the shear surface of the landslide as there is no geophysical contrast either side of the shear surface as it occurs within a single lithological unit. Time lapse images are shown in Fig. 8, with changes in resistivity (%) shown in relation to the baseline image.

Figure 8 demonstrates the seasonal changes in resistivity relating to changes in moisture content, and highlights the evolution of moisture dynamics in the subsurface throughout a single freeze–thaw cycle (Holmes et al. 2020). Generally, the slope is dryer in winter (Fig. 8a), and areas which show a large increase in resistivity at the very-near-surface are likely to be indicative of localized freezing in response to sub-zero temperatures, which are often accompanied by snowfall in this region of British Columbia. During the spring, which marks the on-set of the snowmelt season, there is a large decrease in resistivity, associated with an increase in moisture content as snow melts and water infiltrates into the slope (Fig. 8b). While the whole of the slope surface exhibits a decrease in resistivity, key hydrogeological pathways are revealed. In particular, moisture pathways in the head scarp zone are highlighted where the wetting front propagates

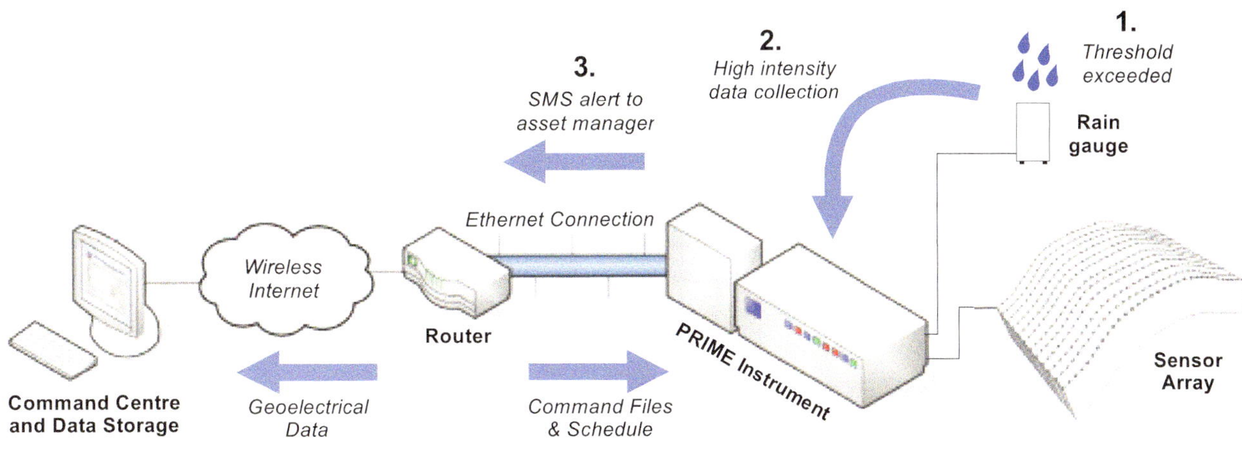

Fig. 4 Schematic describing PRIME operating in an autonomous [*or responsive*] monitoring mode, based on measurement and alert triggers using environmental point sensors (e.g. rain gauge)

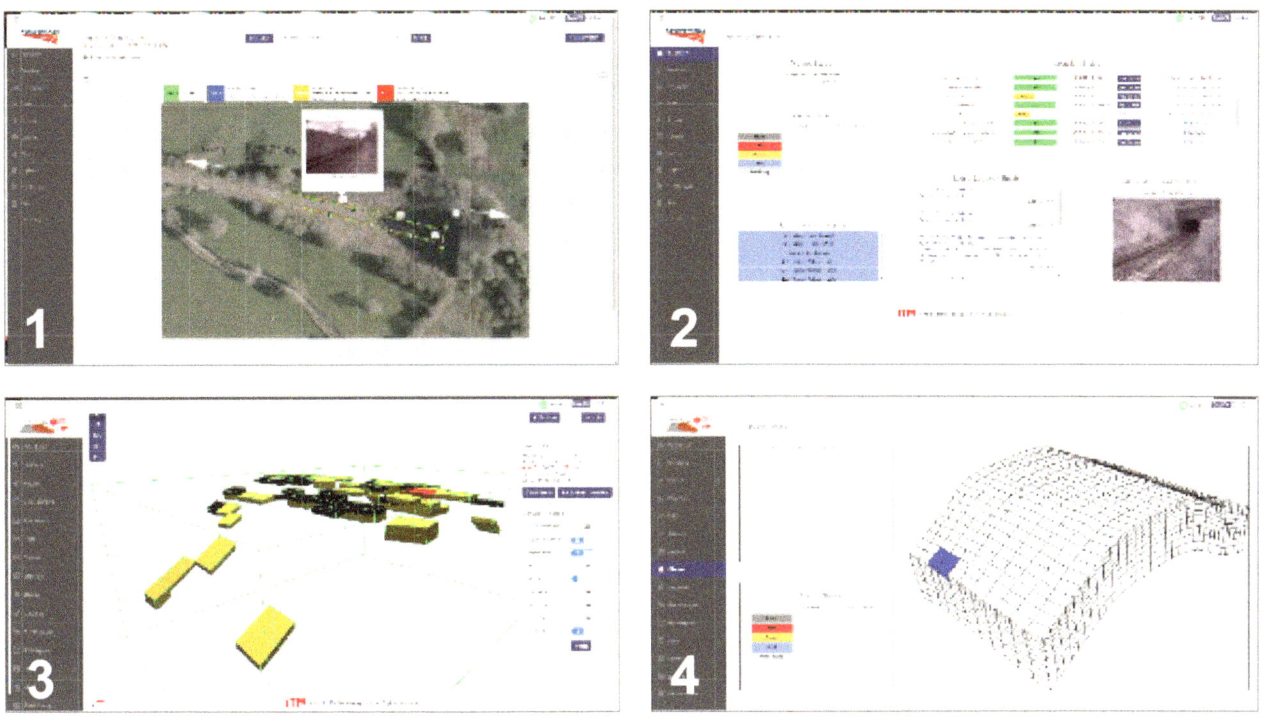

Fig. 5 PRIME-Calyx web-dashboard comprising: (1) GIS front end showing site and sensor locations; (2) Site specific control dashboard showing summary information and alarm states; (3) ERT data viewer enabling alarm thresholds to be set for each voxel/pixel; (4) ERT data viewer enabling time-lapse property changes to be assessed

down the slip face. Towards the end of the snowmelt season (Fig. 8c), the slope surface begins to dry again as moisture levels begin to return to baseline conditions, albeit with some variation owing to annual differences in weather conditions.

The Ripley Landslide case study highlights the utility of PRIME monitoring for contributing to the assessment of slope stability as it enables a detailed understanding of the soil moisture pathways and spatial heterogeneities that cannot be identified using traditional monitoring techniques.

Conclusions

Key Benefits

Spatial/volumetric subsurface information: The use of geophysical ground imaging technology will complement existing landslide monitoring approaches. It extends the information provided by remote sensing (e.g. aerial photographs, LiDAR) by illuminating the internal structure of

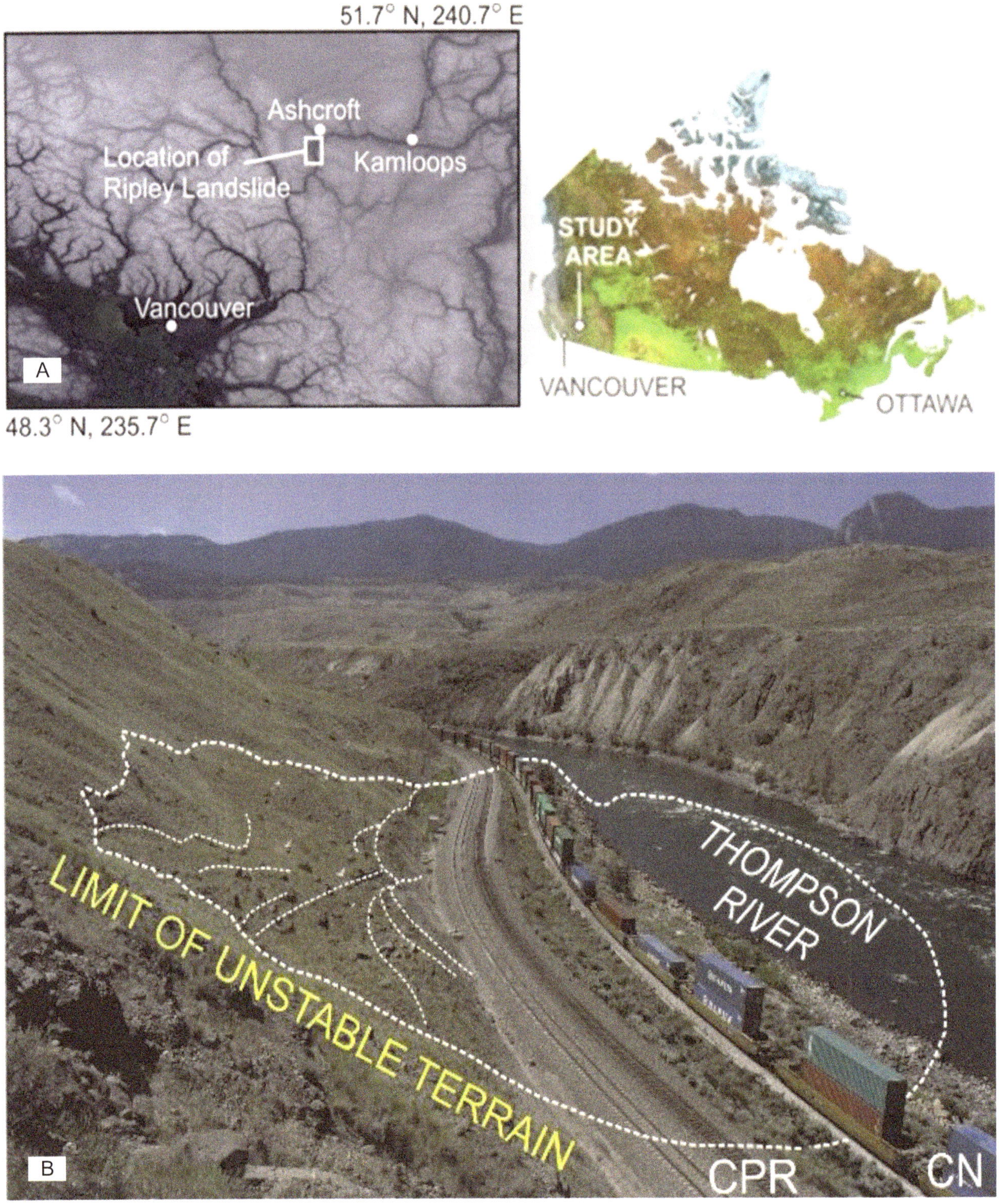

Fig. 6 **a** Location of the Ripley Landslide. **B** Site photograph of the Ripley Landslide showing the surface extent of the landslide on the slope

the earthwork. It also assists in interpreting data from point sensors installed in existing monitoring boreholes by providing spatial information to 'fill the gaps' between intrusive sample/data points. This volumetric approach is important given that unstable slopes can be highly heterogeneous structures, in which small-scale deterioration can rapidly create larger-scale problems, and potentially give rise to catastrophic failures.

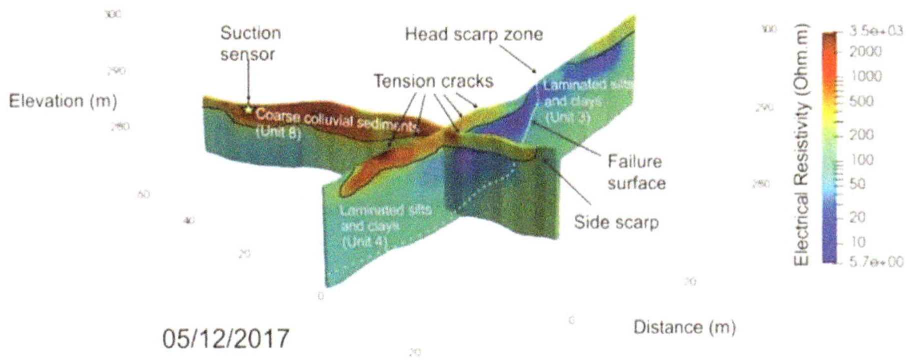

Fig. 7 Baseline ERT image from the PRIME monitoring of the Ripley Landslide. Different lithological units are demarcated by the black lines, and key features of the landslide are highlighted. The dashed white line shows an inferred failure surface; the spacing of the dashes increases with depth as the location of the failure surface becomes less certain

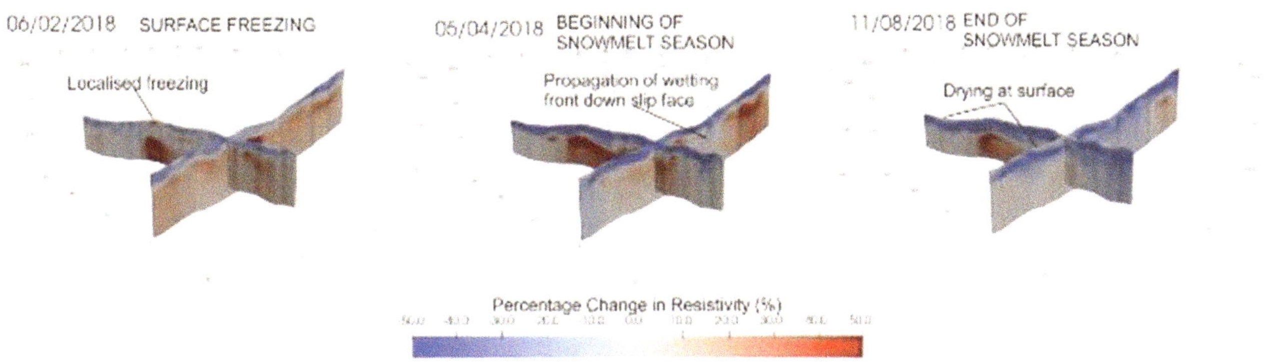

Fig. 8 Percentage change in resistivity from the baseline image showing changes in resistivity through one freeze–thaw cycle from initial freezing (**a**), to the onset of the snowmelt season (**b**), to the end of the snowmelt season (**c**)

Minimally invasive: The deployment of the PRIME system and sensors cause minimal ground disturbance and does not need heavy equipment for installation, which can be advantageous in the context of unstable ground and steep slopes. Sensors can be installed at or just below the ground surface across the slope, thereby avoiding the need for intrusive subsurface installations that could impact on the integrity of the slope.

Near-real-time monitoring: PRIME is fully automated, with condition information accessible through a web-based 'dashboard' or provided through automatic alarms in the case of rapid deterioration. Using this approach, a high temporal resolution (i.e. minutes to hours) can be achieved compared to walk-over or remotely sensed surveys (i.e. days to weeks).

Contribution to Landslide Early Warning

Landslide early warning systems (LEWs) are being widely developed and deployed internationally (e.g. Intrieri et al. 2013 and Guzzetti et al. 2020). Although the specific designs vary, there are generally common features across systems. These include a consideration of: design and sensor deployment; acquisition and processing of monitoring data; forecasting and the development of warning criteria and thresholds; and risk communication and response. We anticipate that PRIME technology can be integrated into the design of existing LEWs at the slope scale. The technology is intended to provide additional subsurface information in relation to moisture driven changes in the subsurface; a key ongoing research goal is to develop thresholds and warning criteria based on these near-real-time observations. Likewise, the facility to deliver monitored data and information via a web-interface is designed to contribute to the communication of risk and response to changing conditions within unstable slopes. By illuminating the moisture dynamics within unstable slopes in near-real-time we anticipate that geophysical imaging technologies, such as PRIME, will play an increasingly important role in the development of LEWs.

Acknowledgements Development of PRIME technology has been assisted through NERC Grants NE/P00914X/1, NE/M008479/1 and NE/N012933/1. This paper is published with the permission of the Executive Director, British Geological Survey (UKRI-NERC).

References

Chambers JE, Gunn DA, Wilkinson PB, Meldrum PI, Haslam E, Holyoake S, Kirkham M, Kuras O, Merritt A, Wragg J (2014) 4D Electrical resistivity tomography monitoring of soil moisture dynamics in an operational railway embankment. Near Surf Geophys 12:61–72. https://doi.org/10.3997/1873-0604.2013002

Chambers JE, Meldrum PI, Wilkinson PB, Ward W, Jackson C, Matthews B, Joel P, Kuras O, Bai L, Uhlemann S, Gunn D (2015) Spatial monitoring of groundwater drawdown and rebound associated with quarry dewatering using automated time-lapse electrical resistivity tomography and distribution guided clustering. Eng Geol 193:412–420

Gunn DA, Chambers JC, Uhlemann S, Wilkinson PB, Meldrum PI, Dijkstra TA, Haslam E, Kirkham M, Wragg J, Holy-oake S, Hughes PN, Hen-Jones R, Glendinning S (2014) Moisture monitoring in clay embankments using electrical resis-tivity tomography. Constr Build Mater 92:82–94

Guzzetti F, Gariano SL, Peruccacci S, Brunetti MT, Marchesini I, Rossi M, Melillo M (2020) Geographical landslide early warning systems. Earth Sci Rev 200:102973

Holmes J, Chambers J, Meldrum P, Wilkinson P, Boyd J, Williamson P, Huntley D, Sattler K, Elwood D, Sivakumar V, Reeves H, Donohue S (2020) Four-dimensional electrical resistivity tomography for continuous, near-real-time monitor-ing of a landslide affecting transport infrastructure in British Columbia. Near Surf Geophys, Canada. https://doi.org/10.1002/nsg.12102

Huntley D, Bobrowsky P (2014) Surficial geology and monitoring of the Ripley Slide, near Ashcroft, British Columbia, Cananda. Geol Surv Can, Open File 7531

Huntley D, Bobrowsky PH, M, Macciotta, R, Elwood, D, Sattler, K, Best, M, Chambers, J, Meldrum, P, (2019) Applica-tion of multi-dimensional electrical resistivity tomography datasets to investigate a very slow-moving landslide near Ash-croft, British Columbia, Canada. Landslides 16:1033–1042

Intrieri E, Gigli C, GN, Nadim, F, (2013) Landslide Early Warning System: toolbox and general concepts. Nat Haz-Ards Earth Syst Sci 13:85–90

Loke MH, Barker RD (1996) Practical techniques for 3D resistivity surveys and data inver-sion. Geophys Prospect 44:499–523

Uhlemann S, Chambers J, Wilkinson P, Maurer H, Merritt A, Meldrum P, Kuras O, Gunn D, Smith A, Dijkstra T (2017) Four-dimensional imaging of moisture dynamics during landslide reactivation. J. Geophys. Res. Earth Surf 122:398–418

Wilkinson P, Chambers J, Uhlemann S, Meldrum P, Smith A, Dixon N, Loke MH (2016) Reconstruction of landslide move-ments by inversion of 4D electrical resistivity tomography monitoring data. Geophys Res Lett 43:1166–1174

UNESCO · ICL · ISC · WFEO · FAO · UNDRR · WMO · UNU
WLF
2020 Kyoto Japan

Geophysical Monitoring of Landslides: State-of-the Art and Recent Advances

Denis Jongmans, Sylvain Fiolleau, and Gregory Bièvre

Abstract

Geophysical monitoring of landslides has developed strongly in recent years with the use of electrical resistivity imaging on the one hand, and the application of techniques based on continuous seismic recordings on the other hand. The paper resituates these developments within the general framework of the application of geophysical methods to landslides and focus on the definition of relevant geophysical parameters that can be precursors to activation or reactivation phases. Four recent case studies from the literature illustrate the potential and limitations of geophysics for landslide monitoring. Development prospects, especially for integration into early warning systems, are discussed.

Keywords

Geophysics • Monitoring • Early warning • Resistivity • Seismic

Introduction

Since the pioneering work of Bogoslovsky and Ogilvy (1977), geophysical methods have been increasingly used to characterize landslides, as evidenced by the growth rate of citations including landslide and geophysics (Jaboyedoff et al. 2019) and increasing number of review papers on the subject since the 90ties (among others, McCann and Forster 1990; Hack 2000; Jongmans and Garambois 2007; Van Dam 2012; Perrone et al. 2017; Whiteley et al. 2019; Pazzi et al. 2019). The main reasons for this growing popularity are known (Jongmans and Garambois 2007): geophysical techniques are non-invasive, low-cost and, unlike geotechnical techniques such as drilling and penetration tests, they allow a large volume of ground to be explored. In recent years, data acquisition and inversion techniques have developed considerably, providing geophysical 2D and 3D images with increasing resolution. As shown in the bibliometric study of Jaboyedoff et al. (2019), the geophysical methods most commonly applied in scientific studies are Electrical Resistivity Tomography and ambient seismic noise techniques which, when interpreted together, can provide information on the nature of soil or rock, moisture and rigidity contrast.

However, as Jongmans and Garambois (2007) point out, geophysical techniques have significant drawbacks: (1) the resolution decreases with depth, (2) the solution obtained by the inversion process is generally non-unique and needs to be calibrated, (3) they provide indirect information (physical parameters) instead of geological or geotechnical properties. These drawbacks, as well as a tendency to over-interpret the data, may explain the reluctance of part of the engineering community to use geophysical techniques (Jongmans and Garambois 2007). Recently, Pazzi et al. (2019) analyzed the efforts of the geophysical community to overcome the limitations identified by Jongmans and Garambois (2007) over the period (2007–2018). They found that many efforts have been made in the geological interpretation of geophysical data but that issues of resolution and penetration are still little discussed.

Recent reviews have also shown that geophysical monitoring studies, particularly over long periods (months to years) are much less common than exploration studies. The main reason is of course the enormous effort required to maintain monitoring systems over long periods in moving terrain. Convinced that complex structures such as landslides cannot be understood without the acquisition of detailed data

D. Jongmans (✉) · S. Fiolleau · G. Bièvre
ISTerre, Univ. Grenoble Alpes, CNRS, IRD, 38000 Grenoble, France
e-mail: denis.jongmans@univ-grenoble-alpes.fr

S. Fiolleau
e-mail: sylvain.fiolleau@univ-grenoble-alpes.fr

G. Bièvre
e-mail: gregory.bievre@univ-grenoble-alpes.fr

© Springer Nature Switzerland AG 2021
N. Casagli et al. (eds.), *Understanding and Reducing Landslide Disaster Risk*, ICL Contribution to Landslide Disaster Risk Reduction,
https://doi.org/10.1007/978-3-030-60311-3_7

sets, several teams have installed observatories measuring various types of data on unstable slopes. Among others, the 4 sites of the French landslide observatory (https://www.ano-omiv.cnrs.fr/), the Salcher observatory in Austria (Stumvoll et al. 2019), the Hollin Hill landslide observatory in England (https://www.bgs.ac.uk/landslides/) and the Tokushima Landslide Observatory in Japan (https://www.dpri.kyoto-u.ac.jp/) are currently in operation. In parallel, numerous sites have been or are being monitored for scientific or operational purposes for months or years in order to gain a comprehensive understanding of the landslide mechanism and to develop early warning systems (e.g. Helmstetter and Garambois 2010; Hibert et al. 2011; Mainsant et al. 2012b; Supper et al. 2014; Gance et al. 2016; Colombero et al. 2018; Bertello et al. 2018).

One major issue raised for the geophysical monitoring of landslides, which are complex and evolving structures with a strong hydro-mechanical coupling, is to define which parameters to measure, where and how. The aim of this paper is to take stock of the geophysical monitoring results published in the literature and to draw some perspectives for the future.

Geophysical Methods

Generalities

Geophysical prospecting is based on physical measurements (data) from which physical parameters, characteristic of rocks and soils, can be deduced. The principle of the different geophysical methods can be found in general works (Reynolds 2011; Kearey et al. 2013; Everett 2013). A geophysical method is characterized by its resolution (ability to detect an object of size D) and its penetration depth (depth beyond which the object sought no longer has any influence on the measurement made). For surface measurements, the resolution generally decreases significantly with depth. The choice of method(s) to be used depends on five main factors (McCann and Forster 1990): (1) the existence of a geophysical contrast. The presence of a lithological, mechanical or hydrogeological contact does not necessarily imply a variation in all geophysical properties, (2) the depth and resolution required, (3) the need to calibrate the geophysical reconnaissance using geological or geotechnical data, (4) the achievement of a sufficient signal-to-noise ratio which depends on site conditions, (5) the cost of the reconnaissance campaign.

Geophysical imaging methods have developed considerably over the past 20 years, generally providing the variation of a geophysical parameter along 2 or 3 spatial coordinates. Inversion of geophysical data is a complex non-linear problem and the obtained solution is generally not unique. The interpretation of the images must then be done critically, considering all the data available on the site (introduced into the inversion if possible) and must be the subject of a critical discussion between geologists, geophysicists and geotechnicians.

Geophysical Parameters

The most commonly used geophysical parameters in engineering geology are P-wave velocity (Vp), S-wave velocity (Vs) and electrical resistivity (ρ). The ranges of variation of these 3 parameters in the most common natural materials are presented in the table in Appendix 1. The knowledge of these three parameters provides initial information on the nature of the material (rock, soil), its rigidity, compactness and its degree of water saturation.

Electrical resistivity ρR in a geological medium varies with different parameters and the most commonly used empirical equation is Archie's modified law (Mavko et al. 2019):

$$\frac{1}{\rho_R} = \frac{1}{a\rho_w \Phi^{-m} S^{-n}} + X \tag{1}$$

where Φ is the porosity, ρw is the water resistivity, S is the saturation degree, X is a clay conductivity term, and a, m and n are constants dependent on the medium. The equation shows that resistivity decreases with the saturation degree, the porosity and the presence of clay and increases with the water resistivity. Another parameter that could significantly influences electrical resistivity is the ground temperature which changes with air temperature at shallow depth. An increase in ground temperature decreases ρR (Hayley et al. 2007).

In terms of seismic methods, S-wave velocity (Vs) has gradually become a geophysical parameter of primary importance in geotechnical engineering because of its sensitivity to variations in stiffness and cracking in geological materials (Barton 2006). In the theory of elasticity, Vs is related to the shear modulus G and density γ by the equation:

$$V_s = \sqrt{\frac{G}{\gamma}} \tag{2}$$

Since G and γ both vary with porosity, empirical relations linking Vs and porosity in various dry and saturated materials have been proposed (e.g. Castagna et al. 1985; Knackstedt et al. 2005; Mondol et al. 2007), showing the decrease of Vs with porosity (increase with compactness).

Application of Geophysical Methods to Landslide Monitoring

Landslides generally lead to changes in surface and internal structure, in terms of mechanical and hydrogeological properties, which may alter the geophysical properties that can then be used to image the ground movement and track its temporal evolution. In recent years, time-lapse geophysical methods have developed considerably in an attempt to track the evolution or triggering of ground movements.

Electrical Resistivity Tomography (ERT) has been particularly used in this sense (e.g. Bièvre et al. 2012; Supper et al. 2014; Gance et al. 2016; Perrone et al. 2017; Uhlemann et al. 2017; Palis et al. 2017) because it makes it possible to identify and monitor water infiltration, which is one of the major causes of the triggering or reactivation of ground movements. Specific methodological developments have been proposed, allowing to monitor the displacement of the electrodes within the ground movement (Wilkinson et al. 2010, 2015).

In clay-rich landslides, field and laboratory experiments have shown that Vs is a parameter very sensitive to clay deconsolidation and the resulting increase in porosity (Jongmans et al. 2009; Renalier et al. 2010a), and that rheological changes in clayey landslides could be tracked by monitoring Vs (Mainsant et al. 2012a). The measurement of Vs on landslides can be performed by different methods: seismic horizontally-polarized shear-wave (SH) refraction (Jongmans et al. 2009; Uhlemann et al. 2016), inversion of surface waves generated by an active source or cross-correlation of seismic noise (Jongmans and Garambois 2007; Renalier et al. 2010b). Temporal monitoring of Vs with an active source is more complex due to the difficulty of repeating the same measurement conditions (source and geophone positions) in a changing environment. Whiteley et al. (2020) explores the effects and challenges that the changing topography of landslides (and consequent changes in source and geophone position) can have on time-lapse seismic (Vp and Vs-SH) refraction measurements. Recently, Bertello et al. (2018) have presented results of periodic and continuous measurements of Rayleigh wave velocity conducted in an active earthflow, showing considerable velocity variations during landslide reactivation phases.

An alternative is to record ambient vibrations and monitor derived seismic parameters (Larose et al. 2015). In the case of rocky sites, a column in the process of decoupling from the massif exhibits resonance phenomena whose frequencies can be tracked over time (Lévy et al. 2010). In a medium that is deconstructing, the cross-correlation of seismic noise between two sensors (Mainsant et al. 2012b) makes it possible to measure variations in the rigidity of the environment based on changes in seismic velocity (surface waves). Landslides can also be monitored by processing endogenous failure signals, the number of which usually increases significantly before a major event (e.g. Amitrano et al. 2005; Tonnellier et al. 2013; Poli 2017; Fiolleau et al. 2020).

We then present four case studies (two seismically and two electrically instrumented sites) illustrating the geophysical monitoring of landslides under very different conditions and allowing to draw conclusions and perspectives for the future.

Geophysical Monitoring of Landslides

Electrical Methods

We present two examples of electrical monitoring of landslides affecting clay-rich slopes. The first study applied a 3D-time lapse (4D) electrical imaging to monitor moisture dynamics during a reactivation of the Hollin Hill landslide (England). At the second site (Super Sauze, France), electrical resistivity measurements were conducted to monitor 2D water circulation in the shallow zone of the landslide, evidencing a significant influence of temperature on the measurements.

Hollin Hill Site

The Hollin Hill landslide is 220 m long, 500 m wide slow moving composite multiple Earth slide/flow (Uhlemann et al. 2016, 2017). It affects a gentle slope shaped in four bedrock formations of Lower to Middle Jurassic age, slightly dipping to the N and covered by superficial deposits of variable thickness. From bottom to top, the slope comprises the Redcar Mudstone (RMF), Staithes Sandstone and Cleveland Ironstone (SSF), Whitby Mudstone (WMF), and Dogger Formation (DF) units, slightly dipping to the North. The Hollin Hill landslide is mainly located in outcropping mudstones and siltstones (WMF, 25 m thick). Underlying the poorly drained WMF is the 20 m SSF unit, which is composed of sandstones and siltstones, and overlies the mudstone RMF unit. The hydrogeological system of the slope is then complex with fluctuating aquifers in the permeable DF and SSF formations. The landslide is superficial (a few m thick) and its mechanism evolves from rotational at the headscarp to translational in the central part of the WMF and to slide/flow-like movements to the toe over the SSF formation. The landslide reactivated in December 2012/January 2013, after a very wet summer and prolonged rainfall in winter. Surface movements of up to 3.5 m occurred in winter 2012/2013 along the back scarp and eastern lobe.

Hollin Hill acts as an observatory for landslide research and has been thoroughly characterized using geotechnical and geophysical investigation over the last fifteen years (Uhlemann et al. 2016). The 3-D ERT monitoring installation is a rectangular grid of 160 electrodes 4.75 m apart, made of 5 parallel lines of 32 electrodes with 9.5 m interline spacing. Electrical data were measured on alternating days. GPS benchmarks were repeatedly surveyed to estimate the electrode movements, providing electrode locations for ERT. The electrical measurement sequence comprised dipole–dipole measurements, providing 2580 apparent resistivity measurements. Data were filtered to remove erroneous measurements. A time-lapse inversion workflow was defined, in which each inversion uses the electrode locations of the corresponding date and is constrained to the resistivity model of the previous time step (Uhlemann et al. 2017). Resistivity values are corrected for temperature (Hayashi 2004) and are converted to gravimetric moisture content (GMC) fitting a Waxman Smit model to laboratory data measured for units WMF and SSF (Uhlemann et al. 2017).

The reference resistivity model (Fig. 1a) was inverted from data acquired on 19 March 2010, while the derived moisture model is shown in Fig. 1b. The SSF and WMF units are clearly imaged as resistive and conductive zones, respectively. The landslide body is well outlined as conductive lobes over the SSF unit, while it cannot be distinguished from the WMF unit. The GMC image shows that the SSF and WMF units are mainly characterized by low and high moisture contents, respectively, resulting from the good and poor drainage in these two formations.

The 43 analysed data sets (one per month) cover a 3-year period (from March 2010 to July 2013), during which the landslide reactivation occurred. At the yearly scale, the slope exhibits characteristic seasonal resistivity fluctuations, with decreasing moisture content in summer and increasing GMC in winter. During reactivation (December 2011; Fig. 1c), ERT shows a preferential flow at a depth of about 3–5 m below the western lobe, which maintains low pore pressure at the critical depth and prevent from movement. In contrast, a significant movement of up to 3.5 m was observed at the eastern lobe where no such flow was detected. Figure 1c, d shows GMC changes in December 2012 (during reactivation) and July 2013 (after reactivation). During the deceleration of the movement (July 2013, Fig. 1d), ERT shows a drainage from WMF to SSF.

4D ERT monitoring of the Hollin Hill slope was shown to be able to image moisture variations and dynamic preferential flow paths in the landslide, providing a better understanding of the relation between movements and water circulations. The reactivation of the landslide can be related to the moment when GMC values are in the range of the liquid limit (LL) (Hobbs et al. 2012) of the ground. This offers perspectives for early warning system.

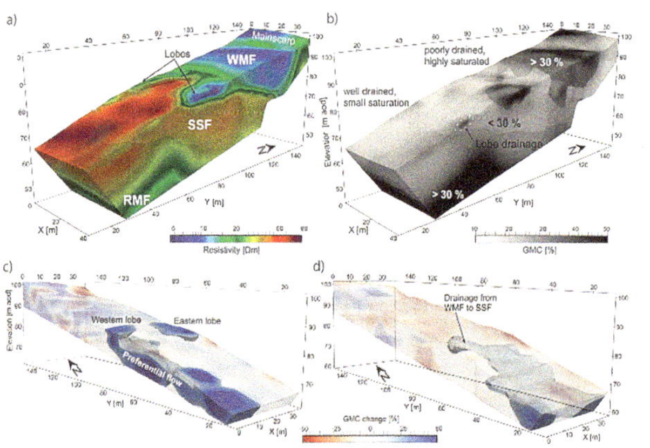

Fig. 1 Electrical imaging of the Hollin Hill site. **a** Reference model (19 March 2010). **b** Derived gravimetric moisture content (GMC) model. **c** GMC change of December 2012 highlighting preferential flow at the western lobe during reactivation of the landslide. **d** GMC change of July 2013 showing preferential flow path from WMF to SSF. Modified from Uhlemann et al. (2017). Copyright J. Wiley (2017)

Super Sauze Site

This case history is a thorough study of resistivity variations generated by rainfalls in the first meters of the Super Sauze landslide. This 900 m long landslide has developed in the weathered black marls since the beginning of the 1960s and affected a volume of about 750,000 m^3 (Malet et al. 2005). The landslide is active (0.05–0.20 m day^{-1}) with a mechanism evolving from sliding in the upper part to flowing in the lower part. The landside affects a few m thick heterogeneous layer made of weathered black marls with the presence of marl blocks and surface fissures.

An experiment of 1 year was conducted on a 113 m long longitudinal profile located in the upper part of the landslide (Gance et al. 2016), with the deployment of 93 electrodes. In this zone, the water table was located at about 1.5 m depth. The objective was to track the water infiltration in the vadose zone. Apparent resistivity sets were measured two times per day with a sequence of 4300 quadrupoles (gradient array). The electrical system was complemented by ground water level and temperature sensors at two plots. Electrode movements were measured with stereo-photogrammetry. Raw electrical data were filtered to remove erroneous measurements and inverted resistivity values were corrected for groundwater conductivity and temperature variations. Figure 1 shows the electrical response to a rainfall that started on 31 May 2011 with a total amount of 40 mm. The air temperature was around 12 °C and the groundwater level was 0.5 m depth. Figure 1 first shows a 5% decrease in resistivity at very shallow depth, which is interpreted as an increase in water content in the vadose zone. Then (t = 24 h), this anomaly disappears and conductive zones appear at about

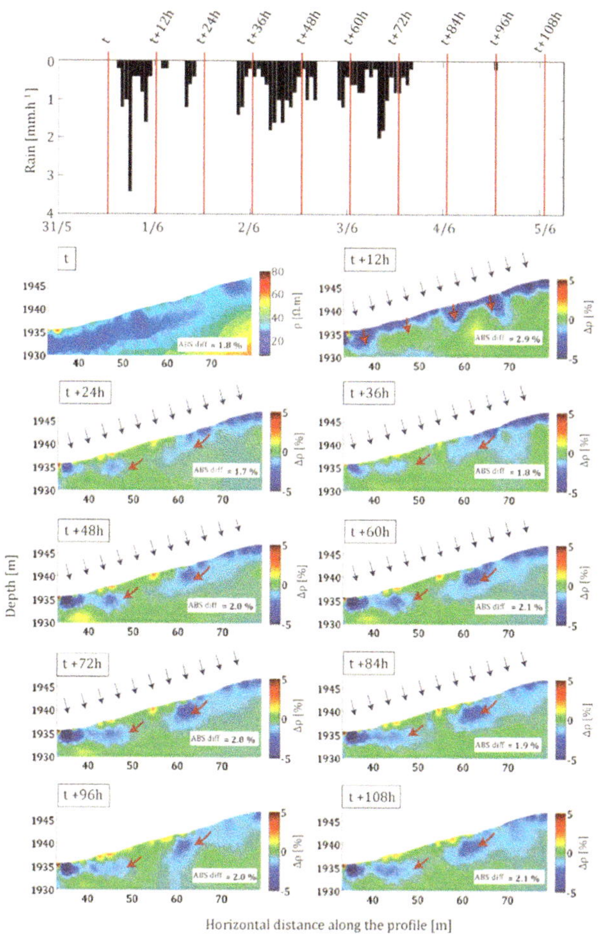

Fig. 2 Electrical response of the super Sauze landslide to the rainfall event shown at the top. The resistivity reference model is shown at t = 0 and the other diagrams show the percentage change of resistivity from t = 12 h to t = 108 h. Red arrows show the interpreted water flows. From Gance et al. (2016). Copyright Elsevier (2016)

2 m depth. These zones move downslope with time. As meteoric water infiltrations are very unlikely to generate a decrease in resistivity below the water table, these anomalies are interpreted as the infiltration of warm water propagating in a fissure network in conditions where the temperature of the groundwater was about 2 °C. For another rainfall occurring in summer with an air temperature of 20 °C, the authors observed that the infiltration of cold rainwater in a warm upper soil produces a 5% increase of resistivity in the vadose zone (Fig. 2).

These results illustrate the effect of temperature and heat exchange between the rain water, the vadose zone and the groundwater, which may hide variations of the soil water content but allow imaging water fluxes in the saturated zone. In the same study, the authors also highlight the sensitivity of apparent resistivity values to electrode movements, 3D effects and superficial fissures, which may generate artefacts, and propose a processing strategy to correct the effects of the first two factors.

These two case studies of shallow electrical monitoring illustrate the potential and difficulties of ERT, in particular the sensitivity of electrical resistivity to temperature. A main difference between the two cases is the presence at the Hollin Hill site of permeable unit (SSF) allowing water flows to be tracked in an unsaturated medium.

Seismic Methods

Continuous seismic recordings are used to derive parameters that have been shown to be precursors to landslides. Several cases have been documented in the literature during rock scale failures with seismic activity (Amitrano et al. 2005) and resonance frequency (Lévy et al. 2010) as parameters, or during mudflow triggering, with seismic velocity variation (Mainsant et al. 2012b).

We present here two recent examples of seismic monitoring of landslides affecting clay material. On the first site (Monte Vecchio, Italy), geophones were deployed on and outside the earthflow and measured the variations in Rayleigh dispersion curves before, during and after rapid movements with solid-to-fluid transition On the second site (Harmalière, France), the monitoring was made using continuous recordings, extracting for precursor parameters derived from ambient vibrations and microearthquakes.

Monte Vecchio Site

The Monte Vecchio landslide is a 700 m long and 30–50 m wide superficial earthflow affecting clayey material (old landslide deposits) over a thickness of about 5–10 m. The landslide may suddenly reactivate with a remobilization of the entire mass, associated to a rheological change from a solid to fluid behavior (Bertello et al. 2018). From February 2014 to June 2015, the landslide experienced three such flows with fluidization of the moving mass at least within the upper 2 m.

The reactivation of this landslide was monitored periodically and continuously using two standard techniques for measuring Rayleigh wave velocities: the active MASW (Park et al. 1999) and the passive ReMi techniques (Louie 2001). The authors used 4.5-Hz vertical geophones 2 m apart to record artificial sources and ambient vibrations (for details, see Bertello et al. (2018)). Periodic measurements were done every 1–2 months along seven seismic lines, four of which were located within the landslide. A cost-effective self-produced monitoring system for continuous measurements (two minutes every hour) of ambient vibrations was installed on the main track of the earthflow after May 16, 2014, allowing an automatic download of the data. During the one-year monitoring period, the system was reinstalled six times due to the strong landslide activity.

All data were analyzed using a cross-correlation algorithm to get the fundamental dispersion curve (phase velocity C(f)). For continuous measurements, C(f) was reliably obtained in a variable frequency range 5–15 Hz. Periodic measurement results show a systematic drop in C(f) and a recovery stage after the three observed flows. Figure 3 shows the variations in C(f) measured by the monitoring system for three frequencies (8, 11 and 15 Hz) during the flow event of 25 May 2015. The investigation depth for these frequencies is restricted to the first meters of the landslide.

In the first 3 weeks of May 2015, the earthflow was moving with a velocity of less than 1 cm/day and the measured velocities varied between 45 m/s (8 Hz) and 60 m/s (15 Hz) with a slight trend to decrease. A first rain event (47 mm) occurred on 22–24 May, which caused an acceleration (4 cm/day) and a drop in C(f) to 35–40 m/s at both frequencies. The mudflow started 16 h after the end of the rain and reached a velocity peak of 5–10 m/day before slowing down to about 1 m/day. A second precipitation (24 mm) occurred in the morning of May 26, provoking a reacceleration of the landslide with a peak velocity of 22 m/day. The cumulative displacement higher than 35 m disrupted the monitoring system, which was reinstalled on June 3, 2015. Phase velocities regularly increase with the deceleration of the landslide until returning to the initial values of 45–60 m/s one month after the first acceleration. The variations in Rayleigh wave phase velocities, which can be related to Vs changes in the superficial layer, show that the earthflow material undergoes a significant change in shear rigidity during each reactivation, with a drop of about 30% in wave velocity followed by a recovery phase.

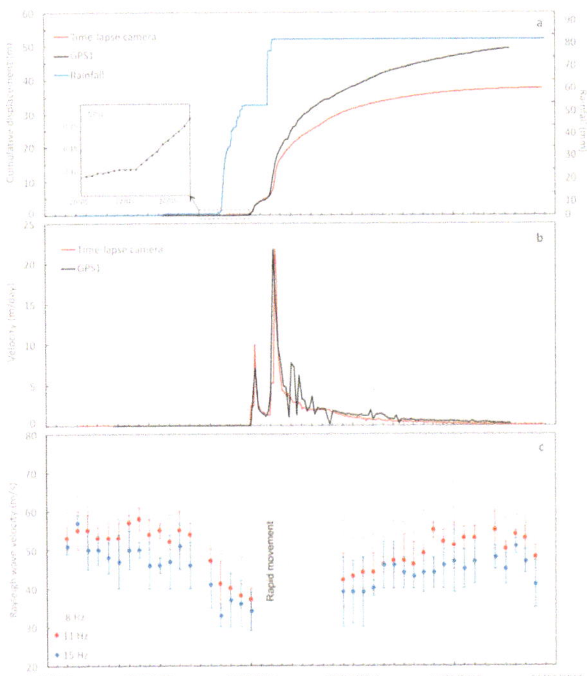

Fig. 3 Comparison between **a** rainfall and cumulative displacement, **b** displacement rate, and **c** Rayleigh velocity measured at different frequencies by the monitoring system before and after the reactivation of 25 May 2015. From Bertello et al. (2018). Copyright J. Wiley (2017)

These observations at Monte Vecchio could be compared to the results obtained at the Pont Bourquin site, where a drop of 7% in C(f) was measured over a few days before the triggering of a mudflow, cross-correlating the ambient vibrations recorded at two sensors placed on the stable flanks of the landslide (Mainsant et al. 2012b). This C(f) variation at 12 Hz was interpreted as a 50% Vs decrease in 2 m layer located at the base of the landslide (11 m depth). Although a decrease in Vs was observed at both sites, several major differences appear between them. The first is the depth of the layer that initially fluidized (superficial at Monte Vechhio, deep at Pont Bourquin). Then, the triggering of landslides at Monte Vecchio is clearly linked to the occurrence of rainfalls, whereas at Pont Bourquin it seems to have resulted from the progressive increase of the shear stress by accumulation of slid material. Finally, the mudflow at Pont Bourquin was preceded in the previous days by a decrease in Vs not related to the rainfall, while at Monte Vecchio, Vs decreased concomitantly with the rainfall before the surge.

Harmalière Site

The Harmalière landslide is located in the Trièves region (French Alps), characterized by the presence of a thick layer of clay of lacustrine origin (Bièvre et al. 2012). The active part of the landslide, 1500 m long and 300 m wide on average, is characterized by a solid behavior at the top evolving towards a fluid behavior at its toe. Its volume is estimated at 25 10^6 m^3. After a major activation in 1981, the main escarpment, which can reach more than 20 m high, has continuously regressed northwards (several m per year on average) with phases of acceleration and deceleration. A major abrupt reactivation of the landslide took place in late June 2016, when the main escarpment retreated in some places by 40 m, with an estimated moving volume of 2 10^6 m^3. Following this event, the main escarpment was affected by numerous small block slides whose volume was between several tens and several hundreds of m^3 (Fiolleau et al. 2020). Seismic equipment comprising two 3 component sensors and an acquisition station was deployed at the end of July 2016, on either side of a rear fracture affecting the main escarpment. The seismic noise was recorded continuously for 4 months until the 70 m^3 block on which the sensor was installed ruptured on November 25, 2016. In addition to the seismic noise, the sensors also recorded thousands of signals generated by microearthquakes caused by the sliding of the block and adjacent blocks. In this study, several seismic parameters were calculated and monitored until rupture: (1) the seismic energy SE, (2) the block resonance frequency f_B, (3) the surface wave velocity variation dV/V in the frequency range 1–12 Hz, (4) the corresponding correlation coefficient CC.

The evolution of these parameters in this order is shown in Fig. 4, together with air temperature and rainfall (Fig. 4c). The parameters showed a different sensitivity to the rainfall preceding the rupture. They all showed a precursor signal before the rupture, but at different times. The dV/V and CC parameters are the first to show significant irreversible variations, about 30 days before the rupture, with a strong decrease of the CC propagating from high to low frequencies. Just before the break, the signals measured on the block and behind the block are totally decorrelated over the whole frequency range, indicating a significant change in the medium. The seismic energy showed an increase at this date and increased sharply 3 days before the rupture. On the other hand, the resonance frequency started to decrease only two days before the block slipped. These results indicate the complexity of the failure mechanism in clayey soils. The interpretation provided by Fiolleau et al. (2020) is that the significant decrease in *CC* and the increase in seismic activity from 25 October onwards is the beginning of the block slip process. The resonant frequency changes little due to the probable presence of a shallow clay bridge under which the fracture surface propagates. This interpretation is supported by numerical modelling.

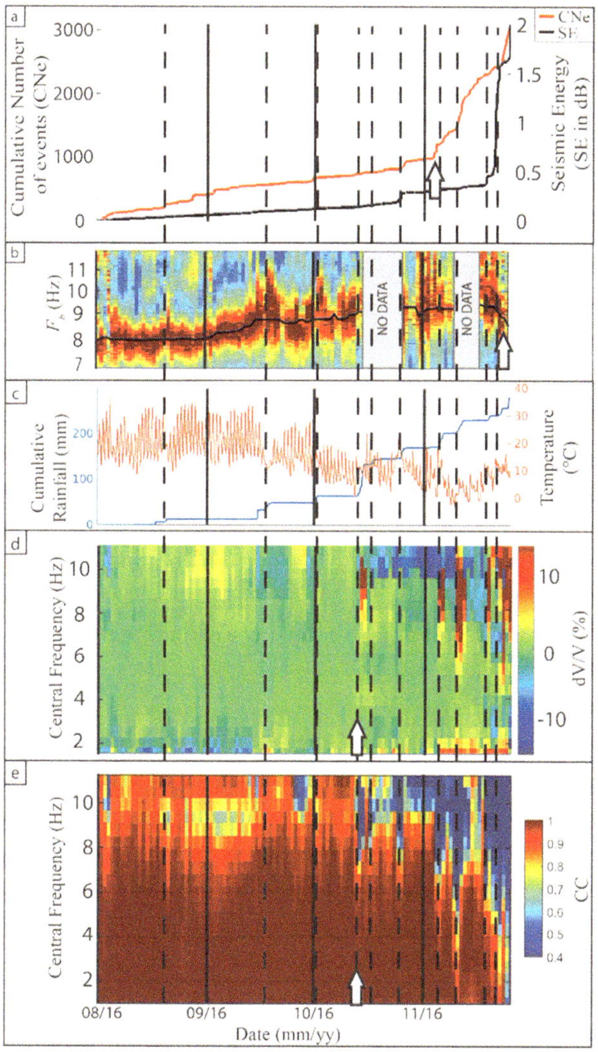

Fig. 4 Summary figure showing the evolution of the five seismic parameters studied. **a** Cumulative number of microearthquakes (CNe, red curve) and the cumulative seismic energy (SE, black curve). **b** Normalized HN0/HN1 ratio analysis zoomed in the frequency band 7–12 Hz. **c** Temperature and cumulative rainfall curves. The dashed black lines highlight the main rain events. **d** Rayleigh wave velocity variation (dV/V) between 1 and 12 Hz. **e** Correlation coefficient (CC). On each subplot, the starting point of the precursory signal for each seismic parameter is indicated by a white arrow. Modified from Fiolleau et al. (2020). Copyright Oxford University Press (2020)

These two cases illustrate the seismic response of two types of landslides (mudflow, slide) stressed by precipitation. At Monte Vecchio, the seismic velocity at shallow depth decreases directly with the rain before the mudflow. At Harmalière the seismic response of the block to rain is more complex but all seismic parameters show significant variations before failure.

Conclusions

The case studies presented show the interest of geophysical imaging and monitoring of geophysical parameters (electrical resistivity and seismic velocity). When installed permanently in the field with continuous data acquisition, these techniques allow the monitoring of hydro-mechanical changes within a landslide. They are complementary to time tracking techniques for surface movements (satellite, UAV, RFID) which are also undergoing spectacular development. They must be complemented by conventional (piezometric level, soil moisture and soil movement) and meteorological data for a complete understanding of the processes. In recent years, geophysical monitoring equipment has greatly improved and has become much more reliable, allowing the acquisition of time series over long periods (several years). These data are essential to improve our understanding of the geophysical response of landslides during their activation and the coupling between geophysical parameters and geotechnical properties. In addition to their spatial coverage, another advantage of geophysical methods is that they can provide information throughout the year, even under snow cover in mountainous areas.

However, the extensive use of geophysical instruments in monitoring systems always comes up against a number of bottlenecks. The first is the cost of the instruments, which remains prohibitive for operational application and still restricts geophysical monitoring mainly in the field of scientific research. The development of robust, if possible wireless and inexpensive devices, remains a challenge for a massive application of these techniques. A second lock is the definition of the relevant geophysical parameters to be monitored. In order to be usable and accepted in a monitoring system, variations in these parameters must be understood and comprehensible in terms of changes in the properties of the sliding mass. In seismic monitoring, several parameters (resonant frequency, seismic velocity variation, correlation coefficient and seismic activity or energy) can be used. The only case study (Harmalière) for which all these parameters were measured shows the interest but also the complexity of the seismic ground response before sliding. In addition, even landslides of the same type can have different causes and mechanisms. The two mudflows (Pont Bourquin and MonteVecchio sites) that were seismically monitored until triggering showed a decrease in surface wave velocity (and thus in S-wave velocity) with, however, a different pattern. In the case of the shallow mudflow at Monte Vecchio, the decrease in surface wave velocity was instantaneous with precipitation, whereas at Pont Bourquin, the S-wave velocity in the layer that liquefied at a depth of 10 m decreased several days before the onset of the mudflow.

For electrical monitoring, by far the most important parameter used is resistivity. The relationship between electrical resistivity and moisture content appears to be well established in the case of Holin Hill and could allow for a possible early-warning when moisture approaches the liquid limit of the superficial material. However, electrical resistivity is a parameter that is sensitive to other factors such as shallow soil temperature (as shown in the Super Sauze case study), porosity and water resistivity. These last two factors are subject to change during a ground movement and any resistivity variation cannot be always interpreted as a change in moisture content. Landslide triggering experiments with simulated rainfall have also shown that the hydrogeological triggering conditions (saturation, fluid pressure) may be more complex than simply reaching a threshold value (e.g. Lehmann et al. 2013).

Finally, the number of landslide triggers or reactivations that have been geophysically monitored remains low and the demonstration of the potential usefulness of geophysical techniques in warning systems will come from new case studies such as those presented in this paper.

Acknowledgements The authors acknowledge financial support from the French VOR federative structure, the French national C2ROP project, the French National Research Agency in the framework of the Investissements d'Avenir program (ANR-15-IDEX-02, CDP-RISK). This work was partly supported by the project SIMOTER 1 funded by the European Union under the ERDF – POIA program and by the French government under the FNADT – CIMA program. The authors are part of LabEx OSUG@2020 (ANR10 LABX56).

Appendix 1

Table showing the ranges of variation of electrical resistivity and P and S wave velocities in the most common natural materials. These values are an order of magnitude and may be modified by weathering processes.

	Resistivity (Ωm)	Vp (m/s)	Vs (m/s)
Soils			
Clay	1–50	1100–2500	30–700
Silt (dry to saturated)	40–200	300–1700	100–500
Dry sand	200–1000	400–1200	100–500
Saturated sand	50–200	1500–2500	100–500
Contaminated sand	1–20	1500–2500	100–500
Dry gravel	500–1500	500–1200	100–500
Saturated gravel	50–250	1500 2500	100–500
Rocks			
Shale	20–200	2000–3500	750–1500
Marl	50–300	1500–3000	600–1500
Chalk	100–400	2000–3000	800–1300
Sandstone	100–1000	3000–4500	1200–2500
Conglomerate	1000–10,000	3000–4500	1200–2500
Limestone	1000–6000	3500–6000	1900–3300
Weathered limestone	*50–1000*	*1700–3000*	*700–1500*
Dolomite	1000–5000	3500–6000	1900–3300
Granite	1000–10,000	4500–6000	2500–3300
Weathered granite	*30–500*	*800–3500*	*300–2000*
Basalt	1000–10,000	5000–6000	2800–3400
Micaschist	300–1000	2800–4000	1200–2000

(continued)

	Resistivity (Ωm)	Vp (m/s)	Vs (m/s)
Gneiss	700–2000	3500–6000	1700–3000
Amphibolite	1000–4000	4000–6500	2000–3500
Fresh water	10–100	1450	0
Salt water	0.2–1	1530	0
Ice	1 10^{4-6}	3400–4000	1700–1900
Concrete	10–2000 (water content—salinity)	3000–4500	1700–2500

References

Amitrano D, Grasso JR, Senfaute G (2005) Seismic precursory patterns before a cliff collapse and critical-point phenomena. Geophys Res Lett 32:L08314. https://doi.org/10.1029/2004GL022270

Barton N (2006) Rock quality, seismic velocity, attenuation and anisotropy. CRC Press

Bertello L, Berti M, Castellaro S, Squarzoni G (2018) Dynamics of an active earthflow inferred from surface wave monitoring. J Geophys Res Earth Surf 123:1811–1834. https://doi.org/10.1029/2017JF004233

Bièvre G, Jongmans D, Winiarski T, Zumbo V (2012) Application of geophysical measurements for assessing the role of fissures in water infiltration within a clay landslide (Trièves area, French Alps). Hydrol Process 26:2128–2142. https://doi.org/10.1002/hyp.7986

Bogoslovsky VA, Ogilvy AA (1977) Geophysical methods for the investigation of landslides. Geophysics 42:562–571. https://doi.org/10.1190/1.1440727

Castagna JP, Batzle ML, Eastwood RL (1985) Relationships between compressional-wave and shear-wave velocities in clastic silicate rocks. Geophysics 50:571–581. https://doi.org/10.1190/1.1441933

Colombero C, Baillet L, Comina C et al (2018) Integration of ambient seismic noise monitoring, displacement and meteorological measurements to infer the temperature-controlled long-term evolution of a complex prone-to-fall cliff. Geophys J Int 213:1876–1897. https://doi.org/10.1093/gji/ggy090

Everett ME (2013) Near-surface applied geophysics. Cambridge University Press

Fiolleau S, Jongmans D, Bièvre G, Chambon G, Baillet L, Vial B (2020) Seismic characterization of a clay-block rupture in Harmalière landslide, French Western Alps. Geophys J Int. In Press

Gance J, Malet J-P, Supper R et al (2016) Permanent electrical resistivity measurements for monitoring water circulation in clayey landslides. J Appl Geophys 126:98–115. https://doi.org/10.1016/j.jappgeo.2016.01.011

Hack R (2000) Geophysics for slope stability. Surv Geophys 21:423–448. https://doi.org/10.1023/A:1006797126800

Hayashi M (2004) Temperature-electrical conductivity relation of water for environmental monitoring and geophysical data inversion. Environ Monit Assess 96:119–128. https://doi.org/10.1023/B:EMAS.0000031719.83065.68

Hayley K, Bentley LR, Gharibi M, Nightingale M (2007) Low temperature dependence of electrical resistivity: implications for near surface geophysical monitoring. Geophys Res Lett 34. https://doi.org/10.1029/2007GL031124

Helmstetter A, Garambois S (2010) Seismic monitoring of Séchilienne rockslide (French Alps): analysis of seismic signals and their correlation with rainfalls. J Geophys Res Earth Surf 115. https://doi.org/10.1029/2009JF001532

Hibert C, Mangeney A, Grandjean G, Shapiro NM (2011) Slope instabilities in Dolomieu crater, Réunion Island: from seismic signals to rockfall characteristics. J Geophys Res Earth Surf 116. https://doi.org/10.1029/2011JF002038

Hobbs PRN, Entwisle DC, Northmore KJ et al (2012) Engineering geology of British rocks and soils. Lias Group

Jaboyedoff M, Del Gaudio V, Derron M-H et al (2019) Characterizing and monitoring landslide processes using remote sensing and geophysics. Eng Geol 259:105167. https://doi.org/10.1016/j.enggeo.2019.105167

Jongmans D, Bièvre G, Renalier F et al (2009) Geophysical investigation of a large landslide in glaciolacustrine clays in the Trièves area (French Alps). Eng Geol 109:45–56. https://doi.org/10.1016/j.enggeo.2008.10.005

Jongmans D, Garambois S (2007) Geophysical investigation of landslides: a review. Bulletin De La Société Géologique De France 178:101–112. https://doi.org/10.2113/gssgfbull.178.2.101

Kearey P, Brooks M, Hill I (2013) An introduction to geophysical exploration. Wiley

Knackstedt MA, Arns CH, Val Pinczewski W (2005) Velocity-porosity relationships: predictive velocity model for cemented sands composed of multiple mineral phases. Geophys Prospect 53:349–372. https://doi.org/10.1111/j.1365-2478.2005.00479.x

Larose E, Carrière S, Voisin C et al (2015) Environmental seismology: what can we learn on earth surface processes with ambient noise? J Appl Geophys 116:62–74. https://doi.org/10.1016/j.jappgeo.2015.02.001

Lehmann P, Gambazzi F, Suski B et al (2013) Evolution of soil wetting patterns preceding a hydrologically induced landslide inferred from electrical resistivity survey and point measurements of volumetric water content and pore water pressure. Water Resour Res 49:7992–8004. https://doi.org/10.1002/2013WR014560

Lévy C, Baillet L, Jongmans D et al (2010) Dynamic response of the Chamousset rock column (Western Alps, France). J Geophys Res Earth 115. https://doi.org/10.1029/2009JF001606

Louie JN (2001) Faster, better: shear-wave velocity to 100 meters depth from refraction microtremor arrays. Bull Seismol Soc Am 91:347–364. https://doi.org/10.1785/0120000098

Mainsant G, Jongmans D, Chambon G et al (2012a) Shear-wave velocity as an indicator for rheological changes in clay materials: lessons from laboratory experiments. Geophys Res Lett 39: L19301–L19301. https://doi.org/10.1029/2012GL053159

Mainsant G, Larose E, Brönnimann C et al (2012b) Ambient seismic noise monitoring of a clay landslide: toward failure prediction. J Geophys Res 117:F01030–F01030. https://doi.org/10.1029/2011JF002159

Malet J-P, Van Asch ThWJ, Van Beek R, Maquaire O (2005) Forecasting the behaviour of complex landslides with a spatially distributed hydrological model. Nat Hazards Earth Syst Sci 5:71–85

Mavko G, Mukerji T, Dvorkin J (2019) The rock physics handbook. Cambridge University Press

McCann DM, Forster A (1990) Reconnaissance geophysical methods in landslide investigations. Eng Geol 29:59–78. https://doi.org/10.1016/0013-7952(90)90082-C

Mondol NH, Bjørlykke K, Jahren J, Høeg K (2007) Experimental mechanical compaction of clay mineral aggregates—changes in physical properties of mudstones during burial. Mar Pet Geol 24:289–311. https://doi.org/10.1016/j.marpetgeo.2007.03.006

Palis E, Lebourg T, Tric E et al (2017) Long-term monitoring of a large deep-seated landslide (La Clapiere, South-East French Alps): initial study. Landslides 14:155–170. https://doi.org/10.1007/s10346-016-0705-7

Park CB, Miller RD, Xia J (1999) Multichannel analysis of surface waves. Geophysics 64:800–808. https://doi.org/10.1190/1.1444590

Pazzi V, Morelli S, Fanti R (2019) A review of the advantages and limitations of geophysical investigations in landslide studies. Int J Geophys 2019:1–27. https://doi.org/10.1155/2019/2983087

Perrone A, Piscitelli S, Lapenna V (2017) Electrical resistivity tomographies for landslide monitoring: a review. Geoelectr Monit 129

Poli P (2017) Creep and slip: seismic precursors to the Nuugaatsiaq landslide (Greenland). Geophys Res Lett 44:8832–8836. https://doi.org/10.1002/2017GL075039

Renalier F, Bièvre G, Jongmans D, et al (2010a) Characterization and monitoring of unstable clay slopes using active and passive shear wave velocity measurements. Advances in near-surface seismology and ground-penetrating radar Society of Exploration Geophysics, Tulsa, pp 397–414. https://doi.org/10.1190/1.9781560802259.ch24

Renalier F, Jongmans D, Campillo M, Bard P-Y (2010b) Shear wave velocity imaging of the Avignonet landslide (France) using ambient noise cross-correlation. J Geophys Res 115:F03032–F03032. https://doi.org/10.1029/2009JF001538

Reynolds JM (2011) An introduction to applied and environmental geophysics. Wiley

Stumvoll MJ, Canli E, Engels A et al (2019) The "Salcher" landslide observatory—experimental long-term monitoring in the Flysch zone of Lower Austria. Bull Eng Geol Environ. https://doi.org/10.1007/s10064-019-01632-w

Supper R, Ottowitz D, Jochum B et al (2014) Geoelectrical monitoring: an innovative method to supplement landslide surveillance and early warning. Near Surf Geophys 12:133–150. https://doi.org/10.3997/1873-0604.2013060

Tonnellier A, Helmstetter A, Malet J-P et al (2013) Seismic monitoring of soft-rock landslides: the Super-Sauze and Valoria case studies. Geophys J Int 193:1515–1536. https://doi.org/10.1093/gji/ggt039

Uhlemann S, Chambers J, Wilkinson P et al (2017) Four-dimensional imaging of moisture dynamics during landslide reactivation. J Geophys Res Earth Surf 122:398–418. https://doi.org/10.1002/2016JF003983

Uhlemann S, Smith A, Chambers J et al (2016) Assessment of ground-based monitoring techniques applied to landslide investigations. Geomorphology 253:438–451. https://doi.org/10.1016/j.geomorph.2015.10.027

Van Dam RL (2012) Landform characterization using geophysics—recent advances, applications, and emerging tools. Geomorphology 137:57–73. https://doi.org/10.1016/j.geomorph.2010.09.005

Whiteley JS, Chambers JE, Uhlemann S et al (2019) Geophysical monitoring of moisture-induced landslides: a review. Rev Geophys 57:106–145. https://doi.org/10.1029/2018RG000603

Wilkinson PB, Chambers JE, Meldrum PI et al (2010) Predicting the movements of permanently installed electrodes on an active landslide using time-lapse geoelectrical resistivity data only. Geophys J Int 183:543–556. https://doi.org/10.1111/j.1365-246X.2010.04760.x

Wilkinson PB, Uhlemann S, Chambers JE et al (2015) Development and testing of displacement inversion to track electrode movements on 3-D electrical resistivity tomography monitoring grids. Geophys J Int 200:1566–1581. https://doi.org/10.1093/gji/ggu483

Whiteley JS, Chambers JE, Uhlemann S, Boyd J, Cimpoiasu MO, Holmes JL, Swift RT et al (2020) Landslide monitoring using seismic refraction tomography–the importance of incorporating topographic variations. Eng Geol 268:105525

Geophysical Monitoring of Landslides—A Step Closer Towards Predictive Understanding?

Sebastian Uhlemann, Jonathan Chambers, Philip Meldrum, Patrick McClure, and Baptiste Dafflon

Abstract

Landslide early warning is still mostly reliant on precipitation thresholds, which can fail to address the subsurface conditions causing slope instabilities. Here we introduce a novel approach combining the latest developments in geophysical and environmental monitoring, with hydrological and geomechanical modelling to provide robust estimates of current and future Factors-of-Safety of slopes, which we propose may be a robust measure for developing early warning thresholds. We aim to develop a methodology that can predict slope instabilities by estimating the causing subsurface conditions in near-real time, thereby allowing for timely early warning to vulnerable communities and implementation of mitigation measures. It is shown here applied to a hillslope with a history of slope failure, using simplified hydrological and geomechanical models. During the monitoring period, precipitation events are shown to give rise to the local water table, thereby reducing the Factor-of-Safety of the slope. It underlines the value of predicting the effect of future storm (or precipitation) events, that imply an additional reduction and hence an increased risk for slope failure. Although applied to a local hillslope, the same approach can be upscaled to regional scales using emerging and established remote sensing and wireless sensor networks.

Keywords

Electrical resistivity tomography • Landslide monitoring • Slope stability • Data-driven hydro-geomechanical modelling

Introduction

Landslides are a major and common natural hazard. They endanger communities and critical infrastructure, and have caused more than 28,000 fatalities and more than $1.8 billion in direct damage within the last decade worldwide. While seismic events are a likely triggering mechanism in many areas, exceptional and prolonged precipitation events triggering landslides are more frequent. This has been shown, e.g., during the 1997–98 El Niño winter storms, which caused more than 300 landslides in the San Francisco Bay Area damaging public and private property, with an estimated direct cost of $158 million (Godt 1999). Climate change, causing more frequent weather extremes, is likely to increase the occurrence of shallow landslides locally and worldwide. In order to mitigate the risk, there is a need to improve our understanding of shallow, rainfall-induced landslide dynamics, which are mostly controlled by hydrological processes (Bogaard and Greco 2016). These processes are known to be spatially and temporally highly heterogeneous. However, the state-of-the-art is still monitoring surface expressions of slope failure (i.e. active failure), either via remote sensing or direct field observations, rather than monitoring subsurface hydrological properties. Landslide early warning systems are also still mostly reliant on precipitation records, rather than soil moisture

S. Uhlemann (✉)
Lawrence Berkeley National Laboratory, Energy Geosciences Division, One Cyclotron Road, Berkeley, CA 94720, USA
e-mail: suhlemann@lbl.gov

J. Chambers · P. Meldrum
British Geological Survey, Environmental Science Centre, Keyworth, NG12 5GG, UK
e-mail: jecha@bgs.ac.uk

P. Meldrum
e-mail: pime@bgs.ac.uk

P. McClure · B. Dafflon
Lawrence Berkeley National Laboratory, Climate & Ecosystem Science Division, One Cyclotron Road, Berkeley, CA 94720, USA
e-mail: ptm@lbl.gov

B. Dafflon
e-mail: bdafflon@lbl.gov

© Springer Nature Switzerland AG 2021
N. Casagli et al. (eds.), *Understanding and Reducing Landslide Disaster Risk*, ICL Contribution to Landslide Disaster Risk Reduction, https://doi.org/10.1007/978-3-030-60311-3_8

measurements, and hence their reliability has been questioned (Marra 2019).

Recent developments of geophysical techniques, geoelectrical monitoring in particular, and wireless sensor networks enable near real-time monitoring of subsurface hydrological properties and processes at unprecedented spatial and temporal resolution. The novelty of this study lies in the integration of this high-resolution data with hydro-geomechanical models. Eventually, this will bring us one step closer towards predicting and reliable early warning of slope instability.

While conventional hydrological monitoring data can readily be implemented in geomechanical modelling (e.g., Bordoni et al. 2015; Springman et al. 2013), and geophysical and hydrological models can be coupled (Hinnell et al. 2010; Irving and Singha 2010; Singha et al. 2015; Johnson et al. 2017), the link between geophysics and geomechanical modelling has not yet been established. Bridging this gap should improve the assessment of Factors-of-Safety (FoS) of slopes by incorporating spatially and temporarily distributed measurements of soil moisture, and should enable us to forecast slope stability by including weather predictions (i.e. recharge rates) and estimating related subsurface flow conditions. It is anticipated that the real-time FoS and its predicted future change can be used as a reliable early warning threshold, e.g. if the predicted FoS becomes close to 1, a warning should be issued.

Here, we present this approach by linking geophysical and environmental data with simplified slope stability and hydrological flow models. We show that including subsurface data into the FoS calculation improves this estimate and that we can use geophysical data to inform hydrological models to derive subsurface flow related to slope stability. This approach is exemplary shown applied to a site in Northern California, which has a long-standing history of slope instabilities due to prolonged and intense rainfall.

Recent Advances in Subsurface Imaging and Modelling

Recent advances in geophysical data acquisition and processing allow for near-real time monitoring of subsurface moisture dynamics at high spatial (10′s of cm's to m's) and temporal (hourly to daily) resolution. By using low-cost, low-powered instrumentation, long-term monitoring at remote locations has become feasible, while data can be analysed at near-real time thanks to wireless data communication technologies (Huntley et al. 2019). This has led to an increased number of geophysical monitoring studies that are recording subsurface dynamics over longer periods of time (Whiteley et al. 2019). While most geotechnical and geophysical landslides studies treat the problem as 2D, Uhlemann et al. (2017) has shown that geophysical data can be used to image 4D spatio-temporal soil moisture dynamics of landslides prior and post failure, and that thresholds derived from such data can be used as predictor for slope instabilities.

Since geophysical data only provide proxies to the parameters controlling slope stability, such as liquid saturation or pore pressures, integrating geophysical monitoring with wireless sensor networks is becoming more frequently applied. Wireless sensor networks employ distributed point sensing to estimate the variability of those parameters and combine it with the latest developments in wireless data transmission to send data to a processing centre, where data are analysed and used, e.g., for early warning (Ramesh and Vasudevan 2012; Ramesh 2014; Watlet et al. 2019).

Similar to the developments in data acquisition, also modelling has progressed. Now we can model subsurface hydrological and landslide dynamics in 3D and at higher resolution than a decade ago, not only assuming fully saturated, but also partially saturated and conditions that can vary in space and time. Hence, real-world failures can be modelled and those tools be used to predict slope instabilities at local and regional scales.

Monitoring and Early Warning Approach

Our aim is to develop a new methodology that takes advantage of those recent developments to provide reliable early warning of landslide hazards to communities at risk. The methodology is outlined in Fig. 1. and combines geophysical and environmental monitoring with hydrological and geomechanical modelling to estimate and predict slopes FoS, which can be used as a threshold for landslide early warning.

The geophysical and environmental monitoring is providing data regarding groundwater level, soil moisture, rainfall and temperature data, but also information on soil characteristics and structure, e.g. clay content. Those data are then used to parameterize hydrological and slope stability models. Parameters like groundwater level can readily be derived from the geophysical and environmental monitoring data and can be included into the slope stability analysis. Hydrological models are used to provide spatial and temporal distributions of parameters such as pore pressures, which will refine the estimation of stresses in the slopes. Weather predictions can be used to define potential recharge volumes, and hence predict groundwater and pore pressure variations, which, when fed into the geomechanical models, allow to provide future predictions of FoS at local and regional scales. Those predictions can then be used as thresholds for early warning and mitigation measures. In contrast to conventional early warning systems, which are

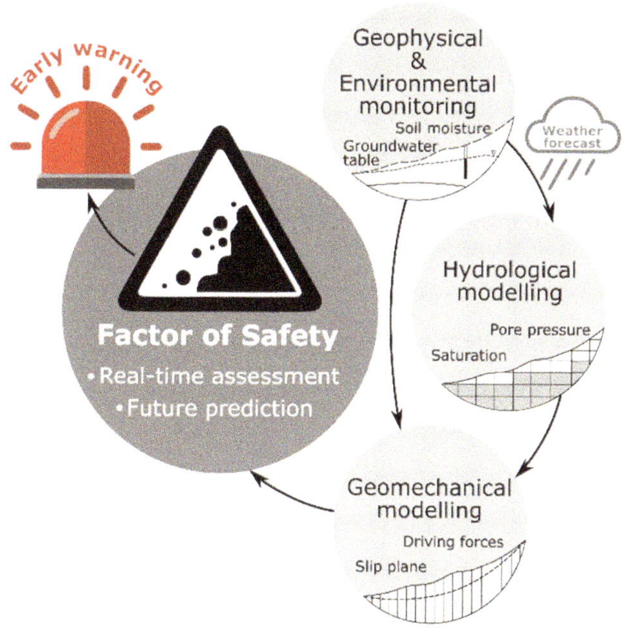

Fig. 1 Methodology for assessment and prediction of slope stability. This workflow integrates the latest developments in geophysical and environmental monitoring with hydrological and geomechanical modelling to provide FoS at local and regional scales that can be used for landslide early warning

mostly reliant on precipitation data only, this approach uses subsurface measurements that are directly linked to the cause of slope instabilities. Hence, we expect these slope stability assessments and predictions to be more reliable.

Study Site

The site to which we apply this methodology is located in the San Francisco Bay Area on the west side of the northwest-trending Berkeley Hills and has a history of slope failures. The landslide investigated here impacts upon a road bridge and has been studied intensively. A ground displacement monitoring system is recording displacements since 2012 and shows movement rates of up to about 10 mm/year, which are correlated with precipitation records (Cohen-Waeber 2018). The landslide can be classified as a very slow moving clay rotational slide (Hungr et al. 2014).

Bedrock geology of the Berkeley Hills is complex and comprises moderately to highly deformed sedimentary, volcanic, and metamorphic rock units. The site itself is located within a mapped landslide deposit (up to 18 m thick) composed of weathered Moraga formation (mainly weathered basalt and andesite flows), which is overlaying Orinda formation, which is composed of partially bedded, non-marine, conglomerate sandstone, and green and red silt- and mudstone. On a regional scale, the Hayward and San Andreas fault are potential sources of seismic activity.

Monitoring Setup

The monitoring setup is shown in Fig. 2. It comprises bi-daily acquisition of Electrical Resistivity Tomography (ERT) data, acquired using a PRIME resistivity monitoring system employing dipole–dipole measurements on 112 electrodes separated by 0.6 m. Hence, subsurface resistivity and chances therein are monitored over a length of 66.6 m and to a depth of approximately 10 m below ground level (bgl). Soil moisture, temperature, and bulk electrical conductivity are measured every 30 min at 20, 40, and 60 m along the ERT monitoring line at 0.1, 0.3, and 0.5 m bgl. Groundwater levels are measured within shallow piezometers (1.83 m bgl) at 20 and 40 m along the ERT line. Precipitation, air temperature and humidity are recorded at a 15 min interval at the toe of the slope. The system is powered using solar energy, and its status is transmitted daily via the Iridium satellite network.

Landslide Structure

Two boreholes in the vicinity of the monitoring site (20–50 m away) indicate weak clay soils of 5.5–7.3 m thickness (paleolandslide deposits) overlying friable sandstones and siltstones of the Orinda formation (Alan Kropp and Associates 2006). This observation is in agreement with the 3D ERT model that was obtained from the site (Fig. 3. recorded at the end of the dry season in November 2019), which shows a low resistive layer (>10 Ωm) above a more resistive

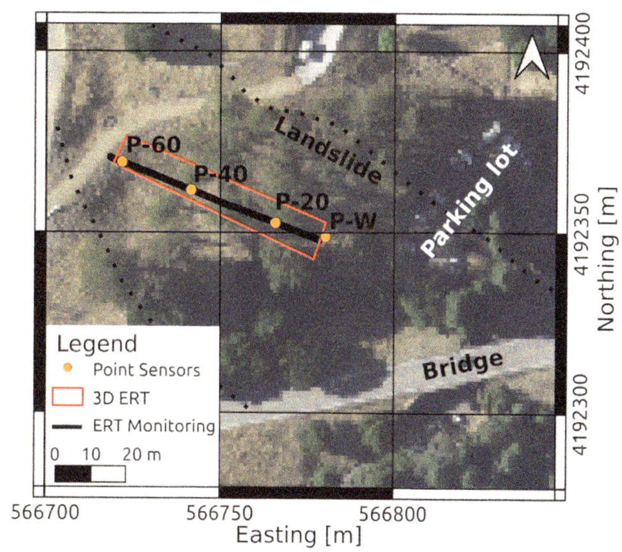

Fig. 2 Map showing 3D ERT data acquisition, ERT monitoring line, and location of point sensors (P-20, P-40, and P-60 soil moisture, temperature and electrical conductivity, P-20 and P-40 include groundwater level monitoring; at P-W precipitation, air temperature and pressure are measured)

layer. On the top of the slope, high resistivities are imaged to deeper depths. This is in agreement with the geological map that indicates a transition from Moraga formation to paleolandslide deposits in this area. The very low resistive feature in the centre of the volume is interpreted to be fully saturated clay, with the vadose zone and dry clay soils above, and the more resistive Orinda formation below. Hence, the groundwater table is expected to be shallow (1.6–2.0 m bgl) towards the toe of the slope. This is in agreement with previous studies that mapped a periodic spring close to the toe. Note that geophysical inversion, which resulted in the model of Fig. 3, is inherently non-unique. To reduce the model space, we applied a L2 smoothness constraint in space (and time for the monitoring data). As described above, the resulting model is in good agreement with independent geological and hydrological information of the site.

Monitoring Results

The monitoring period has covered rainfall events following an about 6 months long dry period. An initial rainfall event occurred November 25, with significant (>20 mm/d) and prolonged rainfall from December 1st to 13th. This caused soil moisture levels to increase, with every rainfall event showing a rapid increase followed by a gradational decrease in soil moisture, with a downward gradient, indicating downward movement of moisture leading to groundwater recharge. This is also confirmed by the rise in groundwater table from >1.83 m to 1.72 m bgl. After this period, rainfall occurred only occasionally, and with daily accumulations mostly below 20 mm/d. The 2D ERT baseline model (Fig. 4c) shows the same observation as were drawn from the 3D model, with a highly resistive surface layer, representing dry soil, overlying a layer of low resistivity, spanning from 1.6 m to about 8 m bgl, which is interpreted to be saturated clay. This layer is underlain by a more resistive layer, representative of the sandstone of the Orinda formation.

The rainfall commencing December 1st led to a decrease in resistivity in the majority of the model (Fig. 4d), indicating an increase in soil moisture. The increase in resistivity between about 1.6–3.0 m bgl is assumed to be caused by an exchange of pore waters, with the rainfall providing an input of more resistive pore water.

The rainfall events of late December to January cause only small changes in the resistivity model, with mostly decreasing resistivity values throughout the model. This is indicative of further increases in soil moisture content following a storm event on January 16th. The shallow resistivities (1–2 m bgl) at x = 20 and 40 m (blue and orange lines in Fig. 4b, respectively) show this resistivity response to the different rainfall events, with step-like decreases following the storms of December 7th and January 16th, while the period of small rainfall between those caused resistivities to increase slightly at x = 40 m, indicating decreasing soil moisture content, as also measured by the point sensor at 0.5 m bgl.

From this data, the location of the groundwater table can be mapped through space and time by extracting the interface between the highly resistive upper layer and the less resistive central layer. This shows that the groundwater table rose by about 0.3 m, particularly in the lower part of the slope, following the rainfall events early December (Fig. 5). The additional rainfall at the end of January 2020 caused an additional rise by about 0.13 m. These observations are in

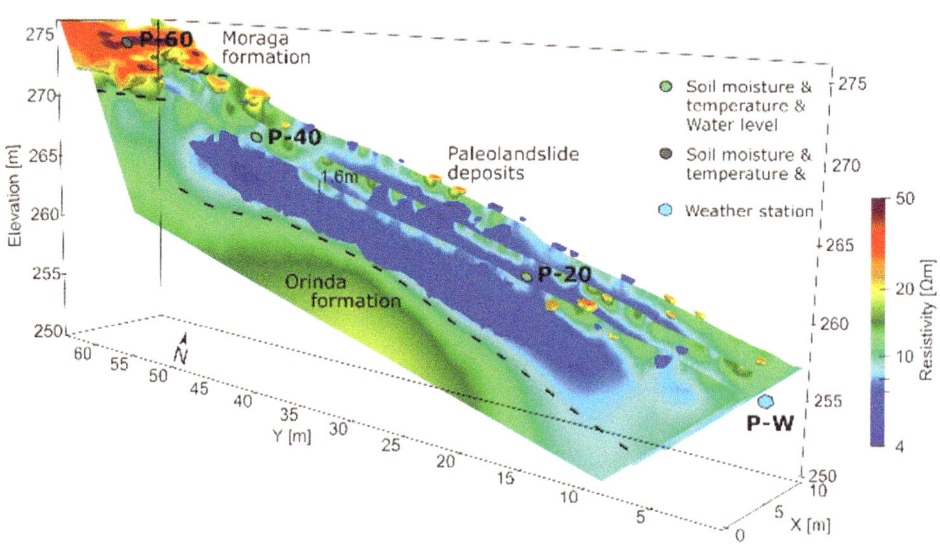

Fig. 3 Interpreted 3D ERT model of the study site, highlighting the different lithological units, and showing the point sensor locations. Inversion RMS = 4.1%. Blue isovolume showing resistivities <6 Ωm, red isovolume resistivities >30 Ωm

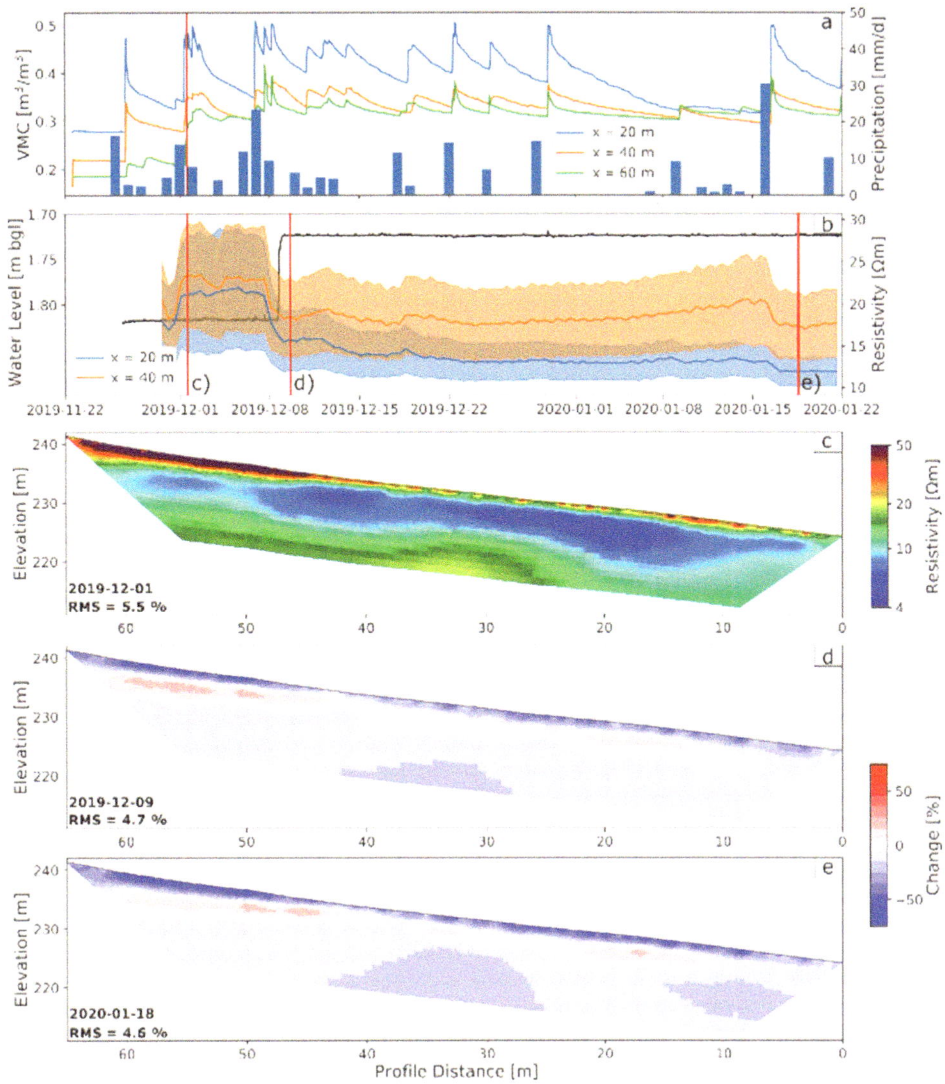

Fig. 4 Geophysical and environmental monitoring data. **a** Precipitation and soil moisture data (at 0.5 m bgl), **b** water level (black line) and average resistivity (shaded area shows the standard deviation) at 1–2 m bgl at two locations, **c** ERT baseline model, **d–e** change in resistivity compared to baseline model

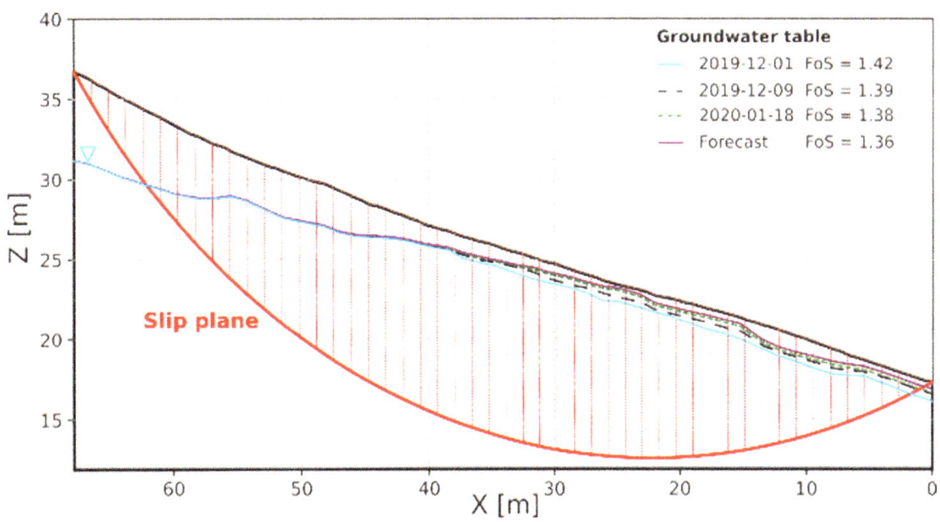

Fig. 5 Simplified slope stability analysis, accounting for the groundwater table variations as mapped from the ERT data. The FoS for a dry slope (i.e. groundwater table below the slip plane) is 2.21, the FoS for the different groundwater conditions is provided in the legend

broad agreement with the groundwater table and soil moisture measurements obtained from site.

Using a groundwater flow model (PFLOTRAN, Hammond et al. 2014), we estimated that a rainfall period similar to the one observed early December would cause the groundwater table to rise on average by 0.21 m from the January 2020 conditions.

Slope Stability Analysis and Prediction

To introduce our approach, we simplify the slope stability problem. Given that the paleolandslide deposits were mapped with a thickness of up to 18 m and that the top of the Orinda formation showed a similar friable characteristic to the overlying soil in soil borings close by, we assume that the slope consists of one soil type only and that the failure type is rotational.

The FoS of the slope is determined for the different groundwater conditions using the General Limit Equilibrium method (Fredlund and Krahn 1977) implemented in the python code pyBIMstab (Montoya-Araque and Suarez-Burgoa 2018), and satisfies both force and moment equilibrium. Previous studies found the soil to have an effective cohesion of 7.75 kPa, a unit weight of 18.85 kN/m^3, and a friction angle of 24° (Alan Kropp and Associates 2006). Analysing more than 5000 possible slip surfaces, Fig. 5 shows the slip surface with the smallest FoS and thus the most likely slip plane. With a maximum depth of about 12 m, this result is in agreement with previous site investigations.

For dry conditions, i.e. the groundwater table is below the proposed slip plane, the FoS is 2.21 and the slope stable. From the piezometer and ERT data we can derive the location of the groundwater table at different times. Including the conditions of December 1st in the analysis results in a FoS of 1.42 (Fig. 5). The prolonged and intense rainfall between December 1st and 13th caused a rise in groundwater table that reduces the FoS to 1.39. The groundwater conditions following the storm on January 16th caused a further reduction of the FoS to 1.38. Continuing from those conditions and simulating another storm with similar precipitation to the one observed early December would cause another rise in groundwater table and hence reduce the FoS to 1.36.

Although simplified and of limited monitoring length, these results show the benefit of including estimated and modelled hydrological subsurface conditions into the slope stability analysis. It allows to provide a well-informed estimate of current conditions. In addition, by modelling the impact of future precipitation events on the subsurface conditions, slope stability in response to these events can be predicted. While we showed the feasibility of this approach, we could not validate an early warning threshold, as the slope remained stable during the monitoring period.

This approach is limited by the accuracy and availability of the geophysical and environmental data, as well as the model validity for both hydrological and geomechanical simulations. Wrong parameterization of the model may provide erroneous, and hence misleading, FoS estimates. Nevertheless, we anticipate that incorporating subsurface information into the early warning methodology will improve our predictions.

Future work will increase the complexity of both hydrological and slope stability models, by including subsurface layers with different physical properties and extending the domain from 2D to 3D. Applying this to a regional scale will increase computational time significantly, and hence new approaches have to be developed to provide estimates and predictions of slope stability in near real-time.

Conclusions

Landslide early warning is still mostly reliant on regionally defined rainfall thresholds. Here, we introduce a methodology that combines the latest developments in geophysical and environmental monitoring, with hydrological and geomechanical modelling to provide robust estimates of slope stability and predict its evolution for future precipitation events by including measured and estimated subsurface conditions.

We introduce this approach using data from a study site in Northern California with a history of slope instabilities. From Electrical Resistivity Tomography (ERT) data and a distributed sensor network, subsurface conditions and variations in groundwater table are estimated and included in the slope stability analysis. Using a hydrological model, the impact of a potential future storm event on the groundwater conditions are derived and included in the geomechanical model to predict its impact on the slope stability. During all conditions, the FoS of the slope is greater than 1, but the changes in subsurface conditions following winter storms show a reduction of the FoS, and hence an increase in potential failure likelihood.

While this approach has been applied to a local hill slope, it could easily be up-scaled and applied to a regional scale using emerging remote sensing products for subsurface characterization (e.g. airborne electromagnetics to map soil–bedrock interface) and monitoring (e.g. radar-based shallow soil moisture measurements), and wireless sensor networks to provide distributed soil moisture data on regional grids. Using 5G networks, this data could be used real-time to update current slope stability models. By including weather forecasts, slope stability could be forecasted and early warning given to communities at risk, well in advance of potentially hazardous conditions. This will ultimately reduce landslide risk and will aid in protecting vulnerable communities.

Acknowledgements This work has been funded by an Early Career Development Grant awarded to Sebastian Uhlemann by the Earth and Environmental Sciences Area of Berkeley Lab. The contribution of Jonathan Chambers and Philip Meldrum is published with the per mission of the Executive Director of the British Geological Survey (NERC).

References

Alan Kropp and Associates I (2006) Initial landslide characterization study East Canyon–Buildings 85 and 85A

Bogaard TA, Greco R (2016) Landslide hydrology: from hydrology to pore pressure. Wiley Interdiscip Rev Water 3:439–459

Bordoni M, Meisina C, Valentino R et al (2015) Hydrological factors affecting rainfall-induced shallow landslides: from the field monitoring to a simplified slope stability analysis. Eng Geol 193:19–37

Cohen-Waeber J (2018) Spatiotemporal patterns of seasonality in landslide deformation from InSAR and GPS. PhD thesis, UC Berkeley

Fredlund DG, Krahn J (1977) Comparison of slope stability methods of analysis. Can Geotech J 14:429–439

Godt JW (1999) Maps showing locations of damaging landslides caused by El Nino rainstorms, winter season 1997–98. San Francisco Bay Region, California

Hammond GE, Lichtner PC, Mills RT (2014) Evaluating the performance of parallel subsurface simulators: an illustrative example with PFLOTRAN. Water Resour Res 50:208–228

Hinnell AC, Ferré TPA, Vrugt JA et al (2010) Improved extraction of hydrologic information from geophysical data through coupled hydrogeophysical inversion. Water Resour Res 46:1–14

Hungr O, Leroueil S, Picarelli L (2014) The Varnes classification of landslide types, an update. Landslides 11:167–194

Huntley D, Bobrowsky P, Hendry M et al (2019) Application of multi-dimensional electrical resistivity tomography datasets to investigate a very slow-moving landslide near Ashcroft, British Columbia, Canada. Landslides 16:1033–1042

Irving J, Singha K (2010) Stochastic inversion of tracer test and electrical geophysical data to estimate hydraulic conductivities. Water Resour Res 46:1–16

Johnson TC, Hammond GE, Chen X (2017) PFLOTRAN-E4D: a parallel open source PFLOTRAN module for simulating time-lapse electrical resistivity data. Comput Geosci 99:72–80

Marra F (2019) Rainfall thresholds for landslide occurrence: systematic underestimation using coarse temporal resolution data. Nat Hazards 95:883–890

Montoya-Araque EA, Suarez-Burgoa LO (2018) pyBIMstab: application software for 2D slope stability analysis of block-in-matrix and homogeneous materials. SoftwareX 7:383–387

Ramesh MV (2014) Design, development, and deployment of a wireless sensor network for detection of landslides. Ad Hoc Netw 13:2–18

Ramesh MV, Vasudevan N (2012) The deployment of deep-earth sensor probes for landslide detection. Landslides 9:457–474

Singha K, Day-Lewis FD, Johnson T, Slater LD (2015) Advances in interpretation of subsurface processes with time-lapse electrical imaging. Hydrol Process 29:1549–1576

Springman SM, Kienzler P, Friedel S et al (2013) A long-term field study for the investigation of rainfall-induced landslides. Geotechnique 63:1177–1193

Uhlemann S, Chambers J, Wilkinson P et al (2017) Four-dimensional imaging of moisture dynamics during landslide reactivation. J Geophys Res Earth Surf 122:1–21

Watlet A, Thirugnanam H, Singh B et al (2019) Deployment of an electrical resistivity monitoring system to monitor a rainfall-induced landslide (Munnar, India). In: AGU fall meeting 2019, 9–13 December, San Francisco, USA

Whiteley JS, Chambers JE, Uhlemann S et al (2019) Geophysical monitoring of moisture-induced landslides: a review. Rev Geophys 57:106–145

Recent Advances in High Spatial Resolution Geophysical Monitoring of Moisture-Induced Landslides

Jim Whiteley, Arnaud Watlet, Sebastian Uhlemann, Philip Meldrum, Paul Wilkinson, and Jonathan Chambers

Abstract

Time-lapse geophysical methods are increasingly used to monitor unstable slopes prone to hydrological destabilisation. Geophysical methods are well suited to this purpose due to the high spatiotemporal resolutions at which monitoring data can be acquired. In particular, geoelectrical and seismic approaches are shown to be particularly beneficial for identifying variations in landslide systems at high spatial resolutions. The integrated use of these approaches, which are sensitive to closely inter-related hydrogeological features and processes driving moisture-induced slope instabilities, can reveal the evolving properties of subsurface materials as they move toward failure. Here, we highlight recent advances in high spatial resolution geophysical monitoring with examples from the Hollin Hill Landslide Observatory, a slow-moving, clay-rich, moisture–induced landslide located in North Yorkshire, UK. We present the details of different high spatial resolution geophysical monitoring arrays deployed at the site, including electrical resistivity, seismic refraction, self-potential, and passive seismic, and consider their relative benefits and weaknesses. Focusing on electrical resistivity and seismic refraction monitoring data, we demonstrate how the integrated analysis of time-lapse data can be used to better understand the key hydrogeological features and processes leading to slope failure.

Keywords

Geophysics • Monitoring • Early-warning • Geoelectric • Resistivity • Seismic • Refraction

Introduction

For over 30 years, geophysical surveys have been used to characterise the subsurface of landslide systems (McCann and Forster 1990; Jongmans and Garambois 2007). For monitoring landslides at risk of failure, geophysical techniques can broadly be split in to two groups: those with high temporal resolution (HTR), and those with high spatial resolution (HSR) (Whiteley et al. 2019a). Passive seismology is the predominant HTR geophysical method used for monitoring landslides, and in a typical deployment, seismic waveforms are recorded from a sparse number of sensors deployed across a landslide surface. Although the cost and power demands of passive seismic sensors is decreasing, the number of seismometers deployed is typically limited by financial and logistical costs, which in turn limits the spatial resolution of the array. On the other hand, HSR geophysical methods acquire data using dense 2D or 3D sensor arrays in which many measurements are acquired during a single deployment, in order to image the subsurface. These methods can produce images of subsurface conditions in the form of two-dimensional (2D) cross-sections and maps, or three-dimensional (3D) volumes. The multi-dimensional subsurface information acquired by HSR methods makes them well suited to identifying subsurface heterogeneity in landslide systems. Where conventional geotechnical point-source monitoring data or observations are sparse or depth-limited, HSR geophysical surveys can be used to identify lithological units, structural discontinuities and heterogeneous moisture distributions and dynamics within the subsurface.

J. Whiteley (✉) · A. Watlet · P. Meldrum · P. Wilkinson · J. Chambers
Geophysical Tomography, British Geological Survey, Nottingham, NG12 5GG, United Kingdom
e-mail: jwhi@bgs.ac.uk

S. Uhlemann
Energy Geosciences Division, Lawrence Berkeley National Laboratory, Cyclotron Road, Berkeley, CA 94720, United States

© Springer Nature Switzerland AG 2021
N. Casagli et al. (eds.), *Understanding and Reducing Landslide Disaster Risk*, ICL Contribution to Landslide Disaster Risk Reduction, https://doi.org/10.1007/978-3-030-60311-3_9

Two types of HSR geophysics are typically used for the characterisation and monitoring of landslides: (i) geoelectrical measurements, including electrical resistivity (ER) and self-potential (SP), and (ii) seismic measurements, including active seismic refraction (SR) and surface wave (SW) surveys, and short-term recording of passive seismic signals to measure the horizontal-to-vertical (HV) ratio of seismic noise.

Acquiring repeat measurements using these different geophysical methods produces time-lapse images of the subsurface, providing a means of both characterising a landslide system, and monitoring changes over time. Due to the large number of readings in a complete dataset from which an image of the subsurface can be produced, HSR methods tend to have lower temporal resolution than HTR methods; conversely, images produced from HSR methods (e.g., using ambient noise tomography) tend to have lower spatial resolution.

There exist many examples of using HSR methods to monitor landslides: slope moisture dynamics have been monitored using 4D ER measurements (Uhlemann et al. 2017), the evolution of surface fissuring has been assessed using SRT and SW surveys (Bièvre et al. 2012), SP surveys have located water flow pathways in an active landslide (Colangelo et al. 2006), and sliding surface depth has been monitored using HV (Imposa et al. 2017). As different HSR methods are sensitive to different features and processes, the use of multiple HSR methods to monitor landslide processes can be highly beneficial. Combining such methods in to a single comprehensive landslide monitoring campaign can therefore provide significant insights in to the processes acting to destabilise slopes at risk of failure.

The sensitivity of geoelectrical methods to moisture content and of seismic methods to elastic properties and structural discontinuities, make them well suited for monitoring landslides. The shear strength (τ_f) strength of a material can be defined as

$$\tau_f = c + (\sigma - u)\tan\phi'_{cv} \quad (1)$$

where c is cohesion, σ is total normal stress, u is pore water pressure, and ϕ'_{cv} is the angle of shear resistance at a critical state. Where shear strength is greater than shear stress (τ_d), the slope has a factor of safety (FOS) > 1, and is in a stable condition, given as

$$FOS = \frac{\tau_f}{\tau_d}. \quad (2)$$

Given the sensitivity of ER measurements to moisture content changes, ER monitoring can provide information regarding the changing state of u in the ground, and relationships between resistivity and u can be established by petrophysical tests (Uhlemann et al. 2017). Similarly, SR measurements can provide information on elastic moduli, which relate to the changing condition of σ in the landslide system (Whiteley et al. 2020).

Geophysical Monitoring at the Hollin Hill Landslide Observatory

The Hollin Hill Landslide Observatory (HHLO) is a slow-moving, clay-rich landslide in North Yorkshire, UK (Fig. 1). The underlying geology comprises the failing Whitby Mudstone Formation (WMF) overlying the stable Staithes Sandstone Formation (SSF). Contact between these units is sub-horizontal, dipping to the north and away from the surface slope angle. An increase in moisture content at the contact between the SSF and WMF in the mid-slope cause plastic failure in the WMF, which moves downslope to form flow lobes. Above the mid-slope, backscarps and rotational failures develop because of the loss of support by material downslope.

The HHLO is a testbed for novel geophysical, geotechnical and geodetic landslide monitoring systems, and hosted a range of HSR geophysical methods for over a decade (see also Whiteley et al. 2019b).

Electrical Resistivity Monitoring

The site is instrumented with a permanent 3D ER monitoring array to monitor slope moisture dynamics (Fig. 1), comprising of five 2D profiles, and orientated parallel to the slope direction. Measurement schedules and data retrieval are facilitated via a GPRS connection, and the system is powered from solar and wind energy generated on site. The ER system typically acquires data every other day, and requires periodic maintenance to repair cables that have suffered damage from landslide displacements (Table 1). The ER measurements provide information on the seasonal patterns of wetting and drying within the landslide system, via the development of petrophysical relationships between the measured resistivity and moisture content of the landslide materials (Uhlemann et al., 2017). The system has given rise to a number of novel applications for ER monitoring of landslides, including contributions to 3D ground model development (Chambers et al. 2011), tracking of electrode displacements from ER data (Wilkinson et al. 2016) and rapid electrical resistance monitoring for detecting landslide movement (Merritt et al. 2018).

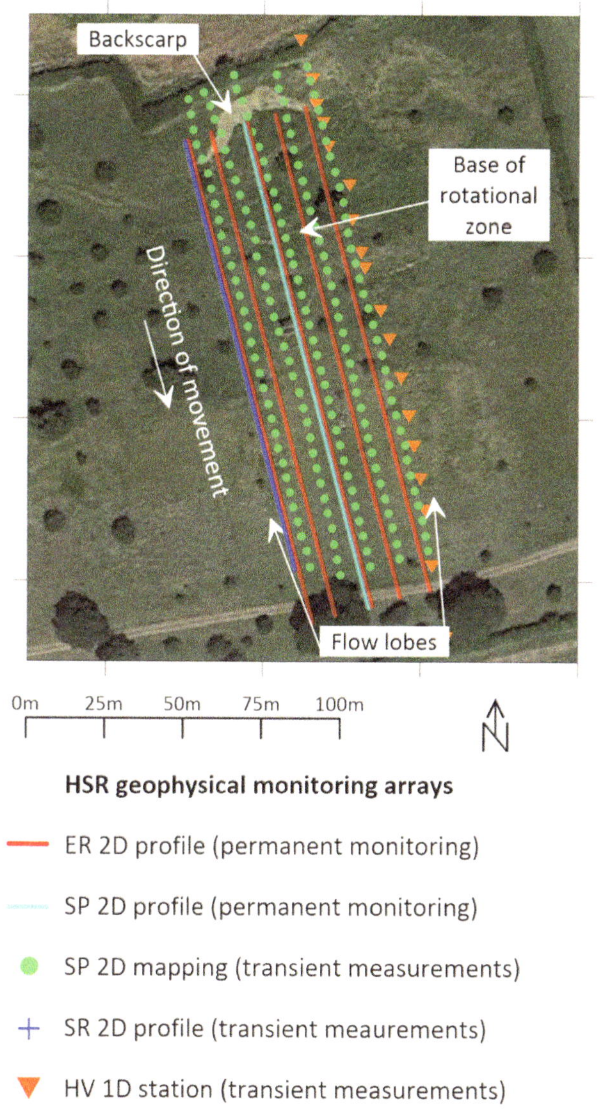

Fig. 1 The location of the HSR geophysical arrays used at the Hollin Hill Landslide Observatory. See Table 1 for details

Seismic Refraction and Passive Seismic Monitoring

Repeat SR measurements (of both P- and S-wave velocity) are acquired at the site every six to eight weeks, along a profile co-located with the western profile of the ER monitoring array. The manually acquired SR data require site visits and re-deployment of geophones at each visit, but the high resolution of the array and sensitivity to variations in elastic properties (which are induced by moisture content variation) provide an excellent complement to the higher temporal resolution ER data (Table 1). Time-lapse changes in seismic velocity have been linked to variations in the moisture content of the landslide system (Whiteley et al. 2020).

In addition, occasional HV measurements have been undertaken at the site to investigate the deeper geological structures underlying the landslide system, and to identify variations in the stiffness of near-surface slipped materials.

Self-potential Monitoring

Self-potential measurements are acquired in two ways at the HHLO; via a permanently installed SP 2D monitoring profile, co-located along the central ER monitoring profile, and by undertaking repeat visits to the site to measure a 2D grid to create a map of SP variation across the landslide surface (Fig. 1).

The SP monitoring data are automatically acquired every hour, providing very high temporal resolution data, whereas the manually acquired mapping data are acquired during site visits every six to eight weeks (Table 1). The SP data are sensitive to variations in fluid flow, with the distinct geomorphological units of the landslide (i.e., rotational movement upslope, flow lobes downslope) showing different patterns of seasonal SP variation.

Geotechnical and Geodetic Monitoring

To complement and support ongoing geophysical monitoring, the HHLO is instrumented with an array of geotechnical and geodetic systems for observing slope hydrology (piezometers, soil moisture sensors, water potential sensors, weather station), surface and subsurface deformation (GPS monitoring array, tilt meters, inclinometers, shape accelerometer arrays, active waveguide sensors) (see Uhlemann et al. 2016). The HHLO surface is also regularly monitored by InSAR and UAV flights (Peppa et al. 2019). HSR geophysical methods can extrapolate the information derived from point-sensors in the landslide subsurface, and remote sensing data are used to update inputs to the geophysical modelling, for example, by providing topographic data for generating inversion meshes.

Integrated Acquisition and Analysis of Electrical Resistivity and Seismic Refraction Data

Between October 2016 and August 2018, 11 co-located ER and SR surveys were acquired from a single profile at the HHLO. By averaging the results of these surveys across the 22 month monitoring period, the major subsurface discontinuities and units can be identified (Fig. 2a). The ratio between P- and S-wave velocity (Vp/Vs) obtained from seismic surveys (Fig. 2b) shows the main contact between

Table 1 Comparative properties of the HSR monitoring methods employed at the Hollin Hill Landslide Observatory, including electrical resistivity (ER), seismic refraction (SR), self-potential (SP) and horizontal-to-vertical seismic noise (HV) measurements. See Fig. 1 for array locations

Survey attribute	ER	SR	SP		HV
Measurement type	Geoelectric	Seismic	Geoelectric	Geoelectric	Seismic
Mode of acquisition	Active	Active	Passive	Passive	Passive
Array dimension	3D	2D (profile)	2D (profile)	2D (map)	1D (interpolated to make 2D profile)
Type of array installation	Permanent	Transient	Permanent	Transient	Transient
Data retrieval	Remote	On site	Remote	On site	On site
Landslide property sensitivity	Moisture content	Elastic properties	Fluid flow	Fluid flow	Elastic properties
Spatial resolution of 2D acquisition	High (4.5 m electrode separation)	Very high (2 m geophone separation)	High (4.5 m electrode separation)	High (5 m electrode separation)	Medium (10 – 25 m station separation)
Spatial resolution of 3D acquisition	High (9 m 2D profile separation)	N/A	N/A	High (9 m 2D profile separation)	N/A
Temporal resolution of acquisition	Days to weeks	Weeks to months	Minutes to hours	Weeks to months	Years
Ease of data acquisition	High (automated)	Very low	High (automated)	Low	Low
Intensity of manual processing	Low	Very high	Low	Medium	Medium
Average number of datasets acquired per year	175	6	8760	6	0.4
Speed of single 2D profile acquisition	Fast (<1 h)	Medium (2 - 3 h)	Fast (<1 h)	Fast (<1 h)	Slow (>4 h)
Length of operation at the HHLO	> 10 years	3 years	2 years	2 years	5 years
Main benefit	Automation of data acquisition and processing	High resolution of survey data	Automation of data acquisition	Ease and speed of acquisition	Ease and speed of acquisition
Main challenge	Array damage by landslide movement	Acquisition and processing intensive	Array damage by landslide movement	Uncertainty of depth sensitivity	Other data (Vs) required for processing

the SSF and WMF at the HHLO. Increased Vp/Vs indicates not only more saturated material, but also indicates compressible material, such as clay-rich WMF. Variations in Vp/Vs, which are primarily located in the central zone of translation at the HHLO, relate to temporal changes in the elastic properties as the plasticity of the sliding material increases (Fig. 2d).

The major features of the averaged resistivity show the difference in moisture content between the saturated WMF and underlying zones of saturated and unsaturated SSF (Fig. 2c). Variations in the resistivity across the monitoring period are concentrated in the flow lobes toward the base of the slope (Fig. 2e), which tend to show preferential wetting and drying over the annual climatic cycle through the formation of preferential flow paths associated with subsurface saturation and drainage (Uhlemann et al. 2017).

Conclusions

Integrated HSR geophysical methods provide a rapid, high resolution and non-invasive means of identifying structural discontinuities and material heterogeneities in landslide systems. The HHLO employs a range of automated and manual HSR geophysical monitoring methods to obtain time-lapse information relating to key landslide features and processes contributing to slope destabilisation. Permanently installed ER and SP arrays provide regular, high temporal

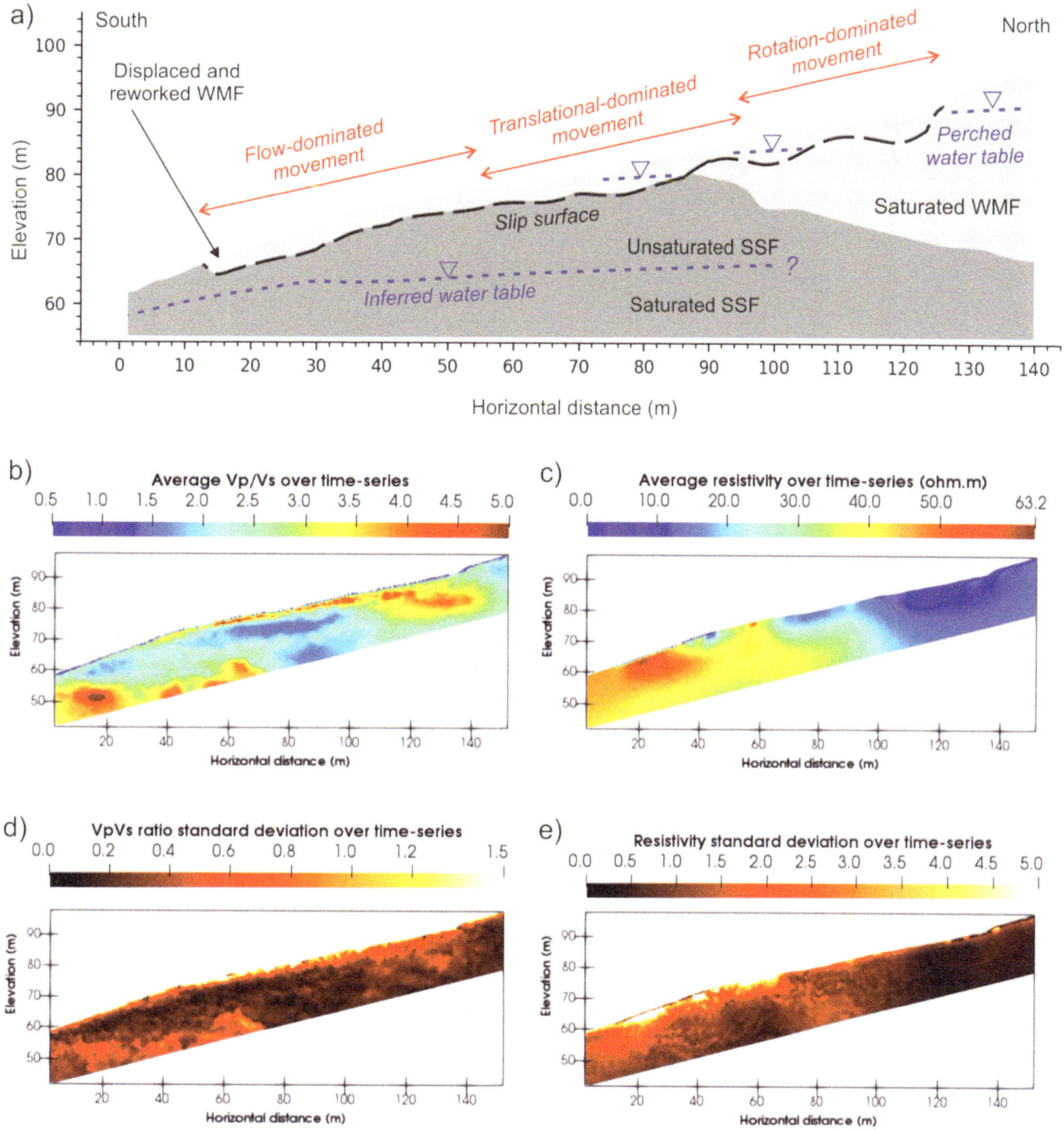

Fig. 2 **a** Conceptual model of the HHLO, developed from geophysical, geotechnical and geodetic data. Below are models of **b** average P- to S-wave velocity ratio (Vp/Vs), and **c** average resistivity, from 11 surveys undertaken between October 2016 and August 2018. The standard deviation of the **d** average Vp/Vs model, and **e** average resistivity model, indicate areas with greatest temporal variation. Modified from Whiteley et al. (2020).

resolution monitoring data that informs the spatial distribution of moisture content and fluid flow throughout the subsurface. Additional manual time-lapse SR, SP and HV surveys, whilst more laborious, provide crucial supplementary data to these permanently installed arrays.

Recent developments in the field of passive seismology, including the continued decreasing financial and logistical costs associated with deploying increasing numbers of seismic sensors in the field, as well as novel developments such as fibre-optic distributed acoustic sensing (DAS) systems, will dramatically increase the resolution of images that can be produced from these near-continuous HTR geophysical methods. The acquisition of increasingly high spatiotemporal resolution data will provide unprecedented insights in to subsurface processes, but will present a new array of challenges for efficient data management, processing and analysis. Automated processes relying on machine learning algorithms and 'big data' management strategies will be needed to handle these increasingly large and complex datasets. However, combining HTR and HSR methods, such as emerging passive seismic systems with the automated ER and SP systems deployed at the HHLO, will allow for unprecedented detail in monitoring unstable slopes at risk of hydrological destabilisation, and will provide important inputs to existing local landslide early-warning systems.

Acknowledgements We thank the current and past members, students and visiting scholars of the BGS' Geophysical Tomography team for their contributions to monitoring Hollin Hill, and thank Josie Gibson, and Frances and James Standen for their continued support. This work was funded by NERC GW4+ UK Doctoral Training Partnership Studentship (Grant NE/L002434/1) and in part by the BGS University Funding Initiative (S337). Jim Whiteley, Arnaud Watlet, Phil Meldrum, Paul Wilkinson and Jonathan Chambers publish with the permission of the Executive Director, BGS (UKRI-NERC).

References

Bièvre G, Jongmans D, Winiarski T, Zumbo V (2012) Application of geophysical measurements for assessing the role of fissures in water infiltration within a clay landslide (Trièves area, French Alps). Hydrol Process 26:2128–2142

Chambers JE, Wilkinson PB, Kuras O, Ford JR, Gunn DA, Meldrum PI, Pennington CVL, Weller AL, Hobbs PRN, Ogilvy RD (2011) Three-dimensional geophysical anatomy of an active landslide in Lias Group mudrocks, Cleveland Basin, UK. Geomorphology 125:472–484

Colangelo G, Lapenna V, Perrone A, Piscitelli S, Telesca L (2006) 2D Self-Potential tomographies for studying groundwater flows in the Varco d'Izzo landslide (Basilicata, southern Italy). Eng Geol 88:274–286

Imposa S, Grassi S, Fazio F, Rannisi G, Cino P (2017) Geophysical surveys to study a landslide body (north-eastern Sicily). Nat Hazards 86:327–343

Jongmans D, Garambois S (2007) Geophysical investigation of landslides: a review. Bull Soc Geol Fr 178:101–112

McCann DM, Forster A (1990) Reconnaissance geophysical methods in landslide investigations. Eng Geol 29:59–78

Merritt AJ, Chambers JE, Murphy W, Wilkinson PB, West LJ, Uhlemann S, Meldrum PI, Gunn D (2018) Landslide activation behaviour illuminated by electrical resistance monitoring. Earth Surface Process Landforms n/a-n/a.

Peppa MV, Mills JP, Moore P, Miller PE, Chambers JE (2019) Automated co-registration and calibration in SfM photogrammetry for landslide change detection. Earth Surf Proc Land 44:287–303

Perrone A, Lapenna V, Piscitelli S (2014) Electrical resistivity tomography technique for landslide investigation: A review. Earth Sci Rev 135:65–82

Uhlemann S, Chambers J, Wilkinson P, Maurer H, Merritt A, Meldrum P, Kuras O, Gunn D, Smith A, Dijkstra T (2017) Four-dimensional imaging of moisture dynamics during landslide reactivation. J Geophys Res Earth Surf 122:398–418

Uhlemann S, Smith A, Chambers J, Dixon N, Dijkstra T, Haslam E, Meldrum P, Merritt A, Gunn D, Mackay J (2016) Assessment of ground-based monitoring techniques applied to landslide investigations. Geomorphology 253:438–451

Whiteley JS, Chambers JE, Uhlemann S, Wilkinson PB, Kendall JM (2019) Geophysical monitoring of moisture-induced landslides: a review. Rev Geophys 57:106–145

Whiteley J, Kendall M, Sujitapan C, Inauen C, Swift R, Watlet A, Raines M, Boyd J, Cimpoiasu M, Holmes J (2019b) An Overview of high spatial resolution geophysical methods for landslide characterisation and monitoring. In: 25th European meeting of environmental and engineering geophysics. The Hague, Netherlands

Whiteley JS, Chambers JE, Uhlemann S, Boyd J, Cimpoiasu MO, Holmes JL, Inauen CM, Watlet A, Hawley-Sibbett LR, Sujitapan C, Swift RT, Kendall JM (2020) Landslide monitoring using seismic refraction tomography: the importance of incorporating topographic variations. Eng Geol 268:105525

Wilkinson P, Chambers J, Uhlemann S, Meldrum P, Smith A, Dixon N, Loke MH (2016) Reconstruction of landslide movements by inversion of 4-D electrical resistivity tomography monitoring data. Geophys Res Lett 43:1166–1174

Characteristic Analysis of the Nayong Rock Avalanche Based on the Seismic Signal

2020 Kyoto Japan

Hao Luo, Aiguo Xing, Kaiping Jin, Shimin Xu, and Yu Zhuang

Abstract

A rock avalanche that destroyed 23 houses and killed 35 people occurred on 28 August 2017, Nayong, SW China. Combined with the dynamic parameters from seismic signal inversion, a discrete element model, MatDEM was used to determine the kinematic behaviour of the rock avalanche. By comparing the velocity evolution process of numerical simulation with that of seismic signal inversion, we are able to find the best fitting parameters. The dynamic process obtained by modelling was compared with the frequency distribution spectrum of the nearest seismometer, showing that the dynamic process is in good agreement with those parameters inverted from seismic signals. The simulation results show that the movement process lasted for nearly 40 s, with a maximum speed of 40 m/s. The selected models and parameters contribute to explain the dynamic processes of similar rock avalanche more accurately and are of considerable significance to the hazard prediction in karst area.

Keywords

Rock avalanche dynamics • Avalanche seismology • Time-series analysis • Discrete element method • Model calibration

H. Luo · A. Xing (✉) · K. Jin · S. Xu · Y. Zhuang
Department of Civil Engineering, Shanghai Jiao Tong University, Shanghai, 200240, China
e-mail: xingaiguo@sjtu.edu.cn

H. Luo
e-mail: a6221709@sjtu.edu.cn

K. Jin
e-mail: king19921130@sjtu.edu.cn

S. Xu
e-mail: xushimin@sjtu.edu.cn

Y. Zhuang
e-mail: zyzhuangyu@sjtu.edu.cn

Introduction

Rock avalanches are the most destructive gravitational instabilities due to their burstiness, high mobility, long runout, and entrainment capacity. In addition to field observation and experiments, many dynamic models and numerical methods have been proposed for predicting the post-failure behavior and movement process of long runout rock avalanches (Hungr 1995; Crosta et al. 2003; Cagnoli and Piersanti 2015). However, it is difficult to obtain the accurate parameters needed for these numerical models. Due to the size effect, the experimental results of rock samples may be inconsistent with the actual material mechanical parameters during large-scale landslide movement, resulting in inaccurate simulation results. The validity of these models and approaches has been verified through comparison with real events. Nevertheless, owing to the lack of direct time-dependent observation evidence for rock avalanches' movement process, few models are considered to be effective.

Recently, the seismic signals recorded by surrounding seismometers provided a possible approach to analyze the energy variation and movement process of rock avalanches (Allstadt 2013; Ekström and Stark 2013; Hibert et al., 2014; Moretti et al. 2015). Therefore, using the dynamic parameters obtained by seismic signal inversion can help to constrain the numerical simulation.

In this paper, through the analysis of seismic signal, more constraints are provided for the dynamic process of the Nayong avalanche. The time frequency distribution spectrum, which can reflect the movement stage of the rock avalanche, can be obtained by Hilbert-Huang Transform. And through the force–time function, detailed dynamic characteristics can be determined. Afterward, the runout behavior of Nayong avalanche was simulated by the 3D discrete element model, MatDEM, and the simulated results are verified by comparing in with field investigation and seismic data. It is expected that this method combining

© Springer Nature Switzerland AG 2021
N. Casagli et al. (eds.), *Understanding and Reducing Landslide Disaster Risk*, ICL Contribution to Landslide Disaster Risk Reduction,
https://doi.org/10.1007/978-3-030-60311-3_10

seismic signal and simulation model could help in better understanding the possible mechanism of rock avalanche and conducting predictive simulations of similar events in bare karst area.

The Nayong Rock Avalanche

Nayong rock avalanche occurred in Southwest China, where there are approximately 62 × 104 km^2 of bare karst topography (Huang and Cai 2007). In the past few years, several catastrophic rock avalanches, especially triggered by mining activities and rainfall, were occurred in the karst areas of China and brought damage to the surrounding infrastructure and facilities (Yin et al. 2011; Xing et al. 2015). Unlike other rock avalanches, there is a clear sign of rock collapse of Nayong rock avalanche and the local residents recorded even the whole process from rock collapse to debris flow with smartphone (Fan et al. 2018). Nevertheless, due to its size and runout distance far beyond the prediction, 35 people died and 23 houses were destroyed (Fig. 1).

Figure 2 shows the 2D longitudinal profile of Nayong rock avalanche. This event involved the displaced materials with a volume of 0.8 Mm^3, and this comprise approximately 0.49 Mm^3 of rock mass from the source area and 0.31 Mm^3 of materials that was entrained along the runout path. The sliding mass travelled about 820 m along the runout path with an elevation difference of 280 m, and eventually deposit at the toe of Pusa village.

Seismic Data and Method

Seismic Data

Six three-component broadband seismometers around the Nayong County have recorded the seismic signals generated from the rock avalanche (Fig. 3). According to the records of the closest station, ZJW station, with a distance of 5.5 km, the seismic signal started at 10:21:48 local time, increasing to peak motion at 10:22:23, within a strong signal lasting for 35 s, and gradually fades into the background noise after 10:24:00. The low-frequency component of the seismic signal is generated by the cycle of loading and unloading of the solid Earth by the bulk acceleration and deceleration of the landslide mass, while the high-frequency component is generated by the friction and collision between rocks and rock bed (Ekström and Stark 2013). In this study, we use Hilbert-Huang transform to obtain the time–frequency distribution spectrum of the seismic signal recorded by ZJW station (Fig. 4).

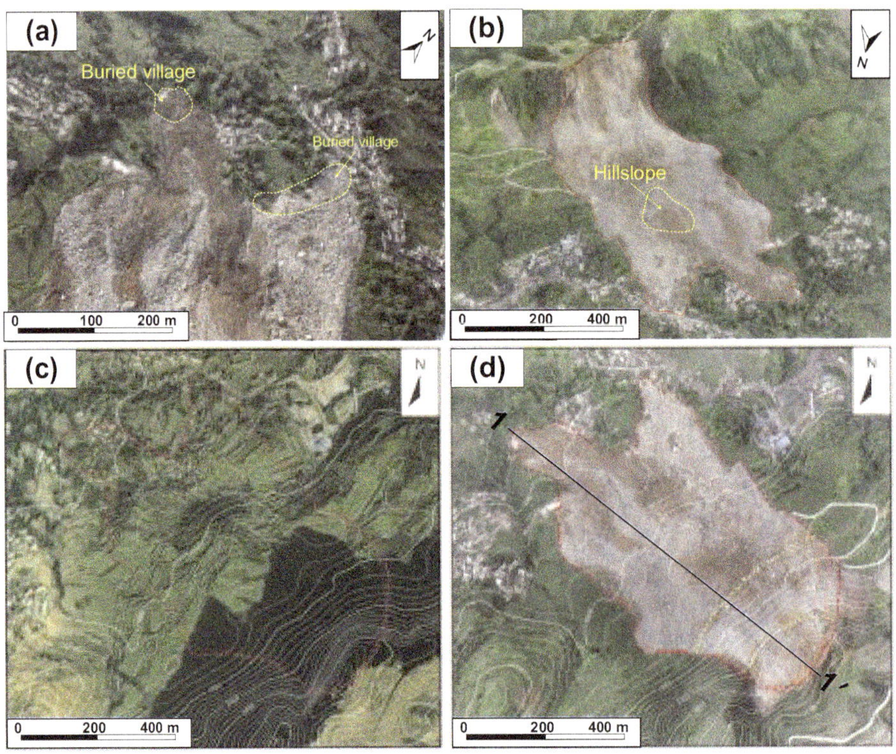

Fig. 1 **a** Buried village in the Nayong rock avalanche. **b** Location of the hillslope. **c** Pre-event image of the Nayong rock avalanche. **d** Post-event image of the Nayong rock avalanche. 1–1′ Cross-section line. (Zhu et al. 2019)

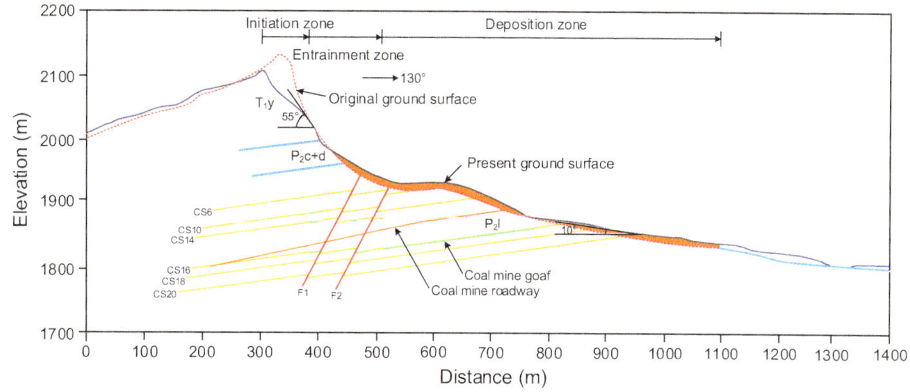

Fig. 2 Longitudinal profile of the Nayong rock avalanche (Zhu et al. 2019)

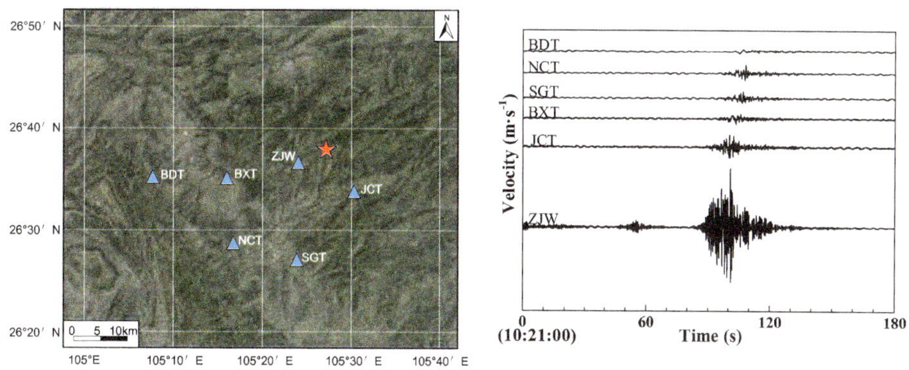

Fig. 3 Locations of the surrounding seismic stations (blue triangles) and recorded seismic signals

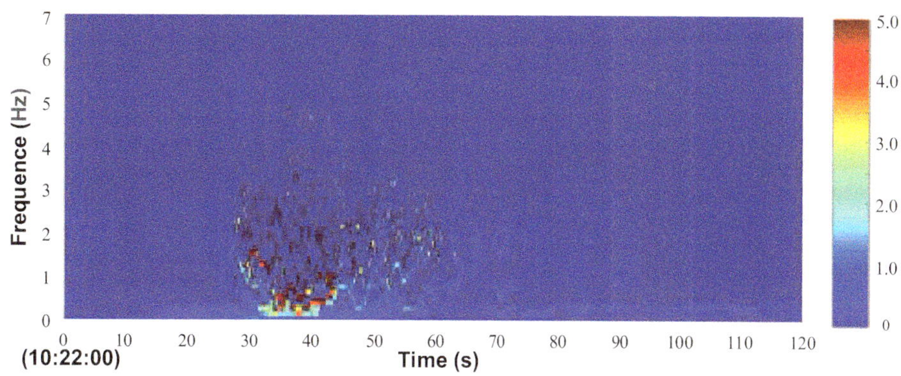

Fig. 4 The amplitude-time–frequency Hilbert spectrum

The time–frequency distribution spectrum (Fig. 4) shows that the low-frequency signal was looming at 10:20:22, and amplitude was getting intense during 10:20:28.3–10:20:44.7. From 10:20:44.7, the amplitude was gradually decreased and finally disappeared in the background noise after 10:20:61.2. On the other hand, the high-frequency component (which greater than 0.5 Hz) starts to appear in 25 s and the amplitude soon reaches the peak. The high-frequency signal appears suddenly in about 25 s, and the large amplitude was maintained at 22–44 s, then started to decrease at 44.7 s, further decreased after 61.2 s, finally disappears in the background at the 80 s.

Method

The earthquake produced by landslides or rock avalanches can be represented by a single-force mechanism (Fukao 1995). Therefore, the solid Earth can be considered a slope and the sliding mass can be a constant mass body sliding above the slope (Ekström and Stark 2013). For the sliding body, the force F_{net} is exerted by the Earth's crust, and there is another force $F_e(t)$ acting on the Earth's crust, both of which are equal in magnitude and opposite in direction. Then the $F_e(t)$ is equal to

$$F_e(t) = -ma$$

$F_e(t)$ is equivalent to the single-force source for generating long-period seismic signals of rock avalanche (Fukao 1995). So $F_e(t)$ is called force–time function this study. $F_e(t)$ can be determined by inverting the low-frequency seismic data (Allstadt 2013). The avalanche source is assumed to be a stationary point source. The impulse response set between the source and each stations pair are known as Green's function, then the seismic displacement records $\boldsymbol{U}(t)$ can be expressed by the convolution of force–time function and Green's function (Stein and Wysession 2003).

$$\boldsymbol{U}(t) = \boldsymbol{F}_e(t) * \boldsymbol{G}(t)$$

For low-frequency signals, Green's functions $\boldsymbol{G}(t)$ can be calculated by the wave number integration method. Because of the insensitivity of low-frequency signals to small-scale heterogeneity, the 1-D generalized Earth model was used in this study, which is from Crust 1.0. To make the models smoother, the force–time function can be inverted by a damped least-squares approach (Allstadt 2013).

$$\boldsymbol{F}_e = (\boldsymbol{G}^{*T}\boldsymbol{G}^* + \alpha^2\boldsymbol{I})^{-1}\boldsymbol{G}^{*T}\boldsymbol{U}$$

where $\boldsymbol{G}^*$ is the convolution matrix of Green's function, $\boldsymbol{I}$ is the identity matrix and α is the damping coefficient. Once the force–time function is obtained, we can calculate the time-varying velocity of the sliding mass v(t) by integrating the force and displacement $\boldsymbol{d}(t)$ by integrating twice.

$$\boldsymbol{v}(t) = -\frac{1}{m}\int_0^t \boldsymbol{F}_e(\tau)d\tau$$

$$\boldsymbol{d}(t) = -\frac{1}{m}\int\int_0^t \boldsymbol{F}_e(\tau)d\tau$$

In this case, the mass of the sliding body can be estimated by the results of field investigation. From Fig. 5, we can see that starting from 23.5 s, the speed of the sliding mass is increasing rapidly, reaching the maximum of 31.8 m/s at 42.5 s. After that, its speed gradually decreases, almost zero at 75.8 s. And the maximum movement distance of mass is 774 m.

The Discrete Element Model

For investigating the detailed mechanisms of the Nayong avalanche, 3D discrete element model, the MatDEM (Liu et al. 2013) was applied to simulate the kinematic behavior of sliding mass. In this model, rock and soil are simulated by a series of tightly packed discrete elements. The motion of elements follows the Newtonian equation of motion, and the elements contact through breakable, linear elastic spring in normal and tangential directions. An artificial viscosity, that can damp the rebound energy of the particles on boundary, is added to the model. In the discrete element model, time-stepping iterative algorithm was developed to model and observed the dynamic evolution of elements (Cundall and Stack 1979). To accurately simulate the model's elastic properties, the time step should be much smaller than the natural vibration period of the system. In this study, the time step is set to 0.02 times of the natural vibration period of the elements system. MatDEM, adopts GPU matrix calculation to support the dynamic simulation of millions of discrete elements. Thus, the entrainable basement layer can also be constructed with discrete elements. The initial model is constructed by simulating the gravitational deposition of discrete elements. The discrete elements with a mean radius of 5 m have a certain initial velocity in a rectangular simulation box, colliding each other under gravity. Then, they deposit in a random position under artificial compression. Deposited elements are shaped according to the digital elevation model. In order to save computation, elements in the lowest layer are defined as wall elements that do not participate in the dynamic simulation process. Based on the geological properties, other discrete elements are divided into four layers, in which the source area is divided separately due to the severe shattering (Fig. 6).

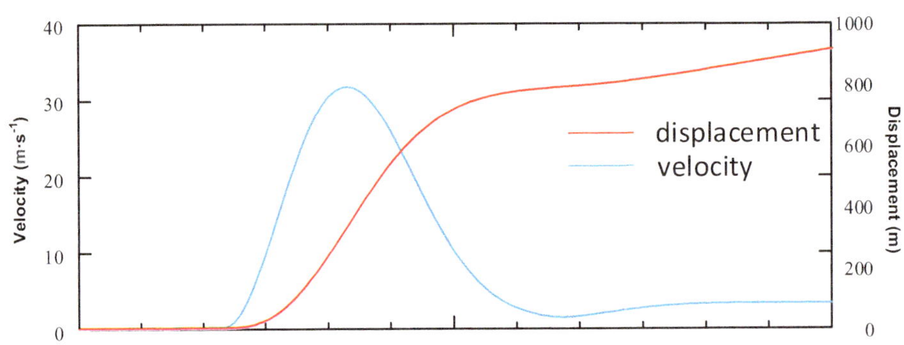

Fig. 5 Kinetic parameters of the sliding mass (velocity and displacement)

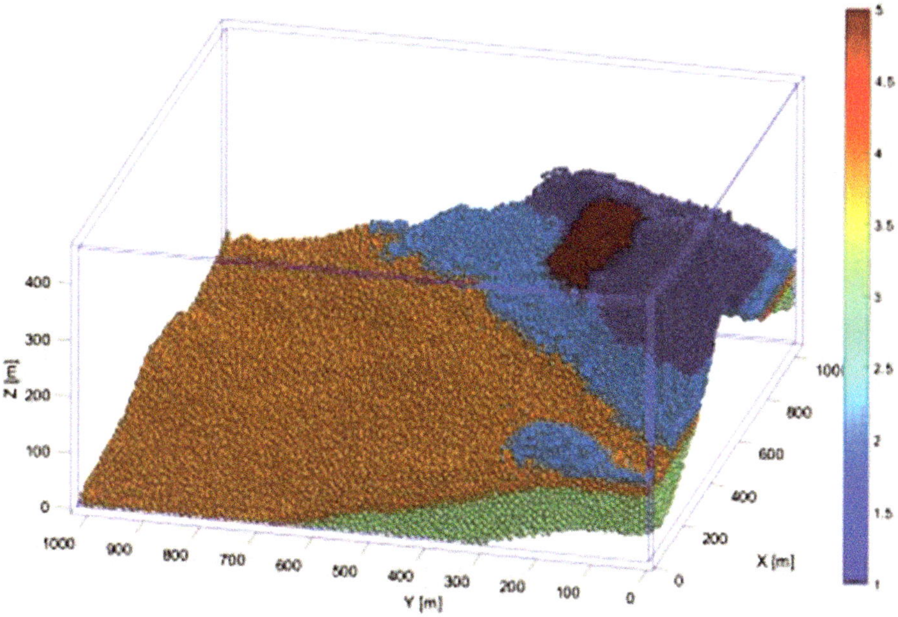

Fig. 6 Basic numerical model of Nayong avalanche

Table 1 The numerical parameters of the avalanche model

Layer	1	2	3	4
Young modulus, E (GPa)	80	30	40	5
Poisson's ratio	0.15	0.25	0.2	0.2
Uniaxial tensile strength (MPa)	8	1	3	0.8
Uniaxial compressive strength (MPa)	80	30	30	5
Intergranular friction coefficient, μ	0.8	0.6	0.6	0.36
Element density, ρ (kg/m^3)	2600	2500	2500	1800

Each layer has different mechanical properties (Table 1), which are obtained by a formula given by Liu that can convert the laboratory mechanical test data into MatDEM parameters (Liu et al. 2017).

Results and Discussion

Simulation Results

Previous studies have shown that the friction coefficient has a tremendous influence on the movement of landslides in numerical model (Lin and Lin 2015). Consequently, this study adopts different intergranular friction coefficient of source area elements (0.4, 0.6 and 0.8) to provide quantitative estimates of the initial conditions. Figure 7 shows the displacement distribution of sliding elements in three scenarios. The final deposition of three different parameters are all about 600 m long and 400 m wide along the sliding direction. Compared with the actual event, the simulation results are roughly the same in length and extends about 100 m to the northeast direction. Similar deposition results are obtained with different intergranular friction coefficients.

This phenomenon may be because the kinetic energy of elements is mainly dissipated by the collision between particles or between particles and the basement. These collisions caused some elements to spread further and more elements entrained into the movement. We can see that in the scenario of low friction coefficient, more elements rushed out of the main deposition body. Moreover, the displacement of elements at the front edge of the landslide mass is smaller than that at the centre of the deposition mass. That means that the elements at the front edge of the landslide mass are not all from the source areas, and there are also many elements that were entrained from the basement. These entrainment phenomena are also more intense in low friction scenarios.

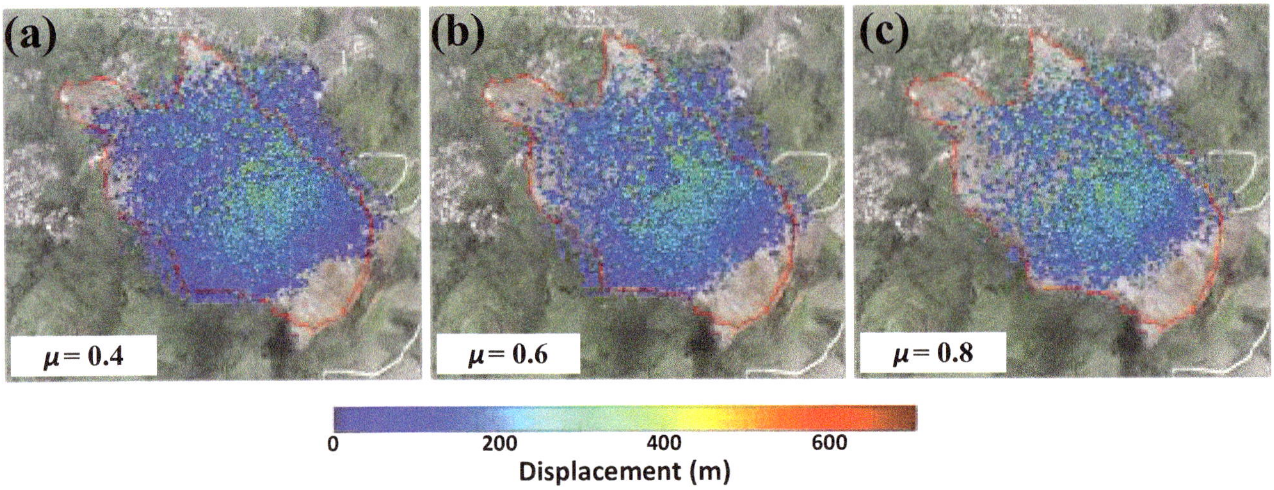

Fig. 7 Simulation of the numerical model with different intergranular friction coefficients

Best Scenario

We can hardly distinguish which simulation result is most consistent with the actual event from Fig. 7. With the dynamic parameters from seismic signal inversion, we can better judge the numerical simulation (Moretti et al. 2015).

Figure 8 shows the velocity-distance pattern obtained from numerical simulation of three scenarios and seismic signal inversion. The velocities of the simulation results are the average velocity of elements at the front edge of the sliding mass, while the velocity inversed from the seismic signal is the centre of the sliding mass. We can see that the velocity of each scenario is increasing rapidly before the horizontal distance reaches 100 m, and except for the data inversed by the seismic signal, the other scenarios reach the maximum speed when the horizontal distance goes to 200 m, i.e., before the 20 m-high hillslope. With the same displacement, velocities of sliding mass in different scenarios are just opposite to the order of the intergranular friction coefficient. The friction coefficient only changes the absolute value of the sliding mass's velocity without influencing its changing law. It indicates that the changing law of the Nayong avalanche's velocity is mainly affected by the terrain and the main obstacle to the sliding movement is the 20 m-high hillslope.

Calculating the area difference between the three curves and the seismic wave inversion curve by integral, we can judge which scenario can best reflect the dynamic process of the seismic signal. Table 2 shows that the numerical model, while the intergranular friction coefficient of source area elements equal to 0.6, is the best scenario closest to the real dynamic process of the landslide.

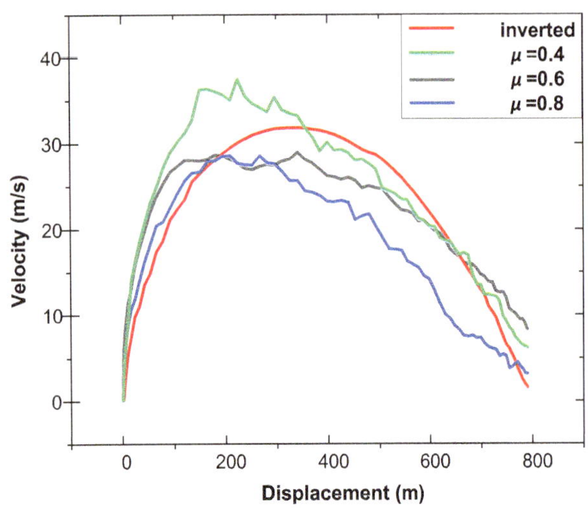

Fig. 8 Avalanche velocity pattern

Dynamic Process

Figure 9 shows the evolution of the Nayong rock avalanche simulated by the best MatDEM scenario. The simulation starts from the overall rock collapse, namely, 10:22:23.5 on the day of the disaster. The primary movement process lasted for about 40 s, and then the sliding body expanded laterally only. In Fig. 9, elements in the source area began to rush, and the front edge of landslide mass reached the toe of the hillslope with a height of about 20 m at 12.6 s. After 31.5 s, the front of the sliding mass almost stopped moving. The final deposition is about 650 m long and 400 m wide along the sliding direction. Compared with the actual event, the simulation results are roughly the same in length and extends about 150 m to the northeast. In the numerical simulation, the front edge of the sliding mass is not divided

Table 2 Area difference of velocity function of the distance between three different scenarios and seismic inversion

Friction coefficient	0.4	0.6	0.8
Area difference	1798.4	−155.8	−3260.9

into two branches since the radius of the elements of 5 m is too large relative to the 20 m-high hillslope.

Figure 9 shows the velocity evolution of sliding mass. Between 0 and 12.6 s, the speed of the avalanche increases rapidly. After 12.6 s, the speed of most elements began to decline. At 18.9 s, only the front edge of sliding mass has a visible speed. After 31.5 s, the speed of most elements is 0, and a small number of elements in the source area continue to fall. The maximum speed of the element is about 40 m/s, which happened at 12.6 s. At this time, the landslide mass is passing through the 20 m-high hillslope.

For a comprehensive understanding of dynamic process, we compare the time–frequency distribution spectrums to the simulations that best fit the observations, and the UAV video (Fig. 10). Both Fan et al. (2018) and Zhu et al. (2019) focus on the analysis of the failure process of the Nayong avalanche through UAV video, while this paper studies the rock fall in the source area after the overall collapse. The magnitude of elements velocity and the number of moving elements can indicate the intensity of energy release. As shown in Fig. 10, the changes in velocity and sliding volume are in good agreement with the seismic signal. At 6.3 s (10:22:29.8), with materials from the source were getting high speed, colliding each other violently, and impacting the basement layer, all the high and low-frequency signal began to be strong. Between 6.3 and 21.5 s (10:22:29.8–10:22:45), a large amount of mass was entrained into the high-speed movement resulting in both the high and low-frequency signal maintaining a strong amplitude in this period. After 21.5 s (10:22:45), most materials had stopped moving, only the front end of the sliding body still spreading forward, so the low-frequency signal was weakened. While the high-frequency signal remained strong due to the constant high-speed falling rocks from the source area colliding with accumulated materials. These rocks impacted the basement severely, which made the low-frequency signal occasionally strong during this period. At 38.5 s (10:23:02) the avalanche movement basically stopped, but the rockfall from source area continued. Until 56.7 s (10:23:20), the rockfall in the source area stopped and the seismic signal faded into the background.

Conclusion

The interpretation of the seismic signal is used to calibrate the parameters of landslide simulation using the 3D discrete element model, MatDEM. Analysis of the intensity and time history of the seismic signals shows that the seismic signals generated by the Nayong avalanche have obvious

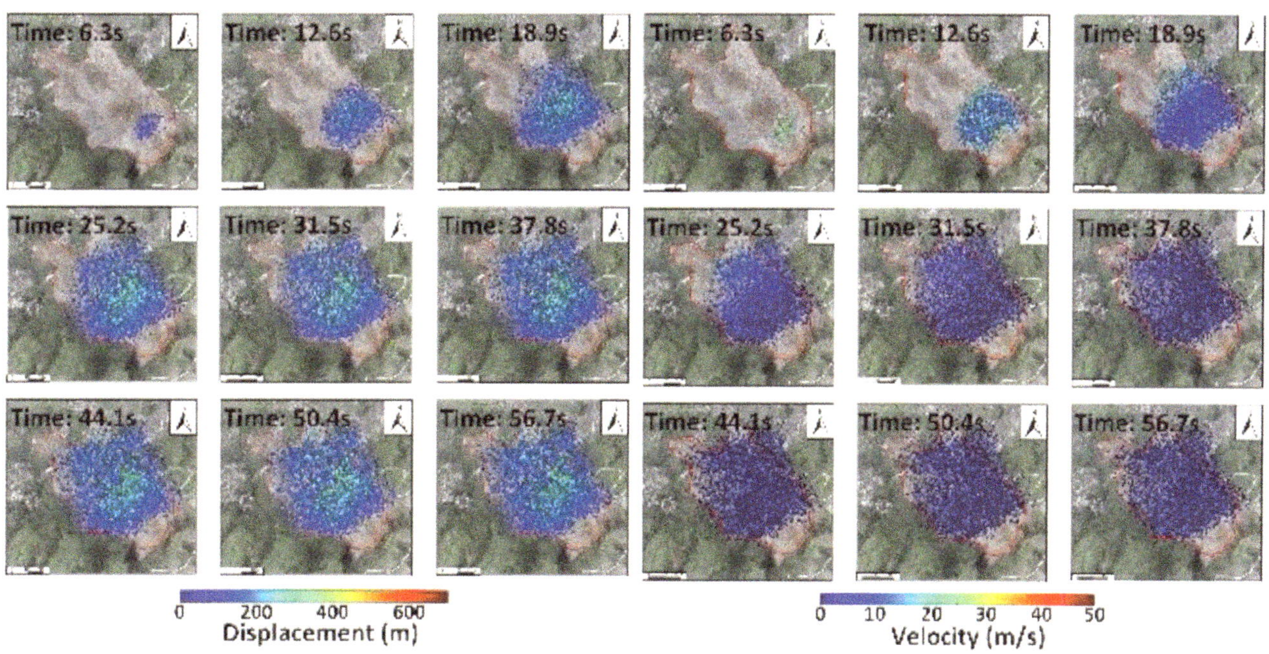

Fig. 9 Evolution of the Nayong rock avalanche simulated by MatDEM

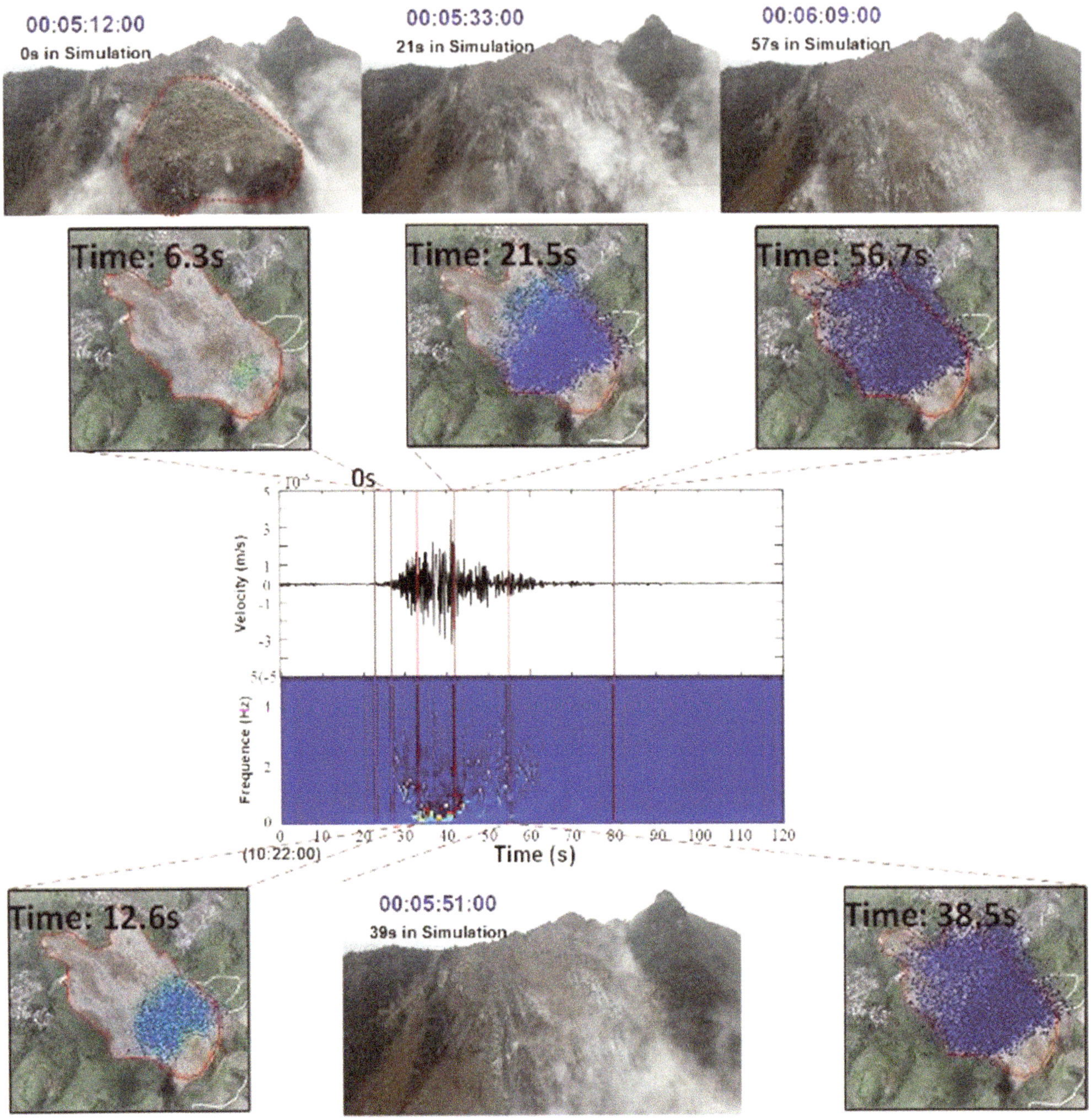

Fig. 10 Comparison of the seismogram and time–frequency distribution spectrum of ZJW station with the simulated velocity evolution of best scenario. Snapshots from UAV video are added

long-period components. Based on the force history, the time-varying velocity and distance of the Nayong avalanche can be obtained by integral. Combined with the results of seismic signal interpretation, the best MatDEM simulation results are selected from three scenarios. The simulation results showed that the movement lasted for about 40 s, and the maximum speed was 40 m/s. The dynamic process is in good agreement with those parameters obtained from seismic signal inversion.

In summary, due to the uncertainty and suddenness of occurrence time, many large landslides have no direct evidence of their dynamic process. Conventionally, the parameter calibration of numerical simulation can only be based on the back analysis of the field accumulation. For rock avalanche, although the dynamic process from seismic signal inversion will be affected by factors such as rockfall from the source area, it is still an effective method to constrain discrete element modelling. Using seismic signals can improve the reliability and accuracy of the results in the modelling, which is useful for predicting and assessing further landslide hazards.

Acknowledgements This study was supported by the National Key R&D Program of China (2018YFC1504804) and the National Natural Science Foundation of China (No. 41530639).

References

Allstadt K (2013) Extracting source characteristics and dynamics of the August 2010 Mount Meager landslide from broadband seismograms. J Geophys Res Earth Surf 118(3):1472–1490

Cagnoli B, Piersanti A (2015) Grain size and flow volume effects on granular flow mobility in numerical simulations: 3-D discrete elements modeling of flows of angular rock fragments. J Geophys Res Solid Earth 120:2350–2366

Crosta GB, Imposimato S, Roddeman DG (2003) Numerical modelling of large landslides stability and runout. Nat Hazards Earth Syst Sci 3(6):523–538

Cundall P, Strack O (1979) A discrete element model for granular assemblies. Geotechnique 29(1):47–65

Ekström G, Stark CP (2013) Simple scaling of catastrophic landslide dynamics. Science 339(6126):1416–1419

Fan XM, Xu Q, Scaringi G, Zheng G, Huang RQ, Dai LX, Ju YZ (2018) The "long" runout rock avalanche in Pusa, China, on August 28, 2017: a preliminary report. Landslides 16(1)

Fukao K (1995) Single force representation of earthquakes due to landslides or the collapse of caverns. Geophys J Int 122:243–248

Huang QH, Cai YL (2007) Spatial pattern of Karst rock desertification in the Middle of Guizhou Province, Southwestern China. Environ Geol 52:1325–1330

Hungr O (1995) A model for the runout analysis of rapid flow slides, debris flows, and avalanches. Can Geotech J 32(4):610–623

Hibert C, Ekström G, Stark CP (2014) Dynamics of the Bingham Canyon Mine landslides from seismic signal analysis. Geophys Res Lett 41:4535–4541

Liu C, Pollard DD, Shi B (2013) Analytical solutions and numerical tests of elastic and failure behaviors of close-packed lattice for brittle rocks and crystals. J Geophys Res Solid Earth 118(1):71–82

Liu C, Xu Q, Shi B, Deng S, Zhu H (2017) Mechanical properties and energy conversion of 3D close-packed lattice model for brittle rocks. Comput Geosci 103:12–20

Lin CH, Lin ML (2015) Evolution of the large landslide induced by Typhoon Morakot: a case study in the Butangbunasi River, southern Taiwan using the discrete element method. Eng Geol 197:172–187

Moretti L, Allstadt K, Mangeney A, Capdeville Y, Stutzmann E, Bouchut F (2015) Numerical modeling of the Mount Meager landslide constrained by its force history derived from seismic data. J Geophys Res Solid Earth 120(2579):2599

Stein S, Wysession M (2003) An introduction to seismology, earthquakes, and earth structure. Blackwell, Malden, MA

Xing AG, Wang GH, Li B, Jiang Y, Feng Z, Kamai T (2015) Long runout mechanism and landsliding behaviour of a large catastrophic landslide triggered by a heavy rainfall in Guanling, Guizhou, China. Can Geotech J 52:971–981

Yin YP, Sun P, Zhu JL, Yang SY (2011) Research on catastrophic rock avalanche at Guangling, Guizhou, China. Landslides 8(4):517–525

Zhu YQ, Xu SM, Zhuang Y, Dai XJ, Lv G (2019) Xing AG (2019) Characteristics and runout behaviour of the disastrous 28 August 2017 rock avalanche in Nayong, Guizhou, China. Eng Geol 259:105154

Electrical Resistivity Tomography (ERT) Based Investigation of Two Landslides in Guizhou, China

Yu Zhuang and Aiguo Xing

Abstract

On 28th June 2010 and 23th July 2019, two long runout landslides occurred in the Guizhou province, Southwest China. These two catastrophic events have a runout distance > 1 km and an estimated maximum velocity > 40 m/s and caused mass casualties in the local area. To investigate the spatial distribution of deposits and better understand the local geological settings of these two landslides, geophysical surveys, including electrical resistivity tomography (ERT) and borehole tests were carried out in this study. The ERT results of the Guanling landslide are roughly consistent with the borehole data, indicating the validity of inversion images. Also, due to the surface and underground river channels, the lithological boundaries of the Shuicheng landslide determined through ERT was not very pronounced. Nevertheless, the location of river channels and deposit characteristics investigated in situ are in rough agreement with the inversion results of ERT. It is expected that the results from the ERT survey are beneficial for the mechanism analysis of these two long runout landslides and proposing more appropriate actions for the landslide mitigation activities.

Keywords

Electrical resistivity tomography • Borehole • Guanling landslide • Shuicheng landslide • Subsurface investigation

Introduction

In recent years, frequent human activities combined with complex geological environments caused a huge number of landslides in the Guizhou, China (Liu et al. 2013). Typically, on 28th June 2010 and 23th July 2019, two long runout landslides triggered by the heavy rainfall occurred in the Guanling and Shuicheng county. To investigate the failure mechanism and disaster-causing process of such disasters, several numerical simulations combined laboratory experiments have been performed (Xing et al. 2014; Liu et al. 2015; Kang et al. 2017).

However, before assessing the stability of sliding mass and planning for remedial measures, a detailed investigation needs to be performed in terms of topographical and geological characteristics of a typical landslide (Ochiai et al. 2004; Lee et al. 2008). The topographic characteristic of a landslide can be obtained based on UAV (unmanned Aerial Vehicle) surveys, while the geological characteristics can be determined through the direct method and indirect method (Holec et al. 2013). Direct methods, including boreholes and associated techniques, are widely used to determine parameters on the lithological, geological, and geotechnical characteristics of a landslide. Nevertheless, the data obtained through a borehole can only represent a specific point, and the spatial characteristics of subsurface landslide requires many tests resulting in a significant increase in cost (Jager et al. 2013).

Alternatively, geophysical surveys provide detailed information about landslide profiles with rapid survey and relative low cost (Jongmans and Garambois 2007). Electrical resistivity tomography (ERT) is regarded as the most successful one to study the internal characteristics of a landslide, as this method could aid in identifying the subsurface characteristics of a landslide, and thus provide valuable data for the landslide analysis (Lee et al. 2008). Therefore, to improve the understanding about the geological settings and subsurface conditions of Guanling landslide and Shuicheng

Y. Zhuang · A. Xing (✉)
Department of Civil Engineering, Shanghai Jiao Tong University, Shanghai, 200240, China
e-mail: xingaiguo@sjtu.edu.cn

Y. Zhuang
e-mail: zyzhuangyu@sjtu.edu.cn

© Springer Nature Switzerland AG 2021

N. Casagli et al. (eds.), *Understanding and Reducing Landslide Disaster Risk*, ICL Contribution to Landslide Disaster Risk Reduction,
https://doi.org/10.1007/978-3-030-60311-3_11

landslide, ERT surveys based on the Wenner electrode array were employed on the central and lower part of these two landslides. Five boreholes were drilled in the Guanling landslide and detailed investigations were conducted to verify the validity of ERT inversion results of these two landslides. The works conducted in this study are expected to provide valuable information for the reconstruction of landslide geometry and propose actions for the landslide mitigation activities.

Electrical Resistivity Tomography Survey

Resistivity measurements are conducted by inputting a current into the ground through two steel electrodes and measuring the potential drop at two other electrodes. The apparent resistivity of subsurface soil can be calculated based on Ohm's law. Technically, various arrays are used for the ERT measurement, and each array has its advantages. For now, many researches have performed the ERT tests on the landslide investigation based on the Wenner array, and the test results indicated that Winner array is robust, high signal-to-noise ratio, and is suitable for the subsurface investigation of a landslide (Lee et al. 2008). Also, the electrical resistivity tomography is derived through the inversion of the apparent resistivity, and RES-2Dinv is one of the most famous inversion algorithms (Akpan et al. 2015). Thus, the Wenner array based ERT measurement was performed in this study. All obvious outliers were removed, and the RES2Dinv software was selected for the inversion work. The main characteristics of the ERT survey are presented in the inverted results of each survey line.

Guanling Landslide

On 28th June 2010, a long runout landslide occurred in the Guanling, Guizhou, China (N25°59′10″, E105°16′50″). The sliding mass has a travel distance of 1300 m with an elevation difference of about 420 m. This catastrophic event killed 99 people, injured 38 people, and caused heavy losses to the property of local people.

Geological and Climate Setting

The study area located in a region of middle-mountain relief with an elevation ranging from 730 to 1642 m. The exposed rocks in the study area range in age from the late Permian to Quaternary. The landslide mainly occurred in the Early Triassic Yelang sandstone, which is underlain by the Late Permian Longtan sandy shale (Fig. 1). The landslide area has a humid subtropical monsoon climate with annual rainfall averages ranging from 1205 to 1657 mm.

Subsurface Characteristics of the Guanling Landslide

To investigate the subsurface characteristics of the Guanling landslide, a detailed field investigation combined with ERT tests were performed. As shown in Fig. 2, seven survey lines were conducted, and the sliding mass was divided into four subzones (zones d-g) according to the surface characteristics of the deposit.

Figure 3 shows the ERT test results of the Guanling landslide. The zones of high resistivity represent the fragmented rocks of Yelang sandstone from the source area and the low resistivity anomaly corresponds to the underlying soft rock of Longtan sandy shale and Quaternary deposits. The estimated distributions of deposits are drawn with the dashed lines based on the comparison of ERT inversion results and field investigation. The discussion of each ERT profile is shown below.

ERT 1 was set at an elevation ranging from 781 to 848 m and passing through zones f and g (Fig. 3a). High resistivity anomalies are identified at the distance of 80–240 m and 300–480 m from the origin of the profile, corresponding to the fractured rock of Yelang sandstone. Zones of low resistivity are observed below the surface at the initial 80 m, and is associated with the mudflow in zone h and loose Quaternary deposits. Also, it clearly reveals the bedrock at the bottom of the image with an increasing electrical resistivity value and is likely corresponds to the Longtan sandy shale. The thickness of landslide deposits ranges from 5 to 20 m with a maximum deposit depth identified at a distance of 180 m from the origin of the survey line.

ERT 2 was installed at an elevation of 870–1000 m, and passed through zones e and f (Fig. 3b). The zone of low resistivity is identified at the initial 140 m of the profile,

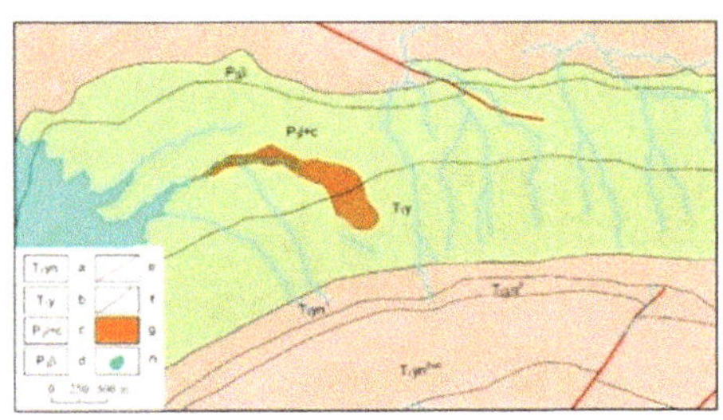

Fig. 1 Geological map of the Guanling landslide. **a** Yongningzhen limestone; **b** Yelang sandstone; **c** Longtan sandy shale; **d** Permian basalt; **e** Stratigraphic boundary; **f** Fault; **g** Landslide area; **h** Guangzhao reservoir (Xing et al. 2014)

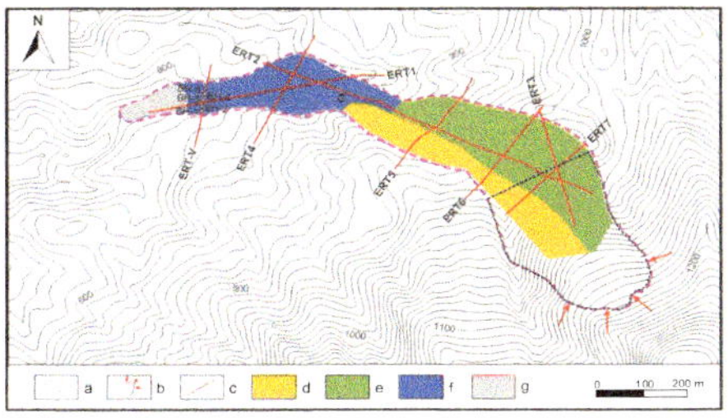

Fig. 2 Detailed topography of the Guanling landslide. **a** Landslide boundary; **b** Source area; **c** ERT survey lines; **d** Boulder-sized debris; **e** Gravel-sized debris; **f** Silty with gravels in small size (<5 cm); **g** Mudflow deposits (Xing et al. 2014)

which is located at the rear part of zone g and corresponds to the Quaternary deposits with few gravels. Also, high resistivity anomalies were observed in shallow depth at the distance of 160–280 m and 360–840 m, and is likely related to the boulder-size Yelang sandstone. Notably, a variation of resistivity value was identified below the high resistivity anomalies and is inferred as the boundary of Yelang sandstone and original deposits overlying Longtan sandy shale. The thickness of landslide deposits in this profile ranges from 4 to 30 m with a maximum deposit depth identified at a distance of 600–700 m from the origin of the survey line.

ERT 3 was installed at an elevation of 965–1020 m and passed through zone e (Fig. 3c). A high resistivity anomaly was observed at the distance of 75–280 m from the origin of the profile with a maximum resistivity value of larger than 400 Ω m, corresponding to the landslide deposits. The thickness of the deposit ranges from 4 to 20 m and increases from the landslide boundary to the central part of the landslide. Also, a boundary was identified between the deposit and underlying bedrock, providing valuable information for estimating the burial depth of Longtan sandy shale.

ERT4 was placed across zone f and installed in the investigated area at an elevation of 828–855 m (Fig. 3d). Deposits in this region are mainly composed of silty with gravels in small sizes. High resistivity anomalies corresponding to fractured Yelang sandstone were identified at the distance of 115–140 m from the origin of the profile. Also, the region at the distance of 140–240 m shows relatively low resistivity and is related to the Quaternary deposits. The boundary between original agricultural soil and landslide deposits was observed at a depth of 5–10 m, and thus the spatial distribution of deposits along this profile was determined.

ERT5 was placed through zones d and e and installed at the elevation of 885–931 m (Fig. 3e). The deposits are mainly composed of debris-avalanche with boulders and gravels. High resistivity anomalies were identified at a distance of 140–190 m from the origin of the survey line, corresponding to the Yelang sandstone. Moreover, the

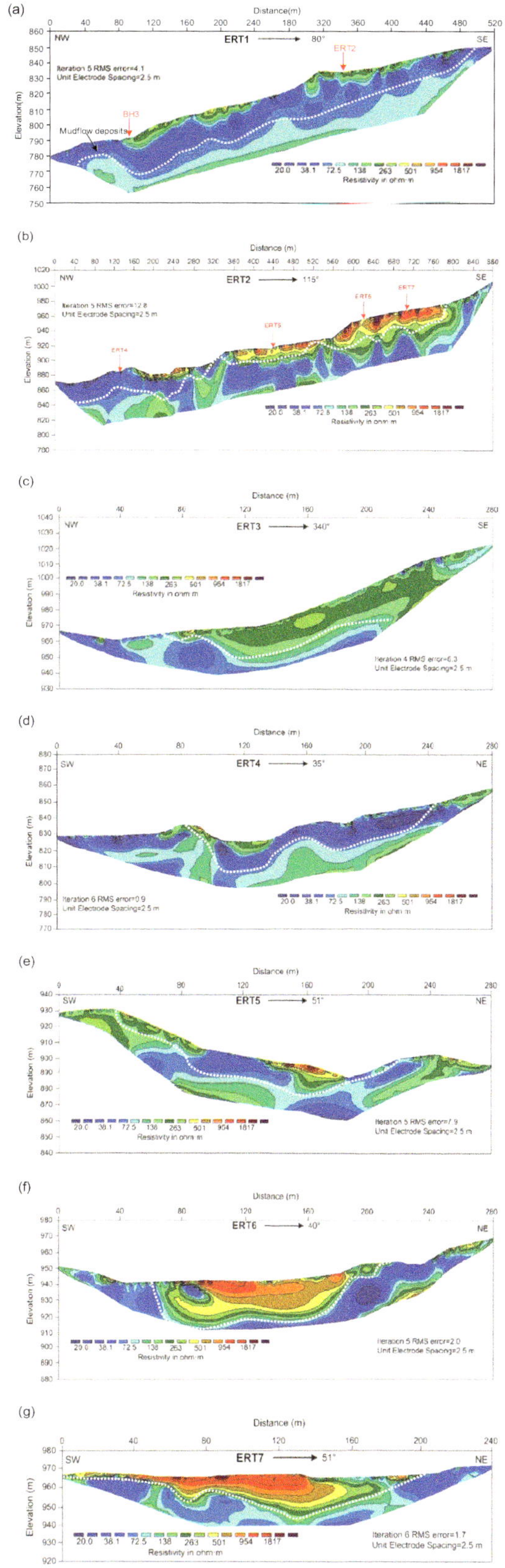

Fig. 3 Inversion results of the Guanling landslide (ERT1–ERT7). White dashed line represents the hypothetical boundary of landslide deposit

regions at the distance of 40–140 m and 190–235 m indicated zones of low resistivity, and was inferred as Quaternary deposits of silts. The deposits along this profile has an estimated thickness of 2–16 m with a maximum deposit depth at a distance of about 160 m from the origin of the survey line.

The inversion images of ERT 6 and ERT 7 show similar results. These two survey lines were both installed across zones d and e with a maximum resistivity value larger than 1500 Ω m. The deposits in this region are mainly composed of Yelang sandstone with particle size larger than 20 cm and Quaternary deposits. Also, low resistivity anomalies were identified below the sliding mass, corresponding to the Longtan sandy shale and possible overlying agricultural soil. ERT 6 was placed at an elevation of 943–966 m (Fig. 3f), and high resistivity anomalies were observed at 60–220 m from the origin of the survey line. A large area of landslide deposits located in this region, with the thickness ranging from 3 to 30 m. For ERT 7, this survey line was placed at an elevation of 966–972 m (Fig. 3g). It gives a wide distribution of deposits with a width of approximately 145 m (at the distance of 35–180 m from the origin of the profile), and a thickness ranging from 2 to 18 m.

To verify the validity of ERT inversion results, five boreholes were drilled along the ERT-V line (Fig. 4). A rough boundary between the landslide deposit and underlying sandy shale can be observed and the thickness of landslide deposits determined by the ERT is roughly consistent with the borehole data, indicating that the ERT method could be used to examine the subsurface characteristics of a typical landslide.

The integration of all the ERT images shows the variation of deposit characteristics along the runout path. The peak value of high resistivity anomalies observed in shallow depth decreases with the runout distance, and this phenomenon is associated with the fragment of Yelang sandstone. With the increase of runout distance, the Yelang sandstone continues to broke, and thus resulted in the increase of rock with small particle size in the debris avalanche. Also, the landslide occurred in the hard rock of Yelang sandstone, which is underlain by the soft rock mass of Longtan sandy shale. The sliding mass deposited loosely after the movement, resulting in the high porosity of the displaced materials. The combination of these factors eventually caused the spatial characteristics of ERT images (high resistivity anomalies in the shallow deposits and low resistivity anomalies in the bedrock and Quaternary deposits).

Shuicheng Landslide

At 20:40 p.m. on 23th July 2019, a rainfall triggered landslide occurred in Shuicheng county, Guizhou, China (N26° 15′27″, E104°40′03″). The sliding mass has a runout distance of 1.3 km and an elevation difference of 465 m. This catastrophic event travelled forward with high velocity and finally caused 52 casualties.

Geological and Climate Setting

The study area is in the middle part of Yunnan-Kweichow, with an elevation ranges from 1180 to 2270 m. The exposed rocks in the study area range in age from Upper Permian to Quaternary, and the Shuicheng landslide mainly occurred in the Emei Mountain basalt (Fig. 5). The study area has a plateau monsoon climate with an annual rainfall ranging from 940 to 1450 mm.

Subsurface Characteristic of the Shuicheng Landslide

To investigate the spatial characteristics of landslide mass of the Shuicheng landslide, a total of 6 survey lines were installed in the entrainment and deposit areas of the landslide (Fig. 6). The electrical resistivity contrasts of the Shuicheng landslide were not very pronounced because of the surface and underground river channels in this region. Nevertheless, combining the field investigation and ERT images, the location of landslide surface, deposit, and river channels are

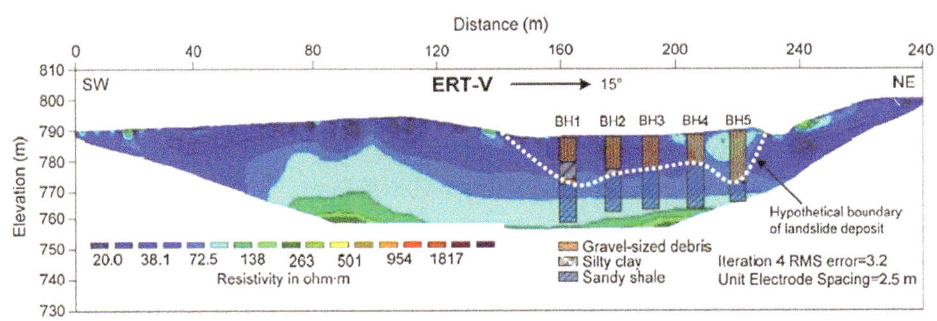

Fig. 4 Comparation results of ERT V and borehole data

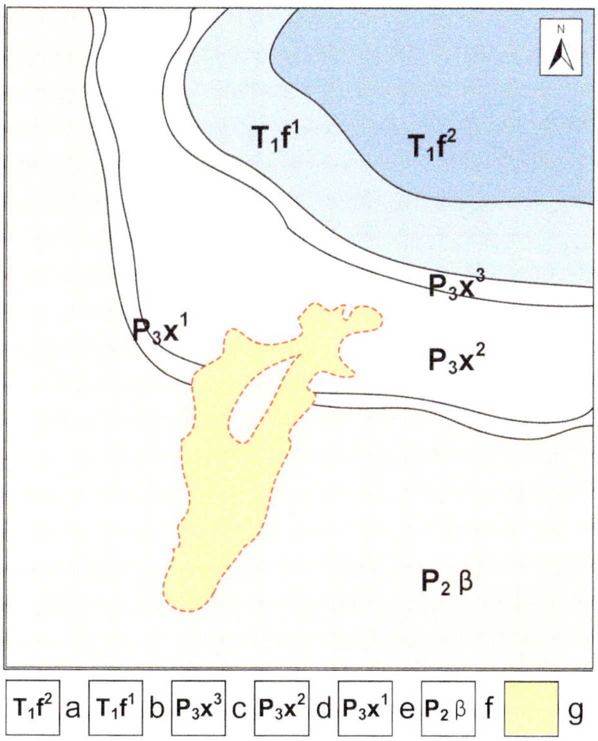

Fig. 5 Geological map of the Shuicheng landslide. **a** 2nd part of Feixianguan formation; **b** 1st part of Feixianguan formation; **c** 3rd part of Xuanwei formation; **d** 2nd part of Xuanwei formation; **e** 1st part of Xuanwei formation; **f** Permian basalt; **g** landslide area

basically identified, and thus the estimated scopes of deposits are outlined by the dashed line. The detailed analysis of ERT images is presented as follows.

The ERT1 was placed parallel to the runout direction in the left gully with an elevation of 1135–1165 m (Fig. 7a). High resistivity anomalies sporadically distributed along the profile surface, and these regions have an average electrical resistivity of about 800 Ω m. Also, a peak value of 5500 Ω m was observed at a distance of 180 m, which is related to the broken Emei Mountain basalt. The deposit along this profile has an estimated thickness of 2–7 m, and a significant decrease of electrical resistivity was observed below, corresponding to the weathered bedrock of Xuanwei formation. Also, a relative low resistivity anomaly was imaged at 80–120 m from the origin of survey line, corresponding to the river channel.

ERT 2 shows similar surface characteristics with the ERT 1. This survey line was placed perpendicular to landslide direction with an elevation of 1150–1165 m (Fig. 7b). High resistivity anomalies were imaged sporadically along the profile with a maximum electrical resistivity value of larger than 6000 Ω m. The landslide deposits have an inferred thickness of 3–7 m. A region with low resistivity value was observed at a distance of 65 m from the origin of the profile, and is inferred as a river channel. Also, an increase of resistivity value was imaged in the bedrock. The surface rock mass was strongly weathered with developed fissures and thus caused the high degree of saturation and relatively low electrical resistivity. Notably, the ERT interpretation of the river channel location is basically consistent with the field investigation results.

ERT3 and ERT4 are a pile of intersecting survey lines installed in the downstream of the landslide. ERT3 was installed at an elevation of 1123–1127 m (Fig. 7c). High resistivity anomalies sporadically distributed along the profile surface, and the peak electrical resistivity of larger than 4500 Ω m was imaged at the distance of 70−90 m from the origin of the survey line. The region with low resistivity anomaly was observed at a distance of 100 m from the profile origin and is associated with a river channel. The deposits in this region have a thickness of 3−10 m with a maximum depth at the distance of 80–100 m from the origin of the profile.

As shown in Fig. 7d, ERT4 was placed at an elevation of 1113–1137 m. High resistivity anomalies were imaged at a distance of 0–50 m and 65–85 m with a maximum resistivity value of 4000 Ω m. Low resistivity anomalies were identified at a distance of 60 m from the origin of the survey line, corresponding to the river channel. The deposits in this profile have a thickness of 3–10 m. Notably, due to the obstruct of a slope located at a distance of 45 m from the origin of the profile, the depth of deposits in the left part (0–60 m) is higher

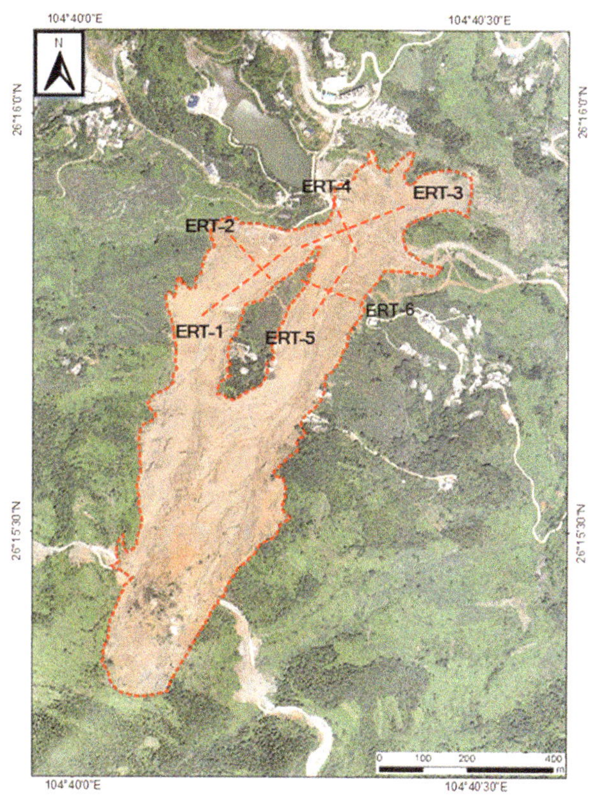

Fig. 6 Location of ERT survey lines

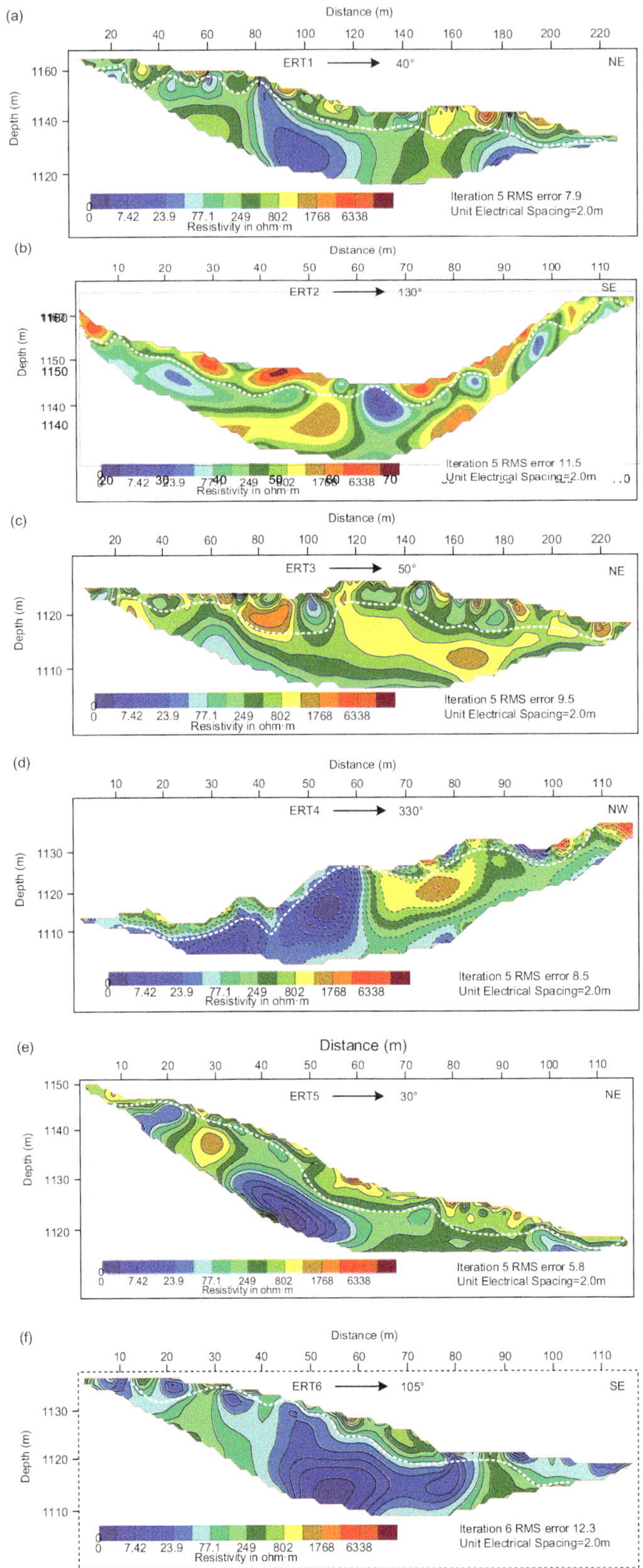

Fig. 7 Inversion results of the Shuicheng landslide (ERT1–ERT6). White dashed line represents the hypothetical boundary of landslide deposit

than that in the right part (60–120 m). The field investigation shows that the deposit of left part of profile is mainly composed of Quaternary deposits and few basalts with thick depth, verifying the validity of ERT inversion results.

ERT 5 was set parallel to the runout direction with an elevation of 1120–1150 m (Fig. 7e). High resistivity anomalies were identified at the distance of 55–100 m from the origin of profile. A maximum electricity value of 5000 Ω m was observed in a narrow region, corresponding to the deposits of silty clay with few broken basalts. Deposits along this profile have an estimated thickness of 3–8 m, and a pronounced decrease of resistivity value was imaged below, corresponding to the sandstone and mudstone of Xuanwei formation. Also, a region with low resistivity anomaly was imaged at bottom of image, and is inferred as an underground river channel.

For the ERT6, this survey line was installed at an elevation of 1123–1136 m, and shows similar subsurface characteristics of ERT 5 (Fig. 7f). High resistivity anomalies were observed at 55–80 m from the origin of profile, and a maximum resistivity value of larger than 5000 Ω m was imaged with a narrow region. The deposits in this region have an estimated thickness of 2–8 m. Specifically, the range of high resistivity anomalies (> 5000 Ω m) in the right gully is pronounced smaller than that in the left gully, and this phenomenon is basically consistent with the field investigation as the amount of basalt rock mass is dominated in the left gully.

All the ERT images of the Shuicheng landslide shows high resistivity anomalies of deposits and low resistivity anomalies of bedrock. Combined with the detailed field investigation, the landslide mainly occurred in the hard rock of Emei Mountain basalt while the bedrock in the deposit area is inferred as Xuanwei formation with soft rock. Meanwhile, the displaced materials deposited loosely after long runout movement, and the strongly weathered bedrock with developed fissures was in a high degree of saturation. Thus, high resistivity anomalies of deposits and relatively low value of bedrock was imaged. Also, a pronounced variation of electrical resistivity was observed in the ERT images along the runout path. The amount of basalt in both gullies is obviously larger than that in the downstream of landslide, resulting in the high electrical resistivity in two gullies. Therefore, it is concluded that the ERT images are in good agreement with the actual condition.

Summary and Conclusion

On 28th June 2010 and 23th July 2019, two long runout landslides occurred in Guizhou, China. To investigate the deposit distribution and better understand the geological settings of these two landslides, ERT surveys combined with boreholes were conducted. The test results indicated that ERT is a cost-effective tool that can easily be performed to provide valuable subsurface data. The main conclusions are made as follows.

(1) All the inverted ERT images illustrated that high resistivity anomalies in shallow depth represent the landslide deposit of hard rock, while the low resistivities anomalies are associated with the Quaternary deposits and bedrock.
(2) Five boreholes were conducted in the Guanling landslide, and the borehole data verified the validity of ERT results. Also, the spatial distribution of landslide deposits determined through ERT images is in good agreement with the actual situation, as the particle size and amount of Yelang sandstone decrease with the increase of runout distance.
(3) The electrical resistivity contrasts of the Shuicheng landslide were not very pronounced because of the surface and underground river channels in this region. Nevertheless, according to the field investigation and ERT images, the location of landslide surface, deposit, and river channels determined are roughly consistent with the actual situation.

Acknowledgements This study was supported by the National Key R&D Program of China (2018YFC1504804) and the National Natural Science Foundation of China (No. 41977215).

References

Akpan AE, Ilori AO, Essien NU (2015) Geophysical investigation of Obot Ekpo landslide site Cross River State, Nigeria. J Africa Earth Sci 109(3):154–167

Holec J, Bednarik M, Sabo M, Minar J, Yilmaz I, Marschalko M (2013) A smallscale landslide susceptibility assessment for the territory of Western Carpathians. Nat Hazards 69(1):1081–1107

Jager D, Sandmeier C, Schwindt D, Terhorst B (2013) Geomorphological and geophysical analyses in a landslide area near Ebermannstadt Northern Bavaria. E&G Quarter Sci J 62(2):150–161

Jongmans D, Garambois S (2007) Geophysical investigation of landslides: a review. Bulletin Société Géologique De France 178 (2):101–112

Kang Y, Zhao CY, Zhang Q, Lu Z, Li B (2017) Application of InSAR techniques to an analysis of the Guanling landslide. Remote Sens 9 (10):1046

Lee CC, Yang CH, Liu HC, Wen KL, Wang ZB, Chen YJ (2008) A study of the hydrogeological environment of the lishan landslide area using resistivity image profiling and borehole data. Eng Geol 98(3–4):115–125

Liu QC, Xiong CR, Ma JW (2015) Study of Guizhou Province Guanling Daz-Hai landslide instability process under the rainstorm. Appl Mech Mater 733:446–450

Liu ZX, Wang X, Shi YH, Zhao ML (2013) Research on mountain landslide WSN monitoring system for Karst area. Appl Mech Mater 325–326:836–841

Ochiai H, Okada Y, Furuya G, Okura Y, Matsui T, Sammori T, Terajima T, Sassa K (2004) A fluidized landslide on a natural slope by artificial rainfall. Landslides 1(3):211–219

Xing AG, Wang G, Yin YP, Jiang Y, Wang GZ, Yang SY, Dai DR, Zhu YQ (2014) Dynamic analysis and field investigation of a fluidized landslide in Guanling, Guizhou, China. Eng Geol 181:1–14

2020 Kyoto Japan

Vibration of Piled Rocks—Which Rock Can Be Removed?

Kiminori Araiba and Shoji Doshida

Abstract

To identify rocks which can be taken from rock-mound without disturbing stability of other rocks, vibration characteristics of rocks in mound were investigated by measuring vibration of each rock in an artificial mound. A mound of sand stone rocks of randomly-shaped, length 30–40 cm was constructed on the ground and was subjected by artificial vibration. The vibration of rock on the top ("free rock") and that of a rock supporting the top rock ("supporting rock") were measured. Their characteristics showed that the dominant frequency was different; 40–50 Hz for the free rock and 50–60 Hz for the supporting rock. After changing the free rock to a supporting rock by means of leaning another rock on it, the dominant frequency became to 50 and 60 Hz. It is considered that waves of frequency range of 40–50 were amplified in the free rock in test conditions of this paper. According to these findings, the authors discussed about possibility of identify free rocks from supporting rocks by vibration characteristics.

Keywords

Vibration • Microtremor • Test • Rescue • Rock mound

Introduction

When someone is affected by a landslide, people wish to rescue him. Evaluation of secondary landslide (Araiba and Sakai 2014; Araiba and Doshida 2017), prompt finding of missing people and quick unearthing are important for safe and effective rescue activity. In this paper, to develop quick unearthing process, preliminary investigation of characteristics of vibration of a rock in piled rocks is investigated.

One of the key factors to survive in a confined place, like deposit of landslide, is the vacant space for breathing (e.g. Haegeli et al. 2011). The deposit of large rocks has larger and much number of spaces than those of fine materials, so searching in rock deposits is considered to have higher priority than fine materials when people are missing by landslide without clue. On the other hand, removing a rock from randomly-piled mound without disturbing stability of other rocks is not easy especially when we do not have enough time for investigation and/or constructing support. Figure 1 is a picture of the site where a boy was rescued from a space between fallen rocks of landslide due to 2004 Chuetsu Earthquake, northeast Japan. Then the body of his sister was found caught between rocks and was diagnosed as instant death on site. The rescuers could not recover her body by means of conventional rescue technique mainly because of difficulties in identifying safe-rock-removing process, then un-manned heavy machines were adopted (Araiba 2014). The approach of machines and recovering took a few weeks. Like this case, it is often the case with us to choose which rock can be removed without disturbing other rocks but there is not effective method for this problem.

To answer this question, the authors investigated characteristics of vibration of rocks in piled rocks. We assume that a "removable rock" can behave (thus oscillate) freely but the "un-removable rock", rock supporting other rock(s), cannot. Whether the difference of constraint of each rock can be detected as a difference in the characteristics of vibration was investigated in this paper by means of model tests.

Ogata et al. (2003) investigated characteristics of vibration of the ground surface and that of the boulder on it by means of model tests as well as in-situ measurements and they found difference between them. Uehan et al. (2012) measured vibration of rock specimen with an artificial notch

K. Araiba (✉) · S. Doshida
National Research Institute of Fire and Disaster, Chofru, Tokyo, 182-8508, Japan
e-mail: araiba@fri.go.jp

S. Doshida
e-mail: sdoshida@fri.go.jp

© Springer Nature Switzerland AG 2021

N. Casagli et al. (eds.), *Understanding and Reducing Landslide Disaster Risk*, ICL Contribution to Landslide Disaster Risk Reduction,
https://doi.org/10.1007/978-3-030-60311-3_12

Fig. 1 A boy was rescued from a space between piled rocks of deposit of landslide caused by 2004 Niigataken-Chuetus earthquake

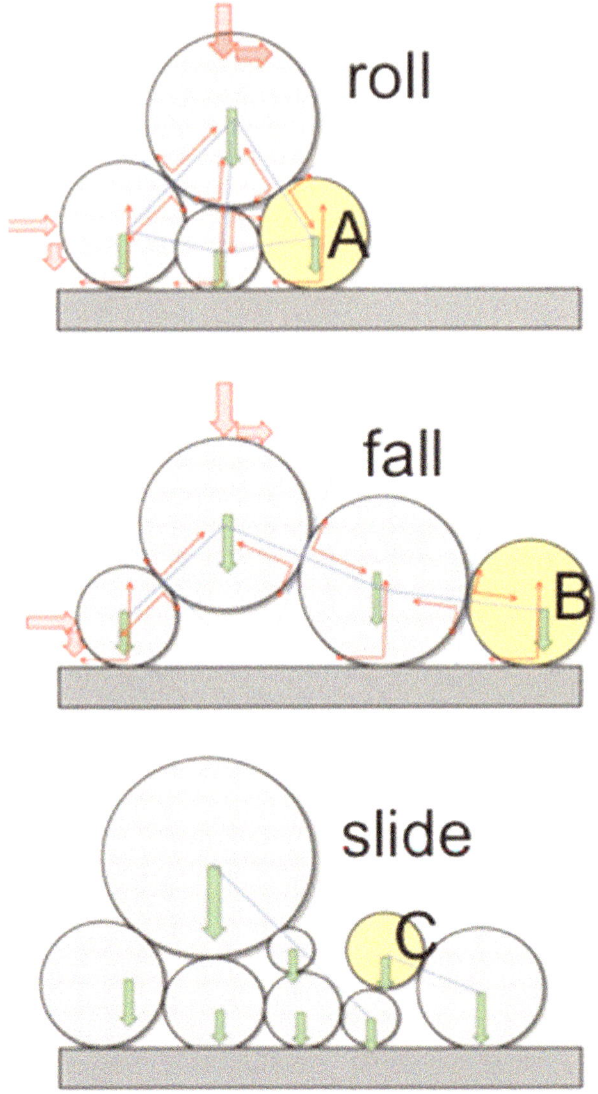

Fig. 2 Schematics of piled rocks. A, B and C cannot be removed without disturbing stability of other rocks. A and B are subjected by force from surrounding rocks, thus their vibrations are considered to be affected by their constrains. C seems impossible to be identified only by vibration and this mode is out of scope of this study

and reported changes in its dominant frequency with depth of the notch. These studies suggest that the characteristics of vibration of rock is enough markedly influenced by its surrounding conditions to be detected. The authors constructed a mound of rocks and measured vibration of rocks.

Figure 2 is a schematic of a mound of rocks. Rocks are simplified as circles. All rocks are supported its position by the balance of gravity, normal force(s) from contacting rocks or ground and friction between them. Removing rock A, B and C will result in roll of the upper rock, fall of the next to the next rock and slide of whole mound respectively. Marked friction and normal forces act on A and B thus they possibly affect vibration characteristics of these rocks. These two modes are discussed in this paper. To identify rock C seems to require consideration not only vibration but also position of each rock in mound thus it will be left for future study.

Test Method

Ten fresh sandstones, randomly shaped, length of 30–40 cm, weight of 20–40 kg were randomly mounded on the ground (Fig. 3). On the top, a rock was placed and named "Rock 1" and one of rocks supporting it was named "Rock 2". Accelerometers was adhered to rocks 1 and 2 using epoxy putty then microtremors of each rocks and artificial tremor caused by impact were recorded. The range of accelerometer was ±2.048 g and sensitivity is 3.9 micro g. Three components of vibration were measured but only vertical data is discussed in this paper as the first step. Data was sampled in 200 Hz using a digital recorder manufactured by Acorn Technologies, Inc. and bandpass filter of 1–100 Hz with box window was applied The artificial tremor was generated by the impact of synthetic resin hammer, head weight 1.1 kg and arm length 30 cm, on a rock on the ground 1 m distant from the rock 1 and 2. The head of the hammer was rotationally free-felt with only support of hand from the height of about 50 cm. Figure 4 is an example of recorded data.

Tests were done with two configurations of rocks. First, the vibrations of rock 1, free, and rock 2, supporting rock 1, were recorded, then another rock was leaned to rock 1 thus rock 1 became "supporting rock" and rock 2 became "supporting two rocks". The data of the first condition will be expressed "Rock 1 (free)" and "Rock 2 (supporting) and those of the second one will be "Rock 1 (supporting) and "Rock 2 (supporting)" in this paper.

Fig. 3 The first configuration of mound of rocks. Another rock will be leaned on Rock 1 for the second test condition

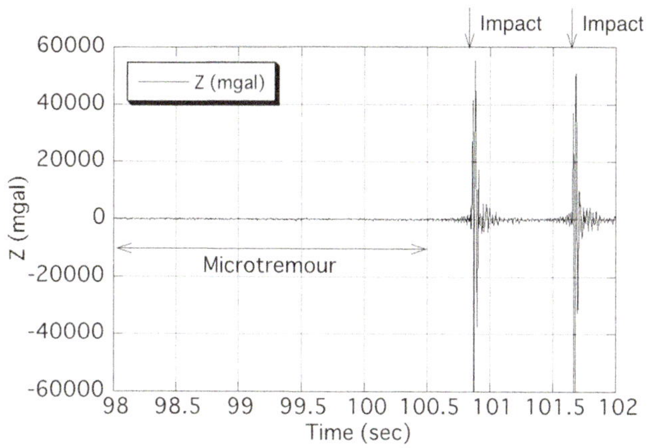

Fig. 4 An example of recorded vibration of Rock 1 (free). The data of microtremor and that of artificial tremor will be analyzed

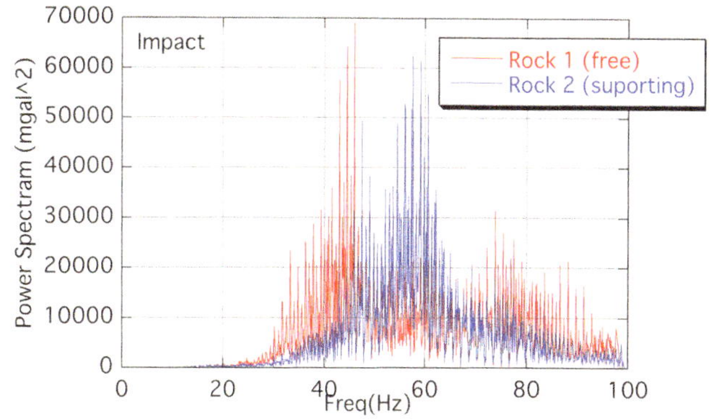

Fig. 5 Power spectrums of vibration for Rock 1 (free) and for Rock 2 (supporting) excited by artificial impact on the ground about 1 m distant

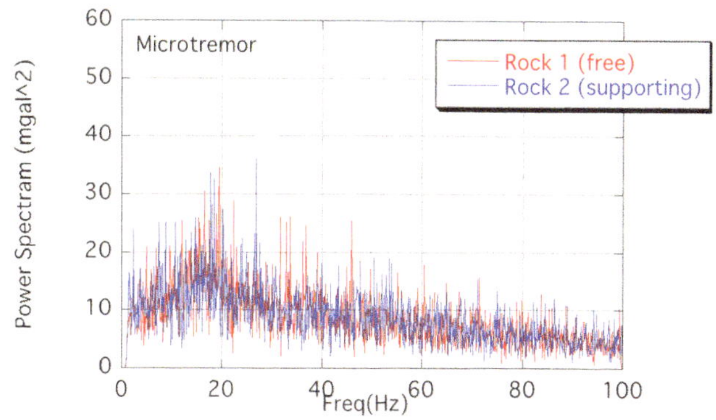

Fig. 6 Power spectrums of microtremor for Rock 1 (free) and for Rock 2 (supporting)

Result and Discussion

Figure 5 shows power spectrums of artificial (impacted) tremor of rock 1 and 2. The dominant frequency for Rock 1 (free) is 40–50 Hz and that for Rock 2 (supporting) is 50–60 Hz. Figure 6 shows power spectrums of microtremor, where marked difference is not seen. It seems that the difference in characteristics of vibration between rock conditions can be observed when marked tremor is excited.

The vibration of each rock is affected not only by contact condition with other rocks and incident wave to it but also by the natural frequency of each rock. The natural frequency of rock is a function of the shape and the elastic wave speed of rock. Considering the wave speed to be 4 km/sec and the representative length of rock to be 40 cm in this experiment, the natural frequency is 5000 Hz, which is out of the measurement range. Therefore, the difference in the natural frequency of each rock is considered to have little effect on the difference in vibration characteristics observed in this experiment.

In real fields, if the wave speed is very slow (for example, 1 km/s; highly weathered rock) and representative length is very large (for example, 10 m), the natural frequency will be close (100 Hz for the example) to the measurement range in this experiment. Further investigation is necessary on whether difference in the rock vibration between supporting/free conditions can be detected in such a large and weak rocks, where the dominant vibration frequency, as well as the natural frequency, of each rock must be different from observed in this experiment.

Figure 7 shows ratio of spectrums in Fig. 5. Considering that vibration of rock 2 (supporting) was incident wave for Rock 1 (free), waves with frequency of 30–50 Hz were markedly amplified besides that dominant frequency was

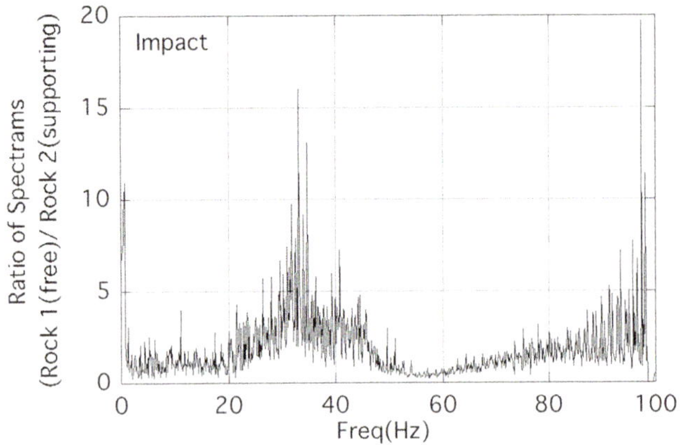

Fig. 7 Ratio of power spectrums of vibration for Rock 1 (free) and for Rock 2 (supporting) (Fig. 5)

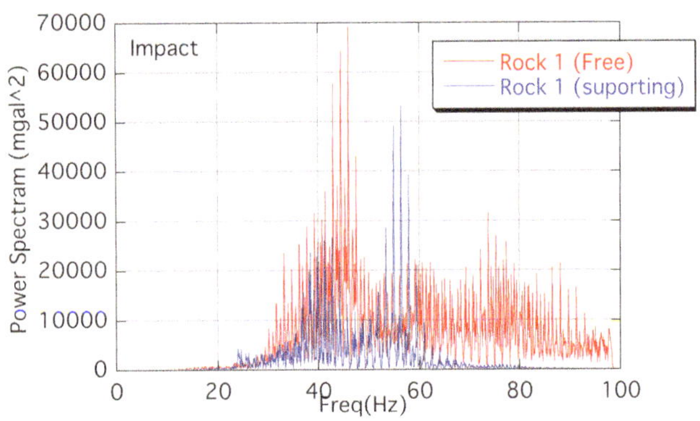

Fig. 8 Power spectrums of vibration for Rock 1 (free) and for Rock 1 (supporting)

50–60 Hz (Fig. 5). Figure 8 shows power spectrums for Rock 1(free) and Rock 1 (supporting). When the rock was free, waves of frequency 30–50 was dominant and those of 50–60 Hz became dominant after it lost freedom. The latter frequency range is similar to the dominant frequency of Rock 2 (supporting) (Fig. 5) so authors think that supporting rocks in a mound have similar vibration and that of free rock (s) may differ from them.

In this experiment, the mound was composed by ten rocks. The vibration of mound as a bulk body is determined by the arrangement of rocks and shape of the mound. Vibration of rocks on the surface of mound is affected by incident waves through contact points of rocks and by boundary conditions. The incident wave is attributed to vibration of whole mound and the boundary conditions are attributed to the contact conditions (free/supporting). Because the incident wave is different for conditions of mound, we need to focus not on the dominant frequency of a certain rock vibration but on the difference in dominant frequency of a certain rock to that of rocks near-by.

The rock mound in this experiment was composed of 10 rocks. Considering the whole mound as one body, its vibration is affected by its shape, composition and incident wave from outside. Vibration of individual rock on its surface is affected by vibration of incident wave transmitted from neighbouring rocks, the natural vibration of the individual rock and boundary conditions (contact with other rocks).

Form these results it can be concluded that vibration of each rock has own characteristics depending on its constraint and it can be observed in waves of frequency 1–100, thus there is a possibility to identify free rock(s) from supporting rocks.

Conclusion

Vibration of rocks in the mound of ten randomly-shaped, length 30–40 cm rocks was measured and difference in frequency characteristics were discussed between a free rock and a supporting rock. Findings are summarized as follows;

(1) The dominant frequency of the free rock was 40–50 Hz and that of the supporting rock was 50–60 Hz, thus the characteristics of vibration of rock was different between the free rock and the supporting rock.
(2) After changing the free rock to a supporting rock by means of leaning another rock on it, the dominant frequency changed its range from 40–50 to 50–60 Hz.
(3) There is a possibility to identify free rocks from supporting rocks by observing vibration characteristics of each rock.

As a preliminary study, the result of experiment indicates the possibility of identifying free rocks from supporting rocks by observing characteristics of vibration. The range of frequency where the difference was observed may change with conditions such as the number, size and shape of rocks. Further investigation is needed to develop a reliable method for identifying which can be used in the field.

References

Araiba K (2014) Technical support for rescue activity in a landslide caused by 2004 Niigataken-Chuets earthquake. In: Proceedings of 17th fire and disaster management seminar (in Japanese)
Araiba K, Doshida S (2017) Mechanisms for secondary slope failure in slope having failed. Adv Landslide Sci 2:1165–1172

Araiba K, Sakai N (2014) Laser scanner application in monitoring short-term slope deformation. In: Landslide science for a safer geoenvironment, vol 2, pp 5–11

Haegeli P et al (2011) Comparison of avalanche survival patterns in Canada and Switzerland. Can Med Assoc J 183(7):789–795

Ogata K et al (2003) Evaluation of the rockfall utilizing the characteristics of vibration. J JSCE 61(749):123–135 (in Japanese)

Uehan et al (2012) Evaluation method of stability of rock slope by means of remote monitoring of vibration. In: Proceedings of 67th JSCE annual conference, vol 242, pp 483–484

Urgent Issues and New Suggestions for Geo-disaster Prevention in Japan

UNESCO · ICL · ISC · WFEO · FAO · UNDRR · WMO · UNU
WLF
2020 Kyoto Japan

Motoyuki Suzuki, Kyoko Kagohara, Kazuyuki Sakaguchi, Hiroaki Matsugi, and Satoru Kataoka

Abstract

Following the devastating damage caused by the heavy rains in western Japan in 2018, the Japan Geotechnical Society compiled and published recommendations for future disasters. This report introduces an outline of the recommendation on slope disasters. In particular, the recommendation emphasises the need for measures to evacuate residents from dangerous places before a disaster occurs. For this purpose, it is noted that it is important to identify potentially dangerous places and their characteristics in the long term based on past disaster information and to inform the residents of the identified areas. As a pioneering effort to achieve this recommendation, this paper describes the results of research conducted in Yamaguchi, Hiroshima, Tokyo, and Kumamoto in Japan, to clarify the frequency of the occurrence of debris flows and its utilisation measures. The main result is that debris flows occur once every few hundred years in weathered granite areas, whereas volcanic ash areas experience even more frequent debris flows than granite areas.

Keywords

Debris flow • Occurrence frequency • Risk assessment • Disaster history

Introduction

In order to maintain a safe living environment in Japan, where heavy rain disasters often occur, it is very important to identify the places where past collapses and debris flows have occurred and the extent of the resulting damage. Slope failures and debris flows occur under the same topographic and geological conditions, hence in the long run they tend to occur repeatedly at the same or adjacent locations. However, it is difficult to understand this over the span of one person's short life. Our predecessors preserved disaster records in the forms of monuments, documentary records, oral traditions such as legends, and local festivals in order to transmit these records to later generations. Unfortunately, modern people have not correctly recognised this information or used it to prevent the next disaster. Therefore, it is important to pass on these results to future generations and to utilise highly reliable data based on many years of history when developing measures to prevent sediment-related disasters.

Under such social conditions, heavy rains in western Japan in early July 2018 caused devastating damage in various places. In the Japanese Geotechnical Society (JGS), the full extent of the damage was investigated at the direction of the president of the JGS, and a special committee was organised to make recommendations based on analysis of the results. The results were published as 'The Issues of Geotechnical Engineering for Heavy Rain Geo-Disasters Based on the Heavy Rain event of July 2018—Recommendations from Geotechnical Engineering' (JGS 2019).

M. Suzuki (✉)
Graduate School of Sciences and Technology for Innovation, Yamaguchi University, Ube, 7558611, Japan
e-mail: msuzuki@yamaguchi-u.ac.jp

K. Kagohara
Faculty of Education, Yamaguchi University, Yamaguchi, 7538513, Japan
e-mail: k-kago@yamaguchi-u.ac.jp

K. Sakaguchi
Public Works Research Center, Institute of Technology, Tsukuba, 3002624, Japan
e-mail: sakaguchi@pwrc.or.jp

H. Matsugi
Fukken Co., Ltd., Hiroshima, 7320052, Japan
e-mail: matsugi@fukken.co.jp

S. Kataoka
Eight-Japan Engineering Consultants Inc., Osaka, 532-0011, Japan
e-mail: kataoka-sa@ej-hds.co.jp

© Springer Nature Switzerland AG 2021
N. Casagli et al. (eds.), *Understanding and Reducing Landslide Disaster Risk*, ICL Contribution to Landslide Disaster Risk Reduction, https://doi.org/10.1007/978-3-030-60311-3_13

This report was then submitted to the Minister of Land, Infrastructure, Transport, and Tourism by the JGS. The first author was the leader of the committee's group on managing slope disaster preparedness. This group of 28 members compiled fourteen recommendations based on the results of several reviews. The titles of the recommendations are given in Table 1.

Of all these recommendations, recommendation 2.13 noted below relates to the effective use of disaster histories and is mainly based on the authors' research results (Sakaguchi et al. 2018; Matsugi et al. 2018).

The recommendation 2.13: Creation and use of hazard maps including information on previous geo-disasters is necessary to develop a hazard map that includes ground history information that cumulatively records past landslides and floods. It is necessary for residents to properly understand and share the importance and urgency of information (sediment disaster alert information, etc.) and to incorporate sociological ideas for the appropriate evacuation of residents.

In this study, we investigated the damage caused by debris flows that occurred as a result of torrential rainfalls in various places during the past 10 years and assessed the long-term risk of sediment-related disasters using a research approach consisting of radiocarbon (^{14}C) dating and literature surveys. Field surveys were also performed for debris flow disasters that occurred in Hofu and Yamaguchi Cities in Yamaguchi Prefecture, Aso City in Kumamoto Prefecture, Oshima Town in Tokyo, Asaminami and Asakita Wards in Hiroshima City and Nagiso Town in Nagano Prefecture. New and old debris flow deposits were identified and stratigraphically classified. To estimate the ages of these debris flow deposits, radiocarbon dating was performed on the organic material in the sediments of the debris flow. The dating results obtained from this study were collated with the disaster events described in historical documents remaining in each area. In addition, we created prototypes of the debris flow chronology in both Hofu and Hiroshima Cities, plotting debris flow generation ages and disaster events (Sakaguchi et al. 2018; Matsugi et al. 2018).

In this paper, the occurrence frequency of sediment-related disasters in Hofu City, Aso City, Oshima Town, and Hiroshima City are reported. Based on the survey results for these four areas, the relationship between topographical and geological conditions and the frequency of sediment-related disasters is discussed. Furthermore, based on the recommendations of JGS (2019), the authors suggest ways to use disaster histories for future disaster prevention.

Table 1 Recommendations from JGS for slope disaster preparedness

No.	Title of recommendation
2.1	Technology in geotechnical engineering to protect human lives and property
2.2	Improvement of accuracy of risk assessment, effective countermeasures, and maintenance and upgrade of existing countermeasures
2.3	Development of geo-disaster prevention system based on management of monitoring and real-time information
2.4	Establishment and disclosure of geotechnical and groundwater information database
2.5	Road/Railway: Efficient disaster preventive inspection by integrated management of design, construction, and disaster information
2.6	Road/Rail: R&D for preventive maintenance and appropriate regulation and deregulation of pre-traffic of road and operation control of train
2.7	Road/Railway: Introduction of concept of performance to resist disasters and realisation of advanced design and countermeasures
2.8	Erosion and mountain controls: Development of a wide-area and efficient geotechnical investigation method for weathered subsoil
2.9	Erosion and mountain controls: Improvement of performance of erosion control facilities throughout mountain streams and promotion of education for disaster prevention
2.10	Erosion and mountain controls: Promotion of comprehensive forest management from landscape and ecosystem to disaster countermeasures
2.11	Residential land: Disclosure of information on residential land safety and promotion of risk communication
2.12	Residential land: Establishment of a system for inheriting geo-information at transaction of real estate
2.13	Creation and use of hazard maps including information on previous geo-disasters
2.14	Strengthen public involvement of land and ground

Debris Flow Disaster in Hofu City, Yamaguchi Prefecture

Overview of Damage

On July 21, 2009, slope failures and large-scale debris flows occurred in various places, mainly in Hofu City, Yamaguchi Prefecture, resulting in 17 deaths in the prefecture. In the city, heavy rainfall on August 2, 1993 caused several slope collapses. It is presumed that sediment-related disasters have been occurred repeatedly in the city in the past. The mountains in this region are mainly composed of Late Cretaceous granite (Hiroshima granites). The granite has been significantly weathered and transformed into sandy and gravelly soil (Masado). There are also exposed core stones from the hillside to the summit. In addition, sheeting joints are densely distributed in the exposed rocks on the collapsed slope. This paper describes the results of on-site surveys of six affected areas: Katsusaka, Matsugatani, Gyokusen Tameike and Mitani River on the right bank of the Saba River flowing south of Hofu City as well as Manao and Ishihara on the left bank of the Saba River. In the field survey, the authors performed stratigraphy of debris flow deposits and collected samples for ^{14}C dating at 41 locations where organic material (carbonized plant material) were found in the sedimentary strata.

Estimation of Debris Flow Frequency Through Dating and Literature Surveys

The samples collected at each site were dated using an acceleration mass spectrometer (Compact AMS, NEC, 1.5SDH). The ^{14}C ages and calendar ages were calculated by correcting for the isotope fractionation effect (carbon isotope ratio $\delta^{13}C$) in the measured results. OxCal 14.1 (calibration curve data: IntCal09) was used for calendar year calibration. Figure 1 is based on the authors' data (a part of the data refers to Sakaguchi et al. 2018); its horizontal axis indicates each district and the vertical axis indicates the historical date (CE). The results of Sugihara et al. (2010) are also plotted in Fig. 1. The measurement results including the error was around 1340 for all districts. The data obtained showed that more than one-third of the data was from the 1300s. This suggests that a large-scale debris flow disaster occurred in each area at this time. In Katsusaka, debris flows have occurred frequently since tenth century. The longest debris flow interval with sedimentary layers coverage was about 300 years and the shortest was about 70 years. Excluding the debris flow in 2009, it is presumed that at least six debris flows have occurred in the Manao area and three or more have occurred in the Katsusaka, Matsugatani, and Ishihara areas simultaneously or individually. On the other hand, the dating data was examined for consistency with disaster records from around 500 CE, which are compiled in 'Yamaguchi-ken Saii-shi' (Shimonoseki Local Meteorological Observatory 1953). Figure 1 shows the dates of the events in which earthquakes and heavy rains are mentioned. The record mentioned heavy rains in 1350. Therefore, this historical data seems to support the occurrence of a debris flow in the 1300s, as mentioned above. In addition, it was confirmed that the dating results and reports in old documents generally corresponded. According to the same report, the intervals between heavy rainfalls were very short, 60 years between 1653 and 1713. The report also noted that earthquakes had occurred frequently since 1677. In particular, the Hoei earthquake, one of the largest earthquakes in Japan, was recorded in 1707, and the Aio area near Hofu City was reported to be severely damaged. The interval between debris flows immediately after the Hoei earthquake was particularly short; this is probably because weathered rocks loosened, and rock masses collapsed due to the effect of the earthquake, and the subsequent heavy rains made it prone for debris flow appearances.

In addition, the debris flow sediments could be divided into gravel-dominated strata and sand-dominated strata based on the geological observations in the field survey. Debris flows mainly composed of gravel (debris flow Nos. 2, 4 to 6, and 8 in Fig. 1) occurred in several mountain streams between 1300 and 1400. It was also found that debris flows mainly composed of sand (debris flow Nos. 1, 3, and 7) occurred before and after the debris flows mainly composed of gravel. This suggests that the debris flow deposits in the region of Hofu are mainly composed of sand, but there is a period when gravel is produced, and that gravel also would be source material for a debris flow.

Although the sediment mainly composed of gravel generally remained in the midstream of the mountain stream, the distribution area was large and the layer was thick, indicating that gravel was produced on a large scale. Debris flows in mountain streams located in areas where granite is distributed are expected to occur with a frequency of once every few hundred years. However, the recurrence interval in the thirteenth and fifteenth centuries was approximately 100 years, while in the eighteenth century, there was a period when the frequency increased to slightly less than 100 years.

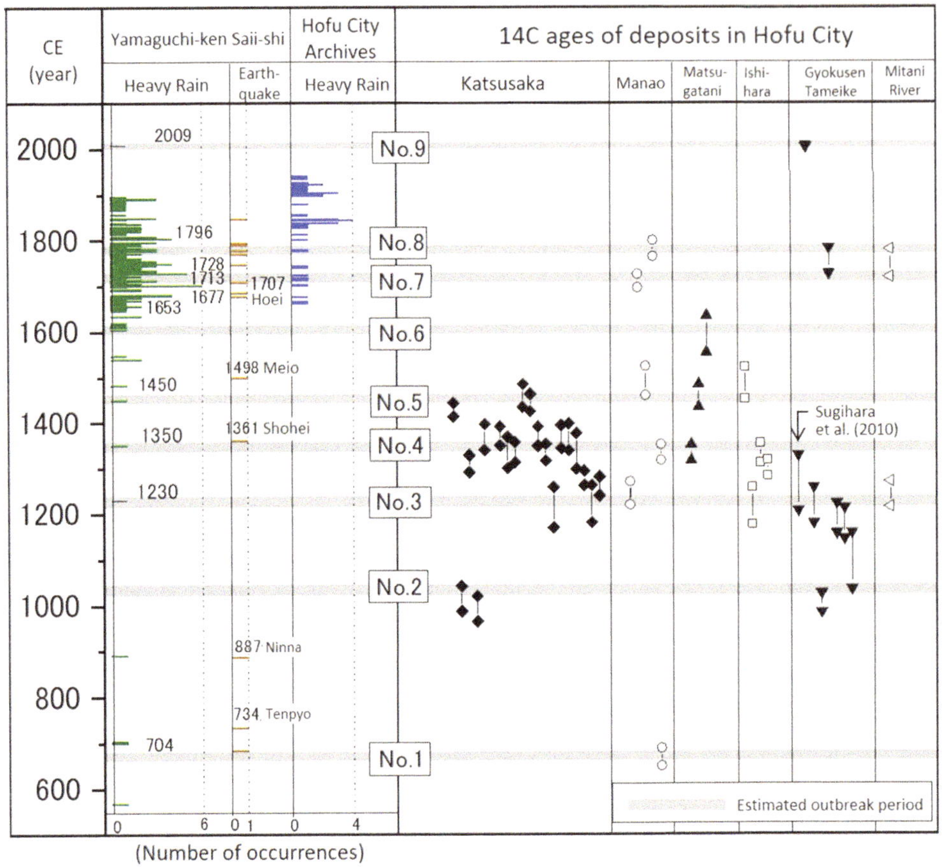

Fig. 1 Chronological table of debris flow occurrence in Hofu City (a part of the data: Sakaguchi et al. 2018)

Topographic and Geological Conditions and Debris Flow Frequency in Other Areas

Asa-Minami and Asa-Kita Wards, Hiroshima City

Heavy rainfalls from August 19–20, 2014, caused a large number of debris flows, mainly in Yagi and Midorii, Asa-minami Ward and Kabe, Asa-Kita Ward, Hiroshima City. A field survey was conducted on the mountain stream where the debris flows occurred. A majority of the mountains in Hiroshima City, as well as those in Hofu City, consist of Late Cretaceous granites (Hiroshima granites). The main lithology of these Hiroshima granites is weathered coarse-grained granite. The Jurassic accretionary complex is distributed along the Ota River from Kabe to Kake. Gentle slopes and alluvial fans are distributed between the mountains and lowlands. Based on the dating of past debris flow deposits shown in Fig. 2, it was estimated that more than five debris flows occurred in the Yagi and more than three occurred in Kabe-Higashi since the beginning of the CE before the disaster in 2014 (Matsugi et al. 2018). Umaki was a place where debris flows occurred due to heavy rain in 2018. The frequency of debris flows in one mountain stream is expected to be approximately once every several hundred years as in Hofu City. Past disaster records around Hiroshima City are described in 'Geihan Tsu-shi' (Rai et al. 1825) and 'Hishiroma Ken-shi' (Hiroshima Prefectural Government 1925). The number of dating materials are limited, and not all events are available. However, event No. 5 shown in Fig. 2 corresponds to a description of the occurrence of a debris flow in 1532 in 'Intoku Taiheiki' (1911). Event No. 6 was also found to correspond to a description of the 1850 landslide in 'Gion Cho-shi' (Gion Town history Compilation Committee 1970). These ages are almost consistent with the dates of the events in each document.

Oshima Town, Tokyo

Heavy rains caused by Typhoon No. 26 on October 16, 2013 caused many shallow slope collapses and mudflows at the western foot of Mt. Mihara, Oshima Town (Izu-Oshima). Mt. Mihara is a Quaternary volcano, and medium to large eruptions have occurred 24 times in the last 1500 years. Tephra was deposited during each eruption. During the period when the eruptive activity paused, loess was accumulated. The current surface is covered by unconsolidated volcanic ash layers such as scoria. As it shown in Fig. 3,

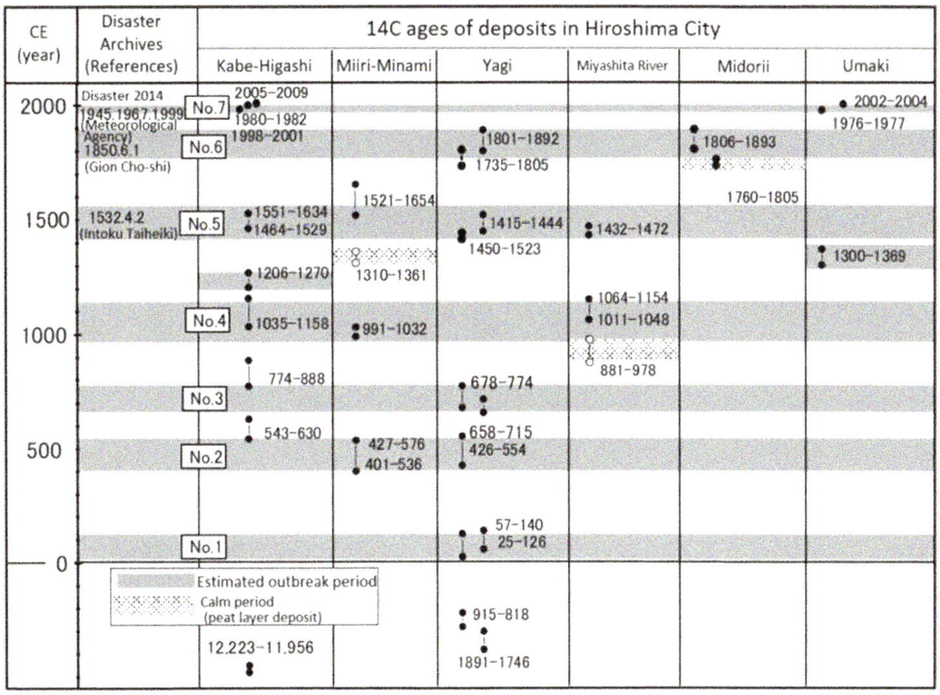

Fig. 2 Chronological table of debris flow occurrence in Hiroshima City (most of the data: Matsugi et al. 2018)

Fig. 3 Condition of sedimentation and geological profile in the downstream area of the survey site (Okanazawa) in Izu-Oshima (JSCE 2015)

mudflow sediments were identified as heterogeneous sediments in which scoria of different colours, lava gravel, and muddy gravel were mixed. In the valley of Okanazawa, where a part of the sediment was stripped off during Typhoon No. 26, it was confirmed that the fourteenth century Motomachi lava flow and tephra from the subsequent eruption activity were stacked.

At that point, it is probable that at least two sediment flows occurred between the time of the Motomachi lava flow to the end of 2013. The intervals were approximately 100 to 150 years. Koyama and Suzuki (2014) stated that the last seven sediment-related disasters (including eruptions) occurred around Motomachi town. According to historical sources, the remarkable sediment-related disaster on Izu-Oshima was caused by heavy rains due to a typhoon such as the Kanogawa Typhoon in 1958. Previous studies have shown that disasters occur approximately every 30–80 years around the Motomachi area, and there is significant evidence that more frequent sediment flows occur in volcanic areas.

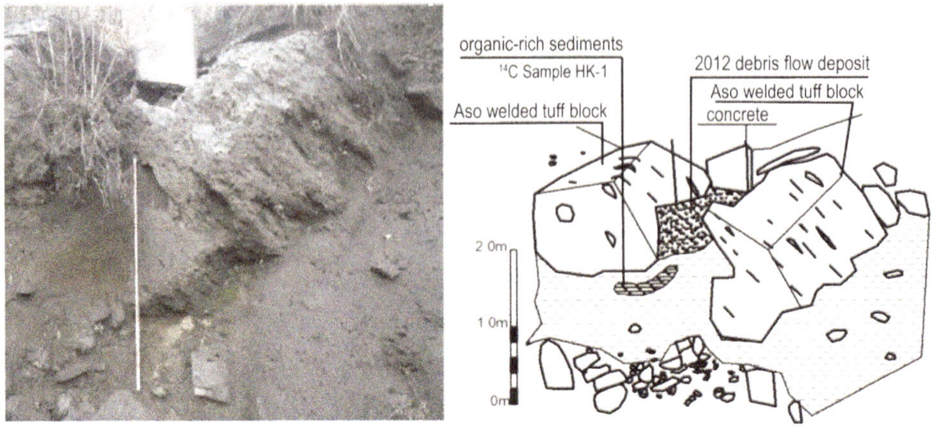

Fig. 4 Left: Debris flow deposits, Right: Overview of the site (JSCE 2015)

Aso City, Kumamoto Prefecture

The heavy rainfall on July 12, 2012 caused a tongue-like flaky slope failure everywhere on the outer rim of Mt. Aso. Mt. Aso is a Quaternary volcano, and the geology of the survey area is covered with tephra fall above the Aso welded tuff; a black volcanic ash layer with humus (Kuroboku) is present near the surface. Three carbon samples collected from debris flow sediments at Ichinomiya-machi Sakanashi and Hakoishi Pass (Fig. 4) had more recent ages since 1970. This suggests that the frequency of debris flows caused by rainfall is relatively high as compared with those caused by granite.

Conclusions

The results obtained from this study are summarised as follows.

(1) In Hofu City and two districts of Hiroshima City where weathered granite is widely distributed, the frequency of debris flows in one mountain stream is estimated to be about once every several hundred years. The reason why the occurrence interval is longer in granite areas seems to be due to the debris production and deposition process in the mountain streams. It is generally understood that the time required for granite up to 1 cm in depth to be transformed to Masado by weathering is on the order of thousands of years. Considering the production volume of Masado, the debris flow interval should be on the order of 1000 years. However, debris flows occurred at shorter intervals of 100–200 years, and the degree of weathering of the granite gravel contained in the debris flow deposits was low. The presence of sheeting jointed rock found on slopes may promote the progress of debris flows.

(2) It is estimated that shallow landslides, mudflows, and debris flows occur approximately once every several decades in one district of Oshima Town and two districts of Aso City, which are Quaternary volcanic areas. The reason for the high frequency of occurrence in Quaternary volcanic regions is thought to be the structure of the volcanic ash layer, which has high permeability on the ground surface, is loosely deposited along the slope and is vulnerable to precipitation.

(3) Finally, the authors would like to state an effective use for the findings of this study. In the studied areas, the risk of sediment-related disasters is very high, and debris flows have frequently occurred in the past. Nevertheless, the victims said, '*I have never heard of such a disaster here before*', and '*I believed here was a safe place*'. The lessons learned from past disasters have not been passed on and used for disaster prevention. For Japan, which keeps extensive historical records, this is regrettably wasteful. In the future, information on when, where, what, and how much has previously happened should be included in hazard maps by finding traces of disasters on the ground and comparing them with disaster facts in historical records stored in the area. If hazard maps are confirmed by the facts, it will raise the awareness of residents.

Acknowledgements The present research was supported by JSPS KAKENHI (Motoyuki Suzuki, 26560185, 15H04038, 17K18954, 19H00785), and the research grants of Japan Society of Civil Engineers

2014 priority research project (Nobuo Anyoji), The River Foundation (Motoyuki Suzuki) and The Chugoku Kensetsu Kosaikai (Motoyuki Suzuki). The authors thank all concerned.

References

Gion Town history Compilation Committee (1970) Gion Cho-shi (in Japanese)

Hiroshima Prefectural Government (1925) Hirosima Ken-shi. Appendix Geibi Chronicles. Imperial Local Administration Association, 86 pp (in Japanese)

Japan Society of Civil Engineers, Geotechnical Committee (2015) 2014 Priority Research Project (Research Grant). Research report on comprehensive disaster prevention and mitigation measures considering occurrence mechanism of sediment disaster in vulnerable volcanic nations Japan and legal systems, 200 pp (in Japanese)

Japanese Geotechnical Society (2019) The issues of geotechnical engineering for heavy rain geo-disasters based on heavy rainfall in July 2018—Recommendations from geotechnical engineering. https://www.jiban.or.jp/file/saigai/houkoku/H30_TeigenSaishu.pdf (in Japanese)

Kagawa M (1911) Intoku Taiheiki. In: Inuyama S (ed) Collection in one volume, vol 1–18, 232 pp (in Japanese)

Koyama M, Suzuki Y (2014) Volcanic mudflow disaster in October 2013 from the eruption history of Izu Oshima. Geography 59–5:34–41 (in Japanese)

Matsugi H, Suzuki M, Kagohara K, Sakaguchi K, Ogasahara H, Kataoka S (2018) Sediment conditions and frequency of debris flows in and around Asaminami-ku and Asakita-ku, Hiroshima City. Japan Geotech J 13(4):403–421 (in Japanese)

Rai K, Kato S, Rai S, Kurokawa K, Tsumura S, Yoshida K, Masaoka G (eds) (1825) Geihan Tsu-shi (in Japanese)

Sakaguchi K, Suzuki M, Kagohara K, Matsugi H, Kanaori Y, Sakurai M, Kataoka S (2018) Occurrence frequency and sedimentation properties of debris flows in Ishihara district, Hofu City, Yamaguchi Prefecture. Japan Geotech J 13(3):237–247 (in Japanese)

Shimonoseki Local Meteorological Observatory (1953) Yamaguchi-ken Saii-shi: Prefectural record of convulsion of nature (in Japanese)

Sugihara S, Kaibori M, Kubota T, Kashiwabara Y, Furukawa K (2010) Consideration of regularity of sediment deposition in Hofu City in Yamaguchi Prefecture by using radiocarbon dating. J Jpn Soc Erosion Control Eng 63(3):27–33 (in Japanese)

WLF 2020 Kyoto Japan

Development of Resident Participation-Type Slope Measurement/Monitoring System in Mountain Region

Tomofumi Koyama, Seiji Kondo, Taizo Kobayashi, Shinichi Akutagawa, Takeshi Sato, Katsuyuki Nakata, and Kazuyuki Shimojima

Abstract

In this paper, the authors proposed the "resident participation-type slope measurement and monitoring system" to detect the premonitory symptoms of sediment disasters in the mountainous area (Takasu Town, Fukui City, Japan). The key concepts of resident participation-type slope measurement/monitoring are (1) visualization, (2) daily routine, and (3) voluntary/proactive involvement. The residents will not only monitor the danger slopes and notice unusual events, but also try to establish a system to grasp the slope deformation/displacement quantitatively. For this purpose, some different types of sensors based on the on-site visualization (OSV) concept were introduced to the dangerous slopes and retaining walls selected by the residents and geotechnical experts. The OSV sensors are low cost and visually excellent measuring device with simple measurement principle. Further improvements of measurement and recording methodology are required to increase the participation of residents to the slope measurement and monitoring.

Keywords

Slope measurement/monitoring • Resident participation • Disaster prevention/mitigation activities • Marginal settlement • Mountainous area

T. Koyama (✉) · S. Kondo
Faculty of Societal Safety Sciences, Kansai University, 7-1 Hakubai-cho, Takatsuki, 569109, Osaka, Japan
e-mail: t-koyama@kansai-u.ac.jp

S. Kondo
e-mail: kondo.s@kansai-u.ac.jp

T. Kobayashi
College of Science and Engineering, Ritsumeikan University, Kusatsu, 525-8577, Shiga, Japan

S. Akutagawa
Graduate School of Engineering, Kobe University, Kobe, 657-8501, Japan
e-mail: cadax@kobe-u.ac.jp

T. Sato
Asano Taiseikiso Engineering, Tokyo, 110-0014, Japan
e-mail: satoh-tks@atk-eng.jp

K. Nakata
Office Himorogi, Kyoto, 603-8347, Japan
e-mail: himorogi@s3.dion.ne.jp

K. Shimojima
Technical Think, Amagasaki, 661-0953, Hyogo, Japan
e-mail: simojima@technicalthink.jp

Introduction

Many villages in mountainous areas in Japan are designated as sediment-related disaster warning areas, and there is a high risk of sediment-related disasters such as debris flows and slope failures. In addition, many villages are becoming vulnerable due to the aging of the population, so-called marginal village.

In order to improve the local disaster prevention ability in mountainous areas, it is necessary to become regional disaster prevention activities as everyday affairs mainly considering "mutual assistance" and "self-help" of each individual (Yamori 2017). For this purpose, experts will discuss with the local residents how to recognize the risks of sediment-related disasters around them correctly and not to rely too much on "public assistance" such as information given from the government. It is necessary for the residents to consider their own safety by themselves.

In this study, in Takasu Town, Fukui City, Japan in order to promote local disaster prevention activities, observations were conducted on slopes that are considered to have a high risk of collapse based on the results of interviews with residents and on-site surveys by geotechnical experts. By installing the measurement/monitoring equipment, the residents will not only monitor the danger points and notice unusual events, but also try to establish a system to grasp the slope deformation/displacement quantitatively called the

© Springer Nature Switzerland AG 2021
N. Casagli et al. (eds.), *Understanding and Reducing Landslide Disaster Risk*, ICL Contribution to Landslide Disaster Risk Reduction, https://doi.org/10.1007/978-3-030-60311-3_14

"resident participation-type slope measurement/monitoring system".

Outline of Takasu Town, Fukui City, Japan

Takasu Town in Fukui City, Fukui Prefecture was selected as research area. Takasu Town is a rural village located on the hillside of Mt. Takasu (elevation of 438 m), about 20 km northwest of the center of Fukui City, with 38 households and 45 (out of 68) people over 65 years old (the percentage of population over 65 years is 66%) according to the demographics of Fukui City (as of March 2020) (Fukui City 2020). Generally, villages over 65 years old with more than 50% of the population are called "marginal villages", and those where the age of 65 and over account for more than 70% of the population are called "critical communities".

There are only two main roads leading to Takasu Town from center of Fukui City, narrow city roads with a width of about 4 m. The road in Takasu Town, which is located at a relatively high altitude of about 200 m above sea level, is a forest road surrounded by slopes. In addition, as shown in Fig. 1, the entire town is in a large landslide terrain, and sediment-related disaster special warning/warning area for steep slopes exist around the village, and some are designated as debris flow alert areas. In fact, collapse of living roads leading to villages, forest roads in the districts, and fields have frequently occurred, and the frequency of torrential rain has increased in recent years, and therefore, residents' awareness toward the risk of landslides is increasing.

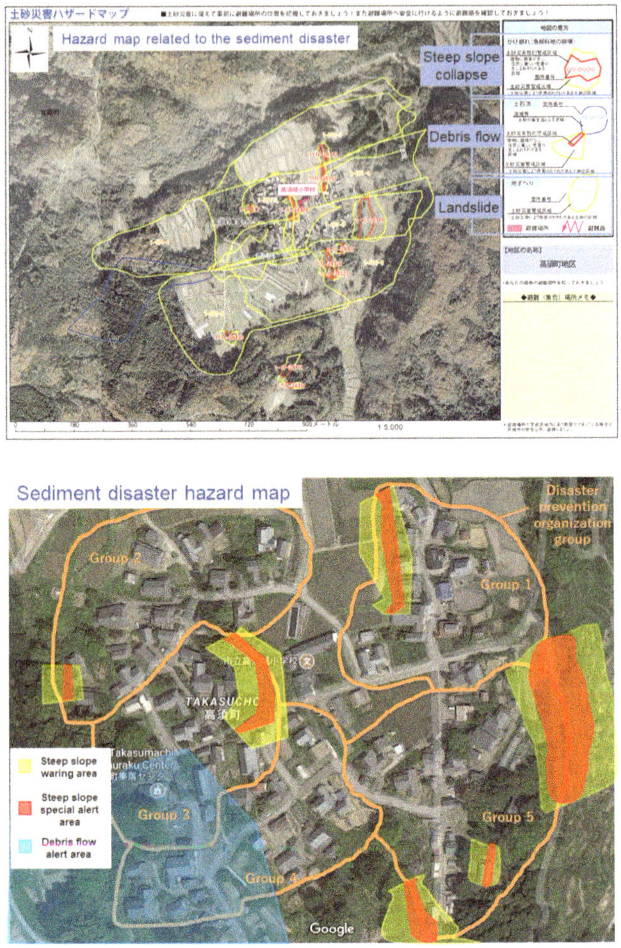

Fig. 1 Hazard map related to the sediment-related disaster of Takasu Town, **a** officially announced by Fukui City and **b** enlarged picture of sediment-related disaster special warning/warning area

Concept of Slope Measurement/Monitoring with Residents' Participation

The following three points should be considered when developing a slope measurement/monitoring system with the participation of residents. In other words, (1) visualization, (2) daily routine, and (3) voluntary/proactive involvement. Firstly, "visualization" refers to visualization of the danger level, and it is easy for residents without specialized knowledge to understand the danger of slopes using simple observation equipment with simple principles of measurement and monitoring. Secondly, "daily routine" means that the residents themselves measure and monitor slopes on a daily basis, rather than conducting them regularly as in disaster prevention drills. To this end, it is necessary to propose simple measurement and monitoring methods so that the residents themselves can easily measure and monitor the slope from their daily lives, and to devise ways to incorporate the act of slope measurement and monitoring as part of their daily lives. Lastly, "voluntary/proactive involvement" means that more and more residents are involved in slope measurement and monitoring daily, raising awareness of "self-help" and disaster prevention. If you notice "unusual events" on a slope different from your daily life, you need to be aware that residents who received unilateral information on disaster prevention may also be able to send information on the contrary.

Monitoring of Dangerous Slopes Using OSV Sensors

Types of Measuring Equipment and its Principle

Since the main purpose was to make the residents themselves involved in measurement and monitoring on a daily basis, and to be aware of "unusual events" that are different from the everyday, the measurement and monitoring system should not be expensive, elaborate, and did not require specialized knowledge. Adopted OSV (On-Site Visualization) sensors (Akutagawa 2017; OSV conthosium 2020) in this study, which is a low cost, simple and visually excellent measuring device.

POCKET (A Pocket-Size Light-Emitting Inclination Sensor)

The "POCKET" is a device that combines a fixed inclinometer and an optical device, and the color of the head changes according to the set criteria (green, yellow and red). In this measurement, the sensor was placed at the position where the inclination of the retaining wall and/or slope was measured, and the slope stability was confirmed by observing the color emitted from the head of the device.

Normally, dry batteries are used as the power supply. However, in this measurement, in order to eliminate the trouble of replacing the power supply and enable long-term measurement, the system was changed to a self-power generation type using a solar battery. POCKET has a built-in data recording mechanism inside the main unit (1 hour pitch, data capacity about 1.5 months). If measurement results are collected, inclination data is stored as digital data in addition to monitoring by light emission.

SOP

"SOP (Single Observation Point)" is a method of visually grasping the movement of the measurement target using a mirror. In the measurement method, a light source is installed in a stable place where no deformation occurs, and the light is visually checked by the observer from the same stable place. The mirror at the measurement point adjusts the direction so that the observer can see the light reflected on the mirror at the initial stage. When the measurement target is deformed/displaced and the mirror rotates or moves, the image of the light source in the mirror moves from the initial state. Therefore, the observer can visually grasp the deformation of the measurement target. Furthermore, when the deformation is large, the reflected light may protrude from the mirror surface and the image may not be visible.

Theoretically, the rotation angle of the mirror can be measured from the distance between the observer and the mirror, the distance between the mirror and the light source. Therefore, in this measurement, since the observers are residents who do not have specialized knowledge, we decided to monitor from the viewpoint of whether the light moved or not.

A See-Through Pole

"A see-through pole" is a method in which a reference pole with a viewing window and several observation poles for measuring displacement are arranged in a straight line, and the movement (gap) of the observation pole is checked from the reference pole. In other words, the displacement of the slope and/or shoulder of roads causes the observation pole to fall and/or shift, so that the line of sight from the reference pole is not straight, and the residents visually notice the deformation of the slope or shoulder of roads easily. If the grasp of quantitative behavior of the observation pole is required, the displacement can be quantitatively confirmed by measuring the mounting position at the beginning of the installation and performing the survey again when the deformation occurs.

Installation Location of OSV Measurement Equipment in Takasu Town

Based on the results of interviews with residents and reconnaissance by geotechnical experts, totally six dangerous slopes and/or retaining walls were selected (two slopes along the access road (city road) to Takasu Town, one retaining wall along the agricultural road, one steep slope/retaining wall behind the elementary school and on three steep slopes behind the houses) and different types of OSV sensors were introduced (see Fig. 2). The positions of the six measurement points selected in this study are shown from ① to ⑥ in Fig. 2.

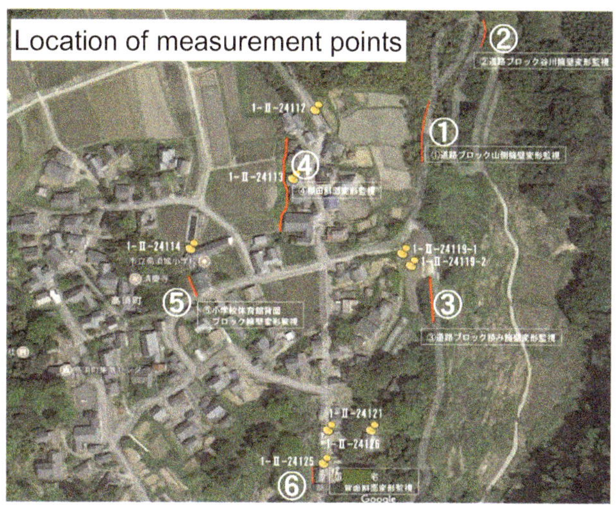

Fig. 2 The location of the six measurement points selected in this study

Fig. 3 "See-through poles" using observation poles and reference pole with a viewing window

Deformation Monitoring of Valley Side Slope Along City Road (①)

This measurement point is along the city road used as community road. On paved roads with a width of 4 m or less, there are concerns about deformation of retaining walls on the mountain side and slope failures on the valley side. Therefore, a measuring device using poles was installed at the shoulder of the city road (see Fig. 3). The shoulder of the road is located near the top of the valley side slope, and this part may collapse due to deformation of the slope and cut off the road.

Since the road alignment at the measurement position was close to a straight line and the street of the pole could be seen, "a see-through pole" was adopted. The observation pole was attached to a guardrail column installed on the shoulder of the road. However, the reference pole, which should not be affected by the slope displacement, was installed off the slope and away from the guardrail support. Slope deformation measurement is performed through a viewing window installed on the reference pole to measure whether the street on the observation poles hold a straight line.

Deformation Monitoring of Valley Side Retaining Wall Along City Road (②)

At the curved part of the city road (extension of the city road in ①), the top of the retaining wall on the valley side has already been deformed so as to fall down to the valley side, and the paved road is partially cracked. (See Fig. 4a–e).

Since the installation location is a curved section, it is impossible to see through the street like an observation pole, so here SOP was adoped to perform deformation measurement. The location of observation point was set on the shoulder of a relatively wide curve so that the reflecting mirror could be viewed from there.

A peephole with a diameter of about 5 mm is installed at the center of the measurement board, and the ovservers check the red and yellow indications of the measuring plate reflected on the mirror through the peephole. At the time of installation, it is adjusted so that the display of the measurement board is reflected on the mirror when looking through the peephole. A displacement confirmation line is also drawn radially around the peephole on the back of the measurement board, and when the retaining wall is displaced, the displacment can be calculated the distance the peephole and eye position to see the display plate reflected on the mirror. The displacement confirmation line drawn on the back of the measurement plate to measure the direction and the amount of the displacement at that time.

Slope Deformation Measurement Near Rice Terraces (④)

The rice terraces are on a hill with a height of 4 to 5 m, and a steep slope continues from the hill. The neighboring houses are built along the steep slope, and the residents living there are concerned about the collapse of slope along the rice terrace. According to the interview surveys, the slope along the rice terraces is recognized as the one of the most dangerous places in the town (see Fig. 1).

Therefore, the SOP was used to measure the deformation of the terrace slope. A ridgeway close to the house under the rice terraces was selected, and three SOP mirrors (reflecting mirrors) were installed. The light source (small LED light) and observation points used for the SOP are set next to the pool at the elementary school and beside the houses under the terraced terraces. From the installed measurement points, the angle of the reflector was adjusted so that the remaining one could be checked from the side of the house under the terraced rice field. Figures 5a-d show the initial setting state of SOP and the loaction of reflecting mirrors and observation points. For the measurement, an LED light is installed at the light source position to check whether the light from the reflector is visible. The measurement accuracy is such that the distance from the observation point to the reflector is about 70 m, so that the angle change of the reflector is about 0.05° and the light source is not reflected on the mirror. The measurements by reseidnets were carried out by recording whether the light reflected by the reflector was visible or not.

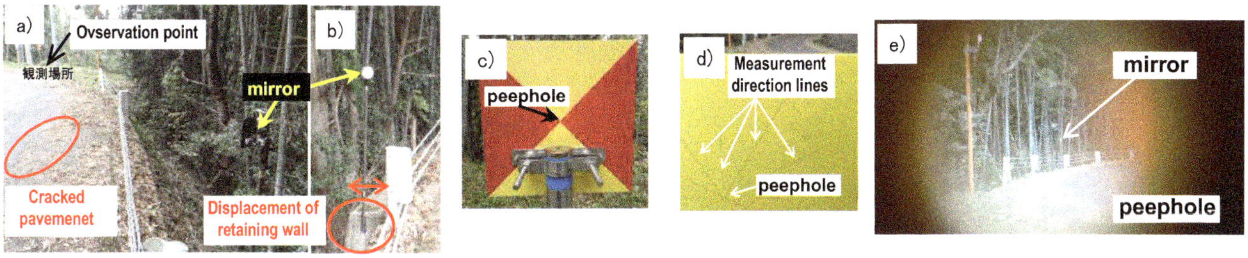

Fig. 4 Slope measurement/monitoring using SOP, **a** retaing wall, **b** mirror, **c** measurement board (front face), **d** measurement board (back face), **e** view from peephole installed at the center of the measurement board

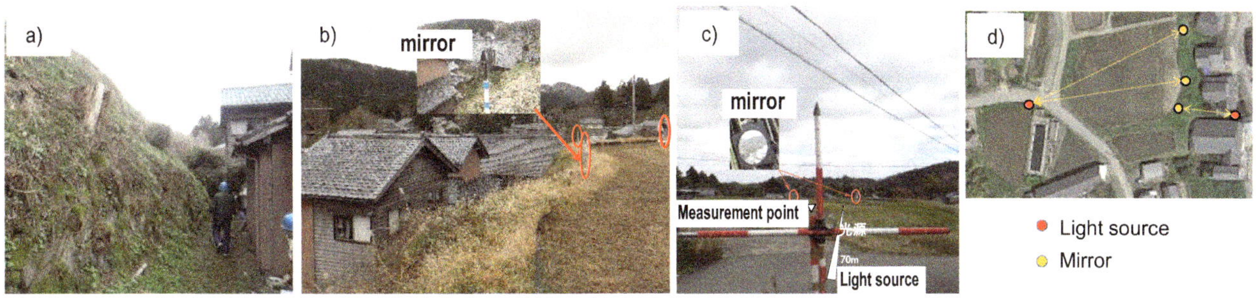

Fig. 5 Slope measurement/monitoring using SOP, **a** rice terraces behind the houses, **b** reflecting mirrors installed at rice terrace, **c** light source and observation points, **d** location of light source and mirrors

Deformation Monitoring of Block Retaining Walls Along Farm Roads, Block Retaining Walls Behind Elementary School Gymnasiums, and Slopes Behind Houses (③, ⑤, ⑥)

The block-retaining wall along the farm road was collapsed once in the past. After the collapse, the slope above the top of the retaining wall has been reenforced by ground improvement. However, it is necessary to monitor the sts-bility of the slope and check the soundness of the retaining wall, and therefore selected as a measurement point (see Fig. 6a).

The only elementary school in the town is currently closed, but its gymnasium is used as a meeting place for residents. In addition, it may be used as a temporary evacuation place at the time of disaster. At present, there is a block retaining wall behind the gymnasium, and a natural slope is approaching above the retaining wall. Groundwater is constantly leaking from the retaining wall. Since the distance between the gymnasium and the retaining wall is about 1.5 m, the gymnasium may be damaged by flooding during torrential rainfall and the collapse of the retaining wall. Therefore, we decided to measure the deformation of the block-retaining wall behind the gymnasium (see Fig. 6b).

In the town, there is a slope on the back of the house that exceeds the two-story roof, and there was a place where the slope collapsed due to past torrential rainfall and sediment flowed into the house. Residents living there were concerned about the slope failure, so this slope was selected as a measurement point, and the deformation of the slope could

Fig. 6 Slope measurement/monitoring using POCKET, **a** block retaining walls along farm roads, **b** block retaining walls behind elementary school gymnasiums, and **c** slopes behind houses

be measured mainly for the residents living there (see Fig. 6c).

For these measurement points, slope inclination measurement using POCKET was performed to measure the deformation of the retaining wall and slope, and the. power supply of the measuring instruments was secured using solar panels at all three measurement points. In POCKET, the color of the light at the head of the measuring instrument changes depending on the inclination, so when displacement occurs on each retaining wall and/or slope, the color can be used to visually judge the situation. In this study, the lighting color of POCKET is set to be green when the tilt angle does not change, and to be yellow and red when the inclination angle changes by 0.2° and 0.4°, respectively. The thresholds for the set color and angle were set with reference to past slope failure experiment results (Toyosawa et al. 2007). When the residents confirmed the change in emission color, the slope was measured and surveyed separately by geotechnical experts to check the stability of the slope. The block retaining wall on the back of the gymnasium has a long depth, so three POCKETs were installed.

Development of Observation System and Establishment of Recording Method of Observation Results

Observation System at the Beginning of Installation of Measurement Equipment

The installation of all OSV measurement equipment was completed in mid-November 2017, and on November 26, a briefing and tour of the use of measurement equipment for members of the town's voluntary disaster prevention organization (the chairman of neighborhood association, vigilante, and group leader). Regarding the measurement/monitoring system (observation members and measurement frequency, etc.), the members of the voluntary disaster prevention organization decided without any instructions or requests from experts. At the beginning of the decision, it was expected that the members of the voluntary disaster prevention organization and the person in charge of the disaster prevention group would be able to manage all of the measurement equipment without any difficulty from the installation of the measurement equipment if the measurement was performed once a month. However, in practice, measurement and monitoring were hardly performed, and the measurement and monitoring system was hardly functioning. One of the reasons for this is that in the method of "measuring/monitoring and recording when the designated person is decided", extra tasks are imposed on the appointed person, although it is a part of disaster prevention activities. This means that even if residents are working on it at first, they will gradually feel burdened, which will lead to reduced motivation.

Door-To-Door Survey for Residents

After installation of OSV measurement equipment and briefing for members of the voluntary disaster prevention organization, on June 24, 2018, along with disaster prevention drills in Fukui City, explanations and tours for slope observation and monitoring were held for residents. Informed the residents of the intention of the "resident participation-type of slope measurement and monitoring". After that, we visited Takasu Town regularly to check the status of measurement and monitoring, and also conducted interviews with residents to extract the current problems of measurement and monitoring.

In particular, in the door-to-door survey conducted on October 12–13, 2018, some positive and negative responses from nine people were obtained as follows;

- "For simple observation equipment, POCKET and line of sight are easy to observe and understand visually."
- "SOP is easy to understand the principle of measurement, but it is difficult to handle."
- "It still don't know if slope monitoring/measurement really helps to reduce sediment-related disaster."

Some women in their 80s found a sense of mission and joy in observation activities. She observed the steep slope behind her home at the SOP every time it rained, and she was "prideful of her grandchildren." This is a good example of a very positive willingness to the newly assigned role. In addition, a mobile co-op sales car visits the village twice a week in front of elementary school. Some residents are observing at the timing. In this way, by grasping the behavior pattern and rhythm of life of the residents in the town in more detail, in addition to the above, daily observations such as watching observation equipment when walking near the equipment in addition to walking dogs and field work. It turns out that there are a several inhabitants who can reasonably observe slope monitoring in their daily lives. In addition, as long as these changes were observed, it was found that the resident-based slope observation system could take root and have the potential to "make habits (daily routine)" in the future.

Management of Observation Records Using IC Card Reader

Regarding the method of recording observation results, some people commented that the handwritten measurement record in ledger at the beginning of the installation. Significant simplification was required so that measurement and monitoring could be performed frequently. Therefore, management of observation records using an IC cards and a card reader was examined (see Fig. 7a).

In the observation, for all OSV measuring equipment, the observer hold IC cards printed "normal" or "abnormal" as observation results over the card readers that placed near the OSV measuring equipment.

As for the POCKET installed behind the elementary school, if at least one of the three POCKETs displayed yellow or red, it was determined that there was an abnormality. A pair of IC cards will be distributed to the members of the "Takasu voluntary slope observation team" as described in the next section. At the same time, an IC card was also placed near the measuring equipment so that residents other than the members of the observation team could observe. A 12-digit ID is engraved on the IC card, and by associating the card ID with the distribution destination, observation records ("who" measured "when") can be stored and managed.

The recording of observation results using an IC card and a card reader started on April 27, 2019. In the future, it is necessary to collect and analyze observation records and examine how to link slope measurement and monitoring to "daily routine" and "voluntary/proactive involvement".

Formation of "Takasu Voluntary Slope Observation Team"

As mentioned in the previous section, there are some residents in Takasu Town who can reasonably observe slope and be involved in monitoring slope in their daily life. They will play a central role in slope measurement and monitoring. Specifically, a total of four women, including the 80-year-old woman mentioned above, a woman of the local welfare officer who is familiar with the situation in Takasu Town, and women who has an interest in measurement and monitoring due to the slope behind their home was formed as a member of the "Takasu voluntary slope observation team". Ask the four women mentioned above to always carry their personal IC cards (a pair of "normal" and "abnormal") and walk around the measuring instruments installed in the town. When they passed, they observed the slope and recorded the observation results using IC card readers.

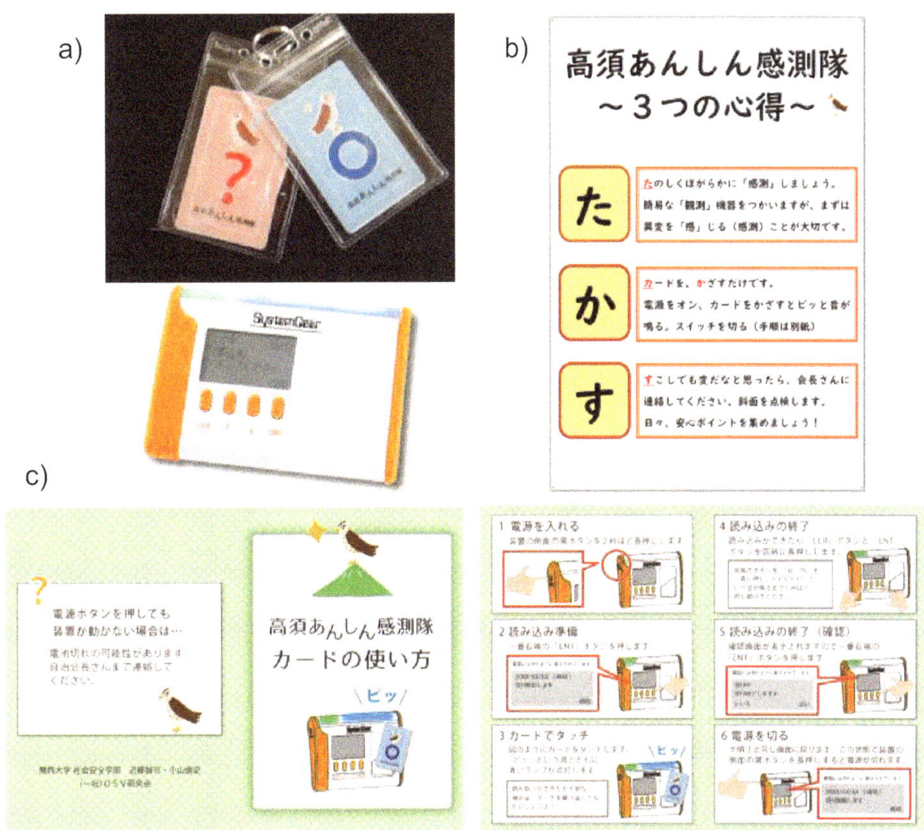

Fig. 7 **a** IC cards printed "normal" and "abnormal" and IC card reader, **b** "Takasu voluntary slope observation team -three hints" and **c** a card reader handling manual

Fig. 8 The formation and appointment ceremony of the "Takasu voluntary slope observation team" was held at the Takasu town village center

On April 27, 2019, the formation and appointment ceremony of the "Takasu voluntary slope observation team" was held at the Takasu Town Village Center (see Fig. 8). Along with the letter of appointment, "Takasu voluntary slope observation team-Three Hints", a card reader handling manual (see Figs. 7b and c) was distributed, and the method of observing slopes and recording the observation results were confirmed on site.

Communication of Disaster Prevention Information by "handmade Newspaper Takasu-Ikusu"

In this study, together with measurement and monitoring of hazardous slopes, university students handed out handmade newspaper called "Takasu-Ikasu" and distributed it to residents about once a month (until the 15th issue as of the end of June 2019). " To improve the awareness of disaster prevention and the ability of local disaster prevention. "Takasu-Ikasu (which means take advantage of the wisdom passed down in Takasu Twon)" contains not only articles related to disaster prevention activities, but also information on the charm of Takasu Town and the wisdom of life rooted in the community. Figure 9 shows the pages of Nos. 14 and 15 issued in June 2019. In the 14th issue, the appointment ceremony of the "Takasu voluntary slope observation team" and the method of confirming the safety of the slope were described. In the 15th issue, the explanation of POCKET installed on the block retaining wall at the back of the elementary school gymnasium and the observation method were described. In the future, we would like to publish other

a)

たかす いかす

第14号

2019年6月

関西大学社会安全学部 近藤研究室・小山研究室

この「たかすいかす」では、自然豊かな高須（たかす）の"暮らしの知恵"を生かしながら（いかす）、防災に関する情報を少しずつ共有していきたいと思います。いっぽいっぽ、たゆまず歩みましょう。

「安心」の記録を残しましょう！！

集落内のあちこちで、崖の様子を観測しています。異変を把握したら、行政に伝えたり、早めに避難したりするためです。

観測機器の近くには、ポストを設置しています。この中には読み取り機器・カード・説明書が入っています。異変がなければ青のカード、異変を感じたら赤のカードをかざして、記録を残してください。みんなの力で、高須に安心を！

ピッ

小学校の体育館裏のポストの様子

「高須あんしん感測隊」任命式（4月27日）

揚部さん、堀川さん、旭さん、橋本さんを「高須あんしん感測隊」の隊員に任命しました。

「高須あんしん感測隊」は、暮らしのなかで異変を感じとる「感測」を目指しています。ぜひ、4人の隊員だけでなくより多くの人の目で、異変（前兆現象）をキャッチしましょう。

感測用のノートをふたりで手作りしました ^o^/

ゆうの

あかね

b)

たかす いかす

第15号

2019年6月

関西大学社会安全学部 近藤研究室・小山研究室

この「たかすいかす」では、自然豊かな高須（たかす）の"暮らしの知恵"を生かしながら（いかす）、防災に関する情報を少しずつ共有していきたいと思います。いっぽいっぽ、たゆまず歩みましょう。

プールの大掃除、がんばりました！

防火水槽に活用できるケロ

小学校裏の観測機器をご紹介します

♪ 小学校の体育館裏には、観測機器が、斜面に3つ設置されています。

♪ 斜面に異常がなければ、「緑色」のランプが点滅しています。そして何らかの異変があれば「オレンジ」・「赤色」に変わります。

♪ 異変を見つけたら自治会長さんまでご連絡ください！

異常なし

これが3台配備

初訪問メンバーも大勢おせわになりました

暑い中で熱い作業になりましたが、高須のパワーをたくさん頂きました！また、うかがいます ^o^/

Fig. 9 Handmade newspaper called "Takasu-Ikasu", **a** Nos. 14 and **b** 15 issued in June 2019

OSV observation equipment (SOP and see through poles) in the same way so that many residents can be aware of it and encourage them to take an active part in measurement and monitoring.

Concluding Remarks

In this paper, the resident participation-type slope measurement/monitoring system using OSV sensors was developed and introduced to in Takasu Town, Fukui City, Japan. Although OSV measurement equipment was able to achieve "visualization" of the danger of slopes, "daily routine" and "voluntary/proactive involvement" were still issues, and more residents worked on measurement and monitoring on a daily basis. Also it is necessary to further improve the environment and mechanism for receiving the information. In addition, if the residents become aware of unusual events, it is necessary to improve communication with specialists and shift to a method of quantitatively measuring slope deformation. In the future, while monitoring the activities of the "Takasu voluntary slope observation team", quantitatively assess how the transmission of disaster prevention information through the "handmade newspaper Takasu-Ikasu" will contribute to the improvement of residents' disaster prevention awareness and local disaster prevention capabilities. We would like to pursue a detailed analysis and finally aim to establish a "resident participation-type slope measurement/monitoring system (Takasu model)".

References

Akutagawa S (2017) A method for visualizing deformation by manipulating light sources and its path. Civil Eng Life 72(9):94–98

Fukui City (2020) Demographics of Fukui City. https://www.city.fukui.lg.jp/sisei/tokei/jinko/jinkoutoukei.html. Accessed 15 March 2020

OSV conthosium (2020) Sensors used in OSV. https://www.osv.sakura.ne.jp/en/index.html. Accessed 15 March 2020

Toyosawa Y, Itoh K, Tamrakar SB, Araki T, Kunimi T, Nishijyoh A, Ohkubo S (2007) A study on prediction of slope failure using highly accurate tilt-sensor. Specific Research Reports of the National Institute of Occupational Safety and Health, JNIOSH-SRR-NO. 35: 91–106

Yamori K (2017) Recommendation of everyday-life preparedness of natural disasters. Nakanishiya-Syuppan, Kyoto. (ISBN-13: 978-4-7795-0570-6). 107

Debris Flow Detection Using a Video Camera

2020 Kyoto Japan

Ko-Fei Liu, Ting-lu Kuo, and Shih-Chao Wei

Abstract

The early warning of natural hazards is an important issue that was raised by the 2015 Sendai Framework for Disaster Risk Reduction. However, early warning systems need complimentary monitoring systems that themselves may be combined with automatic identification and prediction systems. For debris flows, once a disaster has been confirmed to have occurred in an upstream area, early warning of the hazard further downstream may be predicted with relatively good accuracy in time and space. In this study, we use video cameras to identify the arrival times of debris flows using a simple average grey-level method. We show that this method can automatically detect the arrival of debris flow events. This method is tested with both real events video and indoor experiments during the night with moonlight only illumination. All tests have an error of less than 1.3 s. The method is fast and therefore ideal for real-time monitoring and warning.

Keywords

Debris flow • Camera • Monitoring

K.-F. Liu (✉) · T.-I. Kuo · S.-C. Wei
Department of Civil Engineering, National Taiwan University, No. 1, Sec. 4 Roosevelt Road, Taipei, 10617, Taiwan
e-mail: kfliu@ntu.edu.tw

T.-I. Kuo
e-mail: sheep70270@gmail.com

S.-C. Wei
e-mail: stanscwei@gmail.com

S.-C. Wei
Water Hazard Mitigation Center, Water Resources Agency, Ministry of Economic Affairs, Taichung, Taiwan

Introduction

Debris flows are among the most hazardous natural disasters that can occur on sloping land. To provide safety in debris-flow-affected areas, an early-warning system is vital.

The most commonly used early-warning systems are based on indirect warnings using rainfall or hydrology indices with calibrated thresholds (Jan and Lee 2004; Baum and Godt 2010; Thiebes 2012). These rainfall-based methods can issue warnings in an early stage and on a regional scale. This kind of warning indicates there may be debris flows within a large area containing many potential debris-flow streams. However, the specific location of an occurrence cannot be predicted. Confirmation of a debris flow still requires *in situ* monitoring devices or disaster records.

In recent years, debris-flow detection methods have been developed based on direct monitoring systems, such as geophones, video cameras, wire sensors, ultrasonic gauges, radar, etc. (Itakura et al. 2005). These direct detection methods can give precise warnings to affected areas. Among these devices, geophones and cameras have been used most often for debris flows. However, there is currently no automatic detection and warning system using geophones or cameras because the event detection threshold changes in time and still cannot be determined automatically.

Wei and Liu (2019) used the accumulated energy method combined with the characteristic frequency of debris flows to resolve the threshold determination problem and detect debris-flow arrivals. However, since only energy variation was considered, the events detected through geophones could be of a range of types, including granular flow, debris flow, or floods with a high sediment concentration. In these events, the flow energies may be approximately the same but the sediment contents to be very different. Thus, additional information from cameras is essential.

Our research uses a total grey-level method to identify changes in images. The method is based on identifying large

© Springer Nature Switzerland AG 2021
N. Casagli et al. (eds.), *Understanding and Reducing Landslide Disaster Risk*, ICL Contribution to Landslide Disaster Risk Reduction,
https://doi.org/10.1007/978-3-030-60311-3_15

changes in an image's grey level that indicate large changes to the objects in the images. Thus, information from a monitoring video camera can indicate changes in flow conditions and thus the detection of a debris-flow event. This information can in turn be used as part of an early-warning system.

Event Identification Method

Common Image Processing Methods

Past research using camera adopted the particle-tracking method to extract information from the video (Arattano and Grattoni 2000). The same concept has been applied to tracking surface bubbles, debris, and artificial particles in large-scale particle image velocimetry (Fujita et al. 1998; Theule et al. 2018). Many studies used the optical flow method (Horn and Schunck 1981; Lucas and Kanade 1981; Farnebäck 2003) to calculate the velocities of particles in images. After obtaining velocity or particle information, one can use the information to identify debris flows. Chang and Lin (2007) used the movement of a specific target for event detection. They also introduced discrimination of debris flows and floods by using features of contrast, entropy, energy, and homogeneity.

However, these methods usually do not produce accurate results for natural disaster detection, owing to the resolutions of the images and the fact that particles in video are very difficult to identify because of mud/water coverage. The most critical problem is the fundamental assumption used in particle tracking that the same point maintains the same grey level in consecutive images, which is normally not true. Furthermore, tracking analysis is computationally time-consuming, but time is critical for real-time warning.

Therefore, in this research, rather than use a method that is sensitive to variation of particular particles, we will develop a method that does not need to identify particles. Moreover, this method should be usable in the field and in night-time when there is otherwise insufficient illumination.

Therefore, we propose the total grey-level method to identify debris-flow events. We calculate the total grey level in the region of interest (ROI) and if it changes, this indicates there are events occurring within the ROI. This can be used to identify debris flow. After an event is identified, a geophone can be used to confirm if it is debris flow or high-concentration flow. However, the combination of geophones will not be discussed in the present paper.

Total Grey-Level Method

First, video taken in the field is separated into images. Within each image, the ROI is defined and all calculation will be done within the ROI. For each pixel in the ROI, color is transformed to grey level according to the standard from the International Telecommunication Union (ITU-R 1990) code with:

$$Grey = 0.229 \times Red + 0.587 \times Green + 0.114 \times Blue, \tag{1}$$

where each pixel has a grey level from 0 to 255. Then the total grey level for one ROI is calculated as:

$$Total\ Grey\ Level = \sum_{All\ points\ in\ a\ frame} Grey. \tag{2}$$

This total grey level is then divided by the total number of pixels in the ROI to give the average grey level. If it is very dark, the average grey level will be close to 0. If it is very bright, the average grey level will be close to 255. Normal clear water flow will produce very bright and shining images, so the average grey level is around 150 or more. When debris flow or flood occurs, there is granular material and mud and the whole image becomes darker. As one of the characteristics of debris flows is a large amount of granular material concentrated at the front, the average grey level will become darker in a short period of time. An example of real video images taken at Ai-Yu-Zi creek is shown in Fig. 1.

There are boulders of diameter 1 m on top of the flow. A wave of muddy material can be seen in the third photo. There are usually 30–60 frames per second for a standard field camera. Each frame is an image like in Fig. 1, and one average grey level can be calculated. By plotting the variation of average grey level over time, the change can be seen and the rate of change of the average grey level can be calculated. The temporal variation of average grey level for Typhoon Mindulle (the case shown in Fig. 1) is plotted in Fig. 2. For cross referencing, the times corresponding to the three images in Fig. 1 are marked. It can be seen that average grey level remains roughly constant before the debris flow arrives. When the front of the debris flow arrives, average grey level decreased quickly (from brighter to darker). After debris flow passed out of the ROI (time 45–50 s), the grey level roughly returned to its original level. Later than 50 s, there was flood which increased the grey level.

With the temporal variation of average grey level obtained, a condition is required to determine the detection of debris flows. Since average grey level changes for different lighting conditions (sunlight, moonlight, rain), materials, and flow conditions, it seems impossible to find a fixed threshold to determine the arrival of an event (a debris flow or flood). However, before any event, the flow conditions should be steady or slowly varying, so the grey level should remain roughly the same for a long period of time. Hence, the "normal" average grey level for the stable state should be considered as the reference level. Any normal signals have

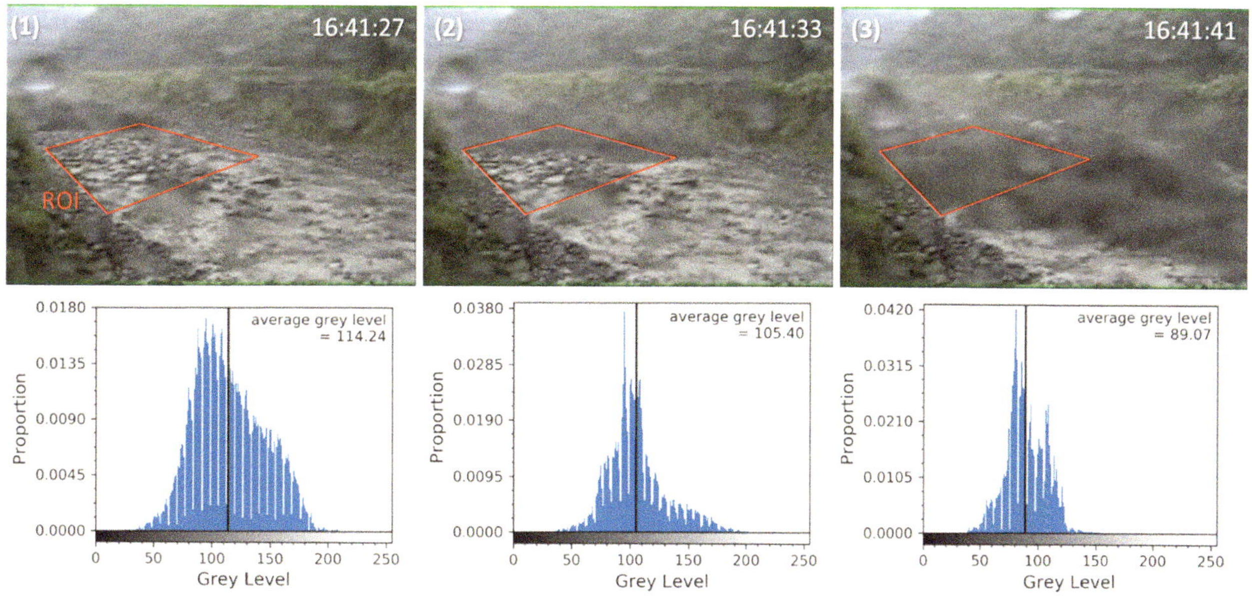

Fig. 1 Grey-level distributions from video of the Typhoon Mindulle debris flow. 'Proportion' denotes the ratio of the number of pixels of a given grey level to total number of pixels in the ROI (region of interest). Next to the grey-level axis, the corresponding brightness is plotted as a grey-level bar. The vertical line indicates the position of the average grey level for that image. (1) Debris flow just entering ROI (from top corner of ROI). The average grey level of the ROI is 114.24. (2) Front of the debris flow reaches the centre of the image. The average grey level of the ROI is 105.4. (3) The front of the debris flow just reaches the boundary of the image. The average grey level of the ROI is 89.07

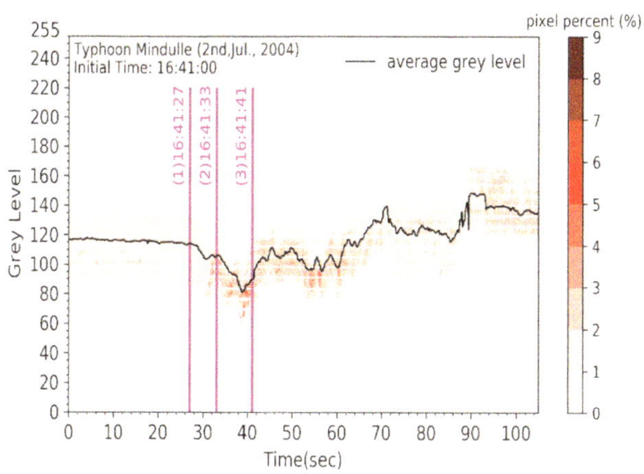

Fig. 2 Temporal variation of average grey level in Ai-Yu-Zi creek for Typhoon Mindulle. The first 25 s is noise before the debris flow arrives. The black line is the average grey level. Darker color indicates a larger proportion of pixels at that grey level. Pink lines indicate the time locations for images in Fig. 1

fast-varying "noises". Therefore, a time average should be taken to remove these noises. Thus, we average the total grey level for every 11 frames (frame from $t_0 - 5$ to $t_0 + 5$) to produce one dataset at each time t_0. These averaged values can effectively remove all pulse irregularities.

To detect a change in average grey level, we use the slope of Fig. 2. The rate of change of the grey level is calculated as:

$$S(t) = \frac{\text{Ave.Grey}(t + \Delta t) - \text{Ave.Grey}(t - \Delta t)}{2\Delta t} \quad (3)$$

For every 10 s, we choose the maximum slope S_{max} within that 10 s as the representative parameter of flow condition. While S_{max} is small, the flow conditions remain the same. However, if at any moment the calculated slope is greater than twice the S_{max} value obtained 2 s before, and this continues to be true for consecutive a 5 points (1.5 s), this indicates something happened within the ROI. Then a debris flow warning will be issued after this detection process.

This simple procedure is tested through laboratory experiments. In practice, the warning triggered by video camera can be checked with detection of geophone or other sensors but this discussion is excluded in this study.

In this paper, the warning given by the proposed detection method will be checked with manual tracking of video images by eyes.

Laboratory Tests

The setup of the experiment is depicted in Fig. 3. The flume is 5 m long and 60 cm wide. A video camera is mounted on top of the flume. Lighting is provided from above.

White Styrofoam balls with a density of 21.4 kg/m^3 are used. We mixed 170 balls of diameters from 1.5 to 10 cm.

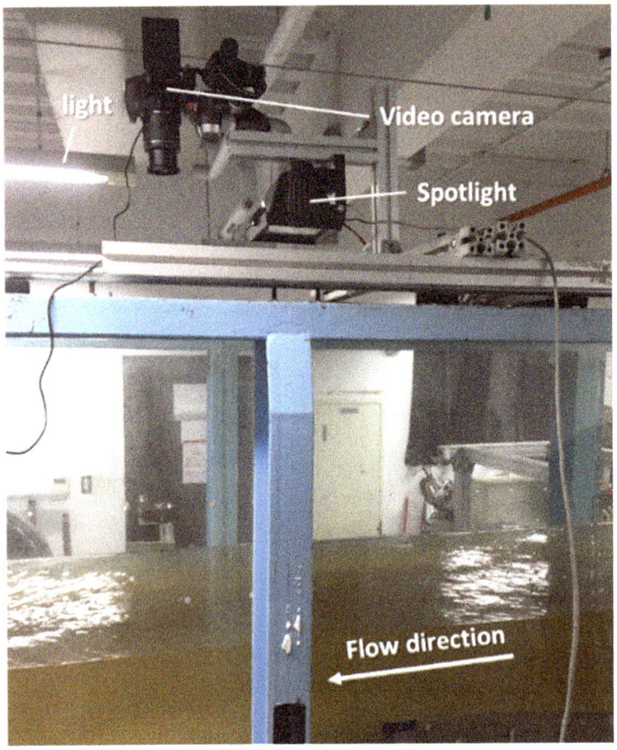

Fig. 3 Flume test setup. Water flow is 20 cm/s to the left. There is strong lighting from the top and normal room lighting

Then we dropped them as a whole into the flume from upstream. The balls flowed down the flume with the water flow at a speed of 20 cm/s. Because the balls float, this is used to simulate granular material on top of the debris flow front. The video camera captured images from directly above. To control the lighting conditions, experiments were conducted at night. We used four different lighting methods to test the influence of lighting: strong artificial lighting (100 W) from on top of the flume, only normal room lighting (two 60 W from 3 m above), no artificial lighting but with all windows open to admit moonlight and starlight, and finally no artificial lighting with all windows closed and curtains blocking all possible light sources. The corresponding sample photos are shown in Fig. 4.

The resulting average grey levels and the calculated slopes are depicted in Fig. 5. The solid blue lines are the average grey levels and the dotted orange lines are the corresponding slopes. In all cases, the grey-level variation before arrival of the balls is small. The slope increases are sufficient to identify the fast change of grey level when the balls arrive. As lighting becomes weaker, the maximum grey level becomes smaller, as do the slopes. For strong lighting, the grey level actually keeps changing, while only the slope remains the same. However, twice the maximum slope for previous 10 min gives a good arrival time.

With no lighting, the maximum grey level is only 2.7 (and cannot be identified by human eyes). However, the slope is still detectable.

The time that the Styrofoam balls enter the ROI is marks the beginning of event detection. The differences between the times determined using the total grey-level method are compared with those identified by eye are listed in Table 1.

The accuracies in all four cases are very good. Since the time difference between two frames is 0.03 s, these values indicate the error is between 1 and 7 frames. The most significant result is for Case 4, where there is no light at all. The human eye can barely distinguish objects in these conditions, yet this method can identify the arrival of the balls. This means that using slope variation enables this method to work under almost no light and still provide an automatic early warning.

Application to Ai-Yu-Zi Creek

Test Area

Ai-Yu-Zi Creek located at Shenmu Village, Nantou County, Taiwan, was selected as the test area. It was identified as having a high potential for debris flow torrents by the Soil and Water Conservation Bureau of Taiwan. The length of the stream is 3.731 km and the watershed area is 405.02 ha. Stream elevation extends from 1200 to 2500 m with an average slope of 39.3°. The river width at the monitored location is approximately 50 m. The study area is mainly located in the Nanchuang and Nankang formations. These formations are composed of sandstone, siltstone, shale, and alternations of sandstone and shale. The landslide area occupies 12–34.2% (1996–2009) of the whole watershed area, and it has had an increasing trend in recent years (Chen et al. 2012). The average annual rainfall is 3054.7 mm with 87% (2644.5 mm) concentrated in the rainy season from April to October. The observed debris flow events in the are involve large boulders of typically 1 to 2 m in diameter, and occasionally boulders of up to 5 m can be observed. All debris flows have high concentration of granular material, but concentration changes from event to event.

Test Result

The total grey-level method was used for the video recordings made during Typhoon Mindulle. The images were those shown in Fig. 1. With the total grey-level method and twice the reference level as the warning criterion, the time of the warning is marked in Fig. 6.

The time differences for this method applied at Ai-Yu-Zi creek are listed in Table 2. We tracked frame by frame and

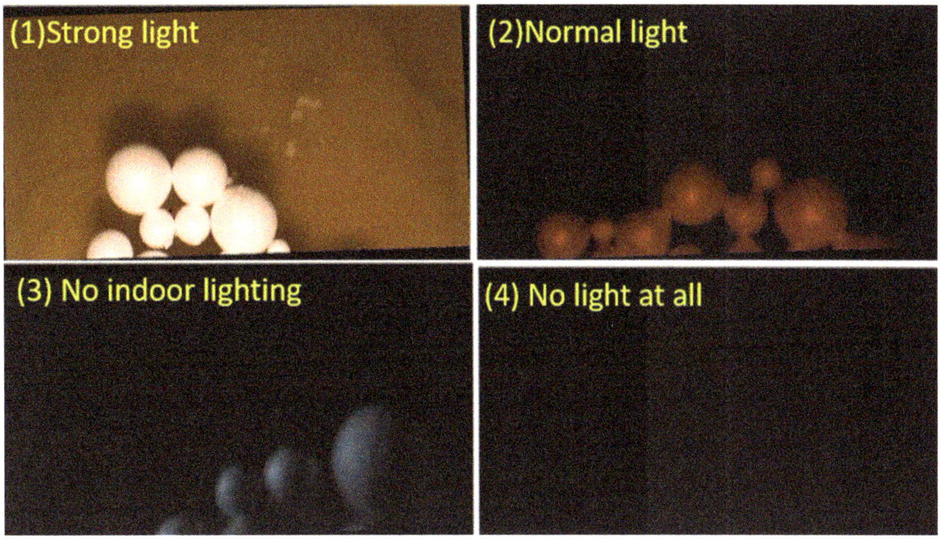

Fig. 4 Images of balls flowing down the flume. The white Styrofoam balls have different grey levels under different lighting conditions: (1) strong lighting provided from on top of the flume, (2) only normal room lighting, (3) no indoor lighting with room lights turned off and only natural moonlight from one small window, and (4) curtains used to block all natural light and no indoor lighting provided

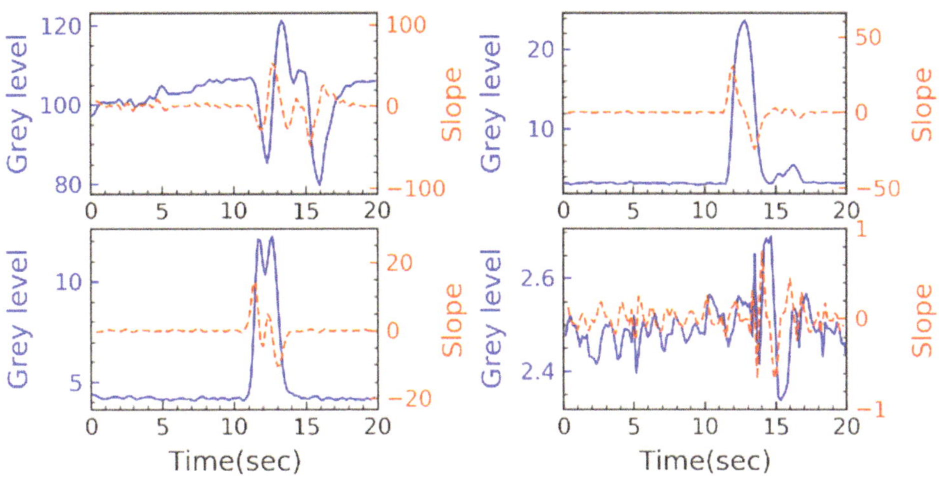

Fig. 5 Temporal variation of average grey level. Solid blue lines are average grey-level variations. Dotted orange lines are slope variations. Graphs correspond to the lighting conditions in Fig. 4

Table 1 Accuracy of using the total grey-level method to detect events in the flume test. Cases are those shown in Fig. 4. Error is the time difference between detections from the total grey-level method and by eye. A negative value indicates detection earlier than the actual time (due to a shadow in front)

Case	1	2	3	4
Error (s)	−0.125	−0.041	−0.308	0.267

when the debris flow front reached the ROI, as judged by eye, we marked that frame as the arrival time for debris flows. Also listed in Table 2 are the results using a higher detection level (five times S_{max}) and averaging over more frames (31 frames). The results for criterion of twice the noise level have a detection time error of less than 1.3 s. This means that the total grey-level method is acceptable for use in the field.

Conclusion

In this study, we extracted a region of interest (ROI) from each frame captured from a video camera setup in both flume and field tests, conducted image processing and calculated the average total grey level of each image. We defined the slope of the temporal variation of the average

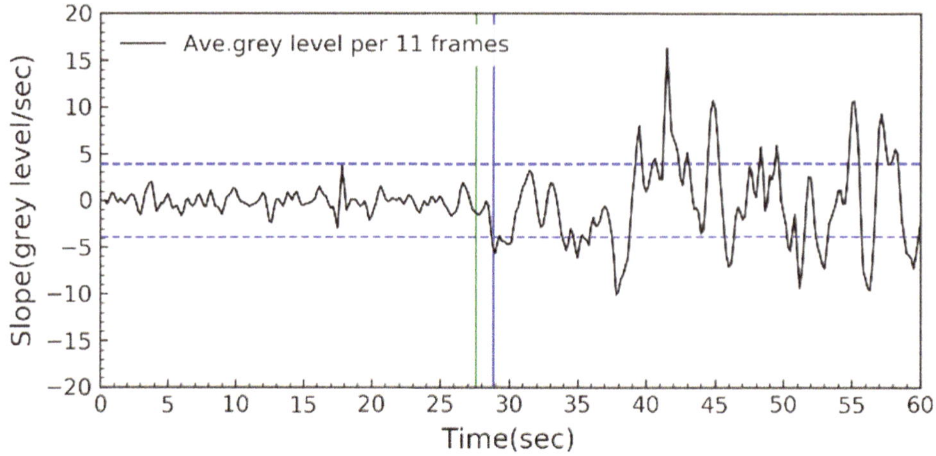

Fig. 6 Temporal variation of the average grey-level slope. The horizontal blue dashed line is twice the noise level S_{max}. The vertical blue solid line denotes warning time with respect to twice the noise level S_{max}. The vertical green line is the reference time estimated by eyes

Table 2 Accuracy of using the total grey-level method to detect debris flows at Ai-Yu-Zi creek during Typhoon Mindulle. Error is between the total grey-level method and estimation by eye. 2T indicates using two times the noise level and 5T is using five times the noise level. 11F denotes results from an average over 11 frames and 31F denotes an average over 31 frames

Case	2 T		5 T	
	11F	31F	11F	31F
Error (s)	1.28	0.95	10.28	1.95

total grey level as the main index for determining changes in flow conditions, and thus detection of debris flows. We adopted twice the reference level, S_{max}, of temporal slope change for detection of events in the video images. Using laboratory experiments and videos of debris flow in the field, we showed that the error of detection method is within 1.3 s of that made by viewing the same imagery with the human eye. Furthermore, in flume tests, we showed that the method can be used in even very dark environments without the need for artificial lighting. This is essential for field monitoring and early warning of debris flows, given that these events can occur at any hour of the day or night. The simple image-processing method used here requires very little computational time compared with other image-processing methods, which is also very desirable for real-time detection and monitoring.

Acknowledgements This research was financially supported by the Soil and Water Conservation Bureau of Taiwan through grants 109AS-10.7.1-SB-S2. The video of the Typhoon Mindulle debris flow was obtained from Soil and Water Conservation Bureau of Taiwan.

References

Arattano M, Grattoni P (2000) Using a fixed video camera to measure debris-flow surface velocity. In: Wieczorek GF, Naeser ND (eds) Debris-flow hazards mitigation: mechanics, prediction, and assessment; Proceedings of the 2nd international DFHM conference, Taipei, Taiwan, August 16–18, 2000, 273–281

Baum RL, Godt JW (2010) Early warning of rainfall-induced shallow landslides and debris flows in the USA. Landslides 7(3):259–272

Chang SY, Lin CP (2007) Debris flow detection using image processing techniques. In: Chen, Major (eds) Debris-flow hazards mitigation: mechanics, prediction, and assessment

Chen SC, Chen SC, Wu CH (2012) Characteristics of the landslides in Shenmu watershed in Nantou County. J Chin Soil Water Conserv 43:214–226

Farnebäck G, (2003) Two-frame motion estimation based on polynomial expansion. In: Bigun J, Gustavsson T (eds) Image analysis. SCIA 2003. Lecture notes in computer science, vol 2749. Springer, Berlin, Heidelberg

Fujita I, Muste M, Kruger A (1998) Large-scale particle image velocimetry for flow analysis in hydraulic engineering applications. J Hydraul Res 36:397–414

Horn BKP, Schunck BG (1981) Determining optical flow. Artif Intell 17:185–203

Itakura Y, Inaba H, Sawada T (2005) A debris-flow monitoring devices and methods bibliography. Nat Hazards Earth Syst Sci 5:971–977

ITU-R Recommendation BT.709 (1990) Basic parameter values for the HDTV standard for the studio and for international programme exchange, Tech. Rep. BT. 709 (formerly CCIR Rec. 709), ITU, 1211 Geneva 20, Switzerland

Jan CD, Lee MH (2004) A debris-flow rainfall-based warning model. J Chin Soil Water Conserv 35(3):275–286

Lucas BD, Kanade T (1981) An iterative image registration technique with an application to stereo vision (DARPA). In: Proceedings of the 1981 DARPA image understanding workshop, April 1981, pp 121–130

Theule JI, Crema S, Marchi L, Cavalli M, Comiti F (2018) Exploiting LSPIV to assess debris-flow velocities in the field. Nat Hazards Earth Syst Sci 18:1–13

Thiebes B (2012) Landslide analysis and early warning systems: local and regional case study in the Swabian Alb. Springer, Berlin Heidelberg, Germany

Wei SC, Liu KF (2019) Automatic debris flow detection with geophones. Landslides 17(2):349–359

Landslide Mapping and Monitoring with Satellite Interferometry

Federico Raspini, Emanuele Intrieri, Davide Festa, and Nicola Casagli

Abstract

The potential of multi-interferometric approach applied to the Sentinel-1 acquisitions for the analysis of slope instabilities is presented and discussed through two different landslides of very different nature. In the first case, Sentinel-1 data, systematically acquired with short revisiting time and promptly processed allows for the quick identification of the acceleration suffered by the Carpineta landslide, a large, active earth slide in the Northern Apennines (Tuscany Region, Italy). In the second case, a post-event analyses of Sentinel-1 data permitted the identification of a clear precursory deformation signal for the Xinmo landslide (Mao County, Sichuan Province, China), a large rock avalanche occurred in the early morning of 24 June 2017. Results suggest that advances in satellite sensors, increase of computing capacity and refinement of data screening tools can contribute to the design of a new paradigm in satellite-based monitoring systems. Sentinel-1 data, systematically acquired over large areas with short revisiting time, could be used not only as a tool for mapping unstable areas, but also for landslide monitoring, at least for some typologies of sliding phenomena.

Keywords

Landslides • Sentinel-1 • Interferometry • Monitoring • Displacement time series

F. Raspini (✉) · E. Intrieri · D. Festa · N. Casagli
Università Degli Studi di Firenze, Department of Earth Sciences, via Giorgio La Pira 4, 50121 Firenze, Italy
e-mail: federico.raspini@unifi.it

E. Intrieri
e-mail: emanuele.intrieri@unifi.it

D. Festa
e-mail: davide.festa@unifi.it

N. Casagli
e-mail: nicola.casagli@unifi.it

Introduction

In many landslide studies, the possibility to monitor deformation and to predict future behaviour is still a major concern. To date, early-warning systems have mostly relied on the availability of detailed, high-frequency data from sensors installed in situ. Methods deducing reliable failure predictions have been largely applied at local scale, where in situ monitoring systems can be installed (Intrieri et al. 2019).

The same purpose could not be chased through spaceborne monitoring applications, as these could not yield information acquired in a sufficiently systematic fashion: the low data sampling frequency of most of the satellite systems hampered the possibility to retrieve the necessary details of tertiary creep characterized by accelerating deformation. So far, the lack of systematic information on ground displacement acquired at regional scale was another serious limit to the application of failure prediction methods at wide scale.

Such limitations can be partially solved through the exploitation of new generation spaceborne platforms. The launch of Sentinel-1 mission opened a new opportunity for InSAR (Interferometric Synthetic Aperture Radar) monitoring applications thanks to the increased acquisition frequency, the regularity of acquisitions and the policy on data access. We demonstrate the potential of satellite InSAR to spot the onset of landslide acceleration and to identify precursors to catastrophic slope failures.

Here we present two sets of Sentinel-1 constellation images processed by means of multi-interferometric approach for the analysis of two very different slope instabilities (a large earth slide in the Tuscany Region, Italy and a large rock avalanche in the Xinmo village in China). The results highlight that satellite InSAR may now be used to support decision making and enhance predictive ability for landslide hazard.

This means that the transition from historical analysis of ground deformation to a continuous monitoring with

© Springer Nature Switzerland AG 2021
N. Casagli et al. (eds.), *Understanding and Reducing Landslide Disaster Risk*, ICL Contribution to Landslide Disaster Risk Reduction,
https://doi.org/10.1007/978-3-030-60311-3_16

prediction capabilities at regional scale using satellite radar data is now possible.

Satellite Interferometry

Two large stacks of C-band SAR images (central frequency 5.405 GHz and wavelength 5.6 cm) acquired by Sentinel-1 constellation have been processed by means of the SqueeSAR algorithm (Ferretti et al. 2011) to generate ground deformation maps for the Tuscany Region and for the area affected by the Xinmo landslide. This algorithm represents the evolution of PSInSAR, the first technique belonging to the PSI (Persistent Scatterer Interferometry) family specifically implemented for the processing of multi-temporal radar imagery (Ferretti et al. 2001). The novelty of PSInSAR algorithm is the capability of identifying a network of coherent radar targets, *i.e.*, point-like-targets (PS, Permanent Scatterers) within the radar image. Such pixels correspond to manmade objects, outcropping rocks, debris areas or buildings that register a steady radar signal over the whole observation time period. The definition of these points enables to reduce decorrelation phenomena and obtain a high signal-to-noise ratio, thus discriminating phase contributions related to displacement from those due to atmosphere, topography and noise.

Over urban fabric, where many stable reflectors can be identified, PS density can be very high (up to hundreds of PS per square km). One limitation of this technique is the impossibility of identifying 'radar friendly' targets outside of urban and peri-urban areas and in highly vegetated zones.

The development of the SqueeSAR techniques contributed to extend the field of application of PSInSAR to natural terrain, overcoming the main limitation of this technique. SqueeSAR estimates deformation rates not only from point-like-targets but also from partially coherent pixels, called Distributed Scatterers (DS). DS points correspond to homogeneous ground surfaces and not to single objects, *i.e.*, uncultivated areas, deserts, debris covered slopes and scattered outcrops.

LOS (Line of Sight) deformation rate can be estimated with an accuracy theoretically lower than 1 mm/yr, at least for very stable PS during a long-time span. PSI analysis is designed to generate time-series of ground deformations for individual reflectors. The accuracy of the single measurement in correspondence of each SAR acquisition ranges from 1 to 3 mm (Colesanti et al. 2003). Each measurement is referred temporally and spatially to a unique reference image and to a stable reference point. Over the last 15 years, satellite interferometry proved to be a valuable technique for slope instability investigations (Raspini et al. 2017; Lu et al. 2019).

The launch of Sentinel-1 mission opened a new opportunity for InSAR applications: it provides, on a regional scale, real-time deformation monitoring and response for geological processes. Developed within the Copernicus initiative, the Sentinel-1 mission is composed of a constellation of two twin satellites, Sentinel-1A and Sentinel-1B. Launched in April 2014 and in April 2016 respectively, they share the same orbital plane and ensure a unique revisiting time of 6 days optimised for SAR interferometry applications.

With respect to previous satellites, Sentinel-1 data couple some favourable characteristics: regional-scale mapping capability thanks to the TOPS acquisition mode), systematic and regular SAR observations and rapid product delivery (typically less than 3 h).

Sentinel-1 SAR products are freely accessible, thus providing the scientific community, as well as public and private companies, with consistent archives of openly available data, suitable for monitoring applications relying on the processing of long series of SAR images.

Sentinel-1 have been used to assess the feasibility (Novellino et al. 2017; Vecchiotti et al. 2017) and to establish nationwide ground deformation services (e.g., Kalia et al. 2017; Dehls et al. 2019; Balasis-Levinsen et al. 2019; Manunta et al. 2019), by mosaicking adjacent SAR data stacks, rather than providing new streamlines of information for monitoring solutions (Raspini et al. 2018). The potential offered by the shorter repeat cycle and regularity of acquisitions of Sentinel-1 with respect to other satellites has been rarely fully exploited.

The Tuscany Region: Identification of Landslide Accelerations with Sentinel-1

Benefits deriving from the systematic exploitation of satellite data for landslides monitoring are presented and discussed through the case study of the Tuscany Region (Central Italy). Being characterized by heterogeneous physical settings, with Apennines flysch ridges marking its northern and eastern parts, Tuscany Region represents a perfect scenario to test, tune and refine this new monitoring approach.

The Tuscany Region results to be a very landslide-prone area. Landslide processes have pervasively shaped the Tuscan landscape (Rosi et al. 2018 mapped more than 90,000 landslide) and are a major issue for regional authorities.

In October 2016, for the initial implementation of the continuous monitoring of the Tuscany Region, the images archives of the Sentinel-1 were acquired and then processed. This phase ended with the creation of ground deformation maps with almost 2 million PS points, covering the whole Tuscan territory, data useful to understand and assess the stability evolution of the territory.

In the second phase of the project, started after the provision of the baseline, the database of ground deformation measurements was updated by continuous processing of Sentinel-1 images. Once a new Sentinel-1 image is available, it is automatically downloaded and added to the existing archive. The new data stack is then entirely reprocessed to generate new ground deformation maps and updated displacement time series. At the end of December 2019, a total of 87 updates have been created for the Tuscany Region, evenly distributed over the monitoring period (29 updates in 2017, as well as in 2018 and in 2019).

In Fig. 1 velocity maps obtained from the processing of all the Sentinel-1 images up to December 2019 are reported. Each velocity map includes more than 900,000 points and provides a synoptic view of the regional displacement field. Measurement point are represented by dots coloured according to their mean velocity along the LOS of the satellite, expressed in millimetres per year, with positive values of mean velocity representing displacements towards the satellite, and negative values representing displacements away from the satellite.

Displacement time series represent the most advanced product of any PSI approach, providing the deformation history over the observed period, and are fundamental for studying the kinematics of a given phenomenon, highlighting any potential changes occurred during the monitoring period, such as sudden accelerations prior to a landslide failure.

Time series, systematically updated with the most recent available Sentinel-1 acquisition, are analysed to detect anomalous points (*i.e.*, points where a change in the dynamics of motion is occurring). Anomalous points are analysed with the support of thematic information to decide if an anomalous pattern can be considered worth reporting to regional authorities (Raspini et al. 2019).

The cluster of anomalous points appeared during winter 2018 in the village of Carpineta (Fig. 1 for location) is a representative case of significant anomalies of movement determining a significant level of risk.

Carpineta (municipality of Sambuca Pistoiese) is a remotely placed village in the Northern Apennines at an elevation of about 800–850 m a.s.l. on an eastward exposed slope. From a geological point of view the area is characterized by the presence of a succession of foredeep turbiditic deposits.

The village is affected by a large, active system of earth-slides. SqueeSAR results reported in Fig. 2, covering the time interval from December 2014 to August 2018, confirmed the presence of active movements. The highest deformation rates (dark red points in Fig. 2 with values above 30 mm/yr) were recorded in the central part of the slope, where the village is located.

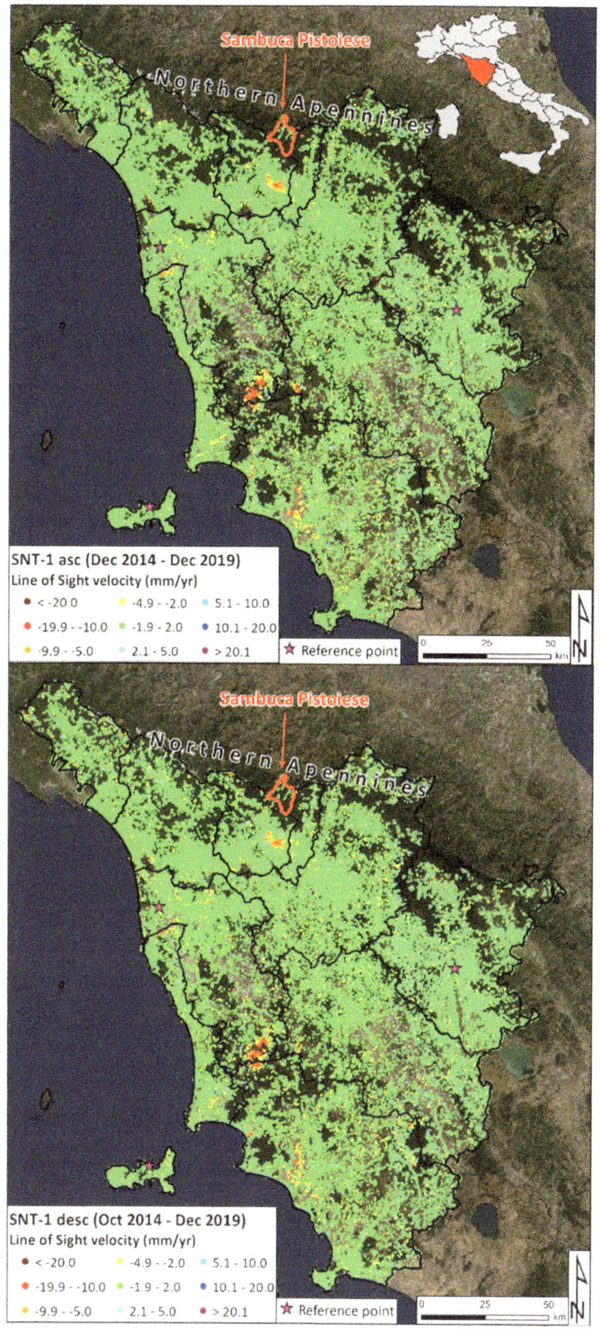

Fig. 1 Ground deformation maps for the Tuscany Region obtained with SqueeSAR processing for ascending (above) and descending geometry (below)

Analysis of time series highlight, during winter 2018, the onset of acceleration which affects, with different timing, different parts of the landslide.

The acceleration of the Carpineta landslide occurred after a period of persistent rainfalls in the area: before the start of acceleration of January 2018, 30-days and 60-days accumulated rainfall of 179.6 and 578.0 mm, respectively, were

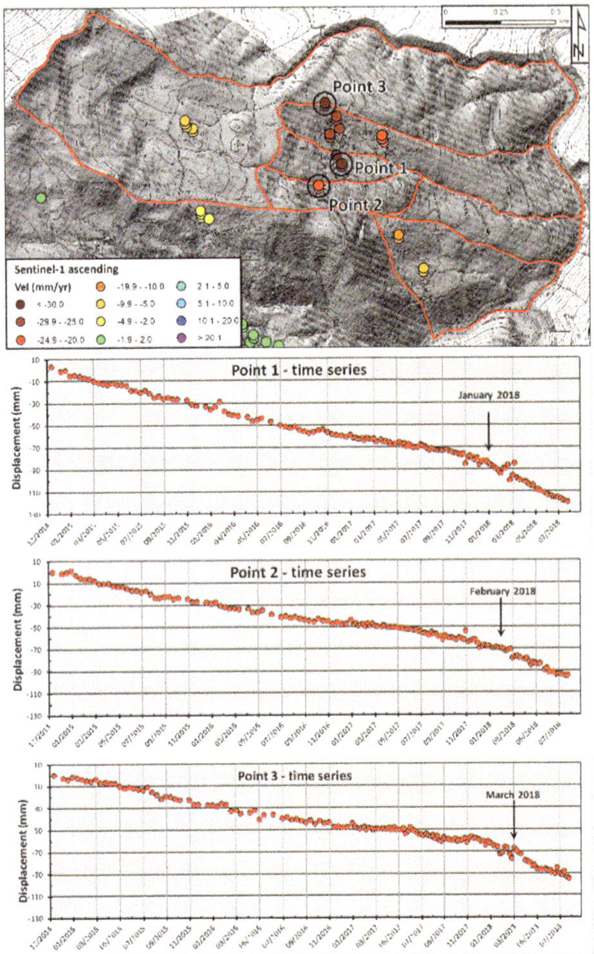

Fig. 2 Ground deformation data for the landslide of Carpineta (municipality of Sambuca Pistoiese)

registered by the closest rain gauge. In addition to the rainfalls, an additional contribution has been given by rapid snow melting, whose effect on the displacement time series is clearly visible in the form of a reduced signal-to-noise ratio from late February to mid-March.

In February 2018 a warning to regional authorities was released in the form of a monitoring bulletin, with the recommendation of further analysis and on-site validation surveys, which confirmed the presence of diffuse factures to buildings induced by the long-term landslide movement.

Landslide Failure Prediction using with Sentinel-1

The Xinmo landslide (Ngawa Prefecture, Sichuan Province, Mao County, China) occurred at 5:38 am on the 24 June 2017. The landslide can be classified as a rock avalanche: it started as a large rockslide in the source area and evolved in an extremely rapid flow-like movement of fragmented rock. About 4.5 million m^3 of rock detached from the top of the mountain ridge at an average elevation of 3.431 m a.s.l. (Fan et al. 2017). Along its travel along the steep slope, the landslide entrained a large amount of pre-existing debris.

The course of the Songping gully, which flows at an elevation of 2280 m from WNW to SSE in the area, was dammed for more than 1 km (Scaringi et al. 2018) by a sliding mass of about 8–13 million m^3 (Su et al. 2017), with a thickness of debris of more than 10 m and a total area of 1.5 km^2. The landslide hit the Xinmo village with a velocity of 250 km/h (Fan et al. 2017), burying 62 houses and killing more than 80 people.

A post-event InSAR (Interferometric Synthetic Aperture Radar) analysis on a stack of 45 descending C-band SAR images acquired by the ESA Sentinel-1 satellites from 9 October 2014 to 19 June 2017 was performed immediately after the event. The purpose of the analysis was twofold:

(i) detect and record any pre-event deformation in and around the village of Xinmo;
(ii) identify potential pre-failure signal for the Xinmo landslide.

Displacement data obtained using the SqueeSAR technique (Ferretti et al. 2011), provided valuable information on ground movements before the event (Fig. 3). The area of the Xinmo village exhibits very low deformation rates, ranging between − 2.0 and 2.0 mm/year, indicating relatively stable ground conditions (green regions).

Results highlight the presence of active movements in a large sector of the slope above the Xinmo village. About 700 measurement points (MPs) are identified, with deformation rates exceeding several millimetres per year. In the upper part of the slope, close to the landslide scarp, velocity values range between—10 and—20 mm/year, with peaks of about —27 mm/year (where the negative sign indicates a movement away from the sensor). Considering the acquisition geometry and the orientation of the slope, the measured deformation rates are consistent with the occurrence of precursory movements over a large sector of the slope affected by the 24 June 2017 landslide. It is worth remarking that this specific sector of the slope was the origin of the sliding event.

Advanced and systematic analysis of displacement time series, which describe the evolution of deformation over the entire monitoring period (from 9 October 2014 to 19 June 2017), is of fundamental importance to understand the dynamic conditions of a specific landslide, as they highlight any changes occurred during the investigated period, such as progressive accelerations before the landslide failure. Deformation time series of points located in the NW sector

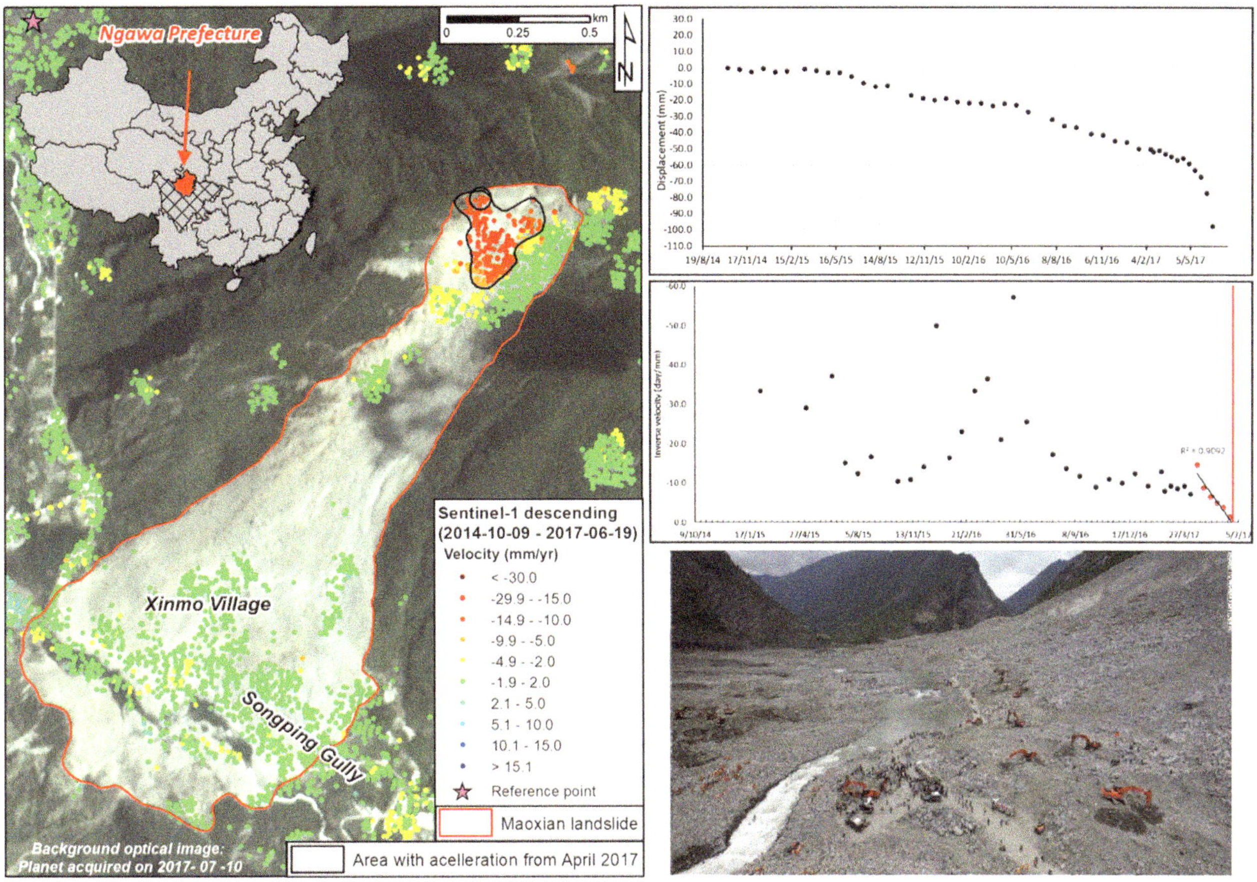

Fig. 3 Pre-failure sign of the Xinmo landslide as seen from Sentinel-1. Ground deformation maps are reported on the left side of the picture, while displacement time series (top) and respective inverse velocity (bottom) of a point taken from the accelerating area are included on the right. The red line indicates the actual time of failure (24 June 2017)

of the source area exhibit an acceleration starting from April 2017. Accelerating areas are the most important to detect for early warning purposes, as they are directly related to slope instability. Anomalous accelerations can indicate that the landslide has entered the tertiary creep and it is approaching collapse (Saito 1969). Specific forecasting methods exist to determine the probable time of failure (Fukuzono 1985; Intrieri et al. 2018).

A retrospective forecast of the time of failure has been performed starting from the accelerating time series. The Fukuzono method for forecasting the time of failure has been applied to the displacement data exhibiting progressive acceleration, pointing out that an accurate estimation of the failure time was already possible since the beginning of June.

Conclusions

The launch of Sentinel-1 constellation, a conflict-free mission, opened the possibility to design of a new paradigm in satellite-based monitoring systems.

Sentinel-1 data, continuously processed and analysed, can be exploited to scan wide areas, to spot unstable zones and to systematically track ground deformation. The main purpose of this application is the early identification of deviations from a linear trend of the displacements as it is a direct sign of a change in the dynamic conditions of a landslide.

Providing continuous displacement monitoring, Sentinel-1 proved the feasibility, although in retrospect, of landslide forecasting, a crucial task in any landslide analysis.

Leveraging the enhanced imaging capabilities of Sentinel-1 and the advances of computing capacities, regional to national scale monitoring systems are now possible, at least for some landslide typologies.

Continuous information on where, when and how fast the ground is moving can be provided, supporting authorities with prioritization of hazards deemed to be most urgent.

Acknowledgements The monitoring system presented in this paper has been founded and supported by the Regional government of Tuscany. The authors also acknowledge TRE ALTAMIRA for having processed the Sentinel-1 data for the Tuscany Region and for the Xinmo landslide.

References

Balasis-Levinsen J, Nissen M, Meister A, Keller K (2019) Linking InSAR-based deformation monitoring with the national geodetic infrastructure in Denmark. In *AGU Fall Meeting 2019*. AGU.

Colesanti C, Ferretti A, Locatelli R, Novali F, Savio G (2003) Permanent scatterers: precision assessment and multi-platform analysis. International Geoscience, Remote Sensing Symposium, Toulouse, France

Dehls JF, Larsen Y, Marinkovic P, Lauknes TR, Stødle D, Moldestad DA (2019) INSAR. No: A National Insar Deformation Mapping/Monitoring Service in Norway–From Concept to Operations. In IGARSS 2019–2019 IEEE international geoscience and remote sensing symposium, pp 5461–5464

Fan X, Xu Q, Scaringi G, Dai L, Li W, Dong X, Havenith HB (2017) Failure mechanism and kinematics of the deadly June 24th, 2017 Xinmo landslide, Maoxian, Sichuan. China Landslides 14 (6):2129–2146

Ferretti A, Fumagalli A, Novali F, Prati C, Rocca F, Rucci A (2011) A new algorithm for processing interferometric data-stacks: squeeSAR. IEEE Trans Geosci Remote Sens 49(9):3460–3470

Ferretti A, Prati C, Rocca F (2001) Permanent scatterers in SAR interferometry. IEEE Trans Geosci Remote Sens 39(1):8–20

Fukuzono T (1985) A method to predict the time of slope failure caused by rainfall using the inverse number of velocity of surface displacement. J Jap Landslide Soc 22:8–13

Intrieri E, Carlà T, Gigli G (2019) Forecasting the time of failure of landslides at slope-scale: a literature review. Earth Sci Rev 193:333–349

Intrieri E, Raspini F, Fumagalli A, Lu P, Del Conte S, Farina P, Allievi J, Ferretti A, Casagli N (2018) The Maoxian landslide as seen from space: detecting precursors of failure with Sentinel-1 data. Landslides 15(1):123–133

Kalia AC, Frei M, Lege T (2017) A Copernicus downstream service for the nationwide monitoring of surface displacements in Germany. Remote Sens Environ 202:234–249

Lu P, Bai S, Tofani V, Casagli N (2019) Landslides detection through optimized hot spot analysis on persistent scatterers and distributed scatterers. ISPRS J Photogramm Remote Sens 156:147–159

Manunta M, De Luca C, Zinno I, Casu F, Manzo M, Bonano M, De Martino P (2019) The parallel SBAS approach for sentinel-1 interferometric wide swath deformation time-series generation: algorithm description and products quality assessment. IEEE Trans Geosci Remote Sens 57(9):6259–6281

Novellino A, Cigna F, Brahmi M, Sowter A, Bateson L, Marsh S (2017) Assessing the feasibility of a national InSAR ground deformation map of Great Britain with Sentinel-1. Geosciences 7 (2):19

Raspini F, Bardi F, Bianchini S, Ciampalini A, Del Ventisette C, Farina P, Casagli N (2017) The contribution of satellite SAR-derived displacement measurements in landslide risk management practices. Nat Hazards 86(1):327–351

Raspini F, Bianchini S, Ciampalini A, Del Soldato M, Montalti R, Solari L, Tofani V, Casagli N (2019) Persistent Scatterers continuous streaming for landslide monitoring and mapping: the case of the Tuscany region (Italy). Landslides 16(10):2033–2044

Raspini F, Bianchini S, Ciampalini A, Del Soldato M, Solari L, Novali F, Del Conte S, Rucci A, Ferretti A, Casagli N (2018) Continuous, semi-automatic monitoring of ground deformation using Sentinel-1 satellites. Sci Rep 8:7253

Rosi A, Tofani V, Tanteri L, Stefanelli CT, Agostini A, Catani F, Casagli N (2018) The new landslide inventory of Tuscany (Italy) updated with PS-InSAR: geomorphological features and landslide distribution. Landslides 15(1):5–19

Saito M (1969) Forecasting time of slope failure by tertiary creep. In: Proceedings of the 7th international conference on soil mechanics and foundation engineering, vol 2. Mexico City, pp 677–683

Scaringi G, Fan X, Xu Q, Liu C, Ouyang C, Domènech G, Dai L (2018) Some considerations on the use of numerical methods to simulate past landslides and possible new failures: the case of the recent Xinmo landslide (Sichuan, China). Landslides 15(7):1359–1375

Su LJ, Hu KH, Zhang WF, Wang J, Lei Y, Zhang CL, Zheng QH (2017) Characteristics and triggering mechanism of Xinmo landslide on 24 June 2017 in Sichuan. China J Mt Sci 14(9):1689–1700

Vecchiotti F, Peduto D, Strozzi T (2017) Multi-sensor a priori PSI visibility map for nationwide landslide detection in Austria. Workshop on World Landslide Forum. Springer, Cham, pp 45–52

Comparison Between PS and SBAS InSAR Techniques in Monitoring Shallow Landslides

Xue Chen, Giulia Tessari, Massimo Fabris, Vladimiro Achilli, and Mario Floris

Abstract

The main aim of this study is to compare the two commonly used multi-temporal interferometric synthetic aperture radar (InSAR) techniques, i.e. permanent scatterers (PS) and small baseline subset (SBAS), in monitoring shallow landslides. PS and SBAS techniques have been applied to ascending and descending Sentinel-1 SAR data to measure the rate of surface deformation and the displacement time series in the Rovegliana area (NE Italian pre-Alps) from 2014 to 2019. As expected, PS results cover only urban areas, while those obtained by SBAS cover up to the 85% of the investigated area. Velocity maps obtained by the two techniques show that some sectors of the investigated slope are affected by active shallow landslides which threaten the stability of buildings, walls and road network. The comparison between ascending and descending velocity maps along the satellite line of sight reveals the presence of a horizontal component in the east–west direction which is consistent with the landslide kinematic. The analysis of the displacement time series shows that, in the case of linear deformation trends, PS and SBAS results are similar, whereas, in the case of high oscillations and non-linear behavior, SBAS technique can provide a better estimation of the displacements. Besides, SBAS provides smoother and less noisy displacement time series. However, both the techniques showed their high capability in monitoring the evolution of the landslides, which is crucial for the implementation of effective risk prevention and mitigation strategies. To deep investigate the differences between the two techniques, other geomatic methodologies, based on global navigation satellite system and terrestrial laser scanning, should be used.

Keywords

DInSAR techniques • PS • SBAS • Sentinel-1 • Shallow landslides • Pre-alps • Italy

Introduction

Differential interferometric synthetic aperture radar (DInSAR) is a powerful remote sensing technique for continuous detection and monitoring of land surface deformation, thanks to its cost-effectiveness and high-precision in the analysis of wide areas. In particular, this technique is capturing the attention of the landslide community in the last decades (Wasowski and Bovenga 2014). DInSAR uses a pair of complex-values SAR images, acquired at different time and from slightly different orbital positions, to generate an interferogram. The phase difference obtained from the two acquisitions can be converted into surface land displacement along the satellite line of sight (LOS) (Zeni et al. 2014). Multi-temporal interferometry methods, i.e. permanent scatterers (PS) (Ferretti et al. 2000, 2001; Crosetto et al. 2016) and small baseline subset (SBAS) (Berardino et al. 2002; Casu et al. 2006), overcome the limitation of DInSAR phase disturbance, such as atmospheric artifacts and topographic inaccuracies, leading to successful applications in

X. Chen (✉) · M. Floris
Department of Geosciences, University of Padua, via G. Gradenigo 6, 35131 Padova, Italy
e-mail: chenxue@cugb.edu.cn

M. Floris
e-mail: mario.floris@unipd.it

G. Tessari
Sarmap SA, Via Stazione 52, 6987 Caslano, Switzerland
e-mail: giulia.tessari@sarmap.ch

M. Fabris · V. Achilli
Department of Civil, Environmental and Architectural Engineering, University of Padua, via G. Marzolo 9, 35131 Padova, Italy
e-mail: massimo.fabris@unipd.it

V. Achilli
e-mail: vladimiro.achilli@unipd.it

© Springer Nature Switzerland AG 2021
N. Casagli et al. (eds.), *Understanding and Reducing Landslide Disaster Risk*, ICL Contribution to Landslide Disaster Risk Reduction,
https://doi.org/10.1007/978-3-030-60311-3_17

landslide investigations (e.g. Colesanti et al. 2003; Hilley et al. 2004).

PS-InSAR technique generates differential interferograms with one common master identifying persistent point-wise reflectors, such as manmade structures and rocks. It is generally applied to analyse deformation affecting urban areas, where the number of persistent scatterers is higher than in natural environments. This technique considers a deformation model (usually a linear model), avoiding phase filtering and unwrapping, simplifying the processing chain compared to the SBAS one.

SBAS-InSAR technique relies on a redundant network of image pairs, with short spatial and moderate temporal baseline, detecting the temporal evolution of the surface deformations and increasing the spatial coverage, especially over non-urban areas. This technique extracts the deformation time series from the observed filtered and unwrapped phases. Considering the much higher number of generated interferograms, this technique is more time-consuming from the computational viewpoint and for the operator intervention too.

In this paper, a comparison between the results obtained from PS and SBAS processing of Sentinel-1 data is reported, in terms of velocity maps and displacement time series, covering the time period 2014–2019. The analysis was carried out in an area affected by shallow landslides, located in the north-eastern Italian pre-Alps. In this area, previous studies (Tessari et al. 2017) have shown how interferometric analysis of several SAR datasets, including Sentinel-1A, represent a useful tool to investigate the instability phenomena.

Study Area

The study area, named Rovegliana, includes 4.2 km^2 wide unstable slopes located in the north-eastern Italian pre-Alps (Fig. 1). Several small agglomerates of houses are placed along the slopes facing to the Agno torrent. Elevation ranges from 800–900 m to 400–330 m a.s.l. and the average slope gradient is about 21 degrees.

The bedrock of the slopes is constituted by two heteropic formations deposited during middle Triassic: *Recoaro* limestone and *Gracilis* Formation. The first one outcrops in the upper part of the slopes and is composed by limestones, marly and dolomitic limestones. The second one outcrops in the middle and lower part of the slopes and consists of an alternance of sandy and marly limestones, interbedded with evaporitic dolomites. These formations are highly fractured due to the tectonic events that occurred during the Upper Triassic-Jurassic and the Alpine orogeny.

The whole area is prone to instabilities of alluvial and colluvial depositions resulting in large quantities of debris material with thickness up to 10 m. The grain size of the debris is very heterogeneous, from millimetric to decametric clasts immersed in a clayey and silty sand matrix. Locally, morphological evidences, such as bumps, dips and sudden changes in the slope, reveal the presence of large boulders in the debris, dislocated from the calcareous formation located at the top of the area.

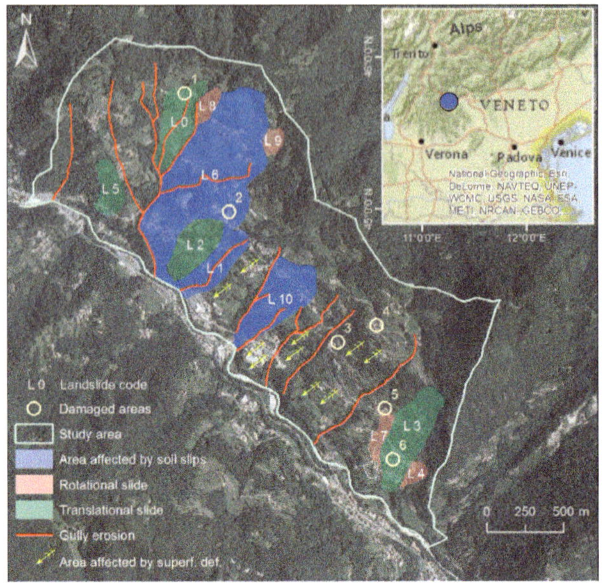

Fig. 1 Location of the study area (inset) and map of the main gravitational and erosional processes. The most damaged areas due to instability phenomena are indicated

The slope instabilities were identified through in situ investigations, aerial photos interpretation and remote sensing surveys (GPS and DInSAR). They consist of translational and rotational slides, soil slips and superficial slow deformations (creep) which involve the debris cover (Fig. 1). Slide phenomena and soil slips have a high state of activity and mainly occur in the wet season (Autumn) after rainfall events (Toaldo et al. 2016; Tessari et al. 2017). Superficial deformations have displacement rates of few millimetres per year estimated by previous remote sensing surveys. They do not show clear geomorphological evidences, but movements result in damages (cracks) to buildings, walls and road network, and upward curvature of trees.

Data and Methods

Ground deformation over the study area has been measured using both ascending and descending Sentinel-1 C-band SAR images, acquired in interferometric wide swath mode, with a 12-day or 6-day revisit time and a spatial resolution of about 15 m. 216 images acquired from ascending track 117 (30 March 2015 to 04 November 2019) and 233 images acquired from descending track 95 (24 October 2014 to 03 November 2019) have been processed.

The multi-temporal process of Sentinel-1 data has been performed through SARscape COTS, using both PS and SBAS algorithms. These approaches provide their best performances on different types of land cover and objects, point targets and distributed targets respectively (Pasquali et al. 2014).

The PS technique analyses the deformation of point scatterers with high temporal stability of the backscattered signal. It establishes a deformation model based on the phase difference of each pixel individually, without performing any phase unwrapping. This leads to preserving the maximum spatial resolution and the total independency of adjacent pixel measurements.

The SBAS technique measures deformations of distributed targets, e.g. sparsely urbanized areas and open fields. In fact, the volume decorrelation typical of natural distributed targets is reduced through an adaptive filtering step. The SBAS processing chain has been applied using an intermittent approach, which consists of extending the analysis to those resolution cells where the information has some temporal gaps because of the signal decorrelation, leading to coherence values smaller then the established acceptable threshold, equal to 0.3 in our analysis. Therefore, SBAS intermittent approach allows to spatially extend the final results. However, the results reliability is guarantee through two parameters establishing the minimum acceptable percentages of interferograms and images, to make sure that most of the deformation temporal information is preserved and directly calculated from the interferograms and using interpolation in the limited decorrelated temporal intervals. In detail, the analyses considered 60% as the percentage of interferograms and 95% as the minimum valid acquisitions, which means that pixels covering at least 60% interferometric connections of the whole connections and 95% acquisitions of all the acquisitions are maintained in the final result. All the pixels which were not respecting this controls have been discarded.

PS connection sets one image as the master, which is usually in the middle of the temporal and spatial distribution of acquisitions, in order to maintain a high coherence with most of the other images. The master image acquisitions for ascending and descending are 25 January 2018 and 25 November 2017, respectively. For SBAS connections, we set 36 days and 100 m as the temporal and spatial baselines constraints. In this case, we had to manually insert additional connections before the launch of Sentinel-1B, because of the low acquisition frequency of 12 days. Then, about 900 pairs were obtained.

We compared the results from PS- and SBAS- InSAR techniques considering spatial coverages, velocity distribution, capability of identifying landslides, and displacement time series.

Results

Ascending and descending velocity maps derived by PS and SBAS techniques are shown in Fig. 2. PS and SBAS results show different spatial coverage and quite similar displacement rates in the coinciding points.

PS points are mainly located in the small urban agglomerates or roads (see Figs. 2a, b and 1). The density of PS points in the entire study area is 262 per kilometres in ascending orbit and 437 per kilometres in descending. In the landslide areas, the density is higher, with 413 and 767 PS per kilometres in ascending and descending orbits, respectively. SBAS results cover most of the study area, providing information not only over anthropic structures but also on non-urban areas.

Results derived by the SBAS processing of ascending dataset cover the 85% (3.7 km^2) of the entire study area, those derived by descending track cover the 76% (3.2 km^2). Landslide areas are almost totally covered by both ascending and descending SBAS results.

The comparison between displacement rates estimated by the two techniques in all the coinciding points is reported in Fig. 3. The mean and standard deviation (std) of difference values (SBAS velocity minus PS ones) are 1.99 and 2.17 for ascending, and -0.91 and 1.39 for descending datasets, respectively.

Considering the result coverage in the landslide areas, just one lanslides (L8) doesn't contain any PS both in ascending and descending orbit, while the others contain 4 or more PS with low variability in the estimated velocities (Tables 1 and 2). In the case of SBAS, more than 80% of landslide areas are covered by the results of the processing. In this case, the estimated velocities present quite high variability for each landslide and the mean values are generally significantly higher than those calculated by PS.

To compare the results from the two techniques, the displacement time series obtained by the processing of descending dataset in the most damaged areas have been considered (Fig. 4). In general, time series trends and shapes are very similar (Fig. 4b, c, and f), but SBAS series appear smoother and less noisy. PS and SBAS time series plotted in areas 1 and 4 (Fig. 4a and d) show differences in the displacement trends due to a divergence in deformation rate in the first part of the monitoring period.

Discussion

As expected, the SBAS technique provided a displacement rate estimation for a larger part of the study area compared to the PS one. However, both techniques provided very interesting information on the behaviour of the landslides

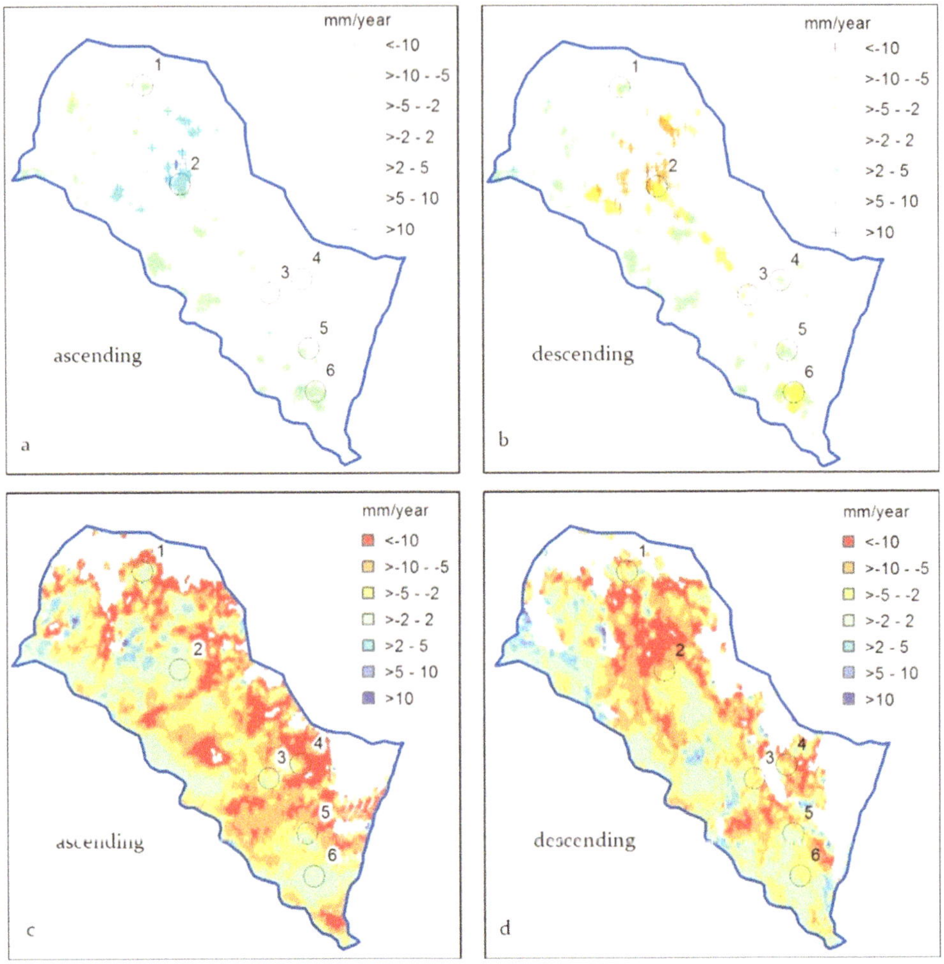

Fig. 2 Velocity maps derived by PS- (**a**, **b**) and SBAS- (**c**, **d**) InSAR processing of Sentinel-1 SAR data acquired in ascending (**a**, **c**) and descending (**b**, **d**) tracks. Black circles indicate the areas most damaged by instability phenomena

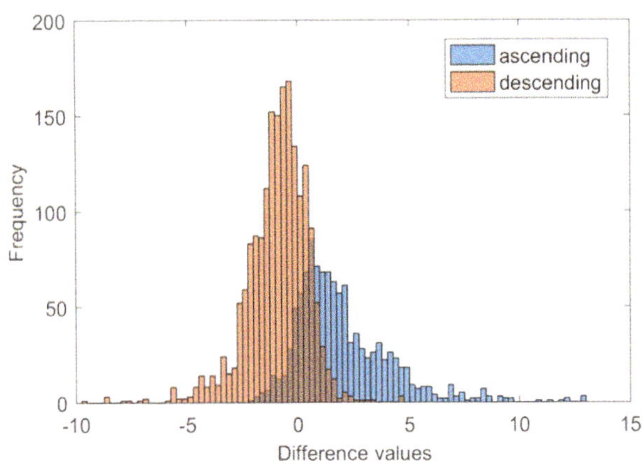

Fig. 3 Difference between velocities obtained by SBAS and PS techniques in all the coinciding points

affecting the investigated slopes. Both PS and SBAS results show that in the most active sectors of the area the displacements measured from ascending dataset are positive, while the descending ones are negative, which means that a horizontal component from east to west is present. These results are consistent with the landslide kinematic which mainly consist of superficial mass movements along the maximum slope direction, which has a dip toward the south-west of about 21°.

Considering the whole study area, the differences between the two techniques in the estimation of the displacements in the coinciding points are quite low (see Figs. 3 and 4) and caused by the different approaches. PS usually considers only a single pixel located in a building having an independent behaviour. Otherwise, SBAS measures a multilooked pixel which mediates the information of building with the surrounding area. In addition, SBAS includes a filtering step that makes the pixels spatially correlated.

Analysing each landslide, we found significant differences in the mean velocity and its variability. In particular, velocities estimated by SBAS are higher and show high variability. In the case of PS, only urban areas which are generally located on flat or gently slope, were detected, while SBAS provides the deformation rates of also non-urban and steepest sectors of the landslides which are generally higher. For this reason, SBAS technique can be considered more effective than PS in detecting and monitoring landslide phenomena.

Table 1 Comparison between velocities estimated by PS and SBAS in landslide areas (ascending orbit)

Landslide		PS asc			SBAS asc		
Code	Area (km^2)	Number of points	Mean vel. (mm/y)	Std	Cover. (%)	Mean vel. (mm/y)	Std
L0	0.100	53	0	0.5	100	−6.2	5.0
L1	0.080	11	0.8	0.6	100	−6.0	3.6
L2	0.080	8	1.7	0.9	100	−4.8	3.9
L3	0.140	117	0.3	0.4	85	−4.1	4.4
L4	0.020	4	0	0.6	77	−1.1	0.9
L5	0.050	19	0.3	0.3	96	−3.1	2.5
L6	0.500	220	3.4	1.3	90	−3.8	5.3
L7	0.020	29	0.8	0.8	100	−2.2	1.5
L8	0.002	0	-	-	100	−12.4	3.7
L9	0.002	7	1.7	0.2	100	−7.9	3.0
L10	0.150	28	0.8	0.3	94	−7.3	4.9

Table 2 Comparison between velocities estimated by PS and SBAS in landslide areas (descending orbit)

Landslide		PS desc			SBAS desc		
Code	Area (km^2)	Number of points	Mean vel. (mm/y)	Std	Cover. (%)	Mean vel. (mm/y)	Std
L0	0.100	55	−0.6	0.4	96	−7.7	5.8
L1	0.080	45	−0.2	1.3	100	−4.6	2.8
L2	0.080	30	−2.7	1.2	100	−6.9	2.9
L3	0.140	174	−2.1	1.1	79	−3.8	3.8
L4	0.020	18	0.3	0.4	83	−0.2	2.4
L5	0.050	39	0.3	0.6	85	−0.6	4.3
L6	0.500	435	−5.2	2.3	89	−9.1	4.5
L7	0.020	38	−0.4	1.7	100	−1.3	1.1
L8	0.002	0	-	-	100	−5.5	4.4
L9	0.002	8	−3.7	0.6	100	−6.5	3.2
L10	0.150	80	−1.7	0.8	98	−2.0	3.3

Regarding the differences in the displacement time series obtained by the two techniques in the case of the damaged areas 1 and 4 (Fig. 4a and d), they can be due to the low frequency of acquisitions before the launch of Sentinel 1B satellite. The low number of SAR images can limit the potential of PS approach in detecting a non-linear trend of the displacement as occurred in the first part of the time series. Therefore, SBAS results should be considered more reliable.

Conclusions

In this study, we compared PS and SBAS InSAR techniques in monitoring shallow landslides affecting an area located in the north-eastern Italian pre-Alps. Both techniques provided very useful information on the landslides. But SBAS has shown better reliability in landslide detection and monitoring because of the larger coverage of the results and the ability to measure non-linear deformation patterns. Mass movements are often characterized by seasonal oscillations or accelerations, in this case, SBAS can provide smoother and more detailed displacement time series, leading to deeper insights on the temporal evolution of instability phenomena. SBAS allows monitoring both the landslides and the deformations of structures and infrastructures due to such phenomena, which is crucial for the implementation of effective risk prevention and mitigation strategies. However, also PS technique can provide information on the main landslide characteristics, but the results are mainly limited to urban areas, so that it can be considered as a very useful tool for the monitoring of the elements at risk.

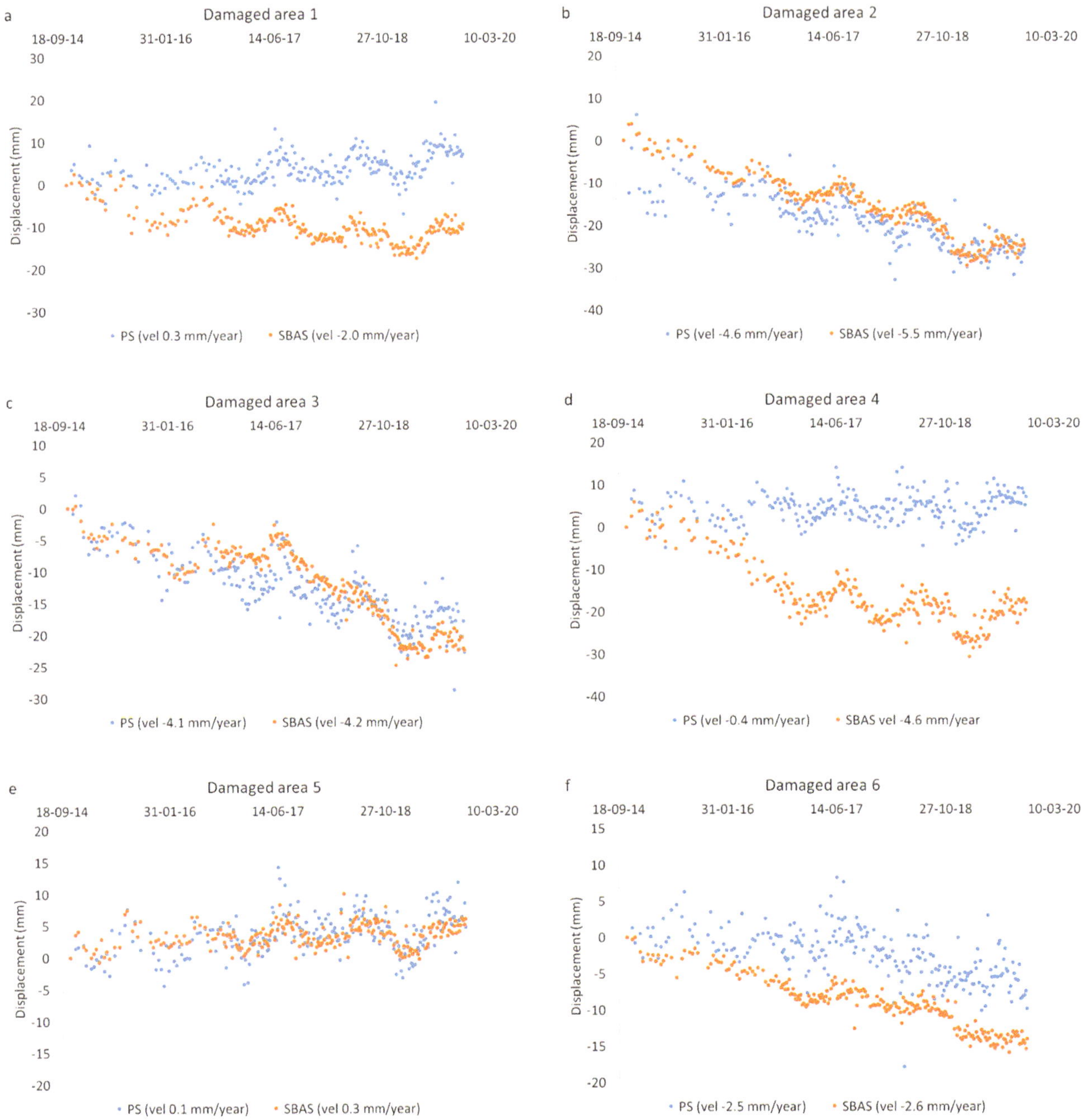

Fig. 4 Displacement time series derived by PS and SBAS processing of descending Sentinel-1 data in the areas most affected by damages (location in Fig. 1)

In the next future, the obtained results will be integrated through ongoing GPS and terrestrial laser scanner surveys, to verify and, eventually, calibrate interferometry data and better understand the relationships between landslides and damages to anthropic structures.

References

Berardino P, Fornaro G, Lanari R, Sansosti E (2002) A new algorithm for surface deformation monitoring based on small baseline differential SAR interferograms. IEEE Trans Geosci Remote Sens 40(11):2375–2383

Casu F, Manzo M, Lanari R (2006) A quantitative assessment of the SBAS algorithm performance for surface deformation retrieval from DInSAR data. Remote Sens Environ 102(3):195–210

Costantini M (1998) A novel phase unwrapping method based on network programming. IEEE Trans Geosci Remote Sens 36(3):813–821

Colesanti C, Ferretti A, Prati C, Rocca F (2003) Monitoring landslides and tectonic motions with the permanent scatterers technique. Eng Geol 68(1–2):3–14

Crosetto M, Monserrat M, Cuevas-Gonzales M, Devanthery N, Crippa B (2016) Persistent scatterer interferometry: a review. ISPRS J Photogramm Remote Sens 115:78–89

Ferretti A, Prati C, Rocca F (2000) Nonlinear subsidence rate estimation using permanent scatterers in differential SAR interferometry. IEEE Trans Geosci Remote Sens 38(5):2202–2212

Ferretti A, Prati C, Rocca F (2001) Permanent scatterers in SAR interferometry. IEEE Trans Geosci Remote Sens 39(1):8–20

Goldstein RM, Werner CL (1998) Radar interferogram filtering for geophysical applications. Geophys Res Lett 25(21):4035–4038

Hilley GE, Bürgmann R, Ferretti A, Novali F, Rocca F (2004) Dynamics of slow-moving landslides from permanent scatterer analysis. Science 304(5679):1952–1955

Pasquali P, Cantone C, Riccardi P, Defilippi M, Ogushi F, Gagliano S, Tamura M (2014) Mapping of ground deformations with interferometric stacking techniques. Land Appl Radar Remote Sens, Holecz F, Pasquali P, Milisavljevic N, Closson D, IntechOpen. https://doi.org/10.5772/58225

Tessari G, Floris M, Achilli V, Fabris M, Menin A, Monego M (2017) Testing sentinel-1A data in landslide monitoring: a case study from North-Eastern Italian pre-Alps. Workshop on world landslide forum, 29 May–2 June 2017. Ljubljana, Slovenia, pp 209–217

Toaldo M, Tessari G, Monego M, Achilli V, Fabris M, Menin A, Floris M (2016) Preliminary study of conditions influencing slope dynamics in the area of Rovegliana (North-Eastern Italian pre-Alps). Rend Online Soc Geol Ital 41:207–209

Wasowski J, Bovenga F (2014) Investigating landslides and unstable slopes with satellite multi temporal interferometry: current issues and future perspectives. Eng Geol 174:103–138

Zeni G, Pepe A, Zhao Q, Bonano M, Gao W, Li X, Ding X (2014) A differential SAR interferometry (DInSAR) investigation of the deformation affecting the coastal reclaimed areas of the Shanghai megacity. In: IEEE geoscience and remote sensing symposium, 13–18 July 2014. Quebec, Canada, pp 482–485

Analyses of Koitash Landslide, Affecting Mailuu Suu Valley, Kyrgyzstan, Through Integrated Remote Sensing Techniques

Giulia Tessari, Loris Copa, Giaime Origgi, Almazbek Torgoev, Lars Uhlig, and Francesco Holecz

Abstract

This work presents an integrated application of remote sensing techniques on monitoring post event conditions of the Koitash landslide, affecting Mailuu Suu valley, Kyrgyzstan. This area is highly prone to landslide affecting the slopes next to the numerous radioactive waste rock dumps and tailings storage facilities disseminated along the valley. In this hazardous context, we propose a methodology to map the location of recent landslides, active slopes and strong surface variation, over large areas. The analysis is tested on the Koitash landslides and combines several remote sensing datasets. First, a high-resolution Digital Surface Model (DSM) is generated using optical tri-stereo acquisitions. The DSM defines the topographic condition before the slope collapse. Secondly, a fast mapping of land cover changes is performed through satellite optical data. Finally, multi-temporal interferometric techniques are used to identify strong topographic variation on the Koitash slope after the collapse and small post event surface deformation affecting the crown area and the landslide deposits. Results highlighted the consequences of the landslide on diverting Mailuu Suu river path, highlighting a strong topographic variation due to the slope collapse. Moreover, ongoing surface deformations have been identified not far from tailings storage facilities containing radioactive deposits. The products are generated mainly from freely available satellite acquisitions, proving the capability of mapping and monitoring wide areas through a cost-effective approach.

Keywords

Mailuu Suu landslide • Radioactive waste tailings • DInSAR techniques • SBAS • Sentinel-1 • Digital surface model • Land cover changes • Landslide monitoring

Introduction

Mailuu Suu valley, located in the southern border of Kyrgyzstan, Central Asia, is known for the dramatic environmental conditions due to the presence of the storage sites of uranium radioactive waste, resulted from the uranium mining and postprocessing completed in the late 1960s. Thousands of tonnes of uranium ore and processing waste are deposited all along this remote valley, in a definitely sensitive setting. In fact, tailings storage facilities are often located on weakly stable rocks and close to landslides-hazardous slopes (Torgoev 2008). Therefore, this context is creating an alarming environmental scenario that could lead to the contamination of the water supply for the Fergana valley, the most densely populated part of Central Asia. This is why this area is considered as one of the most polluted places in the planet. Numerous studies were carried out to analyse the main factors triggering landslides, as Golovko et al. (2017) did in South Kyrgyzstan, generating a landslide inventory for hazard assessment through remote sensing data, to facilitate an analysis at regional scale in case of information scarcity.

G. Tessari (✉) · L. Copa · G. Origgi · F. Holecz
sarmap SA, Via Stazione 52, 6987 Caslano, Switzerland
e-mail: giulia.tessari@sarmap.ch

L. Copa
e-mail: loris.copa@sarmap.ch

G. Origgi
e-mail: giaime.origgi@sarmap.ch

F. Holecz
e-mail: fholecz@sarmap.chz

A. Torgoev · L. Uhlig
WISUTEC Umwelttechnik GmbH, Jagdschaenkenstrasse 50, 09117 Chemnitz, Germany
e-mail: a.torgoev@wisutec.de

L. Uhlig
e-mail: l.uhlig@wisutec.de

© Springer Nature Switzerland AG 2021
N. Casagli et al. (eds.), *Understanding and Reducing Landslide Disaster Risk*, ICL Contribution to Landslide Disaster Risk Reduction, https://doi.org/10.1007/978-3-030-60311-3_18

Mailuu Suu area have been studied as some hundred landslides have affected the region in the last decades (Aleshin and Torgoev 2014). The recent re-activation of most of those paleolandslides has been influenced by mining activities and urbanization combined with geological and climatic main triggering factors. In this work, we focus on the re-activation of the Koitash landslide, affecting Mailuu Suu valley on the 28 April 2017. The landslide was characterized by a sudden and rapid kinematic and the total change of the slope surface. The rapid collapse of the slope caused the failure of a large amount of sediments, which accumulated just next to some of the tailing storage facilities containing radioactive wastes.

Landslide risk management is increasingly exploiting remote sensing techniques, often combining radar and optical remote sensing data, taking advantage of their multispectral characteristics, high revisiting cycle, wide area coverage and high spatial resolution (Metternicht et al. 2005). Several applications show how satellite acquisitions can support a detailed landslide inventory mapping. This can be address to prevention and disaster risk reduction activities but also to emergency response phase to estimate the event extensions, the caused damages and the ground motion situation and evolution (Casagli et al. 2016).

Here, we present an analysis of the post collapse condition of Koitash landslide. We integrated and combined different remote sensing techniques to analyse both optical and Synthetic Aperture Radar (SAR) satellite data. We performed a fast and automatic land cover mapping at large scale through Sentinel-2 data to detect the changes induced by the landslide; we generated a high-resolution digital surface model using tri-stereo optical data to define the slope topography before the event; finally we applied multi-temporal interferometric techniques on Sentinel-1 data, acquired after the event, to estimate the slope topographic changes after the collapse and the eventual ongoing deformation affecting the area. The choice of focusing on the post event allows avoiding the temporal decorrelation affecting the data before and after the collapse (Tessari et al. 2017). The benefit of the proposed approach is here presented, highlighting the cost-effectiveness, accuracy and scalability over larger areas towards an operational system

Study Area

Mailuu Suu valley is located in the western boards of Kyrgyzstan, in Jalal-Abad region (Fig. 1). This area is set in the tectonically and seismically active Tien Shan Mountains, affected by several strong earthquakes in the last century (Havenith et al. 2006). This area is known for the intense mining activities dedicated to radioactive uranium ore extraction, mainly between 1946 and 1968. In fact, in the framework of the Soviet nuclear program, around 9,100 tons of Uranium oxide have been produced. These activities have been abandoned later, as they were no more profitable. More than 23 unstable Uranium tailings storage facilities with unstable dams have been left by the Soviet Union on a tectonically unstable hilly area above the town. Nowadays, Mailuu Suu is one of the most polluted and radioactive places in the world. Moreover, the frequent occurrence of natural hazards, as landslides or floods, make this situation even more dramatic and dangerous.

This environmentally hazardous area was affected by a landslide, Koitash landslide, the 28th April 2017. This dramatic event caused the river blockage and the formation of a natural dam and the consequent migration of Mailuu Suu river, which is flowing at the bottom of the main slope (Fig. 2). The location of the unstable slope is shown in Fig. 3, where two optical satellite acquisitions allowed comparing the slope before and after Koitash landslide.

Methods

Several remote sensing analysis and processing techniques have been integrated in this study, focusing in several datasets and sensors. The conceptual analysis workflow is shown in Fig. 4, where blue boxes refer to the input datasets required to generate the maps (green boxes) and monitoring (yellow boxes) outputs.

More in detail, a high resolution Digital Surface Model (DSM) was generated from stereo or tri-stereo satellite optical data. The DSM was generated using the Space Stereo module of the Opticalscape© software, following a standard photogrammetric workflow. The generated DSM was used as reference for the two additional analyses steps, the Optical and the SAR data multi temporal analysis.

Satellite optical data were analysed to generate Land Cover Map (LCM), to verify the surface condition before and after the landslide event. The series of LCMs allowed identifying eventual changes induced by the slope collapse. The LCM were generated using Mapscape© software to download, process and extract the entire layer needed to be ingested into an own developed knowledge-based classifier. The LCM describes the area of interest in terms of settlements, vegetated area, dense forest, open land and bare soil.

The final steps of the workflow consist of the multi-temporal interferometric analysis of Synthetic Aperture Radar (SAR) data acquired with both ascending and descending geometry to retrieve surface deformation and estimate the collapsed volume. SAR datasets were analysed through the multi-temporal Small Baseline Subset (SBAS) technique (Berardino et al. 2002), implemented in the SARscape© software, to retrieve the evolution of deformations, highlighting potential accelerations, describing the

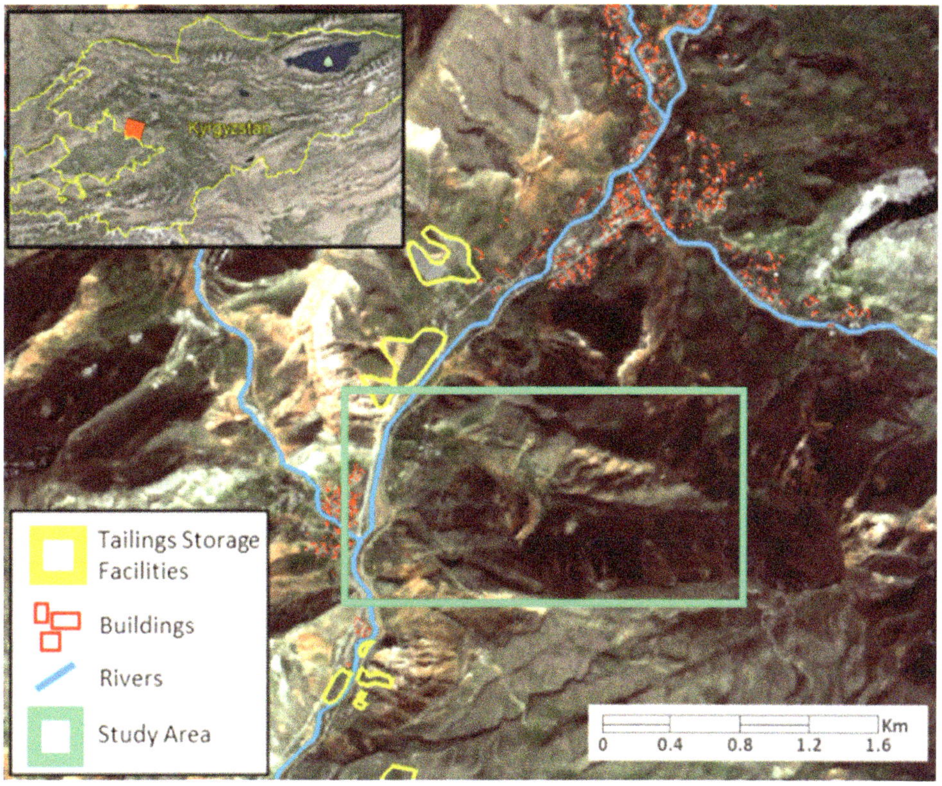

Fig. 1 Optical image of the slope affected by Koitash landslide (green box), showing the location of tailings storage facilities (yellow polygons), buildings (red polygons) and Mailuu Suu river (light-blue polyline)

Fig. 2 Photo of Koitash landslide few days after the collapse, showing the deviation of Mailuu Suu River

ongoing displacement and the potentially unknown instability phenomena affecting the area even after the 2017 collapse. The previously generated DSM was used as a reference in the SBAS processing. The SBAS was run using an intermittent approach, hence accepting some limited temporal decorrelations due to the vegetation seasonal changes, without losing the overall final information. Velocity and sequence of deformations obtained from the ascending and descending datasets were then combined and projected to obtain vertical and east-west velocity maps and time series of deformations, giving a more detailed information about the slope deformation directions. Moreover, the SBAS chain was calculating the topographic correction obtained from the SAR dataset respect to the reference DSM. The topographic variation was used to calculate the collapsed volume on the slope.

Data

The datasets selected and processed were identified trying to take advantage to freely available data. Only in case of the DSM generation, commercial Optical acquisitions were used as the highest resolution of the freely available ones is 30 m.

The DSM was generated from imagery acquired from the PRISM (Panchromatic Remote-sensing Instrument for Stereo Mapping) system, a stereoscopic imaging instrument on-board the ALOS-1 satellite. The instrument provides

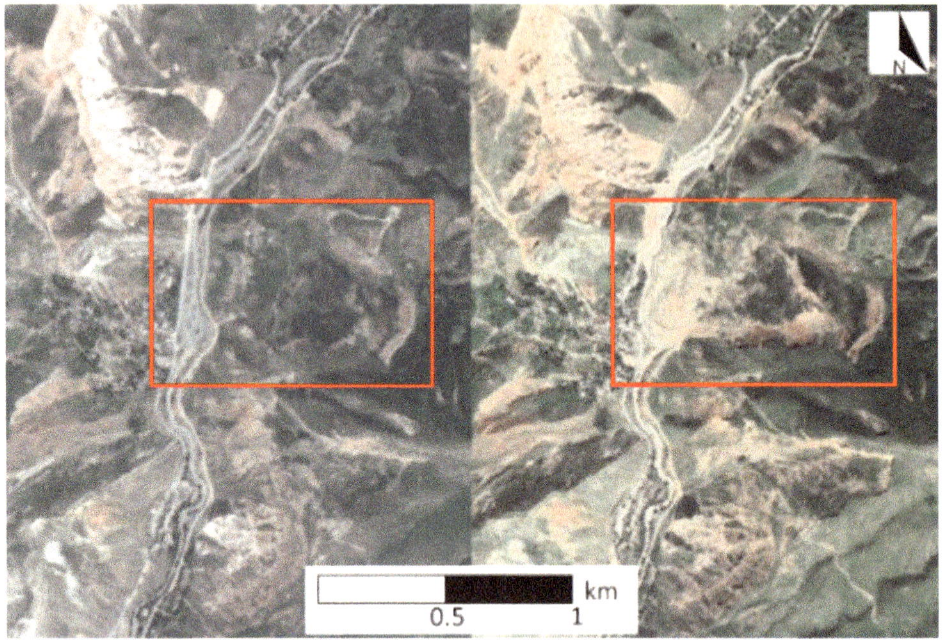

Fig. 3 Koitash landslide, located in Mailuu Suu valley, Kyrgyzstan. Satellite optical views of the study area before (left) and after (right) the landslide event. Red boxes identify the unstable slope (*Source* Google Earth)

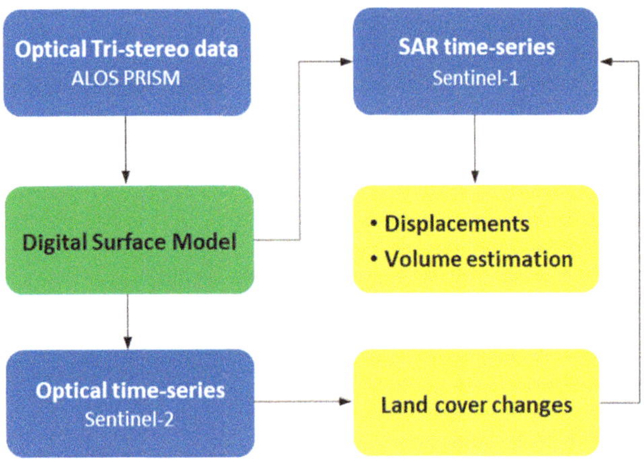

Fig. 4 Methodology conceptual scheme

tri-stereo along-track imagery with a ground sampling distance of 2.5 m at Nadir and a Swath Width of 35 km. The images were acquired in May 2008. The generated DSM has a 5 m resolution. This product is one of the most cost-effective available stereo data, considering their resolution and the cost per square km.

The LCM before and after the landslide event were generating using the freely available Sentinel-2 data. Sentinel-2 data are EC Copernicus spaceborne multi-spectral optical acquisitions, at 10 m resolution. The images were acquired pre and post event, respectively on the 9th April 2017 and on the 8th June 2017.

The multi-temporal SAR interferometric analyses was focused on the freely available Sentinel-1 acquisitions. This EC Copernicus mission provided dense stacks of SAR images, acquired with two different geometries, along ascending and descending orbits, with a 15 m resolution and a 12-day revisiting time. In particular, the analyses considered the post event time period, using the data acquired after the slope collapse, from July 2017 up to January 2020. The analysed datasets consist in 77 ascending images (track 100) and 76 descending images (track 5).

Results

The LCMs generated from Sentinel-2 data are shown in Fig. 5, where black boxes highlight the pre and post landslide condition. Most of the slope, classified as vegetated area and dense forest before the event, changed into bare soil after the collapse. In particular, bare soil on the slope toe corresponds to the collapsed material. The river path was deviated of approximately 120 m because of the landside, causing possible inundation risks for the surrounding inhabited areas. First results allowed to fast mapping large areas, identifying the landslide effects from the land cover variation.

The high-resolution DSM generated from PRISM data extends over 35 × 35 km^2 area. This output has been used as reference topographic layer during the Sentinel-1 SBAS processing. Respect to the existing freely available low resolution DSM, working with a high resolution DSM was helping on estimating and removing the topographic component from each interferogram phase difference, reducing the presence of topographic residual fringes and facilitating the unwrapping step. Moreover, the PRISM DSM refers to 2008 acquisitions, before the landslide event, while

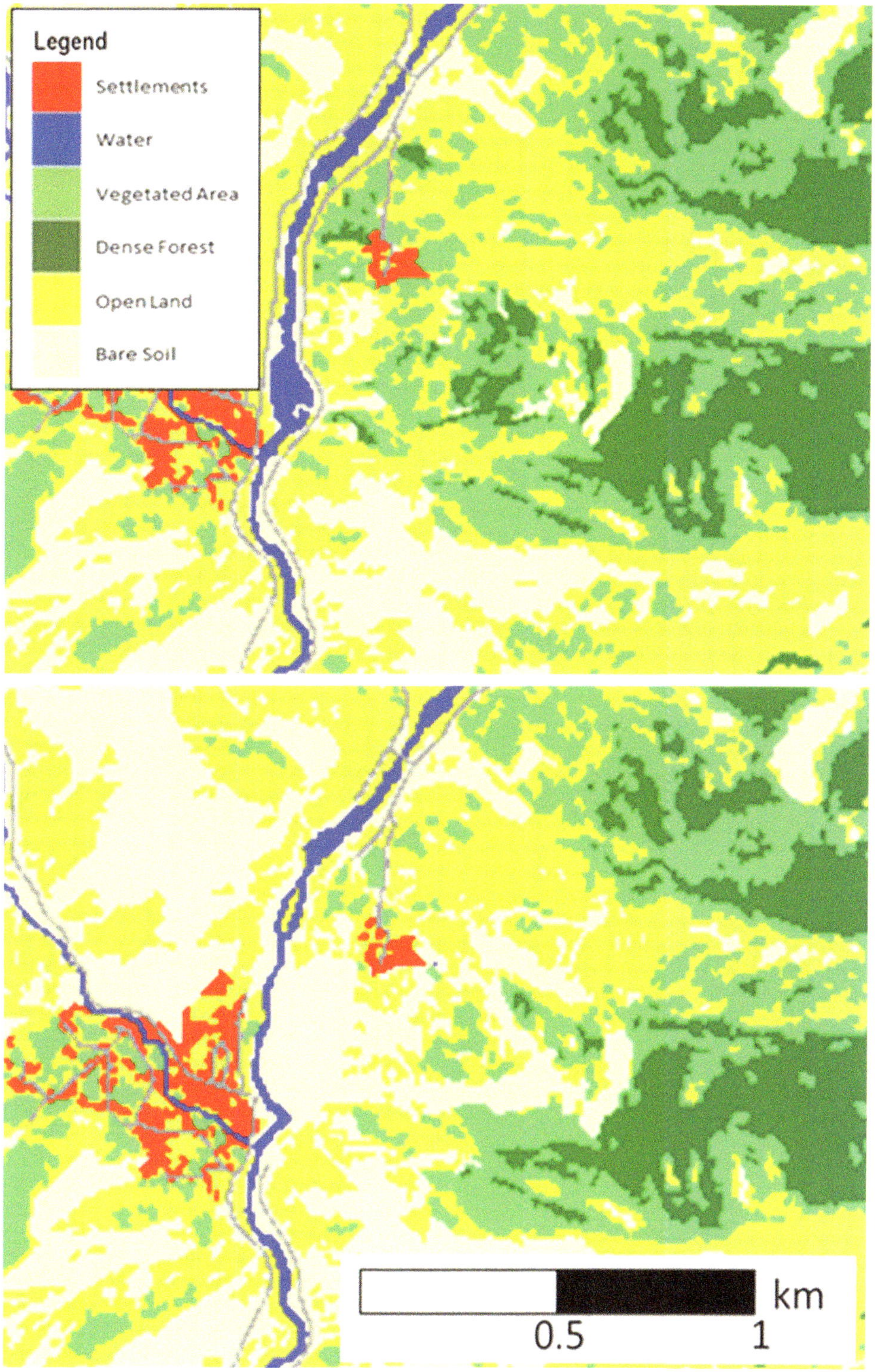

Fig. 5 Land cover maps before (above) and after (below) the Koitash landslide, collapsed the 28 April 2017. Black boxes identify the slope affected by the landslide. Comparison shows the variation of the river flow and increasing of bare soil due to mass movements

Sentinel-1 data span the period 2017–2020, hence a post landslide period. As a consequence, the topographic correction obtained from the SAR data, as one of the output of the SBAS process, measured the dramatic slope morphological change induced by the landslide. In fact, negative elevation corrections obtained in the landslide crown area and the upper part of the slope, correspond to an elevation reduction because of the material collapse. On the contrary, the lower part of the slope was interested by the sediment accumulation, leading to an increase of the elevation, and therefore a positive elevation correction. These results are shown in Fig. 6, where the red polygon highlights the landslide while the red segment shows the A-A′ cross-section which provides the elevation correction along the slope. The A-A′ profile in Fig. 7 shows a maximum sediment accumulation of almost 15 m, while, the depth of the sliding surface in proximity of the crown area is around 30 m. The collapsed along the scarp caused a downward displacement of 4.7 million m^3, as calculated directly from the topography correction in Fig. 6. Moreover, the accumulated material covers the original river path, in agreement with the LCMs in Fig. 5. Additionally, results obtained from Sentinel-1 provide the ongoing deformations affecting the slope after the landslide event. Mean rate of deformation maps, projected along the vertical and east-west directions, are shown in Fig. 8, where the red polygons identify the Koitash landslide.

The deformation colour scale ranges from blue to red, corresponding to subsidence/compaction and uplifting in case of vertical displacements, while in the horizontal

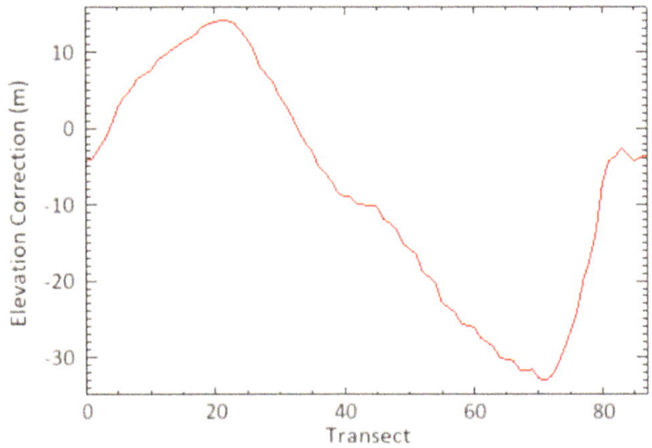

Fig. 7 Cross section of the elevation correction, respect to the reference DSM obtained from PRISM data (2008), obtained from SBAS multi-temporal processing of Sentinel-1 data acquired between 2017 and 2020. The location of the profile is shown in Fig. 6 and it is represented with a red segment

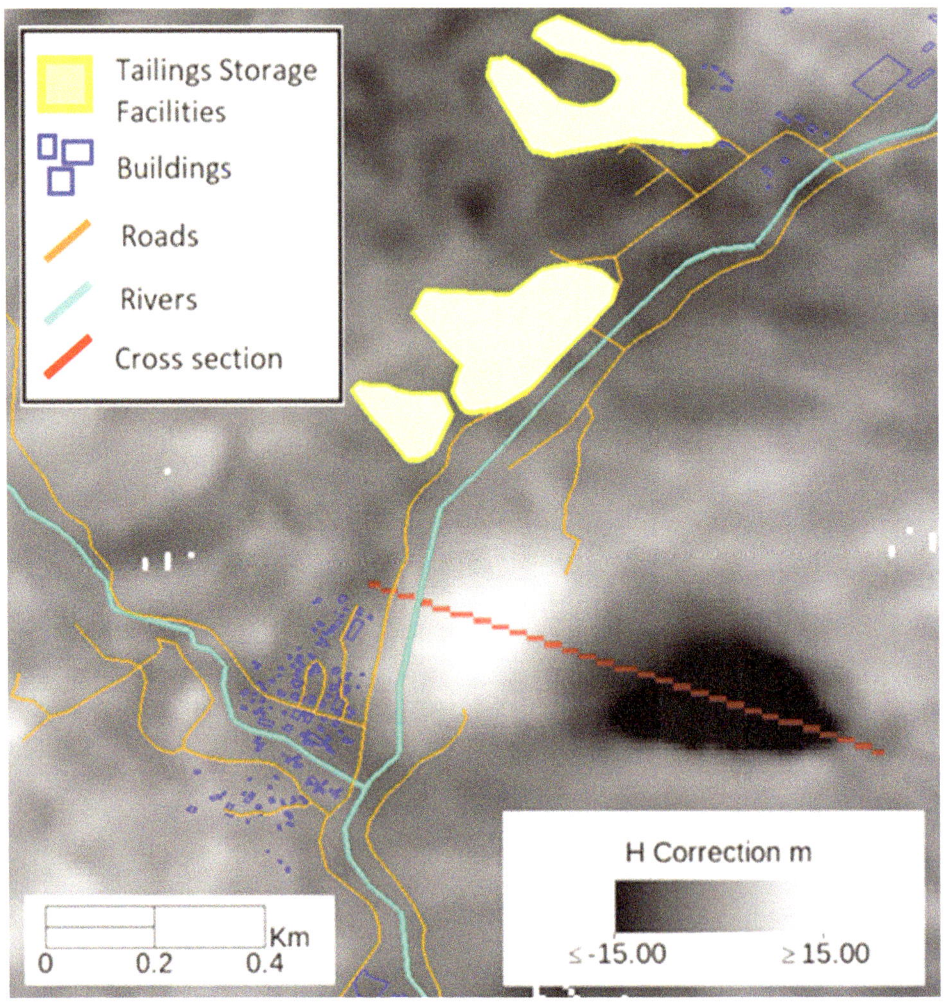

Fig. 6 Elevation correction, respect to the reference DSM obtained from PRISM data (2008), obtained from SBAS multi-temporal processing of Sentinel-1 data acquired between 2017 and 2020. The red polygon highlighted the Koitash landslide

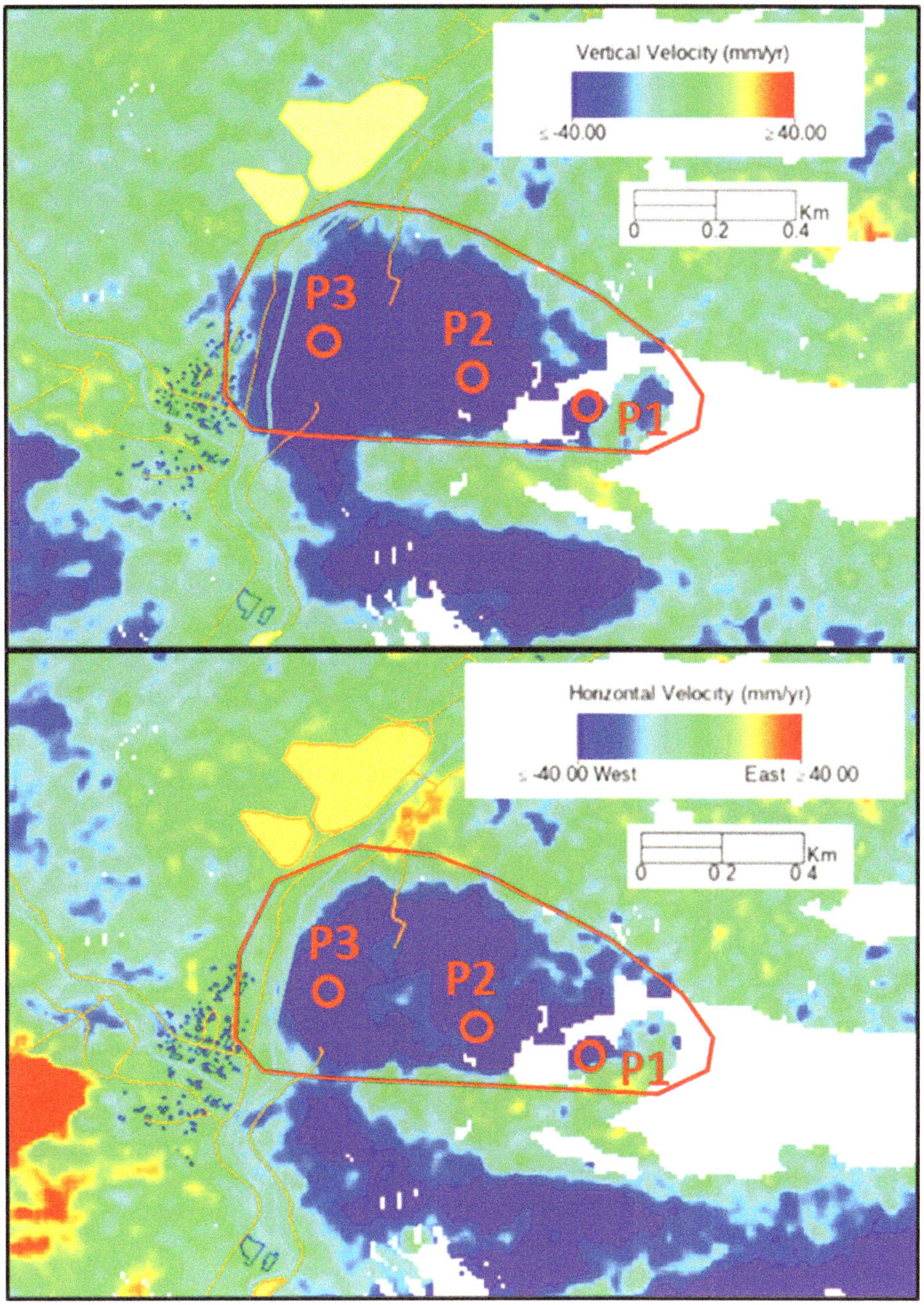

Fig. 8 Vertical (above) and East-West (below) mean velocity maps obtained from the processing of Sentinel-1 data acquired between 2017 and 2020. The red polygons identify the location of the landslide. Yellow polygons correspond to the radioactive tailing store facilities

deformation map, blue areas are moving westward, while red areas are moving eastwards, depending on slope aspect. Results in Fig. 8 show how the slope is clearly affected by vertical deformation as well as an ongoing horizontal displacement along the west direction, both with a rate overcoming 40 mm/year.

Three points along the slope have been selected in Fig. 8. Vertical and east-west time series of deformations are plotted (Fig. 9). It is possible to notice how the crown area is more affected by horizontal deformation than vertical ones, while both the central part and the toe of the slope are showing a higher vertical component. In addition, P2 and P3 are characterized by an almost constant linear trend of deformations, while P1 is showing an initial stability followed by an acceleration, more evident in the vertical component. This describes a re-activation of the deformation some months after the landslide. Moreover, the area affected by ongoing deformation is extending close to the tailing storage facilities, identified by the yellow polygons in Fig. 8, that is located at approximately 200 m far from the collapsed slope.

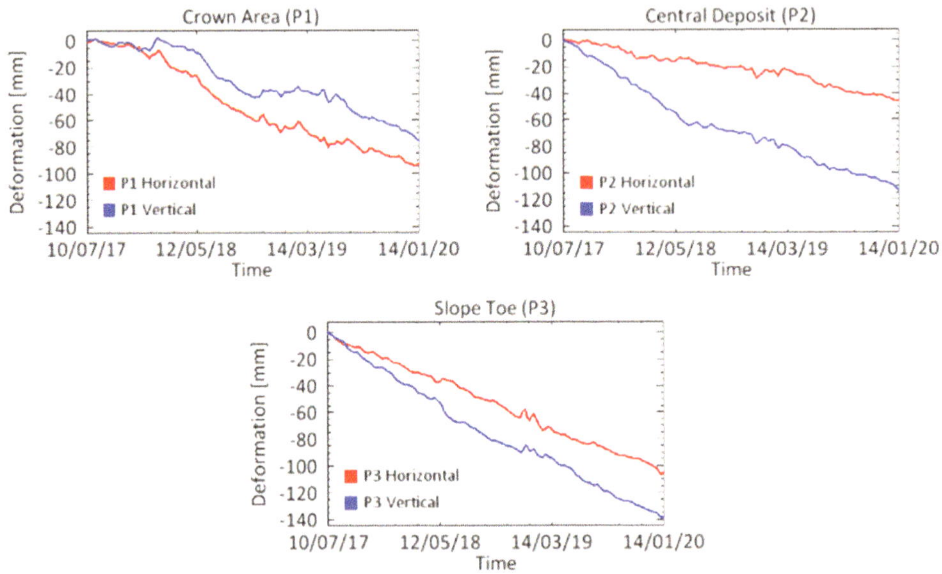

Fig. 9 Horizontal (red curves) and vertical (blue curves) time series of deformations plotted in P1, P2 and P3, located at the crown area, the central portion and the toe of the slope obtained from the processing of Sentinel-1 data acquired between 2017 and 2020. The location of points is shown in Fig. 8

Concluding Remarks

This study analysed the post event condition of Koitash landslide, located in Mailuu Suu area, integrating different remote sensing techniques to highlight the potential of this methodology in case of disaster management, prevention or intervention.

One of the main advantages of the proposed workflow is the capability of providing information over large areas, reducing any possible risk related to field monitoring, particularly relevant after a landslide collapse. In fact, a first fast mapping of satellite optical data allowed identifying the portion of the slope affected by the landslide and tracing the deviation of the Mailuu Suu river path. The generation of a 5 m resolution DSM supported the definition of the topography before the Koitash failure. Moreover, from interferometric SAR data it was measured the ongoing sliding trend, compaction and deformation affecting the slope even after the failure. The landslide crown area re-activated in the spring 2018, one year after the collapse, probably because of the seasonal rainfall events. The lower part of the slope is affected by constant deformation rates after the collapse, probably related to the compaction of the collapsed sediments. The topographic correction in respect to the reference DSM retrieved the morphology and elevation changes due to the landslide giving the necessary information to estimate the displaced volumes. Therefore, despite Sentinel-1 mission has been mainly designed to monitor deformations, results show how the multi-temporal and redundant generation of interferograms having multiple normal baseline values could provide useful information in respect to the topographic changes too.

Future developments of the work could focus on searching for precursor deformations affecting the slope before its collapse. In addition, despite the presented results focused on Koitash landslide, the study was conducted on a much larger area. In fact, taking advantage of the spatial coverage of the generated DSM and the freely available Sentinel-1 and Sentinel-2 datasets, it was possible to extend the proposed analysis to an area, which extends approximately 35 $\times$ 35 km^2. Over the full extension of the analysed data, several slopes showed ongoing deformations and the topographic changes indicated numerous collapses occurred between 2008 and 2017.

Considering the high environmental sensitivity of all Mailuu Suu valley, because of the numerous radioactive material storages, being able to identify even small deformations and dramatic topographic changes, is allowing to have a clearer idea of the most hazardous areas and their proximity to the uranium deposits. In this context, the proposed methodology demonstrated its cost-effectiveness and proved to be a powerful mapping tool to identify and quantify on-going processes in hazardous sites were additional ground measurements are required or where to install monitoring instrumentation.

Acknowledgements This work has been carried out within the ESA Integrated Application Program, Contract No 4000118912/16/NL/US. The authors also kindly acknowledge the European Space Agency (ESA) for making available the Sentinel-1 and Sentinel-2 images in the framework of EC Copernicus activities and the Japan Aerospace Exploration Agency (JAXA) for providing PRISM data.

References

Aleshin Y, Torgoev I (2014) Slope dynamic geomorphology of the Mailuu Suu Area, aspects of long-term prediction. In: Sassa K, Canuti P, Yin Y (eds) Landslide science for a safer geoenvironment. Springer, Cham. https://doi.org/10.1007/978-3-319-04999-1_48

Berardino P, Fornaro G, Lanari R, Sansosti E (2002) A new algorithm for surface deformation monitoring based on small baseline differential SAR interferograms. IEEE Trans Geosci Remote Sens 40(11):2375–2383

Casagli N, Cigna F, Bianchini S, Hölbling D, Füreder P, Righini G, Del Conte S, Friedl B, Schneiderbauer S, Iasio C, Vlcko J, Greif V, Proske H, Granica K, Falco S, Lozzi S, Mora O, Arnaud A, Novali F, Bianchi M (2016) Landslide mapping and monitoring by using radar and optical remote sensing: examples from the EC-FP7 project SAFER. Remote Sens Appl Soc Environ 4:92–108. https://doi.org/10.1016/j.rsase.2016.07.001

Golovko D, Roessner S, Behling R, Wetzel HU, Kleinschmit B (2017) Evaluation of remote-sensing-based landslide inventories for hazard assessment in Southern Kyrgyzstan. Remote Sens 9(9):9436. https://doi.org/10.3390/rs9090943

Havenith HB, Torgoev I, Meleshko A, Alionshin Y, Torgoev A, Danneeld G (2006) Landslide in the Mailuu Suu Valley, Kyrgyzstan—hazards and impacts. Landslides 3:137–147. https://doi.org/10.1007/s10346-006-0035-2

Metternicht G, Hurni L, Gogu R (2005) Remote sensing of landslide: an analysis of the potential contribution to geospatial system for hazard assessment in mountain environments. Remote Sens Environ 98:284–303

Tessari G, Floris M, Pasquali P (2017) Phase and amplitude analyses of SAR data for landslide detection and monitoring in non-urban areas located in the North-Eastern Italian pre-Alps. Environ Earth Sci 76:85. https://doi.org/10.1007/s12665-017-6403-5

Torgoev I (2008) Environmental effects of possible landslide catastrophes in the areas of radioactive waste warehousing in Kyrgyzstan (Central Asia). In: Sassa K, Canuti P (eds) Landslide—disaster risk reduction. Springer Science & Business Media, 31.5.8, pp 604–607

Landslide Monitoring in the Main Municipalities of Sikkim Himalaya, India, Through Sentinel-1 SAR Data

Giulia Tessari, Divya Kashyap, and Francesco Holecz

Abstract

Landslides are a large threat that generate important damage, economic losses and human fatalities worldwide. Identifying terrain deformations related to these phenomena is critical to reduce their impact. In particular, the population of the Indian Himalayas are menaced by landslides, which regularly endanger villages and vital transport axes. Here we present the monitoring of instabilities on some urbanized areas of Sikkim State testing the capabilities of Remote Sensing techniques, and particularly Multi-temporal analysis of Synthetic Aperture Radar data. The main municipalities of Sikkim State have been analysed, covering the most inhabited areas where landslide effects could be more dramatic for the local population. We illustrated the results obtained over the capital, Gangtok, and Namchi. These cases have been analysed thanks to the global availability of Sentinel-1 SAR data. The main multi-temporal interferometric techniques have been applied, Small Baseline Subset (SBAS) and Persistent Scatterers (PS) depending on the extension and density of the urban areas. These techniques allow to map and measure surface deformations with a millimetre sensitivity, providing the temporal evolution of the deformation trends. This is fundamental to identifying activation/re-activation of landslides and analyse eventual correlation with external factors. In case of Gangtok, a strong relation with the rainy season was easily identified. Advantages and limitation of these techniques in such arduous areas are discussed.

G. Tessari (✉) · F. Holecz
sarmap SA, via Stazione 52, 6987 Caslano, Switzerland
e-mail: giulia.tessari@sarmap.ch

F. Holecz
e-mail: francesco.holecz@sarmap.ch

D. Kashyap
Embassy of Switzerland, Swiss Cooperation Office, New Delhi, 110021, India
e-mail: divya.kashyap@eda.admin.ch

Keywords

Sikkim Himalaya • Landslide mapping • Multi temporal InSAR • Sentinel-1

Introduction

Landslides are a large threat to the population of the Indian Himalayas and regularly endanger villages and vital transport axes. Landslide incidences have been of serious concern to the societies of the affected regions because of their potential to cause the loss of life, deteriorate natural resources, damage infrastructural facilities, etc. According to a Geological Survey of India study, 12.6% of India's landmass falls under the landslide-prone hazardous zone and 8% of global landslide fatalities are reported from our country. Out of the total land area prone to landslide, 0.18 million sq km fall in the northeast Himalayas, including Darjeeling and Sikkim. Naithani (1999) has estimated that damages caused by landslides in the Himalayan range cost, on average, more than one billion US dollar per year, besides causing more than 200 deaths every year. Only in 1968, more than 33,000 fatalities were reported in Sikkim because of landslides (Choubey 1992). In the light of these numbers, it is evident that the monitoring of surface displacements and landslide movements becomes an important task.

Traditionally, landslide hazard assessment had to be carried out on-site, relying on ground-based methods (Bhasin et al. 2002). Largely field-based approaches are often made arduous by high risk exposure, as well as observational biases towards objects in easily accessible areas. Observable target areas are local to maximally sub-regional. Due to the large number and extent of the landslide prone areas in the Himalayas ground-based methods are not suited for rapid detection and monitoring of hotspot areas. Remote sensing approaches, on the other hand, enable objective, safe, and spatial continuous observations at different spatial

© Springer Nature Switzerland AG 2021
N. Casagli et al. (eds.), *Understanding and Reducing Landslide Disaster Risk*,
ICL Contribution to Landslide Disaster Risk Reduction,
https://doi.org/10.1007/978-3-030-60311-3_19

scales, covering large areas. Recently, studies focused on landslide susceptibility estimation have been performed on Sikkim cities, e.g. on Gangatok (Kaur et al. 2019). Despite several studies carried out at landslide prone areas, there is lack of in depth scientific study to map the temporal evolution of surface deformation affecting the main municipalities and urbanized areas.

Starting from the Nineties, Space-borne Synthetic Aperture Radar (SAR) data demonstrated to be an effective tool to monitor surface deformations affecting soil, structure and infrastructures. Numerous works were focus on the use of SAR data for detecting and monitoring landslides, clarifying their potentially and limits (Wasowski and Bovegna 2014; Barra et al. 2016; Béjar-Pizarro et al. 2017; Tessari et al. 2017a). Actually, despite numerous successful applications, it is fundamental to consider data decorrelation, which typically affect dense vegetated slopes, layover, shadow and signal distortions due to the acquisition geometry of SAR data and the limitations on the maximum detectable deformation velocities (Cascini et al. 2009). In detail, maximum detectable LOS deformation between two adjacent scatters cannot exceed $\lambda/4$ within one revisiting time interval to prevent ambiguity of phase measurements Therefore, generally slow landslides can be detected while fast kinematic ones, as debris-flow as rock falls cannot be analysed through Interferometric techniques. Some alternatives, as change detection approach, can be used to map fast surface changes (Tessari et al. 2017b). The perspective of landslide detection, mapping and characterization is expected to increase thanks to a new generation of SAR data, Sentinel-1, currently acquired by the European Space Agency, which guarantees a continuous monitoring, covering all over the world surface at least once every 12 days.

Considering the alarming situation daily faced on the main Sikkim municipalities, the Cooperation Office of the Swiss Embassy (SDC) located in New Delhi has been supporting several projects aimed on using new and advanced technologies to provide the necessary information for risk mitigation in the Himalayan Region s in the framework of the action "Strengthening State Strategies for Climate". In fact, the outcomes of the present work have been provided to the Sikkim State Disaster Management Authority (SSDMA).

Study Areas

Sikkim, an Indian small mountainous state in the eastern Himalayas, located between Nepal, Tibet and Bhutan, covers an area of 7069 sq. km. It is also a hilly state consisting of tangled series of interlocking mountain chains rising range above range from the south to the foot of high peaks, which marks the snow line in the north. The state has four districts East District, West District, North District and South District

Fig. 1 Location of the Indian Himalayan state of Sikkim and its main regions, North, West South and East Sikkim with the respective capitals (left side). Satellite optical map of Sikkim (right side)

with their headquarters at Gangtok, Geyzing, Mangan and Namchi respectively (Fig. 1). Sikkim has a very rugged topography and formidable physical features. The whole state is enclosed on three sides by lofty ranges and spurs of Greater Himalaya with varying heights on three sides. Within Sikkim, the main municipalities and villages are almost entirely within the Lesser Himalayas, consisting mainly Daling group of rocks which have undergone several episodes of loading, unloading and uplift during the orogeny. The result is a hazardous combination of weak geology (micro-fractures, joints, fissures, separated foliation planes, faults etc.) and high relief within short distances. The main causative factors of slope instability are weak geology, adverse planar structures in rocks, unstable slope materials, steep slopes or high relative relief. Moreover, intense rainfalls cause surface runoff of water, despite man-made drains are trying to reduce the landslide risk.

In the framework of this project, the main Sikkim municipalities have been analysed: Gangtok, Namchi, Geyzing, Pelling, Soreng, Sombaria and Mangan. Here we present the results of only two cases among these cases: Gangtok, the densely urbanized capital of Sikkim, and Namchi, as representative of all the other cases, characterized by a sparse, low-density urbanisation.

Data and Methods

Sentinel-1 is a Copernicus programme satellite constellation run by the European Space Agency, through a constellation of two polar-orbiting satellite where C-band radar for SAR applications are installed. Two dataset are available over Sikkim, respectively acquired by Sentinel-1 satellite while is flying from South to North (ascending orbit) and from North to South (descending orbit). The main characteristics of the two datasets are described in Table 1.

Table 1 Main characteristics of Sentinel-1 SAR datasets analysed over Sikkim main municipalities

	Acquisition geometry	
	Ascending	Descending
Relative orbit	12	48
Number of Scenes	79	90
Time interval	Mar 15—Sep 18	Oct 14—Sep 18
Revisiting time	12 days	
Resolution	15 m × 15 m	

SAR data record the back-scattered signals of ground surfaces and soil targets. SAR Interferometry consists of the analysis of the phase difference between two SAR images referred to equivalent areas, i.e. relative to the same frame observed from different points of view.

One of the component of the obtained phase difference is due to possible ground deformations, that can be isolated through the unwrapping of the phase difference, the so-called Differential InSAR (DInSAR), to create displacement and velocity maps and identify hazardous phenomena.

Beside the conventional DInSAR, two additional complementary approaches can be used for the measurement of the temporal evolution of deformations. One is the Persistent Scatterers (PS) (Ferretti et al. 2001) and the other is the Small Baseline Subset (SBAS) (Berardino et al. 2002) technique. PS relies on strong stable point scatterers (such as rock or man-made structures) while the SBAS is intended for the analysis of distributed targets. With respect to the PS, the SBAS technique is less sensitive to the number of acquisitions, because it exploits the spatially distributed coherence, instead to exclusively consider single points, as in the PS. However, in general, it is worth mentioning, that the more the acquisitions, the better the resulting product quality because the atmospheric component can be better estimated and reduced. Concerning the displacement, with respect to the PS technique, which is limited to the linear behaviour, the SBAS one can estimate quadratic and cubic models. Moreover, no modelled displacements are derived. On the maximum displacement, while there is no constraints in temporal term, the displacement is limited with respect to the spatial variability, due to the phase unwrapping. In addition, the SBAS approach is more robust than the PS one, because it takes advantage from the higher redundancy of all available cross-interferograms. Depending on each case, the most adequate multi-temporal InSAR algorithm is applied. The availability of two acquisition geometries allows retrieving deformation not only along the satellite Line of Sight (LOS) but also to project the results along the vertical and east–west directions.The main challenges with capturing landslides using these methodologies are both dense vegetation and strong topography, which have a negative impact on the radar signal to properly detect ground movements that could be identified as landslides. These challenges might be more easily addressed in the low density urbanized municipalities, generally located on the top of hills and mountains, as it happens for the considered cases.

Results

The main results obtained over Namchi and Gangtok consist on the mean velocity maps of deformation (Fig. 2) and the time-series of deformation (Fig. 3), available for each pixel covered by the final results, measured along the satellite LOS. For all the study areas, both ascending and descending data have been analysed with PS technique. Therefore, it was theoretically possible to combine the ascending and descending LOS velocity to obtain the vertical and east–west component of deformations. Nevertheless, the composition is possible only for those pixels where the results are available for both the acquisition geometries.

In the considered cases, the dense vegetation and the topographic conditions limited the spatial coverage of the results and the possibility of combining and project the LOS results.

The mean velocities maps of Namchi are shown in Fig. 2 where green points are stable, blue ones are moving far from the satellite while red points are moving towards the satellite with a maximum velocity rate up to 10 mm/yr. In particular, Fig. 2 highlight a portion of a slop oriented West-Est, where the deformation for the ascending and descending datasets are respectively positive and negative, compatibly with the different satellite acquisition geometries. The corresponding time series of deformations are shown in Fig. 3. The different LOS deformation rates are due to a horizontal deformation rate eastwards and a downward component too.

Only in the Gangtok area, also the SBAS technique has been applied thanks to the spatially distributed coverage of the results allowing the projection of deformation velocities and their temporal evolution along the vertical and east–west directions. The total vertical displacements are shown in the left panel of Fig. 4, where blue areas are moving downward. Results highlight that several slopes are affected by deformation exceeding 60 mm in the analysed period (October

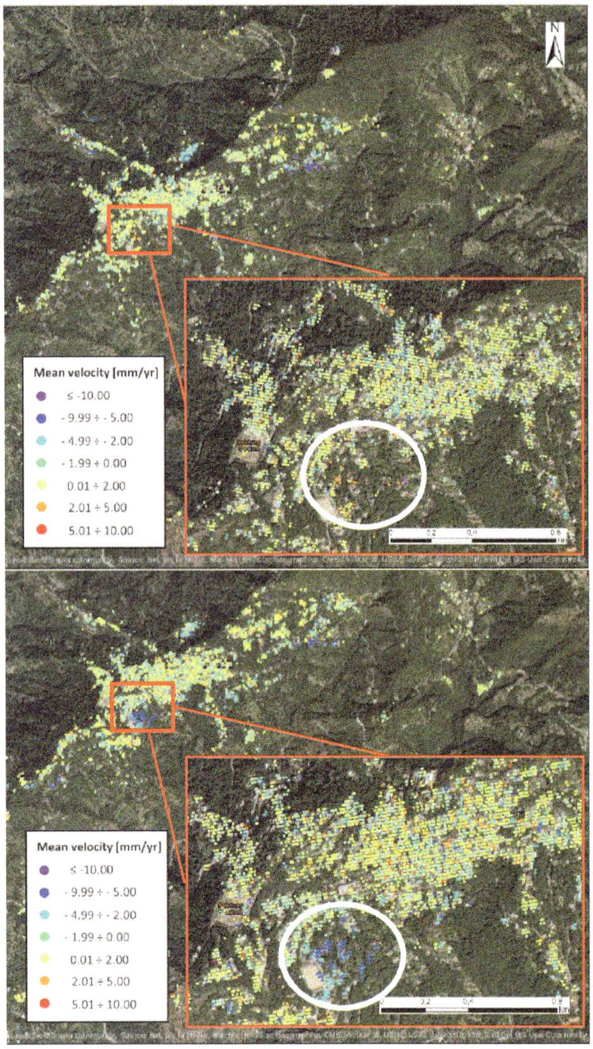

Fig. 2 Line-of-sight mean velocity map of deformation of Namchi obtained through PS technique applied to the Sentinel-1 descending (above) and ascending (below) datasets (time interval October 2014—September 2018). Blue areas are moving far from the satellite, green areas are stable while red areas are moving toward the satellite. The red box show a zoom of an area affected by a landslide

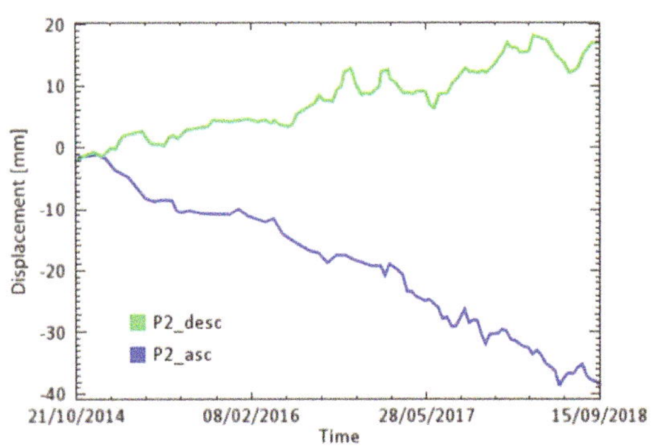

Fig. 3 Time series of deformation of a point selected on the area affected by deformation (inside the white circle in Fig. 2)

2014—September 2018), both on slopes oriented to East and West, on Gangtok city. Horizontal deformations are represented in the right side of Fig. 4 with westward displacements in blue and eastward in red. A significant point have been selected to analyse the behaviour of the ongoing deformations. Figure 5 shows the vertical time series of deformation (red curve) which is characterized by some acceleration and deceleration interval showing a seasonal behaviour. We search a possible correlation between the oscillation affecting the time-series of deformation and the rainfall seasonal distribution. We analysed the cumulative monthly rainfall measured in Gangotok area. Results of the comparison between temporal evolution of deformation time-series and the monthly cumulative rainfalls are presented in Fig. 5.

Comparing the red line, corresponding to the deformation trend, and the blue curve showing the monthly rainfall, it is possible to identify a strong connection between the increasing amount of monthly rainfall (from April/May to August) and an acceleration of the deformation trends, where the maximum gradient correspond to the rainfall peaks, while flat behaviour of the deformation occur in the dry season.

Discussion and Conclusions

The present project has been focused on mapping landslide affecting the main Sikkim municipalities through the capabilities of space-borne Synthetic Aperture Radar (SAR) data

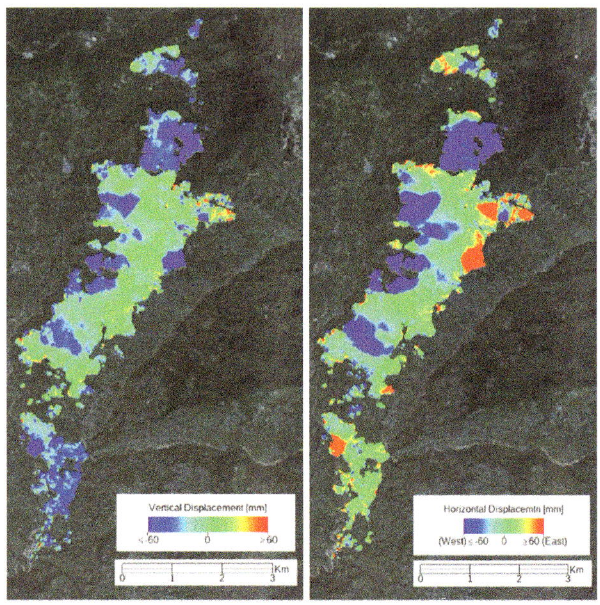

Fig. 4 Vertical (left) and East–West (right) velocity map of deformations of Gangtok obtained through SBAS technique applied to the Sentinel-1 datasets (October 2014—September 2018). Green areas are stable in both the maps, blue areas are moving downwards in the left map and westward in the right one. Red areas in the right frame are moving eastwards

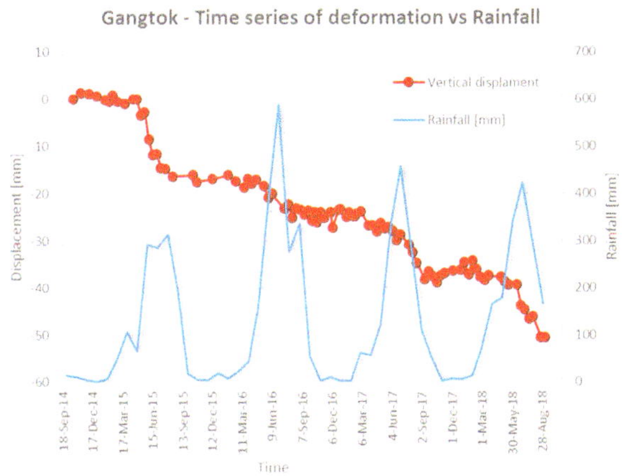

Fig. 5 Vertical time-series of deformation of a selected point in Gangtok area (red line) compared to the cumulative monthly rainfall (light-blue curve)

on monitoring surface deformation. The illustrated analyses have been performed through the processing of Sentinel-1 SAR data, freely available SAR data acquired in the framework of a Copernicus project. All the deformation maps of Gangtok, Namchi, Geyzing and Pelling, Soreng, Sombaria and Mangan have been provided to the SSDMA and here two representative case of Namchi and Gangok have been shown.

They main multi-temporal interferometric techniques, PS and SBAS have been applied to two stacks of data, acquired with two different satellite orbits, ascending and descending. For all the datasets, PS analyses was performed, allowing retrieving information only over the small building and houses widely spread on the mountainous/hilly areas. Because of the localized information, strongly related to the satellite acquisition geometries, the deformation maps obtained from the two datasets are characterized by results overlapping only partially. To avoid losing information, the projection along the vertical and east–west components has not been performed, preferring to extend as much as possible the mapped information.

Only in the case of Sikkim capital, Gangtok, additional analyses have been performed using SBAS technique and the satellite LOS results have been combined to obtain the vertical and east–west components of displacement.

SBAS processing chain showed some restrictions on being applied in Namchi (as well as in all the other municipalities), because of the low density of buildings, generally quite spread and distributed on different crests. This municipality structure is limiting the spatial distribution of detectable targets, creating discontinuities on coherent area on the data processing, and consequently, limiting the SBAS applicability.

Nevertheless, PS technique was able to identify the localized area affected by deformation, possibly connected to landslide characterized by slow dynamics.

Interesting results have been obtained on Gangtok area. In fact, Sikkim capital is definitely the most densely urbanized municipality of Sikkim. This is a favourable condition as it is guaranteeing a backscattered SAR signal stable in time. Consequently, the spatial coverage of the results obtained in Gangtok is much more extended and continues than in the other selected municipalities. Moreover, the good coverage obtained in both the ascending and descending processing allowed combining the results of the two datasets, and project the measured time-series of deformation along the horizontal and vertical directions, providing an additional information related to the direction of the real displacements. The great power of this technique on retrieving the temporal evolution of deformations is particularly interesting as this allows to identify variable trends, acceleration and oscillation, supporting a post-processing stage of result interpretation, trying to identify a correlation with triggering factors.

In the case of Gangtok, an attempt of identify a relation between rainfall season and sliding acceleration has been performed. This analysis showed a strong correlation between monthly rainfall and variation of the deformation behaviour.

Anyway, because of the highly vegetated land surface, the intense rainy season and the seasonal or permanent snow cover, these techniques show their limitation on detecting landslides and deformation in non-urbanized areas.

Nevertheless, some of the limitation could be overcome according to the following observation. Vegetation variation is strongly dependent on seasonal and weather conditions. Therefore, data temporal decorrelation could be more intense during wet seasons than during dry seasons. As a consequence, a possible improvement on the data processing, focused on increasing the spatial coverage of the deformation velocity maps, could be obtained selecting only data acquired during the dry season. Moreover, sensor wavelength, depending on the acquisition band, can strongly influence the sensitivity to surface variation, generally on vegetated surfaces. The capability of penetrate vegetation of radar data is increasing as the wavelength increases. Therefore, using L-band data instead of C-band ones (as Sentinel-1) could help on reducing the sensitivity of vegetated surface changes and the consequent temporal decorrelation. At the moment, the only available space-borne L-band SAR data are ALOS Palsar-2, acquired by JAXA. The main limitation of these data is the poor revisiting time, theoretically 14 day but the acquisition plan is not very dense around the work and there are long temporal gaps, where data are available. The previous point will be solved once the new NASA-ISRO SAR mission, NISAR, will be launch. NISAR will acquired L-band SAR data (and S band only over India), with a world coverage, a revisiting time of 10 days, guaranteeing a constant availability of data. Once this mission will be operational, the applicability of the proposed multi-temporal

DInSAR techniques on vegetated areas will dramatically increase. Finally, more detailed and resolute information could be obtained on urbanized areas, buildings, streets and infrastructure, considering higher resolution X-band SAR data, as the ones acquired by COSMO-SkyMed, TerraSAR-X and PAZ missions. These acquisitions are characterized by a 3 to 1 m resolution, enabling to obtain a much detailed information on the stable scatterers.

All the generated velocity maps should prove their essential value on defining and installing landslide ground monitoring system, identifying and limiting the most dangerous areas. Moreover, thanks to the correlation between slope acceleration and rainfalls, this data could help on setting and calibrating some alert systems based on rainfalls (Floris et al. 2013). Eventually, these data could be used as a reference monitoring system itself, planning to continuously updates the temporal evolution of deformation, every time that a new Sentinel-1 scene is acquired over the study area (Béjar-Pizarro et al. 2017), consciously that, despite dealing with slow landslide, the 12 day revisiting time could not be able to catch in time an eventual acceleration of the slope. Therefore, the integration with future incoming SAR mission will also help on increasing the frequency of SAR measurements in case of monitoring applications.

Acknowledgements The authors gratefully acknowledge Swiss Development Cooperation (SDC) that supported this project in the framework of the projects implemented in India since 2015: "Strengthening State Strategies for Climate Action (3SCA)".

The authors also kindly acknowledge the European Space Agency (ESA) for making available the Sentinel-1 images in the framework of Copernicus activities.

References

Barra A, Monserrat O, Mazzanti P, Esposito C, Crosetto M, Scarascia Mugnozza G (2016) First insights on the potential of Sentinel-1 for landslides detection. Geomat Nat Hazard Risk 7(6):1874–1883. https://doi.org/10.1080/19475705.2016.1171258

Béjar-Pizarro M, Notti D, Mateos RM, Ezquerro P, Centolanza G, Herrera G, Bru G, Sanabria M, Solari L, Duro J, Fernández J (2017) Mapping vulnerable urban areas affected by slow-moving landslides using Sentinel-1 InSAR data. Remote Sens 9:876

Berardino P, Fornaro G, Lanari R, Sansosti E (2002) A new algorithm for surface deformation monitoring based on small baseline differential interferograms. IEEE Trans Geosci Remote Sens 40 (11):2375–2383

Bhasin R, Grimstad E, Larsen JO, Dhawan AK, Singh R, Verma SK, Venkatachalam K (2002) Landslide hazards and mitigation measures at Gangotok, Sikkim Himalaya. Eng Geol 64:351–368

Cascini L, Fornaro G, Peduto D (2009) Advanced low- and full-resolution DInSAR map generation for slow-moving landslide analysis at different scales. Eng Geol 112:29–42. https://doi.org/10.1016/j.enggeo.2010.01.003

Choubey VD (1992) Landslide hazards and their mitigation in the Himalayan Region. In: Proceedings of the sixth international symposium on landslide, 10–14 February, Christchurch, New Zealand, pp 1849–1868

Ferretti A, Prati C, Rocca F (2001) Permanent scatterers in SAR interferometry. IEEE Trans Geosci Remote Sens 39(1):8–20

Floris M, D'Alpaos A, De Agostini A, Tessari G, Stevan G, Genevois R (2013) Variation in the occurrence of rainfall events triggering landslides. In: Margottini C, Canuti P, Sassa K (eds) Landslide science and practice. Springer, Cham, pp 131–138

Kaur H, Gupta S, Parkash S, Thapa R, Gupta A, Khanal GC (2019) Evaluation of landslide susceptibility in a hill city of Sikkim Himalaya with the perspective of hybrid modelling techniques. Annal GIS 25(2):113–132

Naithani AK (1999) The Himalayan landslides. Employ News 23 (47):20–26

Tessari G, Floris M, Achilli V, Fabris M, Menin A, Monego M (2017a) Testing Sentinel-1A data in landslide monitoring: a case study from North-Eastern Italian Pre-Alps. In: Mikoš M, Arbanas Ž, Yin Y, Sassa K (eds) Advancing culture of living with landslides. WLF 2017. Springer, Cham

Tessari G, Floris M, Pasquali P (2017) Phase and amplitude analyses of SAR data for landslide detection and monitoring in non-urban areas located in the North-Eastern Italian pre-Alps. Environ Earth Sci 76:85

Wasowski J, Bovegna F (2014) Investigating landslides and unstable slopes with satellite multi temporal interferometry: current issues and future perspectives. Eng Geol 174:103–138. https://doi.org/10.1016/j.enggeo.2014.03.003

Part II

Landslide Early Warning Systems, Forecasting Models and Time Prediction of Landslides

Definition and First Application of a Probabilistic Warning Model for Rainfall-Induced Landslides

Gaetano Pecoraro and Michele Calvello

Abstract

A methodology for the definition and the performance assessment of a probabilistic warning model for rainfall-induced landslides is proposed and tested in a study area in northern in Italy. To this aim, a database of 513 landslides triggered by rainfall in the period 2010–2018 and satellite-based rainfall data are used. It is worth mentioning that both landslide records and rainfall measurements used for this study are open-access datasets available online. The methodology developed herein can be summarized into several successive steps. First, an automated algorithm is applied for reconstructing the rainfall conditions responsible for the documented landslides in the area of analysis, as well as the rainfall conditions that did not result in any landslide. Then, the conditional probabilities of landslide occurrence are calculated using a two-dimensional Bayesian analysis, differentiating between single landslide events (SLE) and areal landslide events (ALE). Subsequently, several thresholds at different conditional probabilities are evaluated, and different combinations are selected for the activation of two warning levels. For each rainfall combination, the issuing of warning levels is computed by comparing the conditional probability of landslide occurrence with the pre-defined warning level thresholds. Finally, the optimal thresholds combination to be employed, i.e. the one providing the best model performance in terms of success and error indicators, is selected using performance indicators derived from a 3 by 3 contingency table.

G. Pecoraro (✉) · M. Calvello
Department of Civil Engineering (DICIV), University of Salerno, Via Giovanni Paolo II, 132, 84084 Fisciano, (SA), Italy
e-mail: gpecoraro@unisa.it

M. Calvello
e-mail: mcalvello@unisa.it

Keywords

Landslide • Rainfall • Early warning • Probabilistic analysis • TRMM • Franeitalia

Introduction

Rainfall-induced landslides are widespread and destructive natural phenomena occurring all around the world that often cause severe human and economic losses (Froude and Petley 2018). Landslide early warning systems (LEWS) are being increasingly applied as non-structural risk mitigation measures. LEWS can be designed and employed at two different reference scales (Calvello 2017; Pecoraro et al. 2019): local systems address single landslides at slope scale (Lo-LEWS), while territorial systems (Te-LEWS) deal with multiple landslides over wide areas at regional.

Te-LEWS are used to provide generalized warnings over appropriately-defined homogeneous warning zones of relevant extension. Typically, these systems address weather-induced landslides through the monitoring and prediction of meteorological parameters. However, the definition of a regional warning model may be challenging for several reasons: the reconstruction of rainfall events, the absence of a direct relationship between rainfall and landslide initiation, the uncertainty of available landslide catalogues (e.g., Piciullo et al. 2018; Segoni et al. 2018a).

In this study, a conceptual framework for the definition of probabilistic rainfall thresholds for landslides at regional scale is developed. The main steps of the proposed approach are: (i) objective reconstruction of triggering and non-triggering rainfall conditions taking into account their frequency, (ii) probabilistic analysis, (iii) definition and performance evaluation of a two-levels probabilistic warning model. The proposed procedure has been tested by analyzing the reported landslides in the period 2010–2018 within a study area in northern Italy.

© Springer Nature Switzerland AG 2021
N. Casagli et al. (eds.), *Understanding and Reducing Landslide Disaster Risk*, ICL Contribution to Landslide Disaster Risk Reduction,
https://doi.org/10.1007/978-3-030-60311-3_20

Materials and Methods

Study Area and Database

The study area includes 6 of the 158 weather warning zones (*WZ*) defined for hydrogeological risk management in Italy: *Emil-E*, *Emil-G*, *Ligu-B*, *Ligu-C*, *Tosc-L*, and *Tosc-S1* (Fig. 1). Although the selected *WZ* fall into three different Italian regions, all of them are characterized by high susceptibility to the occurrence of rainfall-induced landslides.

Indeed, the study area is one of the rainiest of Italy; moreover, climate change is producing an extraordinary increase of rainfall intensity in there (Libertino et al. 2018). As a consequence, this area is one of the most severely affected by landslides in the last few years in Italy (Battistini et al. 2013).

In particular, thunderstorms characterized by intense and very intense rainfall cause widespread and damaging ground effects, both on the slopes and along the drainage pattern, in *Ligu-B*, *Ligu-C*, *Tosc-L*, and *Tosc-S1* (Roccati et al. 2018). Besides, the frequency of rapid shallow landslides is markedly increasing in the last few years in *Emil-E* and *Emil-G*, as shortest and more intense rainfalls, typically the main triggering factor of shallow landslides and debris flows in the Emilia-Romagna region, became more frequent in the Mediterranean area due to climate change (Segoni et al. 2018b).

The FraneItalia database (Calvello and Pecoraro 2018) reports 540 landslide events that occurred in the study area in the period 2010–2018. 27 records have been excluded from the analysis performed herein as they are reported as human- or earthquake-induced landslides or landslides for which the trigger is not known. Among the 513 landslide events included in the dataset, 353 are classified as single landslide events (SLE, red circles in Fig. 1) and the remaining 160 as areal landslide events (ALE, blue squares in Fig. 1).

The rainfall measurements were derived from the satellite-based Tropical Rainfall Measuring Mission (TRMM) database, which is a joint mission between NASA and the Japan Aerospace Exploration Agency (JAXA) launched in late November 1997 for the study of rainfall for weather and climate research purposes (Huffmann et al. 2007). Precipitation data used in this research have been derived from the TRMM version 3b42, which includes gridded precipitation data collected every 3 h at a $0.25° \times 0.25°$ (~ 25 km $\times$ 25 km) spatial resolution, extending from latitude 50° S to latitude 50° N. It is worth mentioning that the spatial resolution is finer respect to local rain gauge networks usually employed for early warning purposes.

Satellite rainfall data retrieved from TRMM database have been analyzed using Google Earth Engine (https://earthengine.google.com), a cloud-based platform for planetary-scale environmental data analysis. For the purposes of this study, precipitation measurements have been aggregated at 3-hourly temporal resolution and the mean rainfall values over each territorial unit have been calculated.

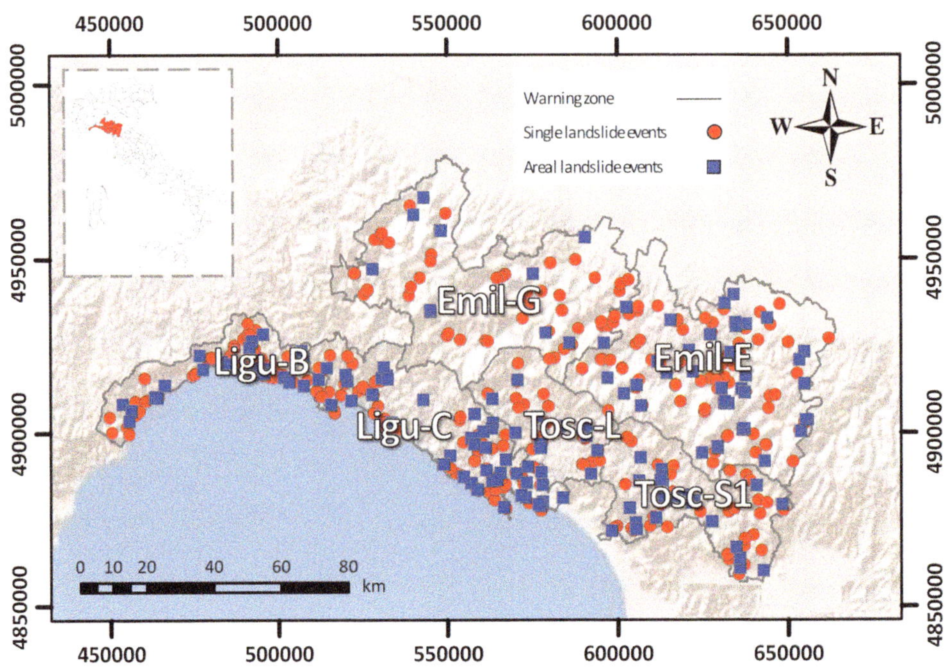

Fig. 1 Shaded relief map of the study area showing the 540 rainfall-induced "FraneItalia" landslide records in the period 2010–2018, differentiated in single (red circles) and areal landslide events (blue squares). The inset shows the location of the six warning zones in Italy

Methodology

The methodology developed for the definition of the probabilistic thresholds of landslides occurrence can be schematized into three main phases: reconstruction of the rainfall events, probabilistic analysis and definition of the probabilistic warning model.

In the first phase, the correlation between landslides and rainfall events in the study area is conducted by reconstructing the rainfall events, in order to convert a series of hyetographs into a point cloud in a graph reporting triggering and non-triggering combinations of rainfall parameters. Duration (D) and cumulated rainfall (E) are identified as the most appropriate rainfall parameters to use. To this aim, a modified version of the "algorithmic" approach developed by Melillo et al. (2016) is applied.

A reduced set of parameters to account for different physical settings and operational conditions has been considered. In particular, all the parameters are differentiated considering the "warm" springer-summer period, C_W, and the "cold" autumn–winter period, C_C (Table 1).

The automated procedure is based on several steps. In the pre-processing step (S_0), the rainfall records lower than a predefined threshold GS are considered noise and are set to $E_H = 0.0$ mm. The remaining steps are differentiated into two main logical blocks. The first block performs the automatic reconstruction of the rainfall events and can be schematized in the following four steps: (S_1) detection of the isolated rainfall events considering a dry interval, R_1 and exclusion of irrelevant events that do not exceed a predefined threshold E_R; (S_2) identification of rainfall sub-events proceeded and followed by dry periods with no rain, R_2; (S_3) exclusion of irrelevant sub-events, whose cumulated (total) rainfall, E_S is lower than a given threshold, R_3; (S_4) identification of rainfall events, constituted either by a period of continuous rainfall or by an ensemble of periods considering a minimum dry period, R_4. Successively, in the second block the algorithm combines information on temporal occurrence of rainfall events and landslide events, performing three additional steps: (S_5) selection of triggering and non-triggering rainfall events; (S_6) reconstruction of multiple aggregations of rainfall sub-events that are likely to trigger landslides; (S_7) reconstruction of multiple aggregations of rainfall sub-events that did not trigger landslides. All the triggering and non-triggering sub-events identified by the algorithm are equally possible.

In the second phase, a probabilistic approach based on a two-dimensional Bayesian analysis, similar to that used by Berti et al. (2012), is developed to calculate the landslide probability associated to the different rainfall combinations. To this aim, the posterior landslide probability is evaluated considering the joint probability of the duration (D) and cumulated rainfall (E), as follows:

$$P(L|D,E) = \frac{P(L) \times P(D,E|L)}{P(D,E)} \quad (1)$$

where: $P(L|D, E)$ is the posterior landslide probability; $P(L)$ is the prior probability; $P(D, E|L)$ is the likelihood; $P(D, E)$ is the marginal probability. The needed probabilities have been determined considering that the triggering and non-triggering rainfall conditions are expressed in terms of multiple combinations, as follows:

$$P(L) = \frac{N_L}{N_R} \quad (2)$$

$$P(D,E) = \frac{\sum_i n_{i,(D,E)} \times f_i}{N_R} \quad (3)$$

$$P(D,E|L) = \frac{\sum_i n_{i,(D,E|L)} \times f_i}{N_L} \quad (4)$$

where: N_L is the total number of landslide events that occurred in the period of analysis; N_R is the total number of rainfall events recorded in the period of analysis; $n_{i,(D,E)}$ is the number of possible rainfall conditions characterized by specific values of D and E; $n_{i,(D,E|L)}$ is the number of rainfall events characterized by specific values of D and E that resulted in landslides; f_i is the relative frequency, defined as

Table 1 Parameters used for the application of the algorithm developed by Melillo et al. (2016)

Step	Parameter name	Parameter value		Unit
		C_W	C_C	
S_0	GS	0.2	0.2	mm
S_1	E_R	0.2	0.2	mm
S_1	R_1	3	6	h
S_2	R_2	6	12	h
S_3	R_3	1	1	mm
S_4	R_4	48	96	h

Table 2 Warning levels defined considering the probabilities of SLE and ALE

Warning level	Correlation law
WL_1	$P(L\|D, E)_{SLE} > P_1$
WL_2	$P(L\|D, E)_{SLE} > P_2$ or $P(L\|D, E)_{ALE} > P_1$

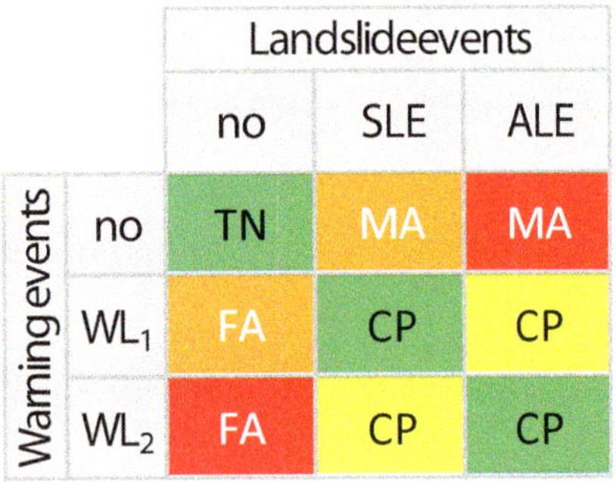

Fig. 2 Contingency matrix used for the performance analysis of the probabilistic rainfall thresholds

the inverse of the total number of possible aggregations of sub-events for a given rainfall event.

In the third phase, a warning model is defined employing two warning levels (WL_1 and WL_2) associated to the exceedance of two thresholds (P_1 and P_2) based on the probabilities of occurrence of SLE and ALE (Table 2).

In the fourth phase, the performance of the warning model is analyzed using statistical indicators, following a procedure similar to that proposed by Calvello and Piciullo (2016). In particular, the performance analysis of a 3 by 3 contingency matrix is based on a set of two performance criteria, both of them assigning a meaning to all the elements of the matrix (Fig. 2). The "alert classification" criterion employs an alert classification scheme derived from a standard 2 by 2 contingency table, and identifies correct predictions (*CP*), false alerts (*FA*), missed alerts (*MA*), and true negatives (*TN*). The "grade of accuracy" criterion assigns a colour code to the components of the matrix in relation to the agreement between a given warning event and a given landslide event. Using this criterion, the elements are classified in four colour-coded classes, as follows: green (*Gre*) for the elements which are assumed to be representative of the best model response, yellow (*Yel*) for elements representative of minor model errors, red (*Ora*) for elements representative of a significant model error and purple (*Red*) for elements representative of a severe model error.

Considering the two performance criteria, several performance indicators can be derived. Table 3 lists the indicators used in this study.

Results

Rainfall Events Reconstruction

1903 rainfall conditions (*D*, *E*) have been identified and plotted in log–log coordinates (Fig. 3). The 207 rainfall conditions responsible for triggering 353 SLE (red circles in Fig. 3) and the 129 rainfall conditions responsible for 160 ALE (blue squares in Fig. 3) are in the range of duration $3 \leq D \leq 915$ h and in the range of cumulated rainfall $1.02 \leq E \leq 243.54$ mm. The non-triggering rainfall conditions, reconstructed in the same period, are 1567 (green circles in Fig. 3). They are in the ranges of $3 \leq D \leq 495$ h and $1.01 \leq E \leq 311.87$ mm.

It is worth mentioning that rainfall combinations characterized by $E < 2$ mm (grey circles in Fig. 3) constitute a negligible amount of rain, thus these combinations have been excluded from the analysis because they are considered irrelevant for the purpose of early warning.

Probabilistic Analysis

The definition of the probabilistic thresholds is based on a two-dimensional Bayesian analysis evaluating the conditional probability of landslide occurrence given the joint probability of *D* and *E*. According to the available data, the

Table 3 Performance indicators used for the performance analysis

Performance indicator	Symbol	Formula
Efficiency index	I_{eff}	$(TN + CP)/\Sigma_{ij}d_{ij}$
Odds ratio	OR	$(TN + CP)/(FA + MA)$
Probability of serious mistakes	P_{SM}	$Red/\Sigma_{ij}d_{ij}$
Probability of serious missed alerts	$P_{SM\text{-}MA}$	MA_{Red}/MA
Probability of serious false alerts	$P_{SM\text{-}FA}$	FA_{Red}/FA
Missed and false alert balance	MFB	$MA/(MA + FA)$

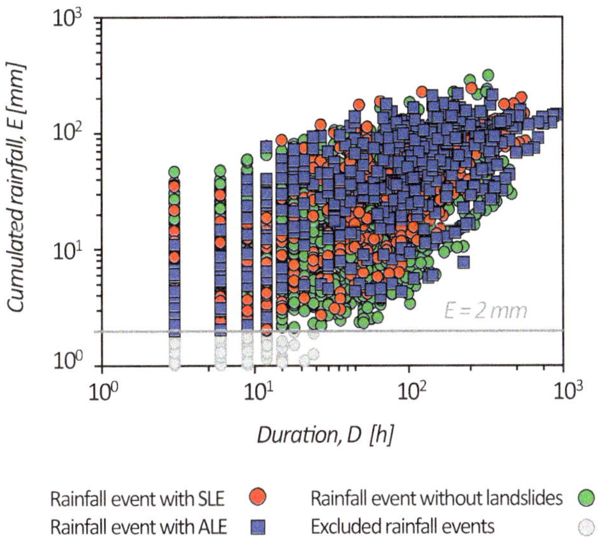

Fig. 3 Rainfall duration (D) versus cumulated rainfall (E) in the study area from 2010 to 2018. Graph plotted in log–log coordinates

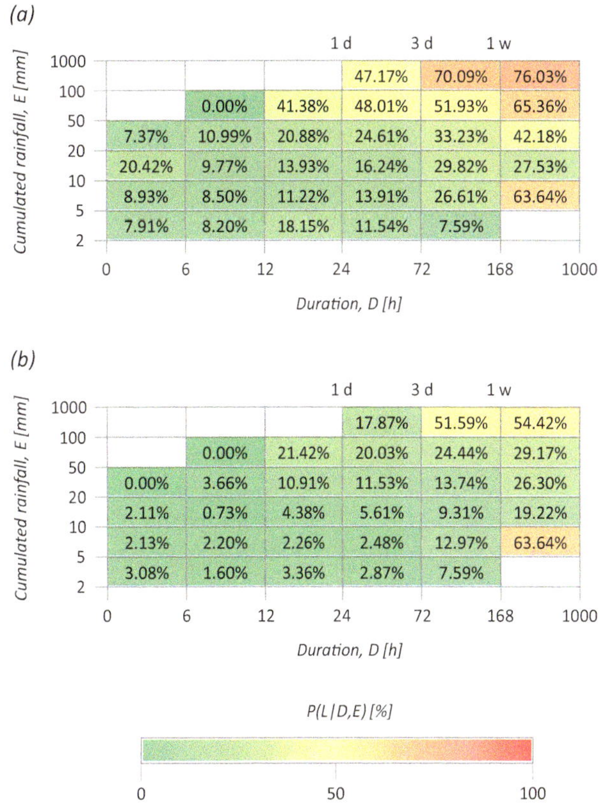

Fig. 4 Posterior landslide probabilities obtained considering SLE (**a**) and ALE (**b**)

prior landslide probability, $P(L)$ has been calculated using Eq. (2) and is equal to 18.79% for SLE and 7.44% for ALE.

Successively, the D, E space reported in Fig. 3 has been divided in 6×6 cells, both for SLE and ALE, and the posterior landslide probabilities, $P(L|D, E)$, have been calculated by applying Eq. (1). Looking at SLE, Fig. 4a displays that long-duration ($12 \leq D \leq 915$ h), high-accumulation rainfall ($50 \leq E \leq 243.54$ mm) events show the highest landslide probabilities ($P(L|D, E) > 40\%$). The only singularity is represented by the combination $5 \leq D \leq 10$ h, $50 \leq E \leq 243.54$ mm, for which $P(L|D, E) = 63.64\%$. However, this can be considered a singularity, as it represents only 0.02% of the rainfall combinations that occurred from 2010 to 2018. The results are substantially confirmed for ALE (Fig. 4b). Indeed, apart for the singularity already highlighted for SLE, the highest values of the posterior probability ($P(L|D, E) > 20\%$) are reached again for $12 \leq D \leq 915$ h and $50 \leq E \leq 243.54$ mm.

Probabilistic Warning Model

A performance evaluation has been conducted in order to identify the optimal two thresholds to be employed in the warning model. Several combinations have been compared, by varying the lower threshold, P_1 and the upper threshold, P_2. As significant differences in the performance evaluation depend only on the variations of P_1, the results are reported grouping the thresholds on the basis of P_1 (Table 4).

Table 5 shows the results obtained for the five combinations considering the elements of the correlation matrix reported in Fig. 2. Higher values of *CP* and *Yel* are obtained when the lower probabilities values are considered to define WL_1 (P_1 from 10 to 20%). In particular, passing from $P_{20,40-50}$ to $P_{30,50}$ results in a reduction of *CP* of about 37%. However, an increase of the *P1* threshold results in a significant reduction of the *FA* and *Red* errors and increasing values of *TN*.

Table 6 shows the results in terms of success (I_{eff} and *OR*) and error (P_{SM}, $P_{SM\text{-}MA}$, $P_{SM\text{-}FA}$, and *MFB*) indicators for the five different thresholds combinations reported in Table 4. Concerning the success indicators and, in particular, the efficiency index (I_{eff}), raising the value of P_1, a general increase is observed, as it is evident when comparing $P_{10,20-50}$ and $P_{12.5,25-50}$ to $P_{20,40-50}$ and $P_{30,50}$. The odds ratio (*OR*), which can be considered a rate between correct and wrong predictions, obviously increases with the reduction of *FA* and the increment of *TN*. However, it should be noted that passing from $P_{20,40-50}$ to $P_{30,50}$ the Probability of serious missed alerts ($P_{SM\text{-}MA}$) shows an increment of about 25%. Besides, the majority of the errors are missed alerts, as demonstrated by the high value of *MFB* (60.72%). For these reasons, $P_{20,40-50}$ can be considered the best-performing thresholds combination of the 5 considered herein.

Table 4 Combinations considered to identify the optimal values of P_1 and P_2

Label	P_1 (%)	P_2 (%)
$P_{10,20-50}$	10	20, 25, 30, 40, 50
$P_{12.5,25-50}$	12.5	25, 30, 40, 50
$P_{15,30-50}$	15	30, 40, 50
$P_{20,40-50}$	20	40, 50
$P_{30,50}$	30	50

Table 5 Number of contingency matrix elements considering the "alert classification" (*CP*, *TN*, *MA*, *FA*) and "grade of accuracy" (*Gre*, *Yel*, *Ora*, *Red*) criteria

Element	$P_{10,20-50}$	$P_{12.5,25-50}$	$P_{15,30-50}$	$P_{20,40-50}$	$P_{30,50}$
CP	249.87	233.66	209.51	185.24	115.85
TN	564.59	692.42	841.70	960.76	1183.58
MA	51.80	68.01	92.15	116.42	185.81
FA	739.21	611.38	462.09	343.04	120.21
Gre	701.24	821.86	956.25	1060.83	1248.63
Yel	113.22	104.21	94.96	85.17	50.81
Ora	434.78	509.22	408.42	346.81	207.70
Red	356.22	170.16	145.83	112.65	98.32

Table 6 Performance indicators computed for the five thresholds combinations considered

Indicator	$P_{10,20-50}$	$P_{12.5,25-50}$	$P_{15,30-50}$	$P_{20,40-50}$	$P_{30,50}$
I_{eff}	50.73%	57.68%	65.48%	71.38%	80.94%
OR	1.03	1.36	1.90	2.49	4.25
P_{SM}	19.92%	9.50%	6.72%	6.66%	6.12%
$P_{SM\text{-}MA}$	25.46%	25.16%	25.43%	26.03%	32.91%
$P_{SM\text{-}FA}$	41.48%	22.16%	18.29%	22.35%	30.92%
MFB	6.55%	10.01%	16.63%	25.34%	60.72%

Conclusions

In this study, a Bayesian approach has been developed for the definition of a probabilistic warning model for rainfall-induced landslides. It has been defined using a landslide inventory retrieved from online news and satellite-based rainfall measurements. Both landslide records and satellite rainfall monitoring used in this study come from open-access datasets available online.

Firstly, the triggering and non-triggering rainfall conditions have been objectively reconstructed. Then, a Bayesian approach has been applied for calculating the posterior landslide probabilities of occurrence of single landslide events (*SLE*) and areal landslide events (*ALE*). Finally, a probabilistic warning model employing two thresholds has been defined and its performance evaluated using performance indicators derived from a 3 by 3 contingency table.

The performed analyses showed that $P_{20,40-50}$ is the best-performing thresholds combination, as it represents the best compromise between the minimization of incorrect landslide predictions and the maximization of the correct predictions. Generally, the probabilistic warning model revealed an overall good performance in predicting landslide events triggered by significantly different rainfall conditions. Although the performance of the model can be further refined considering wider and longer datasets, the preliminary results achieved herein clearly allow to highlight its potential for landslide early warning purposes.

Acknowledgements This work was financially supported by the Project of Relevant National Interest (PRIN) 2015 on "Protecting the Cultural Heritage From Water Soil Interaction" (code: 300393PRN2015CASCI).

References

Battistini A, Segoni S, Manzo G, Catani F, Casagli N (2013) Web data mining for automatic inventory of geohazards at national scale. Appl Geogr 43:147–158

Berti M, Martina MLV, Franceschini S, Pignone S, Simoni A, Pizziolo M (2012) Probabilistic rainfall thresholds for landslide occurrence using a Bayesian approach. J Geophys Res 117:F04006
Calvello M (2017) Early warning strategies to cope with landslide risk. Riv It Geotecnica 2:63–91
Calvello M, Pecoraro G (2018) FraneItalia: a catalog of recent Italian landslides. Geoenviron Disasters 5(13).
Calvello M, Piciullo L (2016) Assessing the performance of regional landslide early warning models: the EDuMaP method. Nat Hazards Earth Syst Sci 16:103–122
Froude MJ, Petley DN (2018) Global fatal landslide occurrence from 2004 to 2016. Nat Hazards Earth Syst Sci 18:2161–2181
Huffman GJ, Adler RF, Bolvin DT, Gu G, Nelkin EJ, Bowman KP, Hong Y, Stocker EF, Wolff DB (2007) The TRMM multi-satellite precipitation analysis: quasi-global, multi-year, combined-sensor precipitation estimates at fine scale. J Hydrom 8(1):38–55
Libertino A, Ganora D, Claps P (2018) Space-time analysis of rainfall extremes in Italy: clues from a reconciled dataset. Hydrol Earth Syst Sci 22(5)
Melillo M, Brunetti MT, Peruccacci S, Gariano SL, Guzzetti F (2016) Rainfall thresholds for the possible landslide occurrence in Sicily (southern Italy) based on the automatic reconstruction of rainfall events. Landslides 13(1):165–172
Pecoraro G, Calvello M, Piciullo L (2019) Monitoring strategies for local landslide early warning systems. Landslides 16:213–231
Piciullo L, Calvello M, Cepeda JM (2018) Territorial early warning systems for rainfall induced landslides. Earth Sci Rev 179:228–247
Roccati A, Faccini F, Luino F, Turconi L, Guzzetti F (2018) Rainnfall events with shallow landslides in the Entella catchment (Liguria, northern Italy). Nat Hazards Earth Syst Sci 18:2367–2386
Segoni S, Piciullo L, Gariano SL (2018a) A review of the recent literature on rainfall thresholds for landslide occurrence. Landslides 15:1483–1501
Segoni S, Rosi A, Fanti R, Gallucci A, Monni A, Casagli N (2018b) A regional-scale landslide warning system based on 20 years of operational experience. Water 10(10):1297

Establishment of an Integrated Landslide Early Warning and Monitoring System in Populated Areas

Nikolaos Depountis, Nikolaos Sabatakakis, Katerina Kavoura, Konstantinos Nikolakopoulos, Panagiotis Elias, and George Drakatos

Abstract

In this work a complete permanent system of timely landslide warning and monitoring in Greece is presented. This system is the first that is designed for a densely residential and mountainous environment. Since 1960's several instability phenomena have been recorded in one of the most traditional settlements in Greece, Metsovo, Region of Epirus. The last major landslide event occured in 2010–2011, and lead to serious damages on the construction and infrastructure within the settlement. The wider geological regime consists of Olonos-Pindos formations with the main appearance of the flysch one of the most critical landslide prone geological formations in Greece. The combinational use of dynamic geotechnical and satellite research methods is discussed as part of this study. In addition, one of the main goals of this investigation is to combine long term monitoring of the parameters connected to the landslide activity with the observation of the landslide kinematics in real time for the planning and realization of a Landslide Early Warning System (LEWS) in Greece.

Keywords

Greece • Real-time • Monitoring • Landslide early warning systems (LEWS)

Introduction

Landslide phenomena are included in the natural hazards with serious socio-economic consequences and with a considerably increasing number of events in the last years, as much internationally as nationally, mainly because of the intense anthropogenic intervention in the geo-environment but also because of the frequent appearance of extreme meteorological events. Characteristically, from 2000 and on, an increase that exceeds 25% has been observed in recorded landslide events in Greece (Sabatakakis et al. 2013).

The monitoring and recording of the kinematics preceding of the activation or reactivation of a landslide, with the application of geotechnical and remote sensing methods, with real time measurements is an increasingly important area in landslide studies (Corsini et al. 2007; Bobrowsky et al. 2015; Casagli et al. 2017) and seems to conclusively contribute to the quick evaluation of its style of activity (WP/WLI 1995). Overall, this knowledge can constitute a basic tool for the creation of a sufficient and reliable evaluation model for the landslide risk, with application in populated areas and in areas of linear or point infrastructures.

In Greece, and in particular on the western part, large landslides have taken place, causing extended destruction to large road axes and to mountainous settlements. Besides, as it results from the landslide susceptibility map of Greece (Sabatakakis et al. 2013), the larger part of Western Greece is classified as "high" to "extremely high" susceptibility.

The pilot application of a complete permanent system of landslide early warning and monitoring is presented in this work, with the combinational use of dynamic geotechnical

N. Depountis (✉) · N. Sabatakakis · K. Kavoura · K. Nikolakopoulos
Department of Geology, University of Patras, 26504 Patras, Greece
e-mail: ndepountis@upatras.gr

K. Kavoura
e-mail: kavoura@upatras.gr

P. Elias
Institute for Space Applications and Remote Sensing, National Observatory of Athens, Athens, Greece
e-mail: pelias@noa.gr

G. Drakatos
Geodynamic Institute, National Observatory of Athens, Lofos Nymfon, Thission, 11810 Athens, Greece

© Springer Nature Switzerland AG 2021
N. Casagli et al. (eds.), *Understanding and Reducing Landslide Disaster Risk*, ICL Contribution to Landslide Disaster Risk Reduction,
https://doi.org/10.1007/978-3-030-60311-3_21

and remote sensing methods. The pilot area corresponds to a densely built mountainous village of the Region of Epirus, known as the Metsovo village. This application is the first that has been realized in a Greek territory and in particularly in a densely residential environment. Preliminary results from the application of a similar combinational system with the use of static geotechnical and remote sensing methods for another region of Western Greece, have been published by Nikolakopoulos et al. (2017).

The differentiation and innovation of the present work lies at the fact that for the first time in Greece, all the known kinematic monitoring techniques (geotechnical, interferometry, satellite geodesy, remote sensing) are combined on a "real-time" level for the evolution of the landslides in residential areas. The measurement data and results of these techniques are transferred on real time to storage stations and with the proper treatment directly on a specially formed internet platform (WebGIS). Then the collected data are ready to be used by a wide spectrum of scientists of competent entities that occupy themselves with landslide risk management issues. This process reflects to the principal data basement and is the first step of planning and realization of a landslide early warning system (LEWS) in Greece.

Study Area

The Metsovo village, Region of Epirus, is built on 1060–1200 m of altitude on the eastern side of the mountain range of Pindos. It is a traditional settlement of about 2500 permanent residents with high touristic activity "Fig. 1".

From the geological point of view, the wider area of Metsovo belongs to the overthrust front of the Olonos–Pindos zone with the main appearance of the flysch formation in various phases (Zouros and Mountrakis 1991). The main geological basement is the medium-thick sandstone horizon of flysch, interbebbed with thin grey horizons of siltstone. Deeper layers consist of clay-sandstone comprising interchanges of fine-grained sandstone and siltstone. Because of the tectonic deformation that the latter has been submitted due to an internal thrust in Flysch, in combination with the intense morphological relief of the study area, serious instability problems are generally caused in various positions around. These instabilities are expressed as individual landslides in the weathering zone of Flysch. Furthermore, the high percentages of rainfall observed in the area with average yearly rainfall of 1472 mm (Koumantakis 2011), mainly during the wet period and the appearance of a considerable number of springs, favour the slope instabilities in such types of formations.

Since the 1960's until today, landslides have been observed in the settlement with direct impact in the infrastructure and the daily life of the residents. In particular, the main landslides examined in the present study are two and are located on the southern part of the residential area at the neighbourhoods of Agios Dimitrios and Agios Charalampos. The most serious activation for both the partial landslides took place on 2010 "Fig. 1" with the appearance of fracturing and subsidence in the streets and infrastructure as well as fracturing in houses, some of which are considered as decrepit. According to the inclinometer measurements taken on both neighbourhoods since 2010 until today, the movement seems to evolve with very low velocity values, thus the landslides can be characterized as "extremely slow" (WP/WLI 1995). More precisely, mean rates of movement ranged from 7–8 mm per year at Agios Charalampos and 10–12 mm per year at Agios Dimitrios.

In the context of the present project, a rich archive of studies, technical reports, scientific projects and reports on the local press was gathered, which was evaluated for the better approach of the problem and the proper organization and planning of the landslide observation system.

Establishment of the Landslide Monitoring System

The observation and analysis of the landslide characteristics on a large scale (site-scale, >1:500) takes into consideration some specific criteria (Corominas et al. 2014; Soeters and van Westen 1996). The selection of the settlement of Metsovo as a pilot area for the installation of a complete observation system of the landslides covers these criteria to a great extent, because: (a) it is a residential area with important infrastructure works at risk, (b) the research areas are smaller than 0.1 km^2 and focus on periodically activated landsliding zones, (c) a relatively sufficient knowledge of the engineering geological and geotechnical conditions of the subsoil and the geometry of the unstable areas is available and (d) the geomaterials participating in the phenomenon belong to geological formations "prone to landslides". Analyses in this research scale permit the subsequent development of landslide early warning systems (LEWS) exploiting the phenomenon monitoring data (Pecoraro et al. 2019).

The installed landslide monitoring system basically provides insight into the evolution of surface and subsurface movements with the combinational use of measurements of

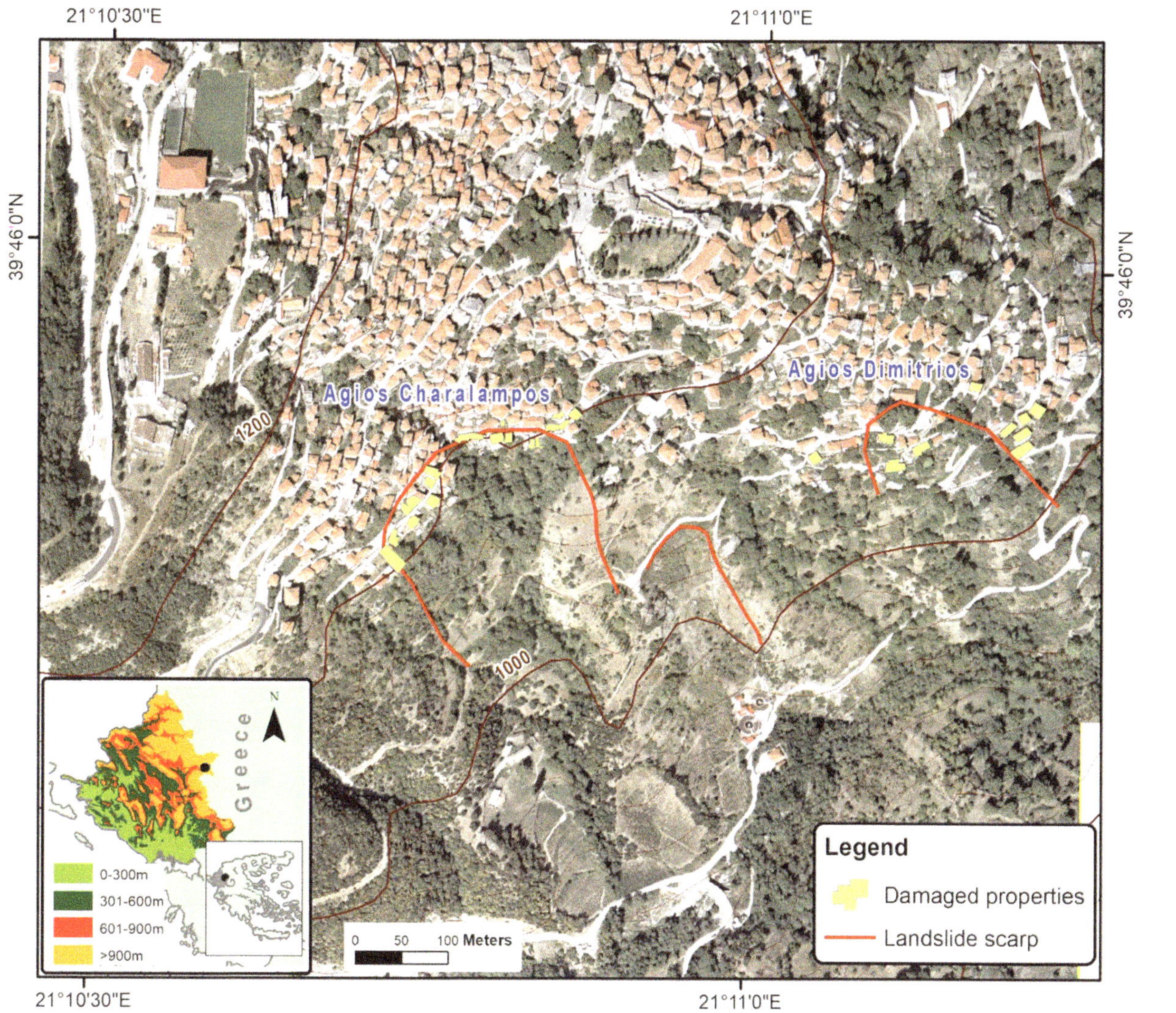

Fig. 1 Location map of the Metsovo settlement and landsliding areas. The colours of the inserted map refer to the different elevation classes in the Epirus region

geotechnics, interferometer, satellite geodesy and remote-sensing with the parallel real-time monitor of rainfall, as it seems that landslides are directly connected with intense and extended rainfall phenomena "Fig. 2".

Persistent Scatterers

Persistent Scatterer Interferometry (PSI) is an approach of interferometry for monitoring Earth's surface, as it is able to measure microscale displacements in terrain surface. In addition, this application permits the measurement of surface deformation over vast areas with at a frequency varying from one month to several days with the earliest satellites (Raspini et al. 2019). For this purpose, five (5) scatterers were installed in several positions of the study area and satellite receptions were programmed with a frequency of eleven (11) days. More specifically, the scatterers are corner retro-reflectors consisting of perpendicularly intersecting flat surfaces and they are mainly made of aluminium, for easy transportation and installation.

Permanent GPS Stations

Permanent GPS stations can monitor the surface deformations (Gili et al. 2000). The GPS network consists of 2 permanent stations. One of them was installed in a stable area, while the other one into the landslide zone. The stations continuously record the land deformation and the data are transferred in real time to data storage station (NAS server). Thus, at any given time, the relative position as much of

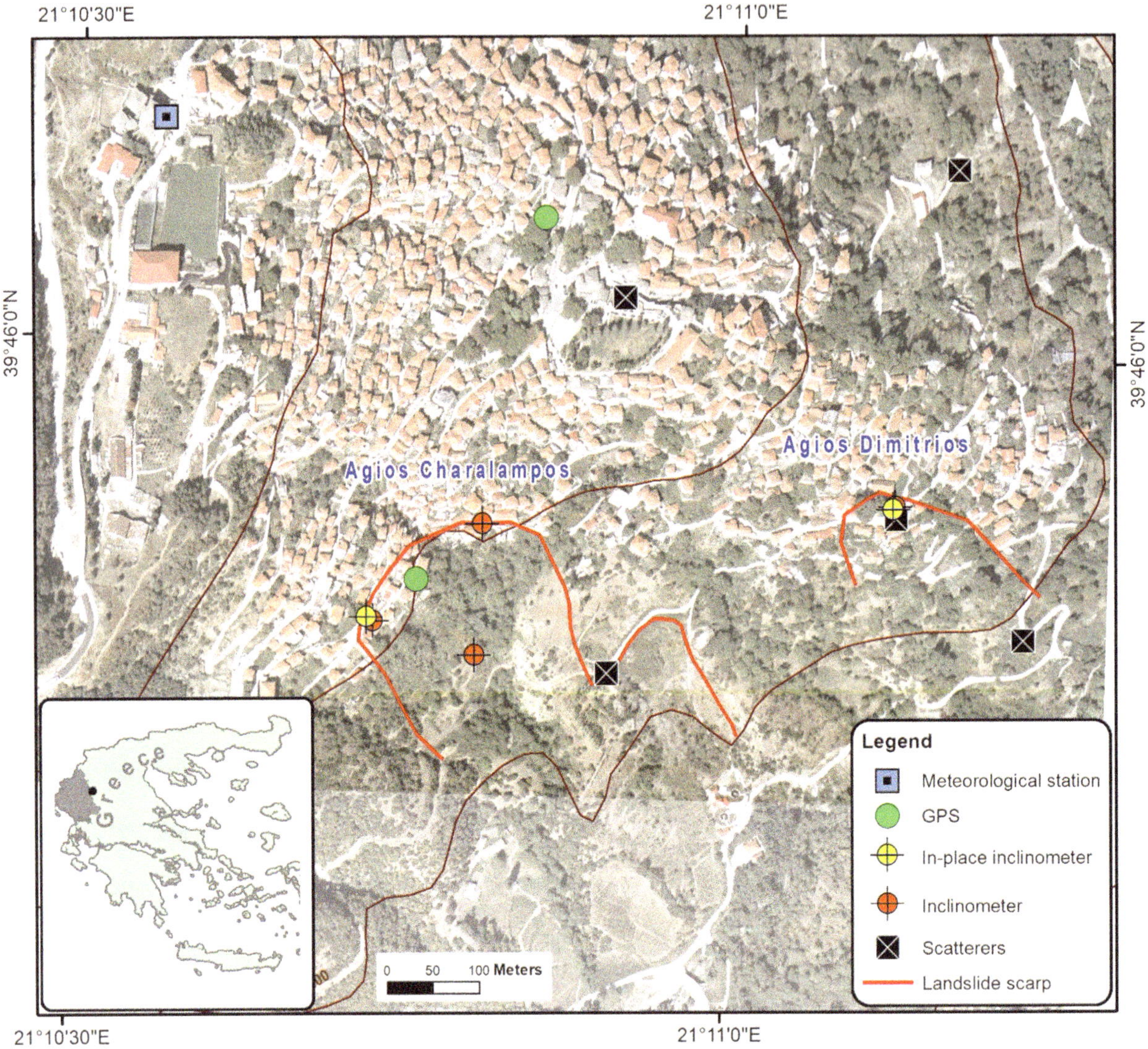

Fig. 2 The permanent landslide monitoring system

scatterers as the position of the GPS station located in the landsliding area will be determined in relation to the stable stations. A characteristic example of application of this method constitutes the observation of slopes in southern Apennines, Italy (Calcaterra et al. 2012).

Unmanned Aerial Vehicle Surveys

They are used for the exact determination of the land topography, the photogrammetric survey on selected points of the pilot area and the creation of photomaps and Digital Surface Models (DSM) of high precision. These data can also be used as much as basemap for the remaining studies as for the exact measurement of changes on the relief in case of land movement. Different studies demonstrating the use of UAV data for landslide monitoring have been published in Rau et al. (2011), Turner et al. (2015) and Nikolakopoulos et al. (2017).

Meteorological Station

The meteorological station is located inside the settlement very close to the landslide area. Rainfall constitutes a basic factor for the initiation of landslides (Sabatakakis et al. 2005; Guzzetti et al. 2008; Lainas et al. 2016) and therefore the pluviometric data are continuously recorded and sent to the data storage station (NAS server) in real time. The recording of rainfall events with the parallel observation of the landslide evolution rate will permit on the following stage, the determination of critical values of rainfall thresholds that activate the landslides, and which will constitute an important factor for the formation of the type of warning of EWS.

Inclinometer Boreholes

A network of eight (8) drilling inclinometers has been installed in the research area that permits the detailed observation of subsurface displacements, for an extended period of time. In two of them inclinometers have been permanently placed (in-place) aiming to the observation in real time of the movement, the data of which are continuously recorded and sent to the data storage station (NAS server). Further analysis of the subsurface displacements time series in combination as much with the surface movement as with the rainfall time series will lead to the full understanding of the evolution rate of the phenomenon. The determination of critical speed thresholds, the exceeding of which can cause noticeable deformations that it is likely to affect the local society, will also constitute a parameter of the EWS.

Discussion–Conclusion

Many mountainous villages have been struck by landslides in western Greece due to growing urbanization and uncontrolled land—use in landslide—prone areas, without considering the engineering geological environment. The presence of the tectonically highly sheared and weathered geological formations of the alpine basement (such as flysch) and the intense geomorphological relief, strongly contribute to the periodically induced instability phenomena mainly triggered by heavy rainfalls and extreme meteorological events.

The study site of Metsovo village, constitutes a typical example of a mountainous, touristic settlement with intense anthropogenic activity, periodically stricken by landslide phenomena that caused severe damages in houses and infrastructures. The installation of a permanent complete system for landslide kinematics observation consisting of persistent scatterers, permanent GPS stations, UAV, meteorological station and in place inclinometers, constitutes the first step for an EWS establishment in Greece.

The obtained results of the system measurements will be appeared in almost real time, in interactive way through a specially designed internet platform (WebGIS). That will constitute a powerful tool for the local authorities and the residents of the area in case of emergency.

References

Bobrowsky P, Sladen W, Huntley D, Qing Z, Bunce C, Edwards T, Hendry M, Martin D, Choi E (2015) Multi-parameter monitoring of a slow moving landslide: ripley slide, British Columbia, Canada. Eng Geol Soc Territory 2: 155–158. https://doi.org/10.1007/978-3-319-09057-3_18

Calcaterra S, Cesi C, Di Maio C, Gambino P, Merli K, Vallario M, Vassall R (2012) Surface displacements of two landslides evaluated by GPS and inclinometer systems: a case study in Southern Apennines, Italy. Nat Hazards 61(1):257–266. https://doi.org/10.1007/s11069-010-9633-3

Casagli N, Frodella W, Morelli S, Tofani V, Ciampalini A, Intrieri E, Raspini F, Rossi G, Tanteri L, Lu P (2017) Spaceborne, UAV and ground-based remote sensing techniques for landslide mapping, monitoring and early warning. Geoenviron Disast 4:1–23. https://doi.org/10.1186/s40677-017-0073-1

Corominas J, van Westen C, Frattini P, Cascini L, Malet J-P, Fotopoulou S, Catani F, Van Den Eeckhaut M, Mavrouli O, Agliardi F, Pitilakis K, Winter MG, Pastor M, Ferlisi S, Tofani V, Hervás J, Smith JT (2014) Recommendations for the quantitative analysis of landslide risk. Bull Eng Geol Env 73(2):209–263. https://doi.org/10.1007/s10064-013-0538-8

Corsini A, Borgatti L, Coren F, Vellico M (2007) Use of multitemporal airborne lidar surveys to analyse post-failure behaviour of earth slides. Can J Remote Sens 33(1–4):116–120

Gili JA, Corominas J, Rius J (2000) Using Global Positioning System techniques in landslide monitoring. Eng Geol 55(3):167–192. https://doi.org/10.1016/S0013-7952(99)00127-1

Guzzetti F, Peruccacci S, Rossi M, Stark CP (2008) The rainfall intensity–duration control of shallow landslides and debris flows: an update. Landslides 5:3–17

Koumantakis I (2011) Landslide phenomena at Metsovo: Proposals for improving conditions. Training seminar "Landslides of Metsovo", 16–18 March 2011, Metsovo, MEDKE/NTUA, p. 6 (in Greek)

Lainas S, Sabatakakis N, Koukis G (2016) Rainfall thresholds for possible landslide initiation in wildfire-affected areas of western Greece. Bull Eng Geol Env 75:883–896. https://doi.org/10.1007/s10064-015-0762-5

Nikolakopoulos K, Kavoura K, Depountis N, Kyriou A, Argyropoulos N, Koukouvelas I, Sabatakakis N (2017) Preliminary results from active landslide monitoring using multidisciplinary surveys. Eur J Remote Sens 50(1):280–299. https://doi.org/10.1080/22797254.2017.1324741

Pecoraro G, Calvello M, Piciullo L (2019) Monitoring strategies for local landslide early warning systems. Landslides 16(2):213–231. https://doi.org/10.1007/s10346-018-1068-z

Raspini F, Bianchini S, Ciampalini A, Del Soldato M, Montalti R, Solari L, Tofani V, Casagli N (2019) Persistent Scatterers continuous streaming for landslide monitoring and mapping: the case of the Tuscany region (Italy). Landslides 16(10):2033–2044. https://doi.org/10.1007/s10346-019-01249-w

Rau J, Jhan J, Lob C, Linb Y (2011) Landslide mapping using imagery acquired by a fixed-wing UAV. In: ISPRS international archives of the photogrammetry, remote sensing and spatial information sciences, XXXVIII-1/C22, pp 195–200.

Sabatakakis N, Koukis G, Mourtas D (2005) Composite landslides induced by heavy rainfalls in suburban areas: city of Patras and surrounding area, western Greece. Landslides 2:202–211. https://doi.org/10.1007/s10346-005-0002-3

Sabatakakis N, Koukis G, Vassiliades E, Lainas S (2013) Landslide susceptibility zonation in Greece. Nat Hazards 65(1):523–543. https://doi.org/10.1007/s11069-012-0381-4

Soeters R, van Westen CJ (1996) Landslides, investigation and mitigation. Transp Res Board Natl Res Council, Special Rep 247:129–177

Turner D, Lucieer A, De Jong SM (2015) Time series analysis of landslide dynamics using an unmanned aerial vehicle (UAV). Remote Sensing 7:1736–1757

WP/WLI (1995) International Geotechnical Societies' UNESCO working party on world landslide inventory. Working group on rate of movement (Chairman: Bonnard C) A suggested method for describing the rate of movement of a landslide. Bull Int Assoc Eng Geol 52 (1):75–78. https://doi.org/10.1007/BF02602683

Zouros N, Mountrakis D (1991) The Pindos trust and the tectonic relation between the external geotectonic zones in the Metsovon: Eastern Zagori area (Northwestern Greece). Bull Geol Soc Greece 25(1): 245–262 (in Greek)

An Integrated WebGIS System for Shallow Landslide Hazard Early Warning

Nguyen Duc Ha, Le Quoc Hung, Takahiro Sayama, Kyoji Sassa, Kaoru Takara, and Khang Dang

Abstract

The landslides are considered as one of the most dangerous natural disasters and can cause catastrophic influence on society. Therefore, improving the effectiveness of landslide early warning systems is an urgent requirement. The heavy and/or prolonged rainfall is the main factor that has triggered most of the landslide events. In this study, we propose the integration of a geotechnical model—LS-RAPID and a hydrological model—RRI in a WebGIS system to enhance the accuracy and efficiency of shallow landslide hazard early warning for a small basin in Ha Long City, Vietnam. LS-RAPID model is applied to determine potential landslide hazard areas and RRI model is employed to identify subsurface water levels. The system utilized real-time rainfall data from an automatic weather station and forecasted rainfall data from the GFS server as input data for the RRI model running inside the WebGIS server. By combining simulated results from LS-RAPID and RRI models, the integrated WebGIS system allows predicting the occurrence of landslide hazard in both location and time. With the ability to deliver highly accurate results in a short time, the system can be very helpful for the authorities at all levels in making early landslide hazard warnings that mitigate disasters in mountainous areas.

Keywords

WebGIS • Landslide hazard • Subsurface water • LS-RAPID model • RRI model • IDV

N. D. Ha (✉)
Graduate School of Engineering, Kyoto University, Kyoto, Japan
e-mail: nh14vn@gmail.com

Vietnam Institute of Geosciences and Mineral Resources, MONRE, Hanoi, Vietnam

L. Quoc Hung
General Department of Geology and Minerals of Vietnam, MONRE, Hanoi, Vietnam
e-mail: hunglan@gmail.com

T. Sayama
Disaster Prevention Research Institute, Kyoto University, Kyoto, Japan
e-mail: sayama.takahiro.3u@kyoto-u.ac.jp

K. Sassa · K. Dang
International Consortium On Landslides, Kyoto, Japan
e-mail: sassa@iclhq.org

K. Dang
e-mail: khangdq@gmail.com

K. Takara
Graduate School of Advanced Integrated Studies in Human Survivability, Kyoto University, Kyoto, Japan
e-mail: takara.kaoru.7v@kyoto-u.ac.jp

K. Dang
VNU University of Science, Vietnam National University, Hanoi, Vietnam

Introduction

Due to the widespread ability and destructive power, landslides have claimed many lives and severely hindered the economic development and urbanization of many countries. To mitigate the landslide disaster, especially human casualties, the most effective method is to identify and early warn the location and time of landslide hazards. Brabb (1991) concluded that if the landslides could be predicted early, more than 90% of the losses could be avoided.

Locating landslide initiation is very important for disaster preparedness and response. However, in order to effectively deal with landslide disasters, processes of both the initiation and the post-failure movement should be predicted (Sassa et al. 2010).

It is well known that subsurface water generated from heavy and/or prolonged rainfalls increases the pore water pressure at potential sliding surfaces and reduces the soil

© Springer Nature Switzerland AG 2021
N. Casagli et al. (eds.), *Understanding and Reducing Landslide Disaster Risk*, ICL Contribution to Landslide Disaster Risk Reduction,
https://doi.org/10.1007/978-3-030-60311-3_22

strength (Reichenbach et al. 1998; Ekanayake et al. 1999; Collins and Znidarcic 2004). Therefore, to be able to accurately simulate and predict the location and time of landslide hazards, the subsurface water level needs to be taken into account (Bogaard and Greco 2014; An et al. 2016).

WebGIS is a combined product of geographic information system (GIS) and internet technologies (Miao and Yuan 2013), and is used to analyze and display the spatial data. With the rapid development of internet and Web technology, sharing geographic information and broadcasting spatial decision in emergencies through the WebGIS system are considered as one of the fastest and most effective methods (Zhang et al. 2011; Li et al. 2015; Chen et al. 2016; Mamai et al. 2017). Hence, WebGIS can play a significant role in preventing and mitigating landslide disasters (Yu et al. 2007; Pessina and Meroni 2009; Li et al. 2010; Miao and Yuan 2013).

In addition to the popular WebGIS systems built to inventory occurred landslides (Devoli et al. 2007; Yanxi et al. 2009; Dahlhaus et al. 2011; Huang et al. 2013; Chen et al. 2016), some WebGIS systems have been established for monitoring and early warning of potential areas of landslides. In most of these early warning systems, landslides are often predicted by empirical methods based on susceptibility maps, empirical rainfall thresholds, and rainfall data (Huggel et al. 2008; Zhang et al. 2011; Hou et al. 2013; Rosi et al. 2017; Artha and Julian 2018). With this empirical method, WebGIS system can only predict about the period but cannot accurately indicate the areas where the landslides will happen. Some warning systems have integrated a landslide model within the WebGIS system for monitoring and warning the failure of single slopes (Li et al. 2010; Thiebes et al. 2013). In all aforementioned systems, the coupling of landslide hazard and hydrological models has not been exploited to improve the effectiveness of disaster mitigation at a basin (or larger) scale.

The capacity of the geotechnical model LS-RAPID (Sassa et al. 2010) and hydrological model RRI (Sayama et al. 2012) for predicting the spatial and temporal occurrence of landslide hazard was proved when verified with the reality (Ha et al. 2020). To improve the effectiveness of mitigating shallow landslide disasters, in this study, we propose a prototype of an integrated WebGIS system combining the LS-RAPID and RRI models into the WebGIS system. The following issues will be resolved:

- How should LS-RAPID and RRI models need to be integrated into the WebGIS system to support the authorities in early warning of shallow landslides?
- How do rainfall forecast and real-time rainfall data need to be acquired and processed to be automatically simulated by RRI model inside the WebGIS system?

Study Area

Located in the tropical monsoon region, Vietnam is one of the countries affected by the global climate change with abnormal weather events and frequent extreme rainfalls. The landslides are a common geohazard in the mountainous areas of Northern Vietnam. Therefore, the development of methods to help minimize losses related to landslide disasters is very important and urgent.

Ha Long City, Quang Ninh Province, Vietnam has a complex and diverse topography including hills, mountains, valleys, coastal areas, and islands. The hilly and mountainous terrain accounts for 70% of the city area and is concentrated mainly in the North and Northeast. The mountain ranges have an average height of 150 m to 250 m, tend to decrease towards the sea and are intertwined by the narrow valleys. Many households are living in high-risk areas affected by the landslides. According to statistics of the Department of Construction of Quang Ninh Province in 2016, 1,278 households living in 174 zones of high landslide susceptibility need to be relocated.

The pilot study area is a small basin (about 0.15 km^2) in Cao Thang Ward, Ha Long City. In this area, 3 shallow landslides were triggered by a historic rainfall in July 2015, in which 1 landslide destroyed 3 houses and killed 8 people. This rainfall event was the heaviest downpour in the city in 40 years with the highest rainfall intensity of 86 mm/h and 257 mm in 5 h. The province's economic damages caused by this rainfall was determined more than 100 million USD (Fig. 1).

Combining LS-RAPID Geotechnical Model and RRI Hydrological Model

In the LS-RAPID model, the effect of pore water pressure on slope stability due to rising water is expressed through the pore water pressure ratio (r_u) based on pore water pressure and total normal stress acting at the potential sliding surface (Sassa et al. 2010).

To identify areas of potential landslide hazards, the LS-RAPID model was applied with a range of r_u values corresponding to different scenarios of the water level (from no subsurface water to when subsurface water reaches to the ground surface). The results of applying LS-RAPID

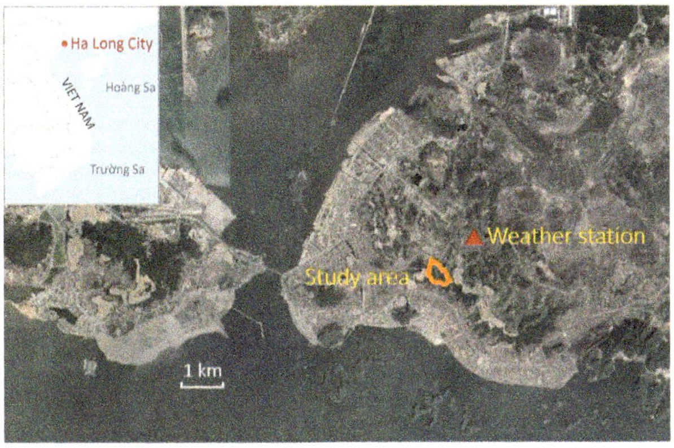

Fig. 1 Google Earth satellite image of Ha Long City with locations of the study basin and the weather station

simulations have identified several potential landslide hazard areas (see areas A1, A2, B1, B2, C, D in Fig. 2). In each forecasted hazard area, three regions: landslide scarp, accumulation zone, and landslide initiation part (locating inside the landslide scarp) were distinguished. In each initiation part of potential landslide hazards, the threshold value of pore water pressure ratio (r_uT) activating the landslide is also estimated. The value of a r_uT is determined as the r_u value when a potential landslide starts to occur in simulations.

The coupling method of the hydrological—geotechnical framework is employing shallow landslide trigger: the subsurface water to connect the simulation results of LS-RAPID and RRI models.

Based on the observed rainfall and subsurface water level, the RRI model was calibrated and then applied to the rainfall data of the 2015 event. By applying the RRI model, subsurface water level maps for all simulation time steps are generated. From there, the various r_u maps for these time steps are created automatically.

To estimate the probability of landslide occurrence, the Risk Index (RI) is determined as $RI = r_u/r_uT$. In each forecasted hazard area, a landslide is likely to happen if the maximum RI is close to or higher 1 (as the maximum value of r_u in the initiation sources approaches or exceeds the r_uT).

For each time step, a RI map is produced by combining the r_u map of that time step and the r_uT map.

The framework coupling LS-RAPID and RRI models is described in more detail in Ha et al. (2020).

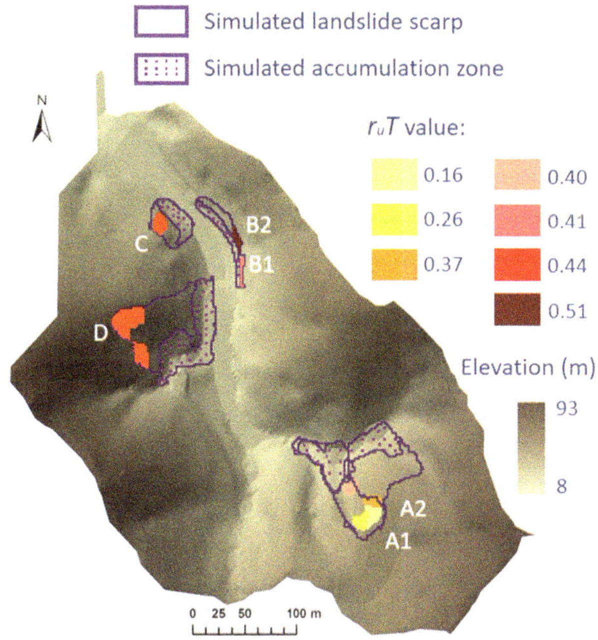

Fig. 2 Potential landslide hazard areas and predicted r_uT values at landslide initiation parts

Integrated System Architecture for Early Warning of Shallow Landslide Hazards

State-Funded Landslide Project (SFLP)

The State-Funded Landslide Project (SFLP) "Investigation, assessment and warning zonation for landslides in the mountainous regions of Vietnam" has been implementing for all mountainous provinces (37/64 provinces) of Vietnam since 2012. In the first phase of the project (ongoing), the WebGIS system has been used as a tool to fill the missing information of the occurred landslide and to update new landslide events. Its spatial database including maps of landslide inventory, controlling factors, and susceptibility zonation would support scientific research as well as contribute to local land use planning to improve the coping capacity of disaster prevention (Hung et al. 2017). In the second phase (not yet implemented), the SFLP has planned to transmit emergency information about the possibility of landslides in critical areas to the government and local authorities through the WebGIS system. Accordingly, through this research, LS-RAPID and RRI models were integrated into the SFLP WebGIS system as a pilot system to support decision-makers.

SFLP WebGIS System

Although the applications to build a WebGIS system are diverse, the general architecture usually consists of three tiers: database tier, application tier, and user interface tier. The SFLP WebGIS system was built with free and open-source software (Fig. 3).

The data tier is built with PostgreSQL database management system and its extension—PostGIS. PostgreSQL is a popular database management system, built with open

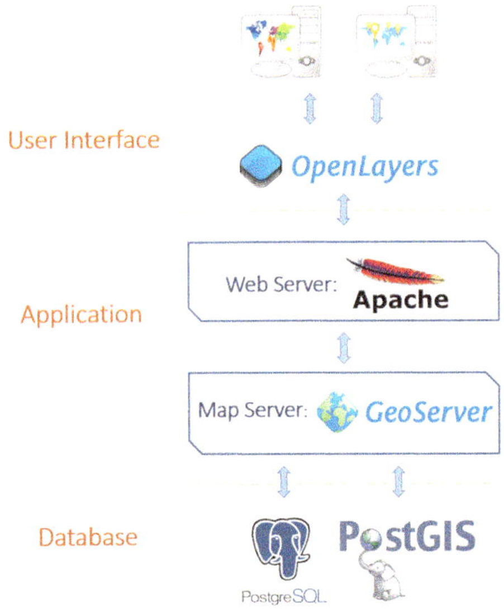

Fig. 3 The three-tier architecture applied for the SFLP WebGIS system

source and can be compatible with most popular operating systems. PostgreSQL is equipped with interfaces to connect to many programming languages like Python, PHP or Java. PostGIS is a spatial database extension of PostgreSQL. Through PostGIS, the location and map information of spatial objects can be easily queried.

At the application tier, to manage the work inside the server, the HTTP web server—Apache and the map server—GeoServer are chosen to use. Web server is a software used to receive and analyze requests from web users and return result information. GeoServer is used to perform services of managing geographic objects over the internet with the OGC (Open Geospatial Consortium) standards. GeoServer is also used to specify the visual styles (colors, shapes, transparency) of geographical objects stored in the PostGIS database.

At the interface tier, OpenLayers communicates with GeoServer via OGC standard. OpenLayers is a JavaScript library that helps to build interactive map tools on the web. With these tools, web users can easily manipulate maps and geographical objects on web browsers.

Integrate the Hydrological-Geotechnical Framework into the WebGIS System

The hydrological-geotechnical framework has been integrated with the SFLP WebGIS system in two separate phases. In the first phase, to identify areas at risk of landslide hazards and the thresholds of pore water pressure ratio (r_uT), the LS-RAPID model is applied to different scenarios of

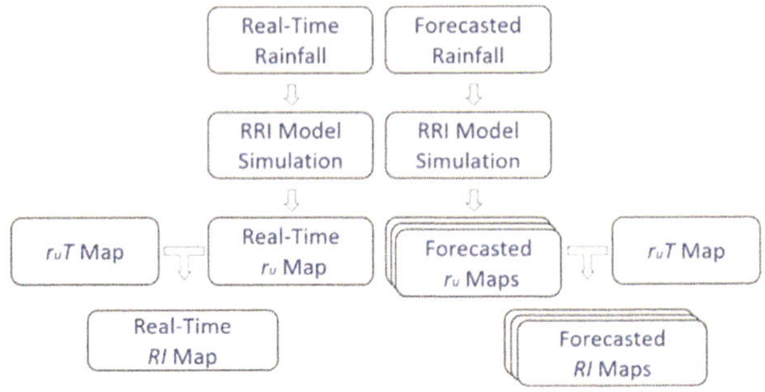

Fig. 4 Method for generating *RI* maps inside the SFLP WebGIS server

subsurface water level. While the simulation with the LS-RAPID model needs to be done before integrating with the WebGIS system, in the second phase, the RRI model needs to be applied automatically regularly within the SFLP server. In order to predict the time at which landslide hazards are at risk, variations in the subsurface water level need to be monitored regularly. From that, the maps of pore water pressure ratio (r_u) according to real-time and forecasted rainfall data can be generated. These r_u maps will be automatically combined with the r_uT map to generate *RI* maps and support warnings about the location and timing of possible landslide hazards in the near future. Figure 4 presents the main steps of the method of combining LS-RAPID and RRI models in the SFLP WebGIS server to identify *RI* maps.

The strategy of applying LS-RAPID and RRI modeling framework into 2 separate phases is very important for emergency situations in disaster management and warning because of the ability to optimize processing time. In addition, the RRI model has the ability to allow the simulation process to continue with newly updated data without having to restart from the beginning of the data series. This feature is very useful for the early warning system since rainfall data (real-time data and forecasted data) are regularly updated (Ha et al. 2020). Specifically for this study area, the time-consuming of acquiring rainfall data and generating a *RI* map for each time step by the author's personal computer range from nearly 1 min to a few minutes depending on the intensity of rainfall.

Connect Real-Time Rainfall Data

A weather station has been installed near the study area (about 1 km Northeast). The weather station consists of 3 main equipment (Vantage Pro2 equipment manufactured by David Instruments Corp): an outdoor station, a console, and a data logger. Rainfall data measured by a rainfall collector

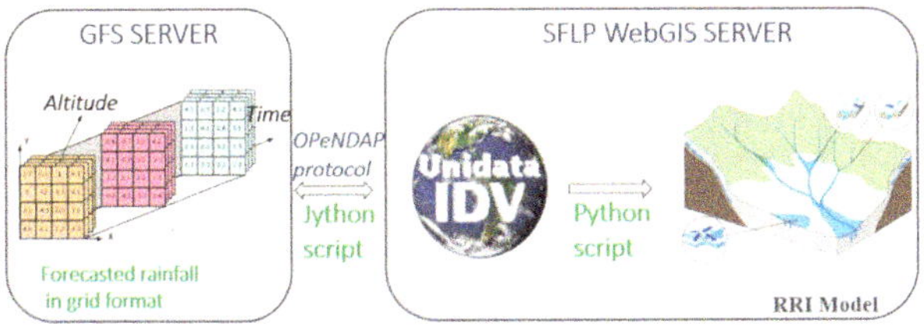

Fig. 5 Conceptual model of utilizing forecasted rainfall data from the GFS server

in the outdoor station are sent regularly to the console every 10 min by the solar-power transmitter. The data logger is employed to link the console and the "WeatherLink Cloud" of David Instruments Corp. With this method, newly updated data can be automatically displayed on the website "www.weatherlink.com" every 60 min.

Inside the SFLP server, every 60 min, the Python script is used to automatically connect to the website "www.weatherlink.com" and collect measured rainfall data. Whenever the rainfall data is updated, the Python script will update the input data file and send requests via the command line to the RRI model to perform simulations. The newly acquired rainfall data is also updated into the database in the PostgreSQL database management system so that it can display the rainfall data in time series to users when required.

Connect Forecasted Rainfall Data

To monitor and forecast of the risk of landslide hazards, real-time rainfall data is a requirement to simulate the hydrological process in the study basin. However, the preparedness is much more effective, and the damages caused by landslide hazards can be significantly reduced if forecasted rainfall data is also applied.

The Global Forecast System (GFS) is a meteorological forecast model produced by the National Centers for Environmental Prediction (NCEP)—USA. The GFS model provides forecasted rainfall data for the world four times a day

Fig. 6 The graphical user interface of the WebGIS on landslides "www.canhbaotruotlo.vn" which introduces the main achievements of the SFLP

Fig. 7 Illustration of three warning levels for potential shallow landslide hazard areas would be shown in the WebGIS "www.canhbaotruotlo.vn"

(at 00, 06, 12, 18 Universal Time Coordinated). The forecasted rainfall data from the GFS model is in a grid format, with a spatial resolution of 28 km and a time resolution of 3 h.

A typical feature of grid data is that the coordinates are stored by longitude and latitude, and time is a dimension of the data structure. These features allow the spatial and temporal relationships of continuous phenomena in the real world to be easily expressed (Di et al. 2003).

The grid data from GFS server is automatically accessible via OPeNDAP protocol. OPeNDAP stands for "Open-source Project for a Network Data Access Protocol". This is a server-client protocol that allows access and sharing of grid data given by Unidata and is widely used in the Earth Science community. A major challenge in using grid data is the huge database set. For the management and warning of natural disasters, the information processing time to make a decision should be shortened as much as possible. Therefore, it is not appropriate to download the entire database and then extract the data of the area and the time required to study. To optimize processing time, the required data of grid data must be determined on the GFS server side. Then the download and processing of data will be much faster.

There are many OPeNDAP client software widely used for connecting OPeNDAP servers and retrieving grid data such as IDV, MATLAB, GrADS, Ferret, and Pydap. Based on selection criteria such as software that needs to support the writing of programming applications, support for building server-client applications, support for reading and analyzing grid data, the IDV software has been selected for use inside the SFLP WebGIS server to address obstacles in identifying and downloading automatically forecasted rainfall data for the study basin (Fig. 5).

The IDV (Integrated Data Viewer) software is developed by Unidata. This is a Java-based software built to support the analysis and display of geoscience data. IDV is able to work with many types of data such as grid data, surface observations, satellite images, and radar data. An important feature of IDV software is the ability to identify and query geo-referenced datasets placed on remote servers. Therefore, by constructing additional functions for IDV with the Jython programming language (an implementation of Python to run on Java platform), a subset of grid data in the spatial range and time of study can be predefined and automatically downloaded to the server of the SFLP WebGIS system.

Results

As part of the SFLP project, a national WebGIS system "www.canhbaotruotlo.vn" has been developed for landslide inventory and early warning with a simple interface for users with diverse backgrounds. This WebGIS system is also considered as one of the interactive interfaces for landslide research among scientists, managers and local people.

In the first phase of the SFLP Project, the main result is the spatial database system for landslides. The database includes maps of landslide inventory and susceptibility zonation at 1:50,000 scale and information of landslide controlling factors. This database has been uploaded to the

WebGIS system to share with local authorities and people about real conditions and potential risks. In addition, the shared information can contribute significantly to scientific research. Local people are encouraged to participate in the WebGIS system to fulfil missing information of the landslide inventory, update the new landslide areas and inform the signs of new landslides as detected. Therefore, WebGIS is also supported with tools that allow local people to easily inform the authorities and staff of the SFLP Project. This information, before verified, will be visualized differently from the style of official data.

In preparation for the second phase of the SFLP Project, the author proposes an early warning system for shallow landslides by integrating the coupled hydrological geotechnical framework in conjunction with the SFLP WebGIS system. This system (still under testing period) will be provided as a reliable reference to assist authorities in shallow landslide hazard early warning (Fig. 6).

The potential shallow landslide hazard areas A2, C, and D are being monitored with the application of the hydrological-geotechnical framework within the SFLP WebGIS system. Area A1 where a landslide occurred and areas B1 and B2 where reinforced by structural measures are not monitored. Figure 7 illustrates three warning levels that will be used to represent the risk status of the area being monitored. Three warning levels: safety (in blue), close monitoring (yellow) and danger (red) are determined based on the calculated Risk Index (*RI*) value range: $0 < RI < 0.2$ (safety); $0.2 \leq RI < 0.7$ (monitoring); $0.7 \leq RI$ (danger). The value of the *RI* index is used for each of the potential hazard area (A2, C, and D) to give a warning that the highest *RI* value is calculated in the landslide initiation parts of those areas.

Discussion and Conclusion

For preventing or mitigating landslide disasters, it is very important forecasting the process of landslide hazards in both location and time. In WebGIS systems utilizing empirical rainfall thresholds, the principal disadvantages are the occurrence (both location and time) of potential landslides are incapable to be determined correctly. In existing systems applying models for early warning landslides at a catchment or larger scales, the possible hazard areas are not identified. In addition, for emergency responses, the processing time for landslide simulating might be also an impediment. By taking advantage of the LS-RAPID and RRI models, the integrated system can support to forecast both the initiations and the runout processes of landslide hazards. Moreover, the application of two models at two separate phases brings a great advantage to the system in terms of the processing time necessary in emergency operations.

For the early warning system, in order to increase the accuracy between predictive and actual results, the uncertainty of the input data needs to be considered and improved. The resolution of the forecasted rainfall data from GFS is not high so that warnings are hard to be issued based solely on this data. However, these data can serve as a reference for the future trend of rain in the study area.

In this study, the geotechnical model LS-RAPID and hydrological model RRI are incorporated into the SFLP WebGIS system as a prototype for the shallow landslide early warning system. The real-time rainfall data from the weather station located near the study area and forecasted rainfall data from the GFS server are automatically extracted and used regularly as input data for the RRI model. From the simulation results, the subsurface water level can be estimated for each location. Therefore, potential area and occurrence time of landslide hazard can be forecasted and support decision-makers in issuing an early warning.

Acknowledgements This study is a part of the State-Funded Landslide Project (SFLP) "Investigation, assessment and warning zonation for landslides in the mountainous regions of Vietnam" and contributes to the Sendai Landslide Partnerships 2015-2025. The research was also financially supported by the Ministry of Education, Culture, Sports, Science and Technology of Japan (MEXT) and the Japan-ASEAN Science, Technology and Innovation Platform (JASTIP). The authors are immensely grateful colleagues from Vietnam Institute of Geosciences and Mineral Resources (VIGMR), MONRE for their support and valuable comments on this study.

References

An H, Tran TV, Lee G, Kim Y, Kim M, Noh S, Noh J (2016) Development of time-variant landslide-prediction software considering three-dimensional subsurface unsaturated flow. Environ Model Softw 85:172–183

Artha Y, Julian ES (2018) Landslide early warning system prototype with GIS analysis indicates by soil movement and rainfall. IOP conference series: Earth Environ Sci 106(1)

Bogaard T, Greco R (2014) Preface "Hillslope hydrological modelling for landslides prediction" . Hydro Earth Syst Sci 18:4185–4188

Brabb EE (1991) The world landslide problem. Episodes 14(1):52–61

Chen W, He B, Zhang L, Nover D (2016) Developing an integrated 2D and 3D WebGIS-based platform for effective landslide hazard management. International Journal of Disaster Risk Reduction 20:26–38

Collins BD, Znidarcic D (2004) Stability analyses of rainfall induced landslides. J Geotech Geoenviron Eng 130(4):362–372

Dahlhaus PG, Miner AS, MacLeod A, Thompson H (2011) A Web-GIS and landslide database for south west victoria and its application to landslide zonation. Aust Geomech 46(2):203–209

Devoli G, Strauch W, Chavez G, Hoeg K (2007) A landslide database for Nicaragua: a tool for landslidehazard management. Landslides 4 (2):163–176

Di L, Chen A, Yang W, Zhao P (2003) The integration of grid technology with OGC web services (OWS) in NWGISS for NASA EOS data. In: Proceedings of the 8th global grid forum (GGF8 & HPDC12 2003)

Ekanayake J, Phillips C, Phillips J (1999) A model for determining thresholds for initiation of shallow landslides under near-saturated conditions in the East Coast region. New Zealand. J Hydrol (New Zealand) 38(1):1–28

Ha ND, Sayama T, Sassa K, Takara K, Uzuoka R, Dang K, Pham TV (2020) A coupled hydrological-geotechnical framework for forecasting shallow landslide hazard—a case study in Halong City, Vietnam. Landslides. https://doi.org/10.1007/s10346-020-01385-8

Hou S, Li A, Han B, Zhou P (2013) An early warning system for regional rain-induced landslide hazard. Int J Geosci 4(3):584–587

Huang R, Huang J, Ju N, He C, Li W (2013) WebGIS-based information management system for landslides triggered by Wenchuan earthquake. Nat Hazards 65(3):1507–1517

Huggel C, Ramirez JM, Calvache M, González H, Gutierrez C, Krebs R (2008) A landslide early warning system within an integral risk management strategy for the Combeima-Tolima Region, Colombia. In: Proceedings of the international disaster and risk conference (IDRC), Davos, Switzerland, pp 273–276

Hung LQ, Van NTH, Son PV, Ninh NH, Tam N, Huyen NT (2017) Landslide inventory mapping in the fourteen Northern Provinces of Vietnam: achievements and difficulties. In: Sassa K, Mikoš M, Yin Y (eds) Advancing culture of living with landslides. WLF 2017. Springer, Cham

Li XG, Wang AM, Wang ZM (2010) Stability analysis and monitoring study of Jijia River landslide based on WebGIS. J Coal Sci and Eng 16(1):41–46

Li B, Wu J, Pan M, Huang J (2015) Application of 3D WebGIS and real-time technique in earthquake information publishing and visualization. Earthq Sci 28(3):223–231

Mamai L, Gachari M, Makokha G (2017) Developing a Web-based water distribution geospatial information system for Nairobi Northern Region. J Geo Inf Syst 9:34–46

Miao F, Yuan Q (2013) A WebGIS-based information system for monitoring and warning of geological disasters for Lanzhou City, China. Adv Meteorol

Pessina V, Meroni F (2009) A WebGis tool for seismic hazard scenarios and risk analysis. Soil Dyn Earthq Eng 29(9):1274–1281

Reichenbach P, Cardinali M, Vita PD, Guzzetti F (1998) Regional hydrological thresholds for landslides and floods in the Tiber River Basin (central Italy). Environ Geol 35(2–3):146–259

Rosi A, Segoni S, Battistini A, Rossi G, Catani F, Casagli N (2017) Definition of a fully functional EWS based on Rainfall thresholds, the case of study of Tuscany Region. In: Mikoš M, Arbanas Ž, Yin Y, Sassa K (eds) Advancing culture of living with landslides. WLF 2017. Springer, Cham, pp 169–174

Sassa K, Nagai O, Solidum R, Yamazaki Y, Ohta H (2010) An integrated model simulating the initiation and motion of earthquake and rain induced rapid landslides and its application to the 2006 Leyte landslide. Landslides 7(3)

Sayama T, Ozawa G, Kawakami T, Nabesaka S, Fukami K (2012) Rainfall-runoff-inundation analysis of the 2010 Pakistan flood in the Kabul River basin. Hydrol Sci J 57(2):298–312

Thiebes B, Bell R, Glade T, Jäger S, Anderson M, Holcombe L (2013) A WebGIS decision-support system for slope stability based on limit-equilibrium modelling. Eng Geol 158:109–118

Yanxi Z, Gangjun L, Erjiang F, Kefei Z (2009) An object-relational prototype of GIS-based disaster database. Procedia Earth Planet Sci 1(1):1060–1066

Yu FC, Chen CY, Lin SC, Lin YC, Wu SY, Cheung KW (2007) A web-based decision support system for slopeland hazard warning. Environ Monit Assess 127(1–3):419–428

Zhang G, Chen L, Dong Z (2011) Real-Time Warning System of Regional Landslides Supported by WEBGIS and its Application in Zhejiang Province, China. Procedia Earth Planet Sci 2:247–254

WLF 2020 Kyoto Japan

The Value of Soil Wetness Measurements for Regional Landslide Early Warning Systems

Manfred Stähli and Adrian Wicki

Abstract

Current regional landslide Early Warning Systems (LEWS) are typically based on rainfall data and rarely use soil wetness information in spite of the direct relationship between soil wetness and landslide triggering. This is partly due to the limited availability of soil moisture measurement networks and the lack of experience in interpreting such data. Here we show how soil wetness measurements could be exploited for LEWS in a region (or country) with variable topography and soil composition. In particular, we highlight the relevance of (steep) topography for the soil wetness observations and its significance for the interpretation in regard to landslide criticality. To this end, a field study has been set up and is currently running in a landslide-susceptible region of central Switzerland (Emmental) where different soil wetness measurement methods are compared on a flat and a steep meadow site. First results suggest that in spite of noticeable topographical effects on soil drying and wetting the informative value of these data for LEWS is not very different for sloped and flat locations. Based on these findings and a comprehensive analysis of the co-occurrence between soil wetness indicators and shallow landslides across Switzerland we attribute soil wetness measurements a significant value for landslide early warning and encourage a wider incorporation of such data in existing warning systems.

Keywords

Soil moisture • Landslide early warning • Monitoring systems

Introduction

In mountainous regions, landslides are a serious hazard for people and infrastructures. In Switzerland, they have caused a total financial damage of 520 million Euros in the period from 1972 to 2007 (Hilker et al. 2009) and the death of 74 people between 1946 and 2015 (Badoux et al. 2016). Shallow landslides are particularly dangerous due to the short interval time between activation and failure and their widespread occurrence.

Regional landslide early warning systems (LEWS) have proven to be a valuable and efficient tool to issue warnings to the public in order to move people and goods at risks to safety (Stähli et al. 2015). Existing LEWS have been mostly based on rainfall exceedance thresholds that empirically relate precipitation characteristics to the occurrence of landslides (e.g. Guzzetti et al. 2007). While precipitation measurements are widely available, rainfall exceedance thresholds bear specific limitations such as the non-representation of antecedent wetness conditions and soil wetness distribution at depth (Godt et al. 2009), or a high spatial and temporal variability of critical threshold values (Aleotti 2004; Baum and Godt 2010).

To overcome these limitations, the need for hydrological-based thresholds has been stated (e.g. Berne et al. 2013; Devoli 2017). In addition, it has been postulated that including information on antecedent saturation conditions could effectively reduce both false and missed alarms (Bogaard and Greco 2018). In fact, recent studies have demonstrated an improvement of the forecast quality of rainfall-based LEWS after the inclusion of soil wetness information derived from measurements (Mirus et al. 2018a, b; Comegna et al. 2016) or models (Segoni et al. 2018).

M. Stähli (✉) · A. Wicki
Swiss Federal Research Institute WSL, Mountain Hydrology and Mass Movements, Zürcherstrasse 111, 8903 Birmensdorf, Switzerland
e-mail: manfred.staehli@wsl.ch

© Springer Nature Switzerland AG 2021

N. Casagli et al. (eds.), *Understanding and Reducing Landslide Disaster Risk*, ICL Contribution to Landslide Disaster Risk Reduction,
https://doi.org/10.1007/978-3-030-60311-3_23

Measuring Soil Wetness for Landslide Prediction

Shallow landslides are typically triggered after the infiltration of rainfall or snow melt water and the consequent rise of the water saturation and perched groundwater tables. The resulting increase in pore water pressure and decrease in matric suction may reach critical values at which point the slope eventually fails (Crozier 1986; Wieczorek 1996). Measurements of soil wetness have thus long been suggested to be valuable for assessing the landslide danger (e.g. Springman et al. 2003; Lu et al. 2010).

Soil wetness is commonly expressed as the volume of pores that is filled with water divided by the total soil volume (volumetric water content, VWC, θ) or as the pressure at which water is bounded to the particles (matric potential). It can be measured or indirectly estimated at different spatial and temporal scales.

At the scale of soil profiles, soil moisture probes (time-domain reflectometry, TDR, or capacitance based) are used to estimate permittivity by electromagnetic sensors. The permittivity values can then be related to the VWC using a specific calibration function (Topp et al. 1980). The matric potential can be measured with tensiometers consisting of a porous cup on a rigid tube which is filled with water and connected to a vacuum gauge that measures pressure levels (Livingston 1908). Both sensor types allow for continuous soil wetness monitoring. While they are representative for a specific measurement volume, multiple probes are usually integrated in depth profiles.

Electrical resistivity tomography (ERT) allows for a distributed 2-D VWC estimation along hill-slope transects of several meters length. Between pairs of equally spaced electrodes, the potential difference is measured, and by variation of the electrode geometry, the 2-D distribution of apparent resistivity is assessed. Specific resistivities are then deducted by the application of an inversion algorithm (e.g. Kneisel and Hauck 2008). While single measurements depend largely on the lithology, resistivity changes can be related to changes in the soil water content (Robinson et al. 2009). This type of measurement allows for a 2-D VWC estimation and is non-invasive, however the longer acquisition time results in a lower temporal resolution.

Soil wetness can also be sensed remotely from space e.g. using satellite-based measurements of passive or active microwave emission of the soil surface (Reichle et al. 2017). However, the spatial resolution of this information is very coarse, the temporal resolution reduced compared to in-situ measurements, and the depth of the represented soil wetness shallow, which strongly limits the potential for LEWS in mountainous landscapes (Thomas et al. 2019).

Dependence of Soil Wetness on Topography

It is well known that the spatial distribution of soil wetness is heavily influenced by topography. There are topographical zones where water accumulates, and such where water can drain (laterally) relatively quickly. In hydrology, a topographic wetness index (TWI; Sørensen et al. 2006) is typically used for discerning these areas.

But not only the lateral water flow in the soil, also soil evaporation or snow cover distribution is significantly influenced by the topography, which in turn can lead to variations in soil wetness.

For LEWS, the question arises to what extent these topographical effects on soil wetness must be taken into account, especially since available measurement networks are often located at flat sites.

Soil Wetness Measurements at a Landslide Prone Site

In order to assess the representativeness of flat measurement sites for critical hydrological conditions at hillslopes, a specific soil wetness observatory has been set up in a landslide prone area of central Switzerland (Emmental, Swiss prealps, Fig. 1). A flat and a sloped location (30° slope angle) were equipped with various soil wetness measurement sensors in spring 2019. To analyse the potential of the different measurement techniques for the use in a LEWS, they were compared in their ability to identify critically saturated conditions in the upper soil layer.

Monitoring Setup

Capacitance-based soil moisture probes (ECH2O 5TE, Meter, Germany) and tensiometers (T8, Meter, Germany)

Fig. 1 Soil moisture site (sloped location) in the Napf area (Switzerland)

were installed in soil pits at three different depths (15, 50 and 100 cm). Perched ground-water levels are measured using a piezometer. Additional measurements include air and ground temperature (temperature probes) as well as precipitation (tipping bucket).

For comparison, the absolute soil moisture measurements are normalized for each sensor by the respective maximum of the time series. Both soil moisture and matric potential records are integrated to depth profiles by calculating the profile mean.

To further assess the 2-D distribution of temporal soil moisture dynamics, an ERT profile line was installed at the sloped site. It consists of 48 electrodes at an 0.25 m electrode spacing, which is small enough to capture small scale hydrological processes. Automated measurements were conducted during precipitation events at hourly time scale, and a time-lapse tomography inversion scheme was applied to relate the resistivity changes to changes in soil water content.

Summer 2019 Results

First results of the soil moisture measurements from April to November 2019 suggest that differences between the flat and the sloped measurement site are generally small (Fig. 2). Differences are smallest near the saturated and the dry end, and they become larger at intermediately saturated conditions and inhibit an S-shaped relationship between the flat and the sloped site.

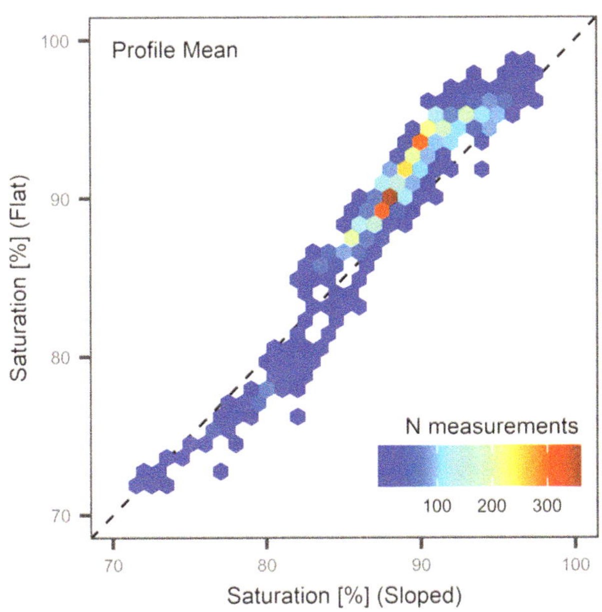

Fig. 2 Profile mean saturation at the sloped site vs. the flat site for the period of April to November 2019

This indicates that during conditions of higher saturation, the sloped site drains faster (more quickly) but saturates slower, which can be explained by the larger gravitational potential (quicker desaturation) and higher surface runoff rates (slower saturation) due to the slope degree. Conversely, during drier conditions, the sloped site dries out slower but saturates faster (more quickly), which we relate to lower evaporation rates due to the SE exposition and shading by trees (slower drying rates) and higher infiltration capacity (quicker saturation).

Potential of an Existing In-situ SWC Network for LEWS

Soil moisture measurements are becoming more frequent within operational monitoring networks. To assess the potential of existing in-situ VWC measurements for regional landslide early warning, a comprehensive soil moisture data base was compiled from existing soil moisture monitoring networks in Switzerland (Fig. 3). In a recent study (Wicki et al. 2020), the temporal variation of the soil moisture measurements was compared to the occurrence of landslides that were recorded in a national landslide inventory (Swiss flood and landslide damage database, WSL, Fig. 3).

Methodology

VWC time series were first homogenized, normalized by the minimum and maximum of each time series, and then integrated to ensembles of sensors at each site. At the ensemble mean time series, the onset and end of individual *infiltration events* (i.e. continuous periods of saturation increase) were identified and various event properties were quantified.

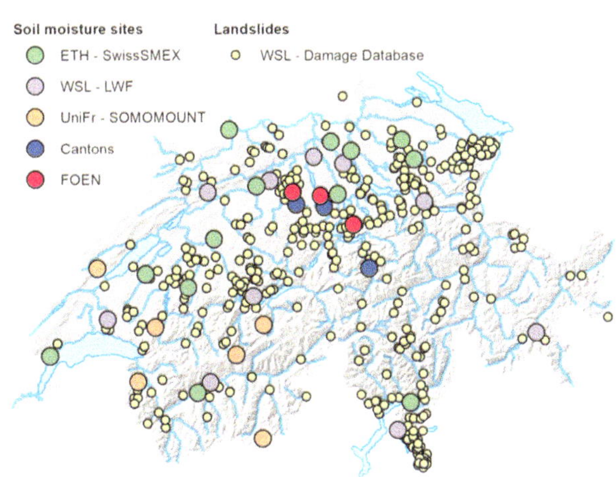

Fig. 3 Map of Switzerland showing all included soil moisture sites and landslide records

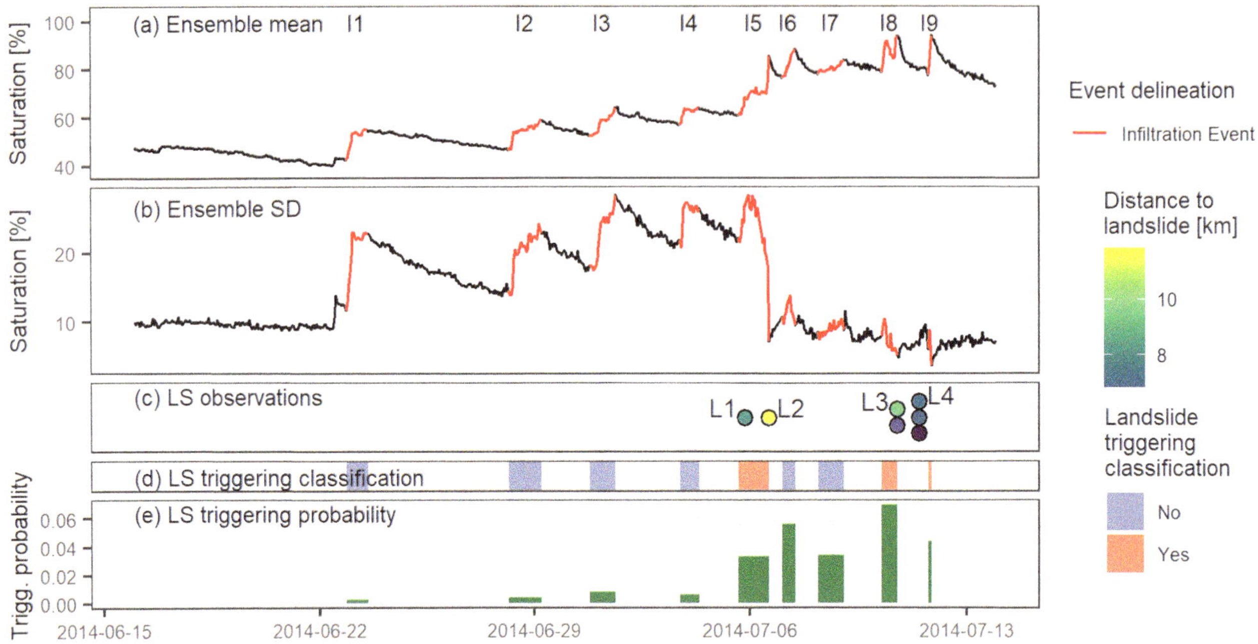

Fig. 4 Temporal evolution for a sample time period at the SwissSMEX Site Plaffeien of **a** the ensemble mean and **b** ensemble standard deviation with infiltration events highlighted in red, **c** recorded landslides within 15 km distance from the site, **d** the respective landslide triggering classification, and **e** the resulting landslide triggering probability from one model fit (all event properties, all sensors, all sites, 15 km forecast distance)

Further, infiltration events were classified as *landslide triggering* or *landslide non-triggering* based on the coincident occurrence of landslides within a specific distance from the site (referred to as the *forecast distance*). To assess the distance sensitivity of the model, the forecast distance was varied in 5 km steps from 5 to 40 km.

Finally, a multiple logistic regression (MLR) function was fit to the infiltration event set to statistically reproduce the landslide probability as a function of the infiltration event properties. To assess the forecast quality of a specific model fit and to compare different model fits, receiver operating characteristics (ROC) analysis was performed.

Results

The components of one specific model fit are illustrated in Fig. 4 for a time period extract from June to July 2014 and at a sample site (SwissSMEX site Plaffeien, Canton of Fribourg, Northern pre-Alps). During this period, a series of rainfall events caused widespread landslide activity in all Switzerland and specifically in the region of Fribourg. The measurements show a continuous increase in the soil saturation (Fig. 4a). It is accompanied by a initial increase of the standard deviation which suddenly decreases at *I5* (Fig. 4b), indicating that the soil profile is homogeneously saturated from then on. It was also during *I5* and in the following days that several landslide events were observed (Fig. 4c). This elevated landslide danger was reproduced by the MLR model as a sudden landslide probability increase from *I5* onwards (Fig. 4e).

Comparison of the different model fits revealed a strong distance dependence of the forecast goodness which increases with short distances between site and landslide. We explain this distance dependence to be mainly driven by variations in precipitation (phase, intensity, duration) that determine the amount of water infiltration, and by variations of irradiance, wind or humidity that mostly drive the drying up of the soil. Some forecast skill persists even at a forecast distance of 40 km. We relate this to spatial persistence of relative soil moisture anomalies as opposed to absolute soil moisture differences, which was found in a previous study for the SwissSMEX sites for much of Switzerland (Mittelbach and Seneviratne 2012).

Main model drivers were identified to be the antecedent saturation (saturation at the onset of an infiltration event), the two-week preceding maximum saturation and the saturation change during an infiltration event. Using a model fit that includes only the antecedent saturation and the saturation change resulted in almost the same forecast goodness as if all event properties are considered. This demonstrates the importance of describing both the medium-term antecedent conditions as well as the short-term event dynamics, as previously proposed by Bogaard and Greco (2016).

Considerable differences were found between landslides triggerd by long-lasting precipitation events compared to landslides due to thunderstorms. If only the first are considered, the forecast goodness improves significantly and event properties that describe the antecedent conditions become more important. If the latter are included in the modelling only, the forecast goodness decreases significantly and event properties that describe the event dynamics become more important.

Finally, improvements of the forecast goodness could be achieved if seasons were considered individually and if the soil moisture sites were grouped by texture. However, the analysis was limited by the size of the dataset. Further, it was shown that most variability could be explained by considering the uppermost sensors (≤ 30 cm depth) only.

Conclusions and Outlook

Our experiences from this work so far suggests that in-situ soil moisture measurements contain specific information on the regional landslide activity and can be effectively used for landslide early warning. Most important soil-wetness information pointing to the imminence of shallow landslides are the antecedent saturation and the saturation change during an infiltration event.

The forecast goodness strongly depends on the distance between the measurement site and the location of the landslide, which is explained to be driven mainly by meteorological variations. The presence of forecast skill at large distance underlines the value of focusing on relative changes as opposed to using absolute soil moisture variations. Finally, the topographic influence is small if the top 1 m is considered only. This supports the use of already existing soil moisture monitoring networks.

Open questions remain about the value of other soil wetness measurement techniques (tensiometers, ERT). Also, in our future work we will test to what degree soil wetness data from measurement networks can be replaced by numerical soil moisture simulations.

Acknowledgements This research project is financially supported by the Swiss National Science Foundation (project number 175785) and is part of the programme *Climate Change Impacts on Alpine Mass Movements* of the Swiss Federal Research Institute WSL. We thank all involved research institutions, private companies and authorities for installation and maintenance of the VWC monitoring sites as well as providing the data for this study.

References

Aleotti P (2004) A warning system for rainfall-induced shallow failures. Eng Geol 73:247–265. https://doi.org/10.1016/j.enggeo.2004.01.007

Badoux A, Andres N, Techel F, Hegg C (2016) Natural hazard fatalities in Switzerland from 1946 to 2015. Nat Hazards Earth Syst Sci 16:2747–2768. https://doi.org/10.5194/nhess-16-2747-2016

Baum RL, Godt JW (2010) Early warning of rainfall-induced shallow landslides and debris flows in the USA.

Berne A, Ferrari A, Frank F, Gruner U, Haeberle J, Hählen N, Huggel Ch, Kos A, Lehmann P, McArdell B, Or D, Sättele M, Schleiss M, Springman SM, Stähli M, Tobler D, Van Herwijnen A (2013) Strategies towards design of next-generation Early Warning Systems (EWS) for rapid mass movements

Bogaard T, Greco R (2018) Invited perspectives: hydrological perspectives on precipitation intensity-duration thresholds for landslide initiation: proposing hydro-meteorological thresholds. Nat Hazards Earth Syst Sci 18:31–39. https://doi.org/10.5194/nhess-18-31-2018

Bogaard T, Greco R (2016) Landslide hydrology: from hydrology to pore pressure. Wiley Interdiscip Rev Water 3:439–459. https://doi.org/10.1002/wat2.1126

Comegna L, Damiano E, Greco R, Guida A, Olivares L, Picarelli L (2016) Field hydrological monitoring of a sloping shallow pyroclastic deposit. Can Geotech J 53:1125–1137. https://doi.org/10.1139/cgj-2015-0344

Crozier MJ (1986) Landslides: causes, consequences & environment. Croom Helm, London

Devoli G (2017) Workshop Regional early warning systems for rainfall- and snowmelt-induced landslides. Need for an international forum? Summary report

Fawcett T (2006) An introduction to ROC analysis. Pattern Recognit Lett 27:861–874. https://doi.org/10.1016/j.patrec.2005.10.010

Godt JW, Baum RL, Lu N (2009) Landsliding in partially saturated materials. Geophys Res Lett 36:1–5. https://doi.org/10.1029/2008GL035996

Guzzetti F, Peruccacci S, Rossi M, Stark CP (2007) Rainfall thresholds for the initiation of landslides in central and southern Europe. Meteorol Atmos Phys 98:239–267. https://doi.org/10.1007/s00703-007-0262-7

Hilker N, Badoux A, Hegg C (2009) The Swiss flood and landslide damage database 1972–2007. Nat Hazards Earth Syst Sci 9:913–925. https://doi.org/10.5194/nhess-9-913-2009

Kneisel C, Hauck C (2008) Electrical methods. In: Hauck C, Kneisel C (eds) Applied geophysics in periglacial environments. Cambridge University Press, Cambridge, pp 3–27

Livingston BE (1908) A method for controlling plant moisture. Plant Wold 11:39–40

Lu N, Godt JW, Wu DT (2010) A closed-form equation for effective stress in unsaturated soil. Water Resour Res 46:1–14. https://doi.org/10.1029/2009wr008646

Mirus B, Morphew M, Smith J (2018a) Developing hydro-meteorological thresholds for shallow landslide initiation and early warning. Water 10:1274. https://doi.org/10.3390/w10091274

Mirus BB, Becker RE, Baum RL, Smith JB (2018b) Integrating real-time subsurface hydrologic monitoring with empirical rainfall thresholds to improve landslide early warning. Landslides 1–11. https://doi.org/10.1007/s10346-018-0995-z

Mittelbach H, Seneviratne SI (2012) A new perspective on the spatio-temporal variability of soil moisture: Temporal dynamics versus time-invariant contributions. Hydrol Earth Syst Sci 16:2169–2179. https://doi.org/10.5194/hess-16-2169-2012

Reichle RH, De Lannoy GJM, Liu Q, Ardizzone JV, Colliander A, Conaty A et al (2017) Assessment of the SMAP level-4 surface and root-zone soil moisture product using in situ measurements. J Hydrometeorol 18:2621–2645. https://doi.org/10.1175/JHM-D-17-0063.1

Robinson DA, Lebron I, Kocar B, Phan K, Sampson M, Crook N, Fendorf S (2009) Time-lapse geophysical imaging of soil moisture dynamics in tropical deltaic soils: an aid to interpreting hydrological and geochemical processes. Water Resour Res 45:1–12. https://doi.org/10.1029/2008WR006984

Segoni S, Rosi A, Lagomarsino D, Fanti R, Casagli N (2018) Brief communication: Using averaged soil moisture estimates to improve the performances of a regional scale landslide early warning system. Nat Hazards Earth Syst Sci Discuss 1–9. https://doi.org/10.5194/nhess-2017-361

Sørensen R, Zinko U, Seibert J (2006) On the calculation of the topographic wetness index: evaluation of different methods based on field observations. Hydrol Earth Syst Sci 10:101–112. https://doi.org/10.5194/hess-10-101-2006

Springman SM, Jommi C, Teysseire P (2003) Instabilities on moraine slopes induced by loss of suction: a case history. Geotechnique 53:3–10. https://doi.org/10.1680/geot.2003.53.1.3

Stähli M, Sättele M, Huggel C, McArdell BW, Lehmann P, Van Herwijnen A, Berne A, Schleiss M, Ferrari A, Kos A, Or D, Springman SM (2015) Monitoring and prediction in early warning systems for rapid mass movements. Nat Hazards Earth Syst Sci 15:905–917. https://doi.org/10.5194/nhess-15-905-2015

Thomas MA, Collins BD, Mirus BB (2019) Assessing the feasibility of satellite-based thresholds for hydrologically driven landsliding. Water Resour Res 55. https://doi.org/10.1029/2019WR025577

Topp GC, Davis JL, Annan AP (1980) Electromagnetic determination of soil water content: measurements in coaxial transmission lines. Water Resour Res 16:574–582. https://doi.org/10.1029/WR016i003p00574

Wicki A, Lehmann P, Hauck C, Seneviratne SI, Waldner P, Stähli M (2020) Assessing the potential of soil moisture measurements for regional landslide early warning. Landslides 17:881–1896. https://doi.org/10.1007/s10346-020-01400-y

Wieczorek GF (1996) Landslide Triggering Mechanisms. In: Turner AK, Schuster RL (eds) Landslides investigation and mitigation. National Academy Press, Washington, D.C, Special re, p 673

Technical Concepts for an Early Warning System for Rainfall Induced Landslides in Informal Settlements

John Singer, Kurosch Thuro, Moritz Gamperl, Tamara Breuninger, and Bettina Menschik

Abstract

In developing and emerging countries informal settlements often develop uncontrollably around major cities. In mountainous regions these low-income settlements frequently are situated in areas subject to high landslide risk. An intermediate solution to reduce landslide risk for the inhabitants is the installation of a landslide early warning system. The Infom@Risk project is developing a socially integrated cost-effective landslide early warning system, that specifically addresses the complex spatial and social conditions of informal settlements. This paper discusses some of the technical concepts implemented in the planned early warning system, such as a low-cost LoRa wireless geosensor network, the measuring system "Continuous Shear Monitor" and the methods to be used in data analysis.

Keywords

Landslide early warning system • Geosensor network • Continuous shear monitor • Informal settlement

Introduction

Although great advances in the recognition, prediction and mitigation of landslides have been made in the last decades, landslide events still claim a high social and economic tribute worldwide. For example, Froude and Petley (2018) have collected and analyzed records of 4862 fatal landslide events around the globe in the years 2004–2016 in which in total nearly 56,000 people were killed. Of these events the majority (79%) were triggered by rainfall. At the same time fatal landslide events cluster around cities and occur most frequently in countries with lower gross national income—usually developing and emerging countries.

This is often linked to the existence of informal settlements around cities in low income countries. These unauthorized, uncontrollably growing settlements often develop due to rural depopulation and migration towards urban centers e.g. in the expectation of better economic opportunities, public services and higher incomes (Hallegatte et al. 2016). In mountainous terrain informal settlements often are built on previously unpopulated steep slopes surrounding the city center and thus are frequently subject to substantial landslide risk. Examples for such a development can be found in Asia, Africa, South and Central America (e.g. Smyth and Royle 2000).

Due to their unclear legal status and the magnitude and extent of landslide hazard present in many informal settlements, municipalities and administrations with limited financial resources are overwhelmed by the task of implementing necessary mitigation measures or by the controlled resettlement of the population to safer areas. Several projects throughout the world have shown that simple to follow construction and development guidelines as well as low cost mitigation measures, e.g. based on bioengineering, can significantly reduce landslide risk, especially if they are conducted in a socially integrated way (e.g. Holcombe et al. 2016). However, even after implementing such measures, the remaining landslide risk often is still far above an acceptable level and the population is still exposed to potentially deadly landslide hazards.

Early warning systems (EWS) can be an effective measure to reduce the landslide risk in these areas until final long-term risk reduction solutions are found. Until now the use of early warning systems in developing countries has been limited due to the high costs and complex operation of such systems. However, with the technological advance and increasing affordability of monitoring devices, for example

J. Singer (✉)
AlpGeorisk, Einsteinstr. 10, 85716 Unterschleissheim, Germany
e-mail: singer@alpgeorisk.de

K. Thuro · M. Gamperl · T. Breuninger · B. Menschik
Chair of Engineering Geology, Technical University of Munich, Arcisstr. 21, 80333 Munich, Germany
e-mail: thuro@tum.de

© Springer Nature Switzerland AG 2021
N. Casagli et al. (eds.), *Understanding and Reducing Landslide Disaster Risk*, ICL Contribution to Landslide Disaster Risk Reduction,
https://doi.org/10.1007/978-3-030-60311-3_24

MEMS (Micro-Electro-Mechanical Systems) sensors and long-range wireless radio transmission techniques, the implementation of landslide early warning systems in informal settlements nowadays is conceivable.

The Inform@Risk research project (2019–2022), which is funded by the German Federal Ministry of Education and Research, plans to develop and implement a socially integrated cost-effective landslide early warning system, that specifically addresses the complex spatial and social conditions of informal settlements. The multidisciplinary project involves landscape architects (Leibniz University Hannover, LUH), geologists (Technical University of Munich, TUM), geotechnical instrumentation experts (AlpGeorisk, AGR), remote sensing experts (German Aerospace Center Oberpfaffenhofen, DLR and Expert Office for Aerial Image Evaluation and Environmental Issues, SLU), and software developers (Technical University Deggendorf, THD).

In the following, after shortly introducing the study area, the technical (monitoring and warning) concepts of the early warning system, which have been jointly developed by TUM and AGR, are presented. The other essential elements of the EWS (UN/ISDR 2006) as e.g. the risk knowledge, warning dissemination and communication and response capability are only covered superficially and will be published in detail elsewhere. A list of publications related to the Inform@Risk research project can be found at https://www.bmbf-client.de/projekte/informrisk.

Fig. 1 Location (white oval) of the project area in the informal settlements above Bello Oriente, Medellin. Base map: Google maps; from Thuro et al. (2020)

Study Area: Bello Oriente, Medellin, Colombia

The city of Medellin, Colombia has been chosen as a test site for the proposed EWS, where according to the municipality currently 200,000 people live in informal settlements at risk of landslides. In an extensive qualitative decisioning process mainly conducted by the municipality of Medellin and local advisors from universities, associations and non-governmental organizations (NGO) the informal settlement above the city district Bello Oriente was chosen as a project site for the implementation of the early warning system.

Bello Oriente is situated on the northeastern slopes above the city of Medellin (Fig. 1). Based on preexisting work from the municipality and especially Werthmann et al. (2012) and Werthmann and Echeverri (2013) as well as findings of a first field campaign conducted in mid-2019, a first assessment of the geology, landslide hazard and landslide exposure are given below.

Geology

In the study area, slopes of 25° ± 5° predominate. The deeper underground consists of dunite, a magmatic rock which is extremely susceptible to weathering due to its high iron content (Fig. 2). Accordingly, as a result of the tropical conditions in the project area, the dunite is mostly covered with saprolite, a clay rich in situ product of weathering, which often still shows the original dunite structure. Depending on the exposure, the saprolite cover has a thickness between zero and several tens of meters. Especially in and below steeper segments of the slope, the dunite and saprolite have already been strongly moved by landslide processes and are therefore present as colluvium with a chaotic block-in-matrix structure. (Thuro et al. 2020).

Landslide Hazard

Most landslides observed in the northeast of Medellín are rotational slides in the deeply weathered soils and colluvium covering the dunite. On the upper parts of the valley, translational slides and sometimes rockfalls are more prominent because the bedrock surface is generally shallower there. Also, flash floods and debris flows occur, which usually follow the predefined morphology of the creeks. Due to its position about halfway up the slope, shallow to medium depth (1–20 m) rotational slides of medium to big size (10–100 m in width) are expected to be the most probable type of landslide to occur in the project area. The area affected by rockfall and flow type landslides is comparably small.

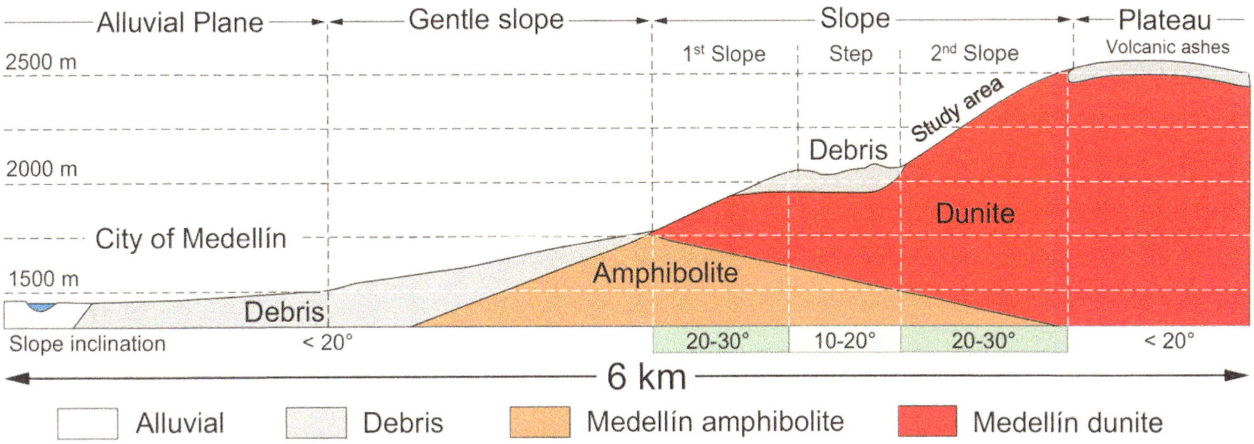

Fig. 2 Geological overview section through the northeastern slope above Medellin; the study area is situated on the upper slope where deeply weathered dunites predominate. Adapted from Werthmann et al. (2012)

The landslides are mostly triggered by rainfall or seismic events. Another frequent landslide trigger is anthropogenic activity like construction or leaky water pipes and septic tanks. When initialized, the slides are expected to have a wide range of possible velocity profiles, from continuously slow creeping movements, to slides with rapid acceleration, complete detachment and depending on the water content possibly liquefaction and very long runouts.

After conducting a geological exploration campaign, which includes exploratory drillings and an extensive laboratory testing program in May/November 2020, a more detailed assessment and differentiation of the landslide hazard will be possible.

Landslide Exposure

Preliminary results show that the project area is characterized by informal settlements built up by simple wooden or low strength masonry (mostly without reinforced concrete frames), one- to two story buildings. In a first analysis DLR counted 836 buildings in the project area, which are populated by estimated 1663 persons.

Most of the project area is endangered by the landslide processes characterized above. As soon as the hazard and exposure have been evaluated in detail, thematic risk maps for population, buildings and infrastructure will be compiled and serve as basis for the detailed planning of the EWS following the concepts described below.

Technical Concepts for a Landslide Early Warning System

To date most EWS for rainfall induced landslides either operate on a regional scale or are developed for a specific, already known landslide (Pecoraro et al. 2019).

Regional landslide EWS are usually based on statistical analyses of historic events or process models developed for geographical information systems (e.g. Marin & Velásquez 2019; Piciullo et al. 2018). These systems can provide a general indication of the current hazard level and can highlight the areas with the highest probability for the development of landslides. While these systems can raise awareness to affected areas, they cannot deliver site specific spatially and temporally precise warnings that allow to move people and assets out of harm's way, which is the goal of the Inform@Risk EWS.

In order to achieve this goal, the planned EWS operates at local scale and incorporates deformation and other geotechnical monitoring techniques comparable to those used in site specific EWS. However, as the exact locations of future developing landslides are unknown and spatially highly resolved area wide observations with conventional monitoring techniques would be very costly, new methodological and technical concepts are needed to implement such an EWS.

General Concept

The general idea is to be able to predict the future behavior of the observed landslide prone area, based on detailed hydrogeological and geotechnical models, which have been calibrated by observational data from hydrogeological field tests, geotechnical laboratory tests and a dense low-cost geosensor network. By including the triggering process in the models (e.g. intense precipitation leading to high ground water levels which trigger landslides), it is feasible to issue first general notifications, several days to hours in advance of a critical phase concerning the stability of the slope. When the onset of slow, but increasing movements is detected, spatially precise early warnings can be issued. In case of further or sudden strong acceleration of the slope, (evacuation-) alarms can be issued hopefully at least hours to minutes prior a catastrophic event, allowing people to leave the endangered area in time.

In order to reliably detect the initiation of landslides—especially in the small order of magnitude described above (rotational slides 10–100 m in width)—spatially highly resolved and accurate deformation observations are required. To achieve this high measurement density, the application of a geosensor network consisting of a combination of horizontally installed Continuous Shear Monitor (CSM) and wire-extensometer measurement systems combined with low cost wireless sensor nodes is planned (Fig. 3). While the CSM and wire-extensometer systems provide continuous, spatially highly resolved deformation observations along measurement profiles, the wireless sensor nodes add punctual observations based on the integrated MEMS sensors and other external geotechnical and hydrological sensors.

In order to achieve an optimal effectiveness of the system in terms of risk reduction, the density of the observations is varied throughout the project area based on the results of the risk assessment. In less risk prone areas, the sensor density is significantly reduced.

While this approach ensures an optimal cost–benefit ratio of the entire system, it must be emphasized, that especially small low intensity and very unlikely events might not be detected.

Geosensor Network

The monitoring system planned for the project site at Bello Oriente will cover about 20 hectares and consist of approximately 1.2 km combined CSM- and wire-extensometer lines and about 75–100 wireless sensor nodes. These communicate with 2–3 data gateways, which are distributed throughout the project area.

Combined CSM and Wire-Extensometer Lines

The Continuous Shear Monitor (CSM) is a specialized advancement of the Time Domain Reflectometry (TDR) technology for geotechnical applications (Singer et al. 2009). TDR is an electrical engineering measuring technique developed in the 1960s for locating cable faults and breaks in coaxial cables (in German-speaking countries often referred to as “cable radar”). With the CSM method, that in addition to the measurement technology itself includes procedures for the cable installation and signal processing, shear deformations (deformations perpendicular to the measuring or cable-axis) along a measuring cable can be monitored.

The CSM can seamlessly monitor measurement lines up to several hundreds of meters in length. While the system can localize and quantify localized shear with high accuracy, the maximum measurement range of shear deformation is limited to about 10 cm, when at this point the measurement cable is severed. In order to accomplish a larger measurement range (1 m and more) and at the same time add the possibility to detect axial extension along the measurement line, wire-extensometers are installed parallel to the CSM cables. These are segmented into elements of maximum 100 m length in order to allow a rough spatial allocation of the detected deformation. This is sufficient, as the location of the deformation usually will be provided by the CSM system.

However, the effort to install these systems is quite high, as they need to be placed in a trench with concrete backfill. Ideally the installation can be carried out in the context of other construction (e.g. streets, sewerage).

LoRa Wireless Sensor Network

The comparably new LoRa (Long Range) wireless technology allows the transmission of small data packets across large distances of several kilometers while only requiring very little power. This makes the development of sensor networks possible, which can be distributed in a wide area with little additional infrastructure required. The nodes can be operated with a few standard AA batteries for a very long period (up to about 2 years) and using small solar panels continuous operation for several years is conceivable. As the required hardware is affordable, the cost of the complete system is low.

For the Inform@Risk EWS new low cost LoRa sensor nodes are developed, which are based on the Arduino MKR WAN 1310 microcontroller. The nodes will include a 24-bit A/D converter allowing to connect precision analog sensors. Each node additionally is equipped with a high-quality MEMS tilt sensor, thermometer and barometer. The developed circuitry and firmware will be made available and distributed as open source.

The LoRa nodes transmit their data to LoRa gateways, of which for redundancy at least 2 are placed in the project area. These require power and internet access. Any data received from the LoRa nodes is directly forwarded to the data server and stored in the project database. The nodes can also receive commands e.g. to change the measurement frequency.

Sensor Placement

While the density of observations is determined by the risk analysis results as described above, the selection and placement of each individual sensor still must be executed with extreme care. The target area should be systematically checked for locations where a sensor most likely will detect changes in the measured value when a landslide develops. The resolution and range of the sensor as well as possible

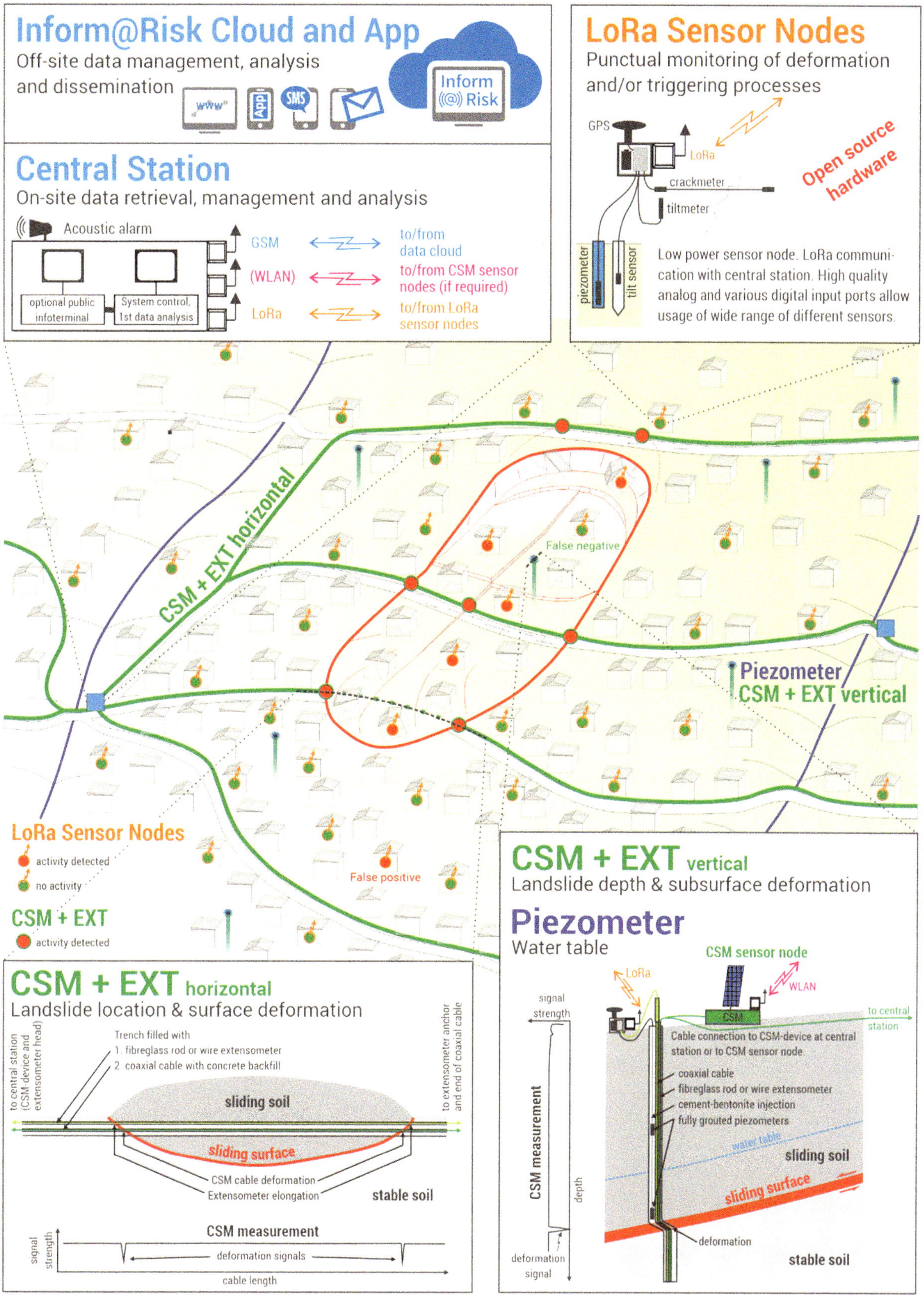

Fig. 3 Schematic layout of the proposed landslide early warning system (Thuro et al. 2020)

outside influences on the measurement should also be considered when placing a sensor. In any case the placement of each sensor needs to be documented in detail, as this will allow a better interpretation of the acquired data. In order to facilitate this process, it is planned to add an interactive sensor installation guide into the Inform@Risk app.

For the predominating rotational slides expected in the project area, MEMS inclination sensors seem to be an appropriate and simple observation method. These will be attached to buildings and infrastructure which are thought to tilt, when the slope starts to move. Additionally, small subsurface inclination probes will be used, which are driven 1–2 m into the ground (see Fig. 3, LoRa Sensor Node inset).

Data Analysis

All data collected from the geosensor network will be immediately transferred to an off-site central server (Inform@Risk Cloud), where it is processed and analyzed in near real time. The data analysis is based on threshold checks for various datasets, time series analyses and sensor fusion methods. Based on the analysis results the short- to medium term hazard level (or probability of failure) is assessed, and—if deformation is detected—early warnings and alarms are issued. The main information dissemination tool will be a newly developed mobile app.

Short- to Medium Term Hazard Level

In order to assess the short- to medium-term hazard level, the triggering factors rainfall and groundwater height are considered. On the one hand time series analyses will be performed to identify causal and temporal relationships between short-, medium- and long-term rainfall and groundwater levels. On the other hand, the hazard level is determined using groundwater level thresholds at e.g. 50, 75 and 90% of the critical water table derived from sensitivity analyses performed using the geological/geotechnical models created during the hazard analysis.

Early Warning and Alarms

Early warnings are issued as soon as significant deformation has been detected. Depending on the amount of deformation observed, different early waring levels are issued. The number and value of the static thresholds used to define these levels will be determined in course of the detailed hazard analysis, but early warning will most probably cover the range from mm per year up to cm to dm per hour. Based on how many/which neighbouring sensor nodes show deformation, the affected area and landslide mass are estimated and reported. If a further or sudden strong acceleration is detected, the system can issue an immediate (evacuation)-alarm using acoustic signals.

Depending on the hazard state, deformation rate and the affected area different actors (experts, trained community members, first responders, whole population) are informed.

Usually warnings are checked by an expert before they are sent to the inhabitants. Only in case at least two neighboring sensor nodes show very strong acceleration at the same time, the warnings are issued without review. The exact definition of the warning levels, warning content and the information dissemination paths will be developed in a participatory process, ensuring that, if possible, each actor gets the required information at the right point in time.

Outlook and Conclusion

It is planned to conduct a first test installation of the Inform@Risk EWS in November/December 2020. Based on these experiences the implementation of the entire system is planned for mid 2021. After a testing and evaluation phase of about 1 year the hopefully fully operational system is going to be handed off to the municipal early warning authority SIATA in 2022.

Although the technical concepts for the Inform-@Risk EWS were designed specifically for the study area of Bello Oriente, the approach can easily be adopted to many other areas in the Andes and worldwide.

Acknowledgements The project is funded by the German Federal Ministry of Education and Research in its program "Client II" (www.bmbf-client.de).

References

Froude MJ, Petley DN (2018) Global fatal landslide occurrence from 2004 to 2016. Nat Hazards Earth Syst Sci 18:2161–2181

Hallegatte S, Vogt-Schilb A, Rozenberg J, Bangalore M (2016) Unbreakable: building the resilience of the poor in the face of natural disasters. World Bank Publications, Washington, p 201

Holcombe EA, Beesley MEW, Vardanega, (2016) Urbanisation and landslides: hazard drivers and better practices. Proc Inst Civ Eng Geotech Eng 169:137–144

Marin RJ, Velásquez MF (2019) Influence of hydraulic properties on physically modelling slope stability and the definition of rainfall thresholds for shallow landslides. Geomorphology 351

Pecoraro G, Calvello M, Piciullo L (2019) Monitoring strategies for local landslide early warning systems. Landslides 16:213–231

Piciullo L, Cavello M, Cepeda JM (2018) Territorial early warning systems for rainfall-induced landslides. Earth Sci Rev 179:228–247

Singer J, Schuhbäck S, Wasmeier P, Thuro K, Heunecke O, Wunderlich Th, Glabsch J, Festl J (2009) Monitoring the Aggenalm landslide using economic deformation measurement techniques. Austrian J Earth Sci 102(2):20–34

Smyth CG, Royle SA (2000) Urban landslide hazards: incidence and causative factors in Niterói, Rio de Janeiro State. Brazil Appl Geogr 20:95–118

Thuro K, Singer J, Menschik B, Breuninger T, Gamperl M (2020) Entwicklung eines Frühwarnsystems für Rutschungen in den tropischen Anden (Medellín, Kolumbien). In press, Geomechanik und Tunnelbau

UN/ISDR (2006) Developing Early Warning Systems: A Checklist. EWC III - Third International Conference on Early Warning. 10 p., Bonn, Germany, March 27-29, 2006, UN Secretariat of the International Strategy for Disaster Reduction, Bonn

Werthmann C, Echeverri A, Vélez AE (2012) Rehabitar La Ladera: shifting ground. Universidad EAFIT Medellín

Werthmann C, Echeverri A (2013) Rehabitar La Montaña: Estrategias y procesos para un hábitat sostenible en las laderas de Medellín. Universidad EAFIT Medellín

Combination of Rainfall Thresholds and Susceptibility Maps for Early Warning Purposes for Shallow Landslides at Regional Scale in Taiwan

An Lu, Wei-Kai Haung, Ching-Fang Lee, Lun-Wei Wei, Hsi-Hung Lin, and Chun-Chi Chi

Abstract

In this study, the development of landslide susceptibility and determination of rainfall threshold for shallow landslides is the main purpose for the prevention of slope disaster. For appliance of the practical operation of disaster prevention, almost 185,000 slope units are used to calculate landslide susceptibility and determine the rainfall threshold of shallow landslides, which are a reference of preparation for slope disasters. Time of occurrence of 941 landslides are collected through field investigation, and rainfall pattern of each landslides are analyzed to clarify the relation between landslide and rainfall. Logistic regression (LR) analysis was performed to evaluate landslide susceptibility and establish the assessment method of the rainfall threshold of landslide by using I_3–R_{24} rainfall thresholds. After verification of rainfall-induced landslide in Typhoon Meranti (2016), the result show that the early warning model is suitable for alerting serious swarm of landslides.

Keywords

Rainfall-induced landslide • Landslide susceptibility analysis • Rainfall threshold • Early warning

A. Lu (✉) · W.-K. Haung · C.-F. Lee · L.-W. Wei
Disaster Prevention Technology Research Center, Sinotech Engineering Consultants, INC, New Taipei City, Taiwan
e-mail: anlu@sinotech.org.tw

L.-W. Wei
e-mail: lwwei@sinotech.org.tw

H.-H. Lin · C.-C. Chi
Central Geological Survey, MOEA, New Taipei City, Taiwan

Introduction

Rainfall-induced landslides number among the most perilous natural hazards, causing severe casualties and economic losses worldwide. Therefore, many efforts have been made to evaluate landslide susceptibility and thereby set criteria for issuing alerts that can save lives and property.

In Taiwan, monsoons and typhoons bring great amounts of rainfall, up to 3,000 mm/year, and numerous landslides cause casualties every year. Therefore, recognizing the areas where rainfall-induced landslides might occur is an urgent issue.

On the other hand, rainfall thresholds for landslides can be categorized as either statistical approaches or deterministic approaches. In the former method, thresholds are decided by collecting historical landslide cases and analyzing their rainfall parameters and the probability lines of rainfall conditions (Guzzetti et al. 2008). In the latter method, thresholds are decided by calculating the safety factors of each slope or grid with geomaterial and rainfall parameters (Kim et al. 2010). Statistical rainfall thresholds for shallow landslides have been well discussed (Guzzetti et al. 2007).

This study focused on shallow landslides of the debris fall, debris topples, debris slide, earthfall, earth topple, and earth slide types proposed by Varnes (1978) and divided slope units according to three different landslide susceptibility levels (high, moderate, and low). After that, we established their rainfall thresholds separately. Furthermore, we set alert levels by adopting a hazard matrix and examined whether differentiated warning thresholds for different degrees of susceptibility existed. Moreover, this study gives an importance of validating the performance of a landslide early warning model, especially the false alarms and missed alarms, to make it feasible for further practical application.

© Springer Nature Switzerland AG 2021

N. Casagli et al. (eds.), *Understanding and Reducing Landslide Disaster Risk*, ICL Contribution to Landslide Disaster Risk Reduction,
https://doi.org/10.1007/978-3-030-60311-3_25

Study Area

Taiwan is located in the western Pacific Ocean, on the convergent plate boundary zone of the Philippine Sea Plate and the Eurasian plate. The orogenic uplift rate is 5–7 mm/year; however, the exhumation rate is also as high as 3–6 mm/year (Dadson et al. 2003) due to the fractured geological materials and the high mean annual precipitation of 2,500–3,000 mm brought by typhoons and monsoons every year. The frequent natural disasters and high population density (23 million people in 36,000 km^2) of Taiwan make it one of the countries most exposed to multiple hazards.

The study area (Fig. 1) includes 157 1:25,000 scale maps (about 24,250 km^2), and covers densely inhabited and landslide-threatening hillslopes. The lithological units are mainly sedimentary rocks composed of sandstone, shale, mudstone, and metamorphic rocks composed of slate, argillite, and metasandstone.

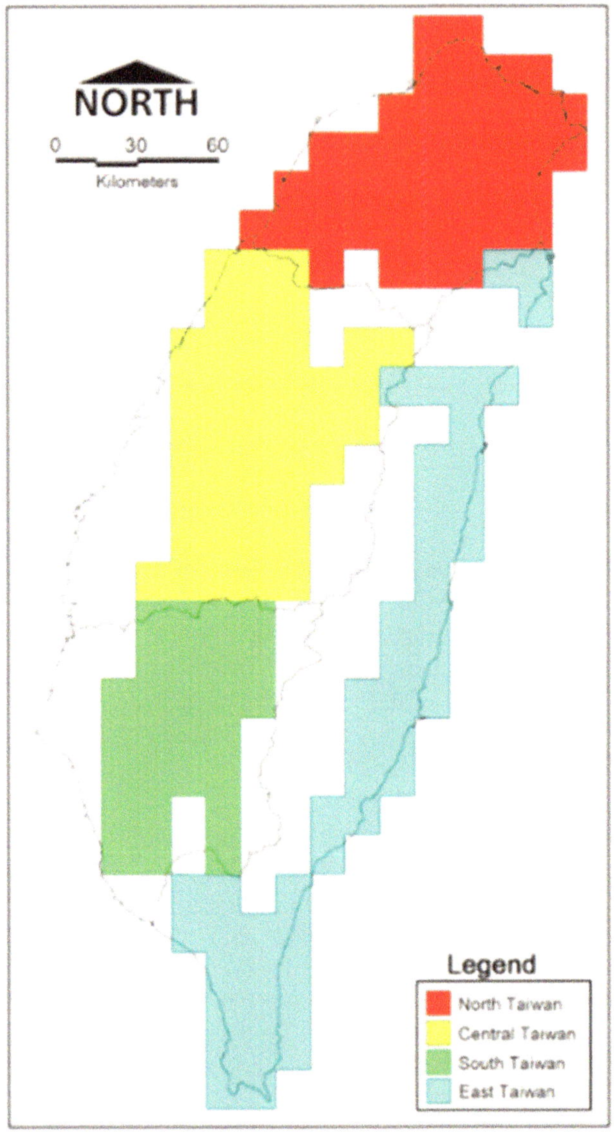

Fig. 1 Study area is composed of 4 regions in Taiwan. Basis of division is different lithological units

Methodology

Slope Unit

Slope units were used for the analysis of landslide susceptibility in this study (Carrara 1988; Carrara et al. 1991, 1995; Guzzetti et al. 1999; Yang 2017). This study followed the method proposed by Xie et al. (2004) in delineating slope units according to gullies and ridges. Almost 185,000 slope units are found in this study for the establishment of an early warning system.

Landslide Occurrence Time and Field Investigation

To analyze the rainfall conditions for each landslide case used in this study, 941 shallow landslide cases, including their occurrence times (date and hour) and the characteristics of the landslides, were gathered for further analysis.

Landslide Susceptibility Analysis

The main purpose of landslide susceptibility analysis is to determine the effectiveness of each predisposing factor and the relative possibility of landslide occurrence in a specific area.

10 × 10 m DEM is adopted to be a basic resolution for slope unit delineation and landslide susceptibility calculation. Several predisposing factors that might lead to landslides were selected initially in this study to construct a landslide susceptibility model for slope units. These factors included rock mass strength-size classification (RMSSC I–VII), dip slope, average slope, variance of slope, ratio of steep slope, total slope height, average elevation, average curvature, variance of curvature, fault density, fold density, average wetness, rainfall intensity, total rainfall, 3-hour mean rainfall intensity (I_3), and 24-hour accumulated rainfall (R_{24}). In this study, we applied logistic regression (LR) to evaluate the susceptibility of each slope unit (Guzzetti et al. 1999). The LR function can be expressed as follows:

$$P = \frac{1}{1 + e^{-z}} \tag{1}$$

$$z = \sum_{i=1}^{m} L_i w_i + \sum_{j=1}^{n} F_j w_{m+j} + C \qquad (2)$$

where P is landslide susceptibility, L_i is RMSSC factor, F_j is other factors, w_i and w_{m+j} are regression coefficients, and C is a constant. Six event-based landslide inventories in this study were used to label whether or not landslides occurred in the slope units. After that, all the slope units were divided randomly into two groups, one for training the model and the other for validation. The index indicating landslide/non-landslide was set as the dependent variable, and all the landslide susceptibility factors were set as covariates in SPSS for training of the model. After iterative training, the regression coefficients of each landslide susceptibility factor, as well as the success rate curve (SRC), the prediction rate curve (PRC), and the area under the curve (AUC), were reported in SPSS. The AUC can be used to examine if the model predicts landslides well, and the regression coefficients can be used for the prediction of landslide susceptibility. The individual landslide susceptibilities of slope units were calculated with this model and classified into high, moderate and low susceptibility levels.

I_3–R_{24} Rainfall Index and Thresholds

Rainfall-induced landslides are triggered by either high-intensity rainfall or high accumulated rainfall (Larsen and Simon 1993; Corominas and Moya 1999; Yu et al. 2006). To identify rainfall indexes responsible for landslides, the triggering rainfall, including the rainfall intensity (I_1, I_2, I_3, I_4, I_5, I_6) and accumulated rainfall (R_6, R_{12}, R_{24}, R_{48}, R_{72}) of different time windows of each landslide case were analyzed according to the landslide occurrence time. The results revealed that 218 landslides occurred within the 3 h following the highest rainfall intensity, and 242 occurred within the 3 h following the 2nd or 3rd highest rainfall intensity (i.e., induced by high rainfall intensity), accounting for nearly 49% of the landslide cases gathered in this study. From these results, it became clear that in Taiwan, I3 is the most important index for landslides induced by rainfall of short duration but high intensity. On the other hand, 481 landslides occurred close to the end of the rainfall events (i.e., induced by high accumulated rainfall), accounting for about 51% of the total cases.

Based on these data and previous studies (Liao et al. 2010), 3-hour mean rainfall intensity (I3) and 24-hour accumulated rainfall (R24) were chosen respectively as the short-term and long-term rainfall indexes for the establishment of the rainfall threshold (Fig. 2). We chose 3-hour mean rainfall intensity here instead of 3-hour accumulated rainfall to focus on the rainfall of short duration but high intensity. Similarly, we chose 24-hour accumulated rainfall to focus on the rainfall of long duration but low intensity. Finally, rainfall thresholds were decided by plotting the I_3 and R_{24} rainfall index of historical landslides in the I_3–R_{24} diagram (Fig. 3). Here we used the ellipse as the threshold line, and the parameters a (semi-major axis) and b (semi-minor axis) of the ellipse were set according to the slope of best fit line obtained from the least square method. Thresholds such as 90, 60, 30 and 15% were determined according to the percentage of historical cases that could be enveloped under the threshold line; e.g., the 90% threshold (T90%) included 90% of the historical cases. A higher threshold indicates a more dangerous condition for the occurrence of landslides. The original warning values of I_3 and R_{24} of the 90, 60, 30, 15% thresholds were equal to the semi-minor axis and semi-major axis of each threshold, respectively. After that, I_3 was rounded to the nearest 5 mm/h, and R_{24} was rounded to the nearest 50 mm for operational purposes, such as the evacuation of residents.

Landslide Early Warning Model

The landslide early warning model in this study considered both landslide susceptibility and rainfall thresholds and alerts were determined by using a hazard matrix. As mentioned above, the LR method was applied to analyze the susceptibility of each slope unit. After that, all the slope units were categorized into high, moderate, and low susceptibility levels. We consequently established rainfall thresholds for each susceptibility level separately and then set alerts of red, orange, yellow and green according to the level of danger.

High-susceptibility slopes might be more susceptible to rainfall. Hence, the alerts were set as red (extreme danger level) for rainfall conditions exceeding the 60% threshold line, orange (high danger level) for those between the 60 and 30% threshold lines, yellow (medium danger level) for those between the 30 and 15% threshold lines, and green (low danger level) for rainfall conditions lower than the 15% threshold line (Fig. 4). For moderate-susceptibility slopes, the alerts were set as red for rainfall conditions exceeding the 90% threshold line, orange for those between the 90 and 60% threshold lines, yellow for rainfall conditions between the 60 and 30% threshold lines, and green for rainfall conditions lower than the 30% threshold line. Low-susceptibility slopes should be less susceptible to rainfall. Hence, the alerts were set as orange for rainfall conditions exceeding the 90% threshold line, yellow for those were between 90 and 60% threshold lines, and green for those lower than the 60% threshold line (Fig. 5).

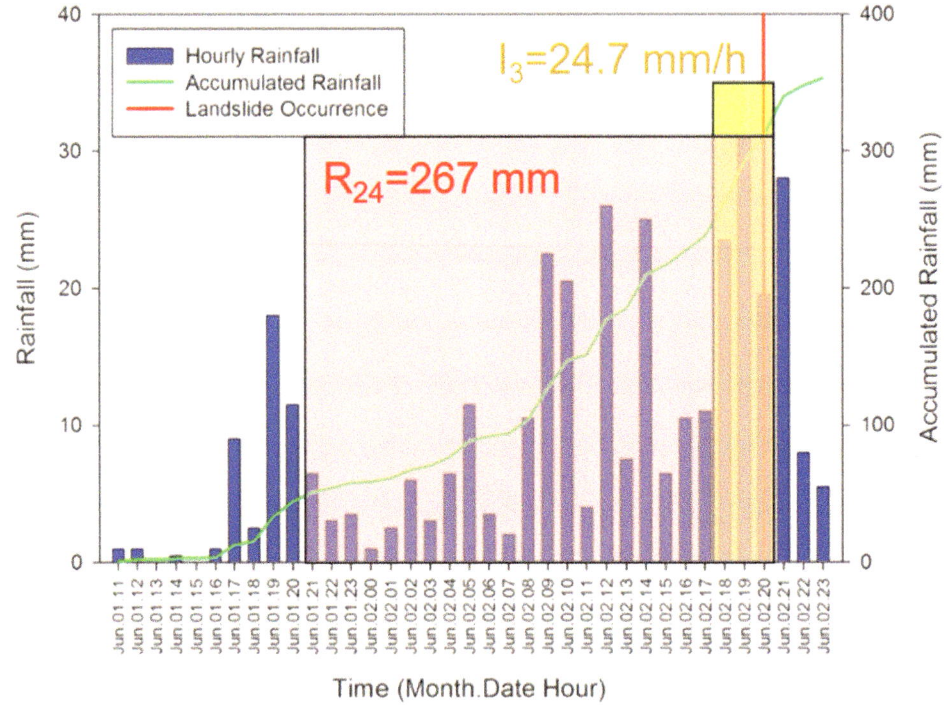

Fig. 2 3-hour mean rainfall intensity (I_3) and 24-hour accumulated rainfall (R_{24}) were used as short-term and long-term rainfall indexes for the establishment of rainfall thresholds

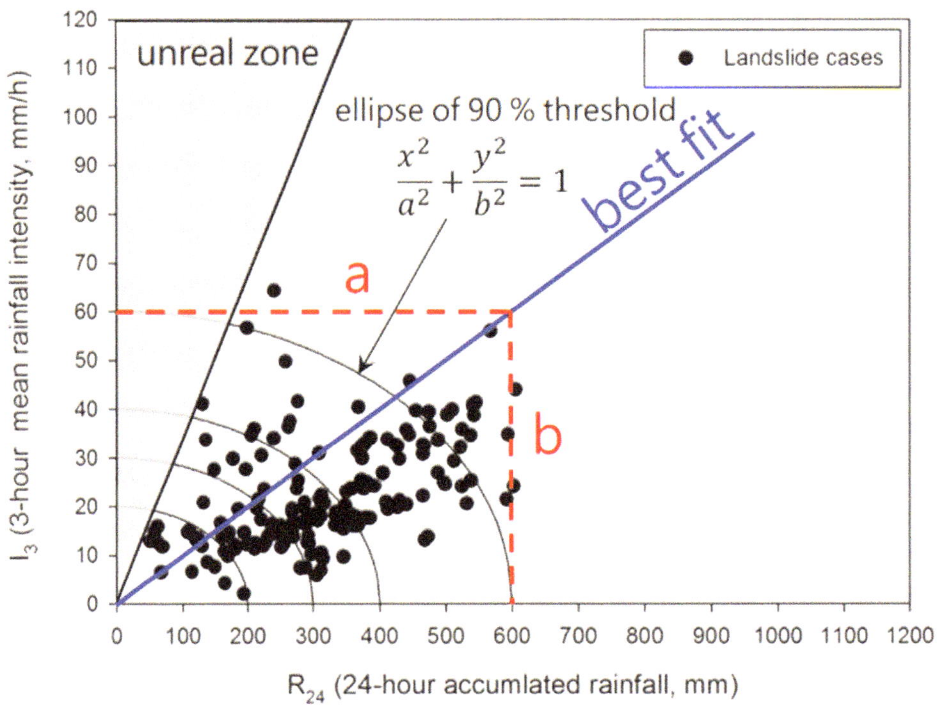

Fig. 3 Establishment of I_3–R_{24} rainfall thresholds for shallow landslides. The best fit line was derived by the least square method, and the ratio of a and b was used as the ratio of the semi-major axis and semi-minor axis in the ellipse threshold line

I_3–R_{24} Rainfall Threshold

We gathered a total of 941 landslide cases in this study and used 240 cases located in southern Taiwan, consisting of 110 high-susceptibility cases, 84 moderate-susceptibility cases, and 46 low-susceptibility cases, to establish a susceptibility-based regional landslide early warning model. For practical use, the original threshold values of I_3 and R_{24} (as shown in the parentheses in Fig. 6) were separately rounded to the nearest 5 mm/h and the nearest 50 mm for an operational purpose. It was found that the threshold values gradually

Fig. 4 Landslide early warning model and alert considering both landslide susceptibility and rainfall thresholds

		Rainfall threshold (T)							
		$T_{90\%}$		$T_{60\%}$		$T_{30\%}$		$T_{15\%}$	
		I_3	R_{24}	I_3	R_{24}	I_3	R_{24}	I_3	R_{24}
Landslide susceptibility	High susceptibility	Extreme danger level		Extreme danger level		High danger level		Medium danger level	
	Moderate susceptibility	Extreme danger level		High danger level		Medium danger level		Low danger level	
	Low susceptibility	High danger level		Medium danger level		Low danger level		Low danger level	

Fig. 5 Rainfall thresholds for southern Taiwan. The values were calculated as 90, 60, 30, and 15% of the original threshold, respectively. The original values are shown in parentheses

		Rainfall threshold (T)							
		$T_{90\%}$		$T_{60\%}$		$T_{30\%}$		$T_{15\%}$	
		I_3	R_{24}	I_3	R_{24}	I_3	R_{24}	I_3	R_{24}
Landslide susceptibility	High susceptibility	70 (68)	750 (745)	55 (56)	600 (610)	40 (40)	450 (438)	30 (27)	300 (291)
	Moderate susceptibility	70 (61)	650 (657)	45 (46)	500 (498)	35 (34)	350 (368)	20 (22)	250 (236)
	Low susceptibility	50 (50)	550 (539)	40 (40)	450 (430)	30 (29)	300 (316)	15 (15)	200 (167)

increased as the susceptibility of slope units decreased for the same alert level, indicating that greater rainfall amounts would be needed when issuing alters on less susceptible slope units. These results showed that establishing rainfall thresholds according to different landslide susceptibilities and then setting alert levels by adopting a hazard matrix provided differentiated not only thresholds but also avoided the over- or underestimation of the thresholds for slopes.

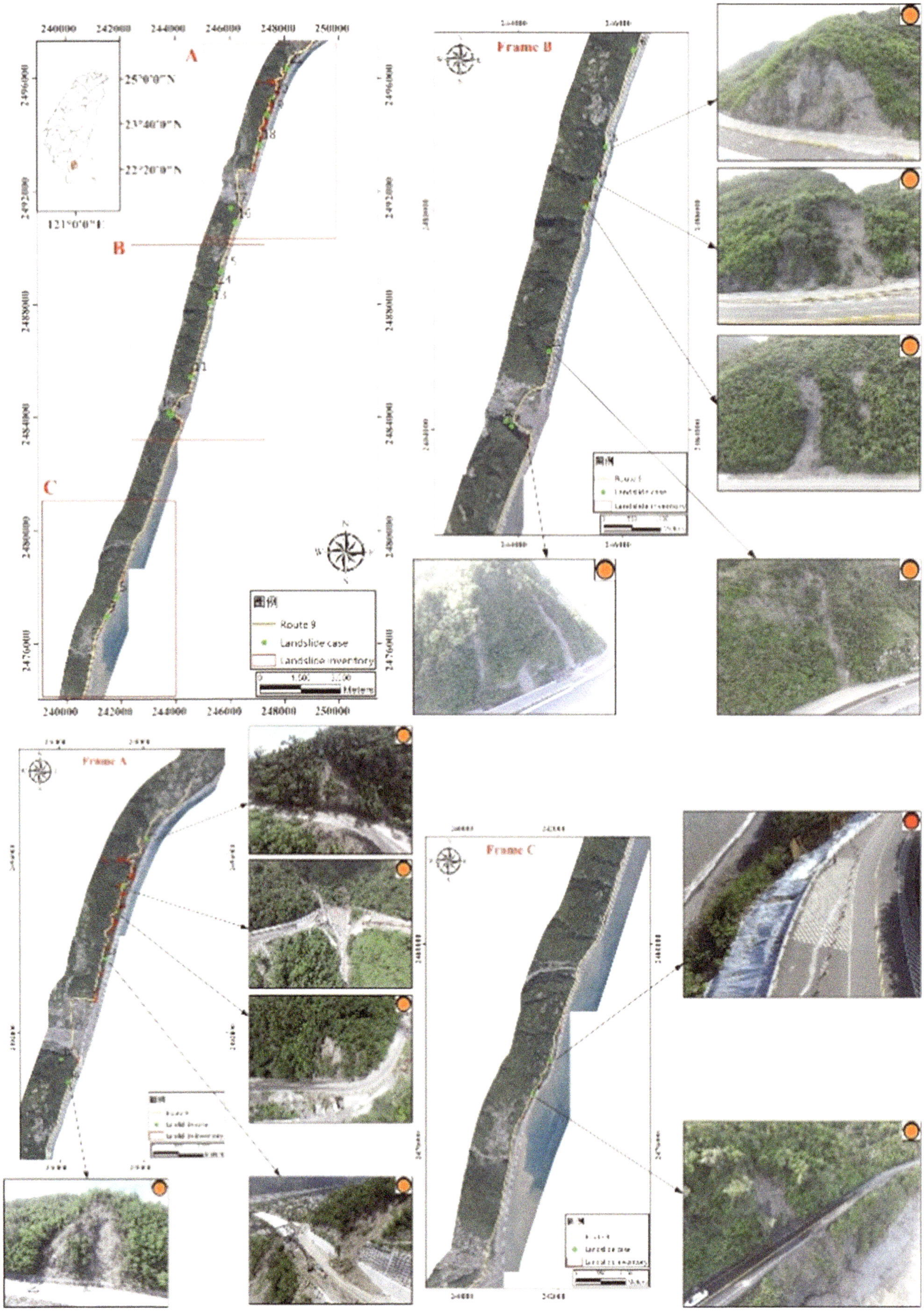

Fig. 6 Location and alert level of 12 representative landslides near route 9 caused by Typhoon Meranti in 2016

Fig. 7 Photo of route 9 410.8 k landslide

Validation of Early Warning Model

During Typhoon Meranti in 2016, 178 shollow landslides occurred in 29 slope units, including one high-susceptibility slope, two moderate-susceptibility slopes, and 26 low-susceptibility slopes. According to newspaper reports, most of the landslides occurred at 04:00 on September 15, 2016. Alert level and photo of 12 shollow landslides, which are representative from 178 shollow landslides, are shown in Fig. 7. One of them was a red alert level when the landslide occurred; the others are orange alert level.

The Route 9 410.8 k landslide (Fig. 8) occurred on a low-susceptibility slope in southern Taiwan. The rainfall path (Fig. 9) showed that on September 15, the alert was raised to yellow at 02:00. Then the landslide occurred at 05:00 on September 15. The Route 9 413.5 k landslide (Fig. 10) occurred on a low-susceptibility slope in southern Taiwan. The rainfall path (Fig. 10) showed that on September 15, the alert was raised to yellow at 00:00 and then to orange at 02:00 during the downpour. Then the landslide occurred at 04:00 on September 15.

Conclusion

This study attempted to establish regional rainfall thresholds for shallow landslides according to their landslide susceptibility levels and set alerts with a hazard matrix to provide more detailed results for disaster mitigation.

This study also examined the relationships between rainfall indexes and the occurrence of landslides. From 941 landslide cases we gathered, it was found that 3-hour mean rainfall intensity (I_3) and 24-hour accumulated rainfall (R_{24}) was the most dominant short-term and long-term parameters responsible for rainfall-induced landslides in Taiwan. There were 460 cases (about 49%) that occurred within the 3 h following the highest, 2nd, and 3rd rainfall intensities, while 24-hour accumulated rainfall had the lowest coefficient of variation of the long-term rainfall indexes. The I_3-R_{24} rainfall index was therefore used to establish rainfall thresholds.

Slope units are categorized into three landslide susceptibility levels (high, moderate, and low) and then separately established a susceptibility-based regional rainfall threshold.

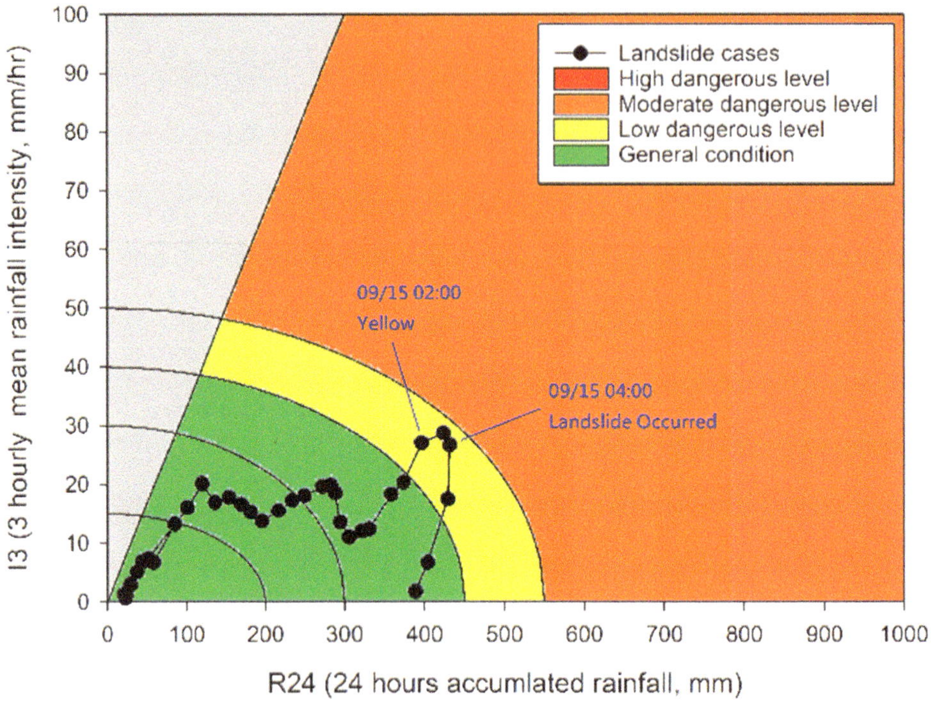

Fig. 8 Rainfall path of route 9 410.8 k landslide in the I_3–R_{24} diagram

Fig. 9 Photo of route 9 413.5 k landslide

We also set three alert levels, including red (extreme danger level), orange (high danger level), and yellow (medium danger level), by adopting a hazard matrix for application to evacuation decisions.

Validations using landslide cases in Typhoon Meranti in 2016 showed that, for the representative landslide cases in Typhoon Meranti, orange or red alerts could have been issued before the landslides occurred.

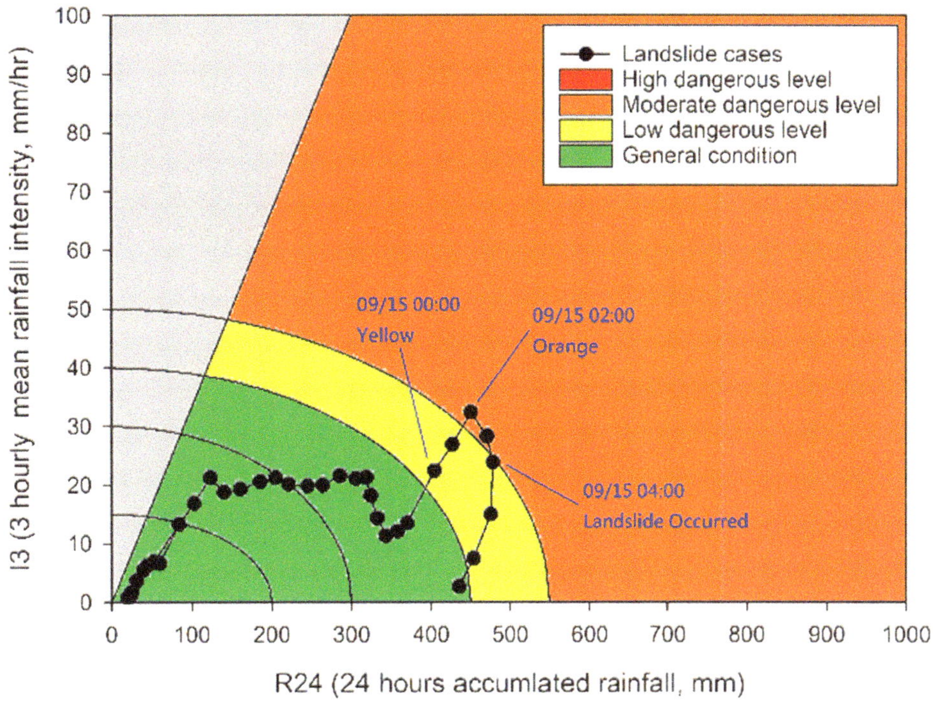

Fig. 10 Rainfall path of route 9 413.5 k landslide in the I_3–R_{24} diagram

It can be concluded that classifying landslide susceptibility and establishing rainfall thresholds separately not only provides refined thresholds but also avoids over- or underestimation of the thresholds for slopes, especially when considering the application to disaster prevention.

Acknowledgements The authors would like to thank the Central Geological Survey, Taiwan, for supporting this research financially and for providing helpful comments on the research.

References

Carrara A (1988) Quantitative analysis of mineral and energy resources. Springer, Dordrecht, pp 581–597

Carrara A, Cardinali M, Detti R, Guzzetti F, Pasqui V, Reichenbach P (1991) GIS techniques and statistical models in evaluating landslide hazard. Earth Surf Proc Land 16(5):427–445

Carrara A, Guzzetti F (1995) Geographical information systems in assessing natural hazards. Springer, Netherlands, pp 135p–175p

Corominas J, Moya J (1999) Reconstructing recent landslide activity in relation to rainfall in the Llobregat River basin, Eastern Pyrenees. Spain Geomorphol 30(1):79–93

Dadson SJ, Hovius N, Chen H, Dade WB, Hsieh ML, Willett SD, Hu JC, Horng MJ, Chen MC, Stark CP, Lague D, Lin JC (2003) Links between erosion, runoff variability and seismicity in the Taiwan orogeny. Nature 426(6967):648–651

Guzzetti F, Carrara A, Cardinali M, Reichenbach P (1999) Landslide hazard evaluation: a review of current techniques and their application in a multi-scale study. Central Italy Geomorphol 31 (1):181–216

Guzzetti F, Peruccacci S, Rossi M, Stark CP (2007) Rainfall thresholds for the initiation of landslides in central and southern Europe. Meteorol Atmos Phys 98(3–4):239–267

Guzzetti F, Peruccacci S, Rossi M, Stark CP (2008) The rainfall intensity-duration control of shallow landslides and debris flows: an update. Landslides 5(1):3–17

Kim D, Im S, Lee SH, Hong Y, Cha KS (2010) Predicting the rainfall-triggered landslides in a forested mountain region using TRIGRS model. J Mt Sci 7(1):83–91

Larsen MC, Simon A (1993) A rainfall intensity-duration threshold for landslides in a humid-tropical environment, Puerto Rico. Geogr Ann Series A, Phys Geogr. 75:13–23

Liao Z, Hong Y, Wang J, Fukuoka H, Sassa K, Karnawati D, Fathani F (2010) Prototyping an experimental early warning system for rainfall-induced landslides in Indonesia using satellite remote sensing and geospatial datasets. Landslides. 7(3):317–324

Xie M, Esaki T, Zhou G (2004) GIS-based probabilistic mapping of landslide hazard using a three-dimensional deterministic model. Nat Hazards 33(2):265–282

Yang SR (2017) Assessment of rainfall-induced landslide susceptibility Using GIS-based slope unit approach. J Perform Constr Fac 31 (4):04017026

Yu FC, Chen TC, Lin ML, Chen CY, Yu WH (2006) Landslides and rainfall characteristics analysis in Taipei City during the Typhoon Nari event. Nat Hazards 37(1):153–167

Development of Landslide Early Warning System Based on the Satellite-Derived Rainfall Threshold in Indonesia

Agus Setyo Muntohar, Olga Mavrouli, Victor G. Jetten, Cees J. van Westen, and Rokhmat Hidayat

Abstract

Landslide is a common natural disaster occurring in Indonesia during the rainy season from November to February. Attempts have been made to develop an early warning system based on the rainfall derived from satellite observation. It is essential to verify the accuracy level of the rainfall threshold in predicting the occurrence of rainfall, causing landslides and non-landslides to model the lower limit that can be used as an early warning device of the landslides. In this analysis, modelling was carried out with an empirical (intensity—duration/ID) approach using 220 data of rainfall that triggered landslide with satellite-based TRMM in Indonesia territory. The intensity and duration of antecedent rainfall were utilized in rainfall threshold modelling. The rainfall threshold was validated with ROC analysis. This method used seven statistics indices and ROC curve to determine the accuracy rate of the rainfall threshold. The results showed empirical equation $I = 7.83D^{-0.328}$ within the interval time 2–18 days. The results of the analysis of the ROC on the rainfall threshold indicate that the model has a good accuracy rate and can be used in an early warning system of landslide even though it still has a fairly high error rate.

Keywords

Rainfall • TRMM • Threshold • Warning system

Introduction

Landslides are one of the natural disasters that frequently occur in one tropical country in Indonesia.. High precipitation can cause unstable soil conditions and cause slope collapse. Rainfall triggerring landslides can be predicted using rainfall thresholds used in early warning systems. In this work, rainfall modeling is generated by evaluating rainfall intensity and duration of rain from Tropical Satellite Rainfall Measuring Mission (TRMM) data based on Multi-Satellite Precipitation Analysis (TMPA) (Mathew et al. 2013). Each threshold model created has a different level of accuracy in predicting landslides. Several methods are frequently used to evaluate the empirical threshold models.

Rainfall triggering landslides can be divided into two categories, namely critical and antecedent rainfall. Figure 1 shows the occurrence of rain that can trigger landslides (Aleotti 2004). Critical rainfall is the amount of rainfall that has increased drastically and can trigger landslides. Critical rain can be a trigger for landslides if the rainfall generated is high, and exceeds the maximum limit. Meanwhile, the antecedent rain is a successive rainfall event measured prior to the critical rainfall event until the beginning of the critical rainfall. Both critical and antecedent rainfall can be used as parameters in determining the rainfall threshold model. Measuring antecedent rainfall threshold modeling is simpler than the critical rainfall. The common empirical model of rainfall threshold is determined by intensity—duration curve

A. S. Muntohar (✉)
Department of Civil Engineering, Universitas Muhammadiyah Yogyakarta, Jl. Brawijaya Tamantirto, Bantul, D.I. Yogyakarta 55183, Indonesia
e-mail: muntohar@umy.ac.id

O. Mavrouli · V. G. Jetten · C. J. van Westen
Geoinformation Science and Earth Observation (ITC), University of Twente, Hengelosestraat 99, Enschede, 7514 AE, Netherlands
e-mail: o.c.mavrouli@utwente.nl

V. G. Jetten
e-mail: v.g.jetten@utwente.nl

C. J. van Westen
e-mail: c.j.vanwesten@utwente.nl

R. Hidayat
Ministry of Public Works and Housing, Balai SABO, Water Resources Research Centre, Jl. Sabo No. 1, Maguwoharjo, Sleman, Daerah Istimewa Yogyakarta 55282, Indonesia
e-mail: rokhmathidayat33@gmail.com

© Springer Nature Switzerland AG 2021
N. Casagli et al. (eds.), *Understanding and Reducing Landslide Disaster Risk*,
ICL Contribution to Landslide Disaster Risk Reduction,
https://doi.org/10.1007/978-3-030-60311-3_26

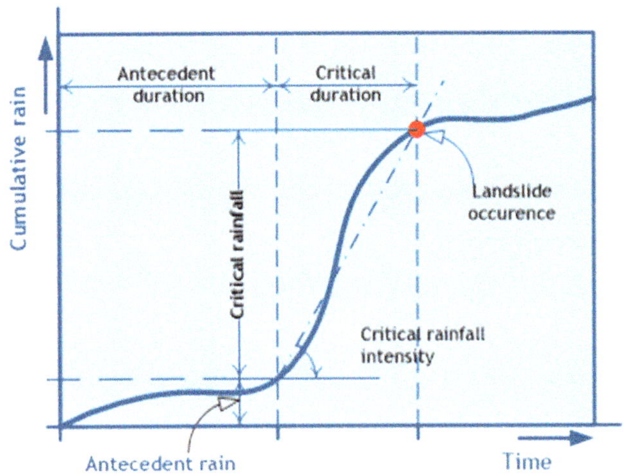

Fig. 1 Definition of rainfall parameters for threshold determination (modified after Aleotti 2004)

(I-D curve) (Aleotti 2004; Guzzetti et al. 2007; Reichenbach et al. 1998; Muntohar 2008).

The developed-rainfall threshold models for early warning system must be evaluated to determine the level of accuracy in predicting landslides during the rainy period. This level of accuracy can usually be evaluated by Receiver-Operating Characteristic (ROC) analysis. This method applies statistical index values and ROC curves that represent the level of accuracy of the rainfall threshold model (Fawcett 2006; Zou et al. 2007). The main purpose of this study is to establish the empirical model of the rain threshold based on antecedent rainfall intensity and duration. The accuracy of the empirical model was evaluated by the ROC method. Thus, the rainfall threshold model can be implemented in a landslide early warning system.

Research Method

Landslides Inventory in Indonesia 2010–2018

An empirical model of rainfall threshold to predict landslides requires data such as location, time, and rainfall. Several landslides in Indonesia have been documented through the website of the National Disaster Management Agency (BNPB), the Crisis Centre of the Ministry of Health, and the Geology Agency. However, many landslides were barely well recorded at government agencies but reported by newspapers and online sources. The distribution of landslide locations is shown in Fig. 2. Total landslides that can be recorded are 220 locations. The most recorded locations of landslides were in Java and Sumatra islands, while, in the regions of Kalimantan, Sulawesi, Papua, and Maluku, landslides rarely occur. Therefore, the model developed in this study is valid for the Java and Sumatera islands.

Rainfall Records

The rainfall records for subsequent landslide location was obtained by the satellite data of the TRMM. Some studies on the validation of TRMM data over Indonesia have been

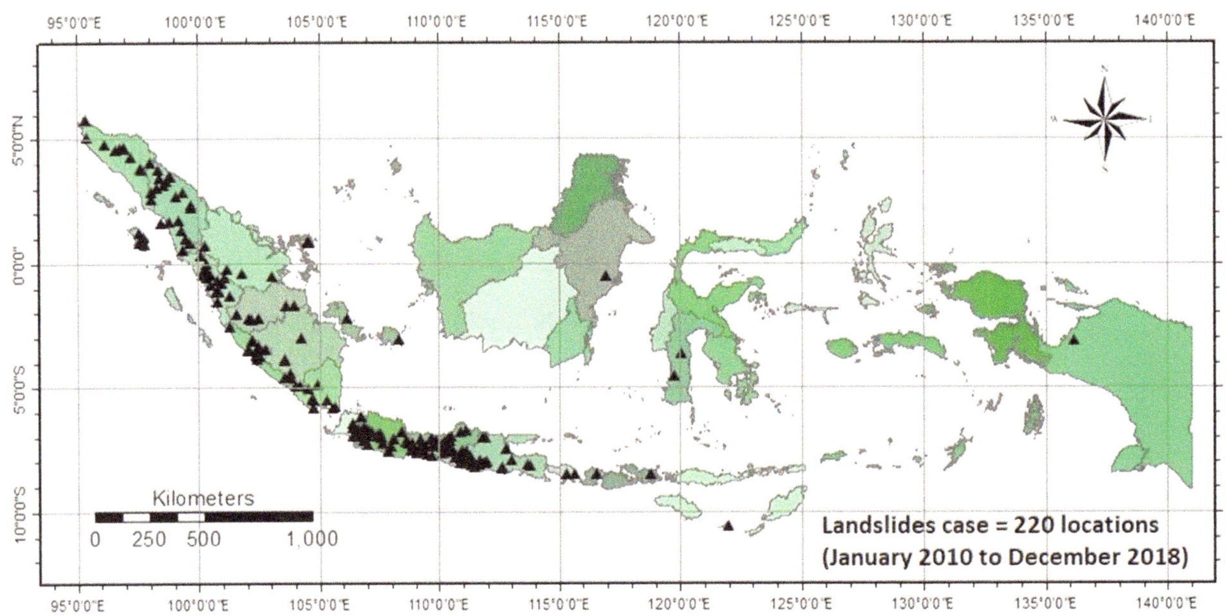

Fig. 2 Landslides occurrence in Indonesia during January 2010–December 2018

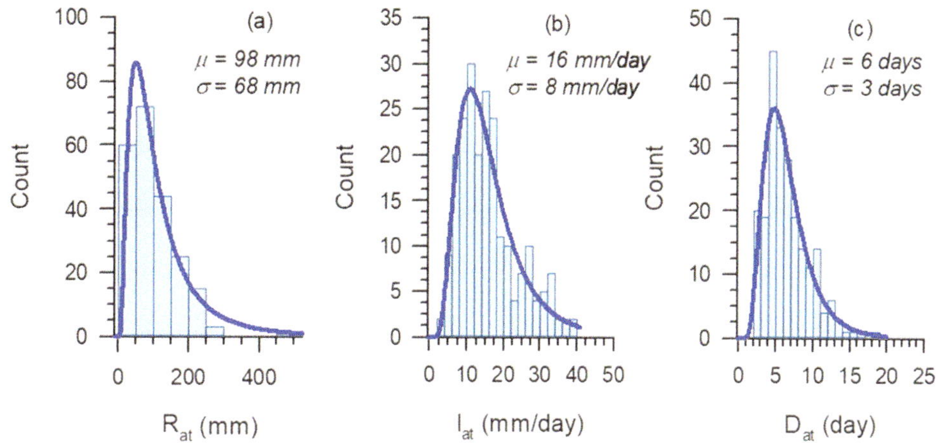

Fig. 3 The statistical distribution of antecedent rainfall, R_{at} (**a**), average intensity, I_{at} (**b**), and duration, D_{at} (**c**) from the 220 landslide

conducted by As-Syakur et al. (2011), Sipayung et al. (2014), Pratama et al. (2016), Giarno et al. (2018), and Fatkhuroyan et al. (2018). The studies concluded that the rainfall derived from the satellite was over predicted during the wet season, but the rainfall was applicable for an early warning system. The statistical descriptor (mean μ, and deviation standard σ) and distribution of the antecedent rainfall are shown in Fig. 3. The mean value of the duration of antecedent rainfall is about six days, and the average antecedent rainfall intensity is 16 mm/day, with the mean antecedent rainfall is 99 mm.

Empirical Model

Empirical models of landslides were analyzed from 220 daily rainfall records at each landslide location. The rainfall threshold model was made based on the relationship between rainfall intensity and duration (I-D). Empirical I-D curve was developed by regression analysis by determining the lowest limit on the curve based on the distribution of data points. Furthermore, empirical equations are approached with power equation models such as Eq. 1 (Guzzetti et al. 2007).

$$I = aD^{-b} \tag{1}$$

where I_t is rainfall intensity (mm/day), D is he duration of the rainfall event that triggered each landslides (day), a and b is the constants obtained from the best fit.

In this study, a model developed was based on the antecedent rainfall triggering landslides. The definition of antecedent rainfall and respective duration, as explained by Aleotti (2004), is presented in Fig. 1. Parameter I in Eq. (1) is defined as average antecedent rainfall intensity (I_{at}) determined by Eq. (2).

$$I_{at} = \frac{R_{at}}{D_{at}} \tag{2}$$

where R_{at} is the antecedent rain (mm), and D_{at} is the duration of the antecedent rainfall (days).

Performance Analysis of the Empirical Model

In this study, the only first-time landslides at a single location was counted. The accuracy of the model was tested by a contingency table, skill scores, and ROC curve. Rainfall events that are not triggering landslide (no-landslide) were also collected to evaluate the empirical model. The rainfall events that are not triggering landslides were defined by the method proposed by Muntohar and Liao (2008). The rainfall events with no-landslide were defined at the same location in previous years, for instance, a landslide occurred at location L1 on November 21st, 2013 (T_n) during a rainfall event, which indicates that the slope has not failed on November 21st in previous three years 2012 (T_{n-1}), 2011 (T_{n-2}), and 2010 (T_{n-m}) (see Fig. 4). Thus, the antecedent rainfall and duration in previous years are defined as no-landslide. A total of 5468 rainfall events were observed for performance analysis.

Contingency Table

The contingency table or confusion matrix considers two classes of a classification model (classifier) and instances (Fawcett 2006; Frattini et al. 2010; Piciullo et al. 2016). The classification model and examples used in this study are observed landslide and predicted landslide. The contingency table correlates the observed landslide and predicted landslide. The predicted landslide is positive if a rainfall is positioned at or upper the threshold, whereas the negative is below the threshold. The positive observation is considered

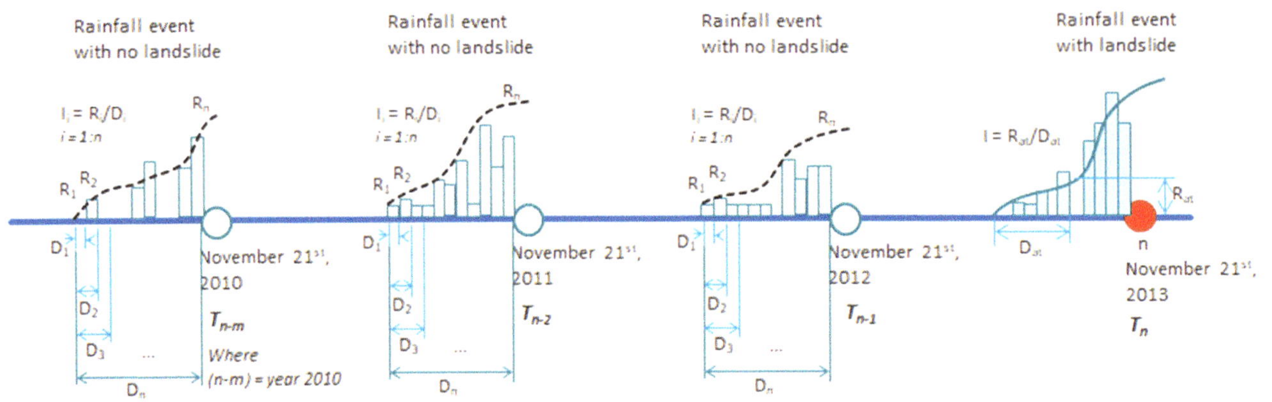

Fig. 4 A schematic definition of rainfall events with no landslides

Table 1 Confusion matrix for accuracy analysis of empirical model of rainfall threshold

Predicted	Observed	
	Landslide	No landslide
At or above threshold	True positive, TP	False positive, FP
Below threshold)	False negative, FN	True negative, TN

as true positive (TP), and is considered as false negative (FN) below the threshold. The related definitions for negative results above and below the threshold are false positives (FP) and true negative (TN). As a consequence, there are two main cases of prediction: (1) correct prediction: true positive (TP) and true negative (TN), and (2) wrong prediction: false positive (FP) and false-negative (FN) as presented in Table 1.

Statistical Indices

Frattini et al. (2010) used statistical indices to assess the accuracy of empirical models obtained through statistical analysis between the results of the model and the observed data. Seven statistical indices were used to analyze the accuracy of the rainfall threshold model (Fawcett 2006; Frattini et al. 2010), as presented in Table 2. The indices are

(i) True Positive Rate or hit rate is the ability of the threshold to identify rainfall events that trigger landslides;
(ii) False Positive Rate states the level of error in identifying rainfall events that do not trigger landslides;
(iii) True Negative Rate or specificity measures the ability to identify rainfall events that do not trigger landslides;
(iv) False Negative Rate states the error rate in identifying rainfall events that triggered landslides;
(v) Positive Prediction Power or precision is to determine the probability of a rainfall event which triggers a landslide;
(vi) Negative Prediction Power measures the probability of a rainfall event do not trigger a landslide; and
(vii) True Skill Statistics is the ratio between the True Positive Rate and False Positive Rate.

Table 2 Statistical indices for measuring the level of accuracy

Index	Equation	Range	Best value
True positive rate or hit rate	$TPR = \frac{TP}{TP+FN}$	(0;1)	1
False positive rate	$FPR = \frac{FP}{FP+TN}$	(0;1)	0
True negative rate or specificity	$TNR = 1 - FPR$	(0;1)	0
False negative rate	$FNR = 1 - TPR$	(0;1)	0
Positive prediction power or precision	$PPP = \frac{TP}{FP+TP}$	(0;1)	1
Negative prediction power	$NPP = \frac{TN}{FN+TN}$	(0;1)	1
True skill score	$TSS = TPR - FPR$	(0;1)	1

(After Fawcett 2006; Frattini et al. 2010; Piciullo et al. 2016)

ROC Curve

The ROC analysis method can be applied to measure the accuracy of the rainfall threshold. The ROC curve is a relationship between the True Positive Rate and the False Positive Rate indices, which at each point shows the level of the ability of the rainfall threshold to predict landslides. The level of accuracy is high if the ROC curve is close to the perfect classification point; for instance, the point with TPR = 1 and FPR = 0. The area under the curve (AUC) is an area that shows the level of accuracy of the empirical model. AUC is area whose value is always between 0 and 1. Random Performance results in an AUC value of 0.5 since the curve obtained is a diagonal line between point (0,0) and point (1,1). If the AUC is < 0.5, then the statistical model has a deficient level of accuracy and indicates the worst prediction when applied (Fawcett 2006; Zou et al. 2007).

Result and Discussion

Empirical Rainfall Threshold

The proposed empirical threshold in this study is presented in Fig. 5. The red line is the best-fit line of threshold. The threshold was defined by Eq. (3). The I-D threshold of this study is compared to the other regional thresholds by Caine (1980), Larsen and Simon (1993), Guzzetti et al. (2007), Mathew et al. (2013), and Rosi et al. (2017) as illustrated in Fig. 6. The figure shows that present threshold is lower than the other empirical equation. The figure also plots the landslides in Indonesia from year 2000 to 2004 that was provided by Muntohar (2008). The rainfall was obtained from the gauge measurement. The plot alludes to show that the established empirical rainfall threshold has a good performance to predict the landslides.

$$I = 7.83D^{-0.328} \tag{3}$$

However, Rossi et al. (2017) argued that satellite-derived rainfall could not be used directly in analytical or calibrated hydrogeological models if they were derived from gauge-based data unless the estimating satellite are correctly scaled locally. In terms of locally scale area, Sipayung et al. (2014) and Pratama et al. (2016) found that rainfall from gauge measurement was 0.36–0.44 of the rainfall derived from TRMM. It indicates that the rainfall threshold from satellite-based rainfall estimation can be applied to predict the landslide initiation. Furthermore, Kirschbaum and Stanley (2018) and Guzzetti et al. (2020) stated that satellite rainfall projections might be useful in determining empirical rainfall thresholds for landslides initiation or in general for other hydrogeological models to forecast phenomena of instability over the large territorial area. Therefore, the selection of the empirical threshold model is the key to issuing early warning system (EWS) for rainfall-induced landslides (Monsieurs et al. 2019; Piciullo et al. 2018).

Performance Evaluation of the Rainfall Threshold Model

Table 3 presents the confusion matrix to evaluate the performance of the empirical rainfall threshold. The total number of events of 5468 was evaluated in the performance analysis. The True Positive was counted as many as 1389 events, while the True Negative, False Positive, and False Negative were 1572, 121, and 2386 events, respectively. Further performance analysis for statistical indices of the model is presented in Table 4.

The empirical model proposed in this study showed a higher True Positive Rate (TPR = 0.92) but a moderate True

Fig. 5 The proposed I-D curve as rainfall threshold for landslides warning

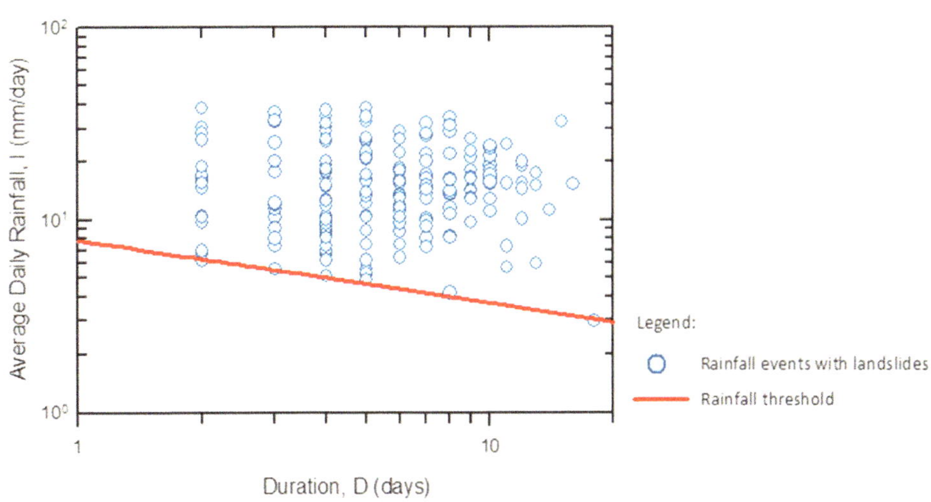

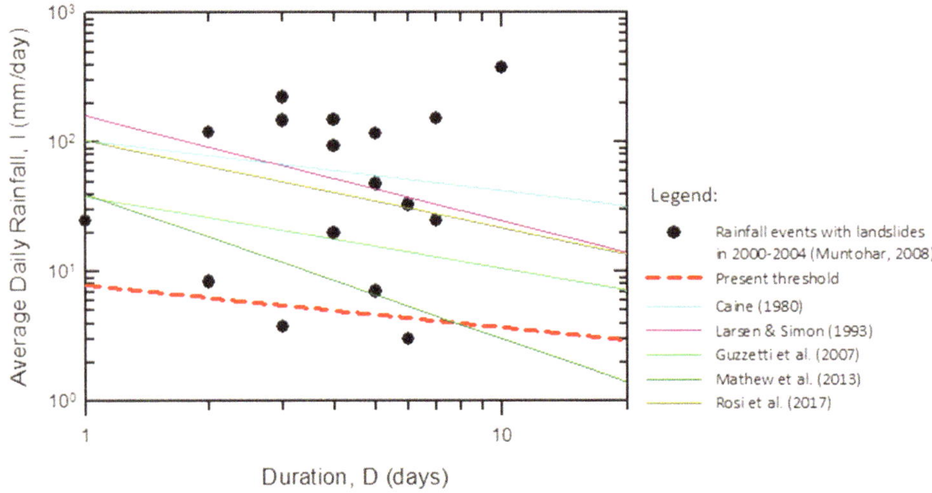

Fig. 6 Comparison the threshold with the worldwide threshold for rainfall events with landslide in 2000–2004

Table 3 The confusion matrix for performance evaluation of the model

Predicted	Observed	
	Landslide	No landslide
At or above threshold	TP = 1389	FP = 2386
Below threshold)	FN = 121	TN = 1572

Negative Rate (TNR = 0.40). The result implies that the model was right in identifying the occurrence of rainfall-triggered landslides, but it presented low accuracy in predicting not-failure case. Estimated TNR values > 0.90 stated that the rainfall threshold is perfect for avoiding alarm errors in the early warning system (Rosi et al. 2017). However, the TNR value (TNR = 0.40) of the threshold could be categorized as low to a moderate level in identifying rainfall events that did not trigger a landslide, which could cause a high alarm error rate in the warning system. The FPR value of the threshold was highly moderate (FPR = 0.60), indicating that the rainfall threshold was suitable enough to predict the rain event that did not trigger landslides. The FPR is strongly related to the uncertainty of rainfall intensity and causing the condition of FP (Guzzetti et al. 2007).

Table 4 Statistical indices of the model

Index	Value
True positive rate (TPR)	0.92
False positive rate (FPR)	0.60
True negative rate (TNR)	0.40
False negative rate (FNR)	0.08
Positive prediction power (PPP)	0.37
Negative Prediction power (NPP)	0.93
True skill statistic (TSS)	0.32

The PPP value (PPP = 0.37) was relatively low for an empirical model. The result indicates that the ability to classify precisely the rainfall-triggered landslide was minimal. Furthermore, there was still a high possibility that the rain threshold was incorrect in classifying the occurrence of rainfall-triggered landslides despite having a high TPR value. However, NPP values > 0.90 indicates that the rainfall threshold model has an excellent prediction of rainfall events that do not trigger landslides (Rosi et al. 2017). The NPP value obtained in this study was 0.93, showing that the ability of the rainfall threshold in classifying rainfall events that did not trigger landslides was very high, despite the low TNR value.

The TSS index is expressed as an interval number (0,1), if TSS = 0 then TPR = FPR and TSS = 1 for a perfect prediction level if TPR = 1 and FPR = 0. The high value of False Positive Rate caused the low value of True Skill Statistics (TSS). The analysis of this study revealed a TSS value of 0.32. It means that the rain threshold still has a low predictive level (Peres and Cancelliere 2014). Overall statistical evaluation was considered in ROC analysis. The result is presented in a ROC curve, as in Fig. 7. The diagonal line is a random performance value of 0.50 (AUC = 0.50), assuming the true and false are equal (TPR = FPR). The AUC of the threshold shows that the level of accuracy in detecting rainfall triggering landslides and non-landslides was 0.66. Figure 7 shows the AUC and the rain threshold produce a pretty good level of accuracy since the results obtained exceed the value of random performance.

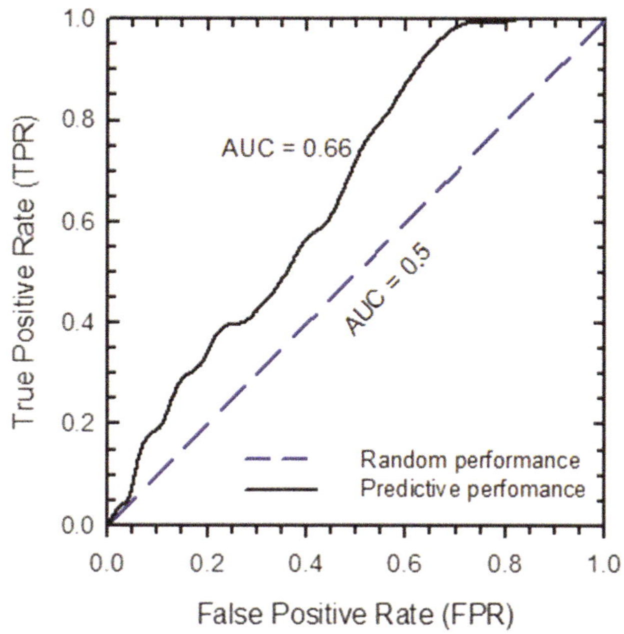

Fig. 7 Calculation for AUC of ROC curve

Application of the Rainfall Threshold Model

The developed rainfall threshold has been implemented in a prototype of Indonesia Landslides Early Warning System (ILEWS) by Research Centre of Water Resources of the Ministry of Public Works and Housing (PUSAIR 2017). The rainfall threshold model was coupled with the Delft-FEWS platform. The platform was available to manage the forecasting and save the time series data (Werner et al. 2013). The satellite forecasting rainfall data was automatically programmed to download the TRMM precipitation data and forecasting from Weather Agency (BMKG). The territory of analysis was based on water catchment area which provided by Directorate General of Water Resources of the Ministry of Public Works and Housing. The warning scenario is categorized into three levels as presented in Table 5. Figure 8a and b illustrate the forecasting rainfall-derived from TRMM for a day and three days rainfall respectively, while the predicted landslide is shown in Fig. 8c. The warning has been displayed on website of SABO Research Centre (http://202.173.16.248/status_longsor.html). The warning system can predict landslides with next three days based on the rainfall forecasted by BMKG. However, it should be noticed that the ILEWS prototype has limitation on the resolution. The resolution 0.25° × 0.25° is quite coarse and covers several river catchment area in one grid (Hidayat et al. 2019).

Conclusion

The research has successful developed a rainfall threshold based on TRMM satellite observation for landslide prediction. The performance of the empirical threshold model has been evaluated by several accuracy method including statistical confusion matrix, skill scores, and ROC analysis. The application of the model has been also adopted by government authority to develop a prototype of Indonesia Landslide Early Warning System. Thus far, it can be concluded that the empirical rainfall threshold $I = 7.83D^{-0.328}$ was valid for an interval time 2 to 18 days. The results of the analysis of the AUC on the rainfall threshold indicate that the model has a good accuracy rate and can be used as an early warning system of landslide even though it still has a fairly error rate. The rainfall threshold model showed a good level of accuracy in predicting rainfall triggered landslides and non-landslides events. Thus, it can be implemented in a landslide early warning system.

Acknowledgements The authors thank to Dio Fairus Akhbar, Abidah Uli Rohmaniah, and Gayuh Aji Prasetyaningtyas for their assistance for collecting the landslides and rainfall records. The Ministry of Research and Technology, the Republic of Indonesia funded the research through the international collaborative research grant under contract number 227/SP2H/LT/DRPM/2019. The supports from SABO Technological & Research Centre, Ministry of Public Works and Housing, and Faculty of Geoinformation Science and Earth Observation (ITC), University of Twente, the Netherland are highly appreciated.

Table 5 Warning scenario of prototype of ILEWS

Warning level	Criteria	Remark
Green	Daily rainfall and average rainfall for three days is *below* the threshold	No landslides
Yellow	Daily rainfall and average rainfall for three days is *at* the threshold	Potential to landslides (Low susceptibility)
Red	Daily rainfall and average rainfall for three days *exceeds* the threshold	Landslides (High susceptibility)

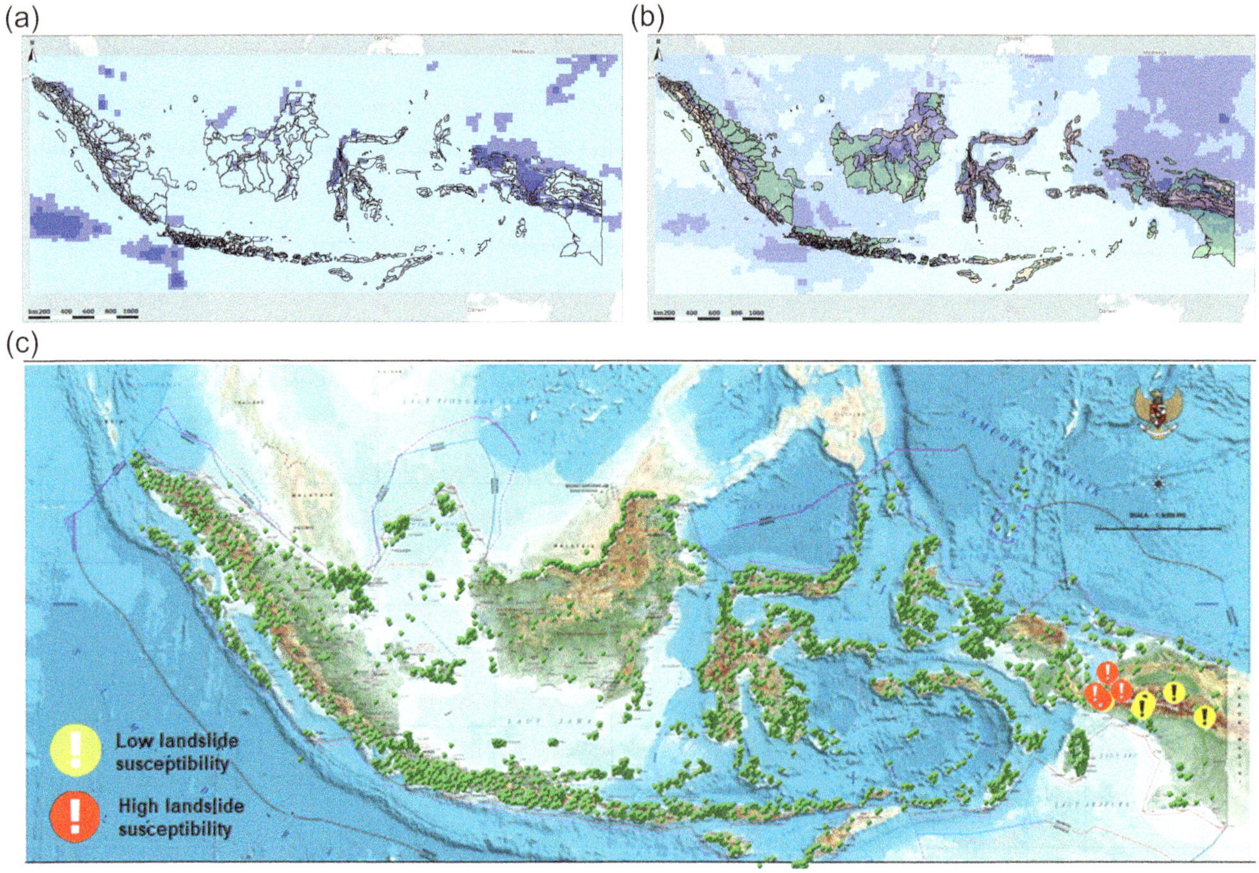

Fig. 8 **a** A day rainfall forecasting, **b** three days rainfall forecasting, **c** prediction of landslides

References

Aleotti P (2004) A warning system for rainfall-induced shallow failures. Eng Geol 73(3–4):247–265. https://doi.org/10.1016/j.enggeo.2004.01.007

As-Syakur AR, Tanaka T, Prasetia R, Swardika IK, Kasa IW (2011) Comparison of TRMM multisatellite precipitation analysis (TMPA) products and daily-monthly gauge data over Bali. Int J Remote Sens 32(24):8969–8982. https://doi.org/10.1080/01431161.2010.531784

Caine N (1980) The rainfall intensity: duration control of shallow landslides and debris flows. Geogr Ann Series A, Phys Geo 62(1–2):23–27

Fatkhuroyan Wati T, Sukmana A, Kurniawan R (2018) Validation of satellite daily rainfall estimates over Indonesia. Forum Geogr 31 (2):170–180. https://doi.org/10.23917/forgeo.v31i2.6288

Fawcett T (2006) An introduction to ROC analysis. Pattern Recogn Lett 27(8):861–874. https://doi.org/10.1016/j.patrec.2005.10.010

Frattini P, Crosta G, Carrara A (2010) Techniques for evaluating the performance of landslide susceptibility models. Eng Geol 111(1–4):62–72. https://doi.org/10.1016/j.enggeo.2009.12.004

Giarno Hadi MP, Suprayogi S, Murti SH (2018) Distribution of accuracy of TRMM daily rainfall in Makassar Strait. Forum Geogr 32(1):38–52. https://doi.org/10.23917/forgeo.v31i2.5774

Guzzetti F, Gariano SL, Peruccacci S, Brunetti MT, Marchesini I, Rossi M, Melillo M (2020) Geographical landslide early warning systems. Earth Sci Rev 200. https://doi.org/10.1016/j.earscirev.2019.102973

Guzzetti F, Peruccacci S, Rossi M, Stark CP (2007) Rainfall thresholds for the initiation of landslides in central and southern Europe. Meteorol Atmos Phys 98(3–4):239–267. https://doi.org/10.1007/s00703-007-0262-7

Hidayat R, Sutanto SJ, Hidayah A, Ridwan B, Mulyana A (2019) Development of a landslide Early warning system in Indonesia. Geosciences 9(10):451–461. https://doi.org/10.3390/geosciences9100451

Kirschbaum D, Stanley T (2018) Satellite-based assessment of rainfall-triggered landslide hazard for situational awareness. Earths Future 6(3):505–523. https://doi.org/10.1002/2017ef000715

Larsen MC, Simon A (1993) A rainfall intensity-duration threshold for Landslides in a humid-tropical environment, Puerto Rico. Geogr Ann Series A Phys Geogr 75(1–2):13–23

Mathew J, Babu DG, Kundu S, Kumar KV, Pant CC (2013) Integrating intensity–duration-based rainfall threshold and antecedent rainfall-based probability estimate towards generating early warning for rainfall-induced landslides in parts of the Garhwal Himalaya. India Landslides 11(4):575–588. https://doi.org/10.1007/s10346-013-0408-2

Monsieurs, Dewitte, Depicker, Demoulin (2019) Towards a transferable antecedent rainfall—susceptibility threshold approach for landsliding. Water 11(11). https://doi.org/10.3390/w11112202

Muntohar AS (2008) Toward regional rainfall threshold for landslide occurrence in Yogyakarta and central of java. Jurnal Teknik Sipil 3 (1):40–47

Muntohar AS, Liao H-J (2008) Analysis of rainfall-induced infinite slope failure during typhoon using a hydrological–geotechnical model. Environ Geol 56(6):1145–1159. https://doi.org/10.1007/s00254-008-1215-2

Peres DJ, Cancelliere A (2014) Derivation and evaluation of landslide-triggering thresholds by a Monte Carlo approach. Hydrol Earth Syst Sci 18(12):4913–4931. https://doi.org/10.5194/hess-18-4913-2014

Piciullo L, Calvello M, Cepeda José M (2018) Territorial early warning systems for rainfall-induced landslides. Earth Sci Rev 179:228–247. https://doi.org/10.1016/j.earscirev.2018.02.013

Piciullo L, Gariano SL, Melillo M, Brunetti MT, Peruccacci S, Guzzetti F, Calvello M (2016) Definition and performance of a threshold-based regional early warning model for rainfall-induced landslides. Landslides 14(3):995–1008. https://doi.org/10.1007/s10346-016-0750-2

Pratama GN, Suwarman R, Junnaedhi IDGA, Riawan E, Anugrah A (2016) Comparison landslide-triggering rainfall threshold using satellite data: TRMM and GPM in South Bandung area. In: Sakakibara M, Saepuloh A, Kurniawan IA (eds) 2nd Transdisciplinary research on environmental problems in Southeast Asia science. Bandung, Indonesia, 20–22 September 2016. IOP Conference Series: Earth and Environmental. https://doi.org/10.1088/1755-1315/71/1/012003

PUSAIR (2017) Modelling of rainfall induced landslide for prediction of prone area in Indonesia. Research center of water Resources, ministry of public work and Housing. Bandung, Indonesia

Reichenbach P, Cardinali M, De Vita P, Guzzetti F (1998) Regional hydrological thresholds for landslides and floods in the Tiber River Basin (central Italy). Environ Geol 35(2–3):146–159. https://doi.org/10.1007/s002540050301

Rosi A, Peternel T, Jemec-Auflić M, Komac M, Casagli N (2017) Definition of rainfall thresholds triggering landslides in Slovenia. In: Mikoš M, Arbanas Ž, Yin Y, Sassa K (eds) Fourth World landslides forum Ljubljana, Slovenia, 29 May 2 June 2017. Springer, Netherland, pp 177–182. https://doi.org/10.1007/978-3-319-53485-5_19

Rossi M, Kirschbaum D, Valigi D, Mondini A, Guzzetti F (2017) Comparison of satellite rainfall estimates and rain gauge measurements in Italy, and impact on landslide modeling. Climate 5(4). https://doi.org/10.3390/cli5040090

Sipayung SB, Cholianawati N, Susanti I, Maryadi E (2014) Pengembangan Model Persamaan Empiris Dalam Memprediksi Terjadinya Longsor di Daerah Aliran Sungai (DAS) Citarum (Jawa Barat) Berbasis Data Satelit TRMM. Jurnal Sains Dirgantara 2(1):12–21

Werner M, Schellekens J, Gijsbers P, van Dijk M, van den Akker O, Heynert K (2013) The Delft-FEWS flow forecasting system. Environ Model Softw 40:65–77. https://doi.org/10.1016/j.envsoft.2012.07.010

Zou KH, O'Malley AJ, Mauri L (2007) Receiver-operating characteristic analysis for evaluating diagnostic tests and predictive models. Circulation 115(5):654–657. https://doi.org/10.1161/CIRCULATIONAHA.105.594929

Establishing Soil Moisture and Rainfall Intensity-Duration Thresholds for Initiation of Mass Movements Along the National Higway-58 in the Chamoli District of Uttarakhand

WLF 2020 Kyoto Japan

Shobhana Lakhera, P. K. Champati Ray, Michel Jaboyedoff, and Harshita Tiwari

Abstract

The Rishikesh-Badrinath National Highway 58 (NH-58), in the district of Chamoli, Uttarakhand, India records several incidences of landslides and related casualties every year during intense rainfall periods. Mass movements that occurred along this highway during monsoon, in the period 2015–2019 have been studied to arrive at meaningful rainfall and soil moisture thresholds. Data from the Border Road Organization (BRO) was compiled for 166 mass movement events, including 67 debris slides, 33 rock slides, 24 debris flows and complex movements. The relationship between rainfall maximum intensity (I_{max}) and total duration (D) was analysed for each site and I_{max}-D thresholds were derived using Global Precipitation Mission (GPM) half-hourly data. Significant differences were observed between the derived I_{max}-D thresholds, which are: $I_{max} = 59.06D^{-1.31}$ (all mass movements), $I_{max} = 21D^{-0.96}$ (debris slides), $I_{max} = 11.55D^{-0.6}$ (rock slides) and $I_{max} = 4.45D^{-0.44}$ (debris flows). For short-duration rainfall events (<24 h), higher maximum intensities were required to trigger debris flows. In contrast for long duration rainfall events, similar maximum intensities triggered both landslides and debris flows. Additionally, the relationship between soil moisture and rainfall was analysed by deriving maximum soil moisture (SM_{max}) based rainfall thresholds (SM_{max}-I_{max}D) using Global Land Surface Model (GLDAS) 3-hourly data, which makes this study the first attempt towards development of SM_{max}-I_{max}D thresholds for the region. The thresholds hence obtained are $SM_{max} = 677.7 I_{max}D^{-0.96}$, $SM_{max} = 104.6I_{max}D^{-0.38}$, $SM_{max} = 307 I_{max}D^{-0.68}$ and $SM_{max} = 878.7ID^{-0.64}$ for mass movements, debris slides, rock slides and debris flows, respectively. Notable differences pertaining typology of mass movements were observed likewise the I_{max}-D thresholds. Also, it was found that a large amount of rainfall is required to cause an increment in maximum soil moisture, and any rainfall event which leads to increment in maximum soil moisture value above 30 mm is likely to cause a failure. Further, for long duration rainfall events, soil moisture value (>47 mm) can trigger both landslides and debris flows.

Keywords

Mass movements • debris slides • Rock slides • Debris flows • I_{max}-D thresholds • SM_{max}-I_{max}D thresholds • Rainfall • Soil moisture

S. Lakhera (✉)
Department of Earth Sciences, CERG-C, University of Geneva, 13 Rue des Maraichers, 1205 Geneva, Switzerland
e-mail: shobhana303@gmail.com

P. K. Champati Ray · H. Tiwari
Geoscience and Disaster Management Group, Indian Institute of Remote Sensing (ISRO), 4-Kalidas Road, Dehradun, 248001, India
e-mail: champati_ray@iirs.gov.in

H. Tiwari
e-mail: harshita.iirs@gmail.com

M. Jaboyedoff
University of Lausanne, Geopolis - 3793, CH-1015 Lausanne, Switzerland
e-mail: michel.Jaboyedoff@unil.ch

Introduction

Slope failure and related secondary hazards are common geological hazards in mountainous regions characterized by high slopes and complex geo-tectonic settings (Aleotti and Chowdhury 1999). These slope failures often lead to losses in terms of both life and property. They include mass movements of all types namely: falling, sliding and flowing and often occur in combination with triggering agents like cloudburst, heavy rainfall, earthquakes, and floods. The factors that control slope instability in such regions involve geology, geomorphology and hydrology along with intricate

© Springer Nature Switzerland AG 2021
N. Casagli et al. (eds.), *Understanding and Reducing Landslide Disaster Risk*, ICL Contribution to Landslide Disaster Risk Reduction, https://doi.org/10.1007/978-3-030-60311-3_27

tectonics, geo-dynamics and climatic factors. Increased urbanization, accompanied by expansion of roads and deforestation creates an increasing pressure on the landscape, and leads to higher degrees of vulnerability to the occurrence of mass movement activity in such high altitude regions.

The Rishikesh-Badrinath NH-58, in the Chamoli district, Uttarakhand, India has been made by excavations in the rocks, fluvio-glacial material and talus deposits on the slopes of valleys. The townships are generally established alongside the road. The expansion of this road together with rapid urbanization has rendered these unstable hill slopes, apparently more vulnerable to slope failures. This highway has socio-economic and cultural importance as it is the only motorable route connecting Badrinath, an important Hindu pilgrimage center and other hill cities to the rest of the nation. Hence slope failures along this route lead to disruption of traffic leaving the pilgrims, tourists and inhabitants grounded for several hours to days. Destruction of the highway has deeper impacts as many towns, villages and hamlets are often completely cut-off from the rest of the country. Also, the landslides and the involved secondary hazards such as landslide dams and subsequent flash floods often turn into major disasters and cause destruction in the downstream areas. Given these disastrous impacts many studies have been carried out to derive rainfall triggering empirical thresholds in the region based on historic analysis of relationship between rainfall and landslide occurrence. These triggering thresholds are predominantly focused on rainfall parameters like rainfall intensity, duration, cumulative rainfall and antecedent rainfall. However rainfall intensity-duration based thresholds are most widely used (Kuthari 2007; Guzzetti et al. 2008). Many studies have also found that peak rainfall intensity and the initiation of mass movements are often concurrent (Aleotti 2004; Chien-Yuan et al. 2005; Guzzetti et al. 2008). Other studies have showed that antecedent rainfall has also played an important role in the initiation of mass movement (Crozier 1999; Guzzetti et al. 2008; Mathew et al. 2013). However the simplicity of empirical approach often neglects important hydrological and geological controls, but offers a straight forward means for issuing regional-scale mass movement warnings based solely on rainfall data.

The processes by which mass movements occurs can be complex, as a heavy or long duration rainfall event may not always cause a slope failure. Studies have showed enhanced soil moisture and rainfall prior to major landslide events in landslide prone regions of California, U.S.; Leyte, Philippines; and Dhading, Nepal (Ray and Jacobs 2006). One case study for EL Salvador derived a two-dimension threshold curve consisting of depth-integrated soil moisture and rainfall to predict landslides (Posner and Georgakakos 2015). Recently, Irawan et al. (2019) and Segoni et al. (2018b) highligted the integration of soil moisture for improving landslide early warning. Hence the use of soil moisture as another variable coupled with rainfall can give a more reliable threshold which can incorporate soil properties, terrain properties, geology and hydrometeorology to an extent (Segoni et al. 2018a, b; Krøgli et al. 2018). However the integration of soil moisture data with rainfall is still uncommon for establishing threshold in the Himalayan region, therefore, it was explored in the present study.

Study Area

The study focused on the mass movement activity that occurred along the NH-58 highway in the district of Chamoli, Uttarakhand, India. This part of the National Highway runs around 110 km, aligned alongside the river Alaknanda from Srinagar to Badrinath and extends in the North from 30.23° N to 30.83° N and in the east from 78.13° E to 79.76° E (Fig. 1). The highway is constructed by excavating highly jointed friable rocks and is currently under expansion under the all-weather road project funded by the central government. This has led to a concern that it may further increase the number of mass movement events along this highway. The area around the highway is characterized by deep gorges and resilient peaks with a maximum elevation of 5826 m masl (Fig. 2a). Primarily the area lies in the moderate altitude zone with mostly areas under 2500 m releif. The topography is highly rugged, immature and characterized by moderate to steep slopes controlled by both structural and lithological factors (Fig. 2b). Geologically, the study area is divided into three main zones—upper, middle and lower, which are separated from each other by major faults disecting the entire region (Validiya 2010) (Fig. 1). The major drainage in the area constitutes the four main tributaries of Alaknada namely Dhauliganga, Nandakini, Pindar and Mandakini all of which rise from high hills forming sub-parallel drainages. Consequently, the hill slopes are gullied and dissected. Several streams of Alaknanda like the Patal Ganga show contribution from deep underground sources which indicate conditions suitable for landslides. Accoring to the Central Ground Water Board (CGWB), India, climate of the district varies from Sub-tropical monsoon type (mild winter, hot summer) to Tropical upland type (mild winter, dry winter, short warm summer). The average maximum and mininmum temprature for the district varied between ~40 °C to around −1.5 °C for the period 2014–2018 (Fig. 2c). The study area witnesses heavy rainfall varying from 1000–2500 mm/year. Of this total rainfall about 50–80% of the rainfall is received during the months of June to mid-September and extreme rainfall events like cloudburst are also reported from the region (Fig. 2c). The region is mostly rain fed, hence landslide activity is mainly

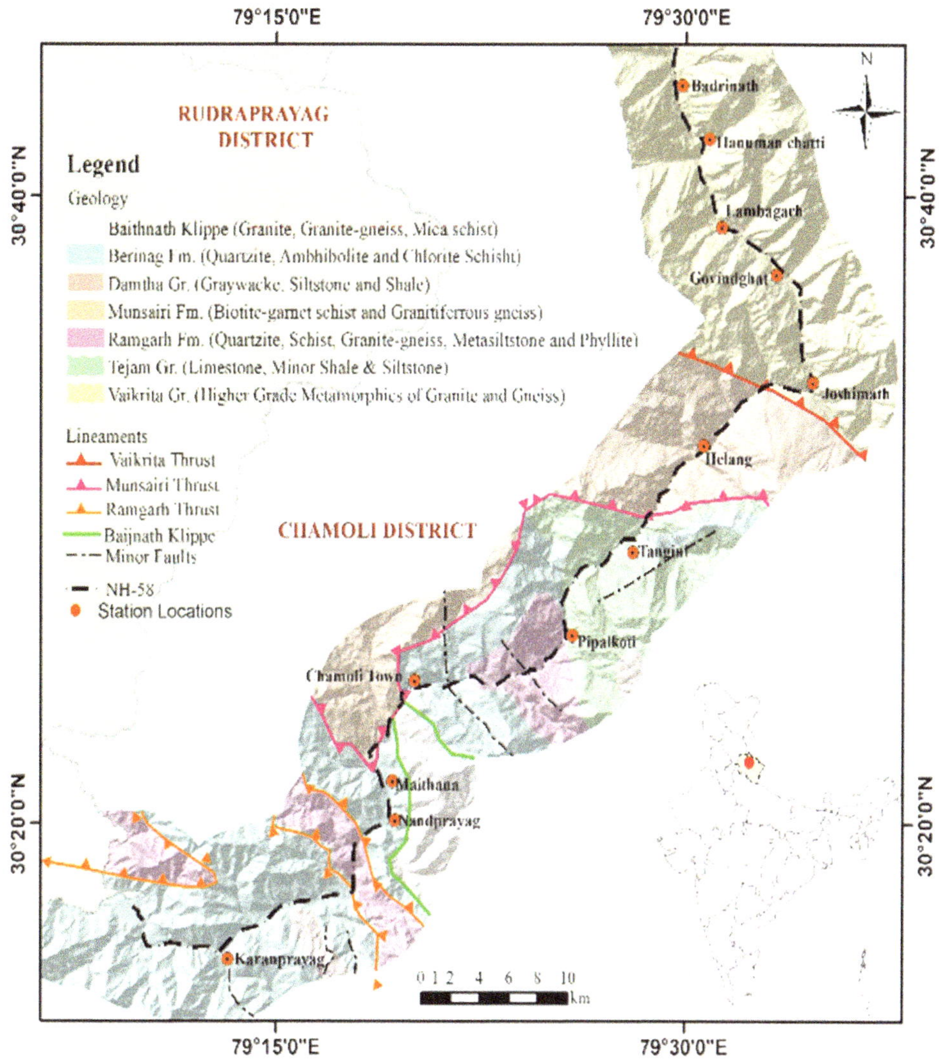

Fig. 1 Geological setting along the NH-58 highway

triggered by rainfall and the role of snow melt is very marginal in the summer and summer monsoon period.

Methods

Mass Movement Data

The Border Road Organisation (BRO) data for the years 2015–2019 was sorted for twelve major landslide prone locations along the NH-58 highway. The location points were chosen based on three criteria which are landslide history of the location, socio-economic importance and availability of AWS (Automated Weather Stations). A total of 166 mass movement events were used for the creation of database. This included 67 debris slides, 33 rock slides, 24 debris flows. Apart from this 20 events recorded both landslides (rock slides and debris slides) and debris flows, 12 events recorded both rock slides and debris slides, 6 events recorded both rock fall and landslides and 4 individual rock fall events were also recorded in the area. For each event, the database defines the location and day of the occurrence of the mass movement. The typology for mass movement was determined based on the field survey and ancillary reports from the Uttarakhand Secretariat, Dehradun. Further it was important to note that majorly shallow debris slides were present in the study area along with few deep ones.

Rainfall and Soil Moisture Data

The GPM (Global Precipitation Measurement Mission)/IMERG (Integrated Multi-satellite Retrievals) (level 4, 0.1° —30 min gridded, 90° N–90° S, 60° N–60° S full, March 2014–March 2019 near real time) data was used for the study. The GIS extension/product of the GPM data has been used. The GPM data was then compared with AWS rain gauge data to determine the correlation between the two datasets for five functional AWS stations for 2015 (June to

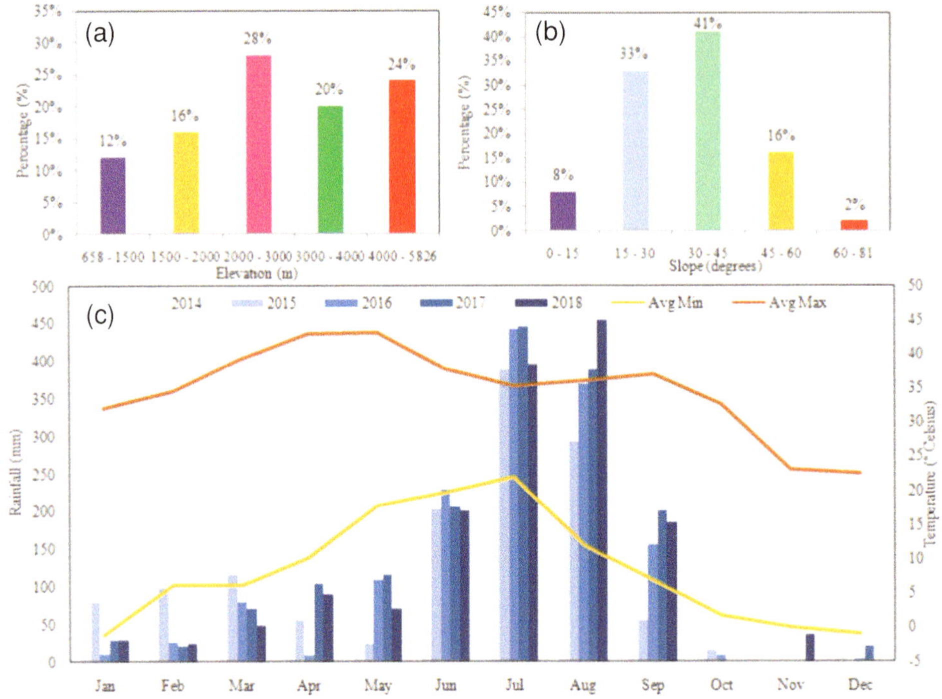

Fig. 2 **a** Elevation and **b** slope around the 5 km buffer zone of NH-58 highway. **c** Monthly rainfall and average temperature distribution in the Chamoli district from 2014 to 2018 (*source* Indian Meteorological Department (IMD))

September) and an overall good correlation of nearly 76% was observed. The soil moisture up to depth of 0–40 cm as estimated in mm sourced from GLDAS (Global Land Surface Model) version 2.1 data products (GLDAS_-NOAH025_3H_2.0). These datasets have a temporal resolution of 3-h and spatial resolution of 0.25° and are available in netcdf format. The soil moisture data was extracted GIS layer wise, for the depths up to 0–10 cm and 10–40 cm for each event day and then stacked together for both the depths using ArcGIS software, to get a single day file. The soil moisture values corresponding to both the GIS layers (daily stack) were then added to get soil moisture up to 0–40 cm. Soil moisture mean, median and maximum values for the depth up to 0–40 cm were analysed against the maximum rainfall intensity values, for the monsoon months i.e. June to September for five major locations to see how well the two variables correspond and it was observed that the maximum soil moisture up to 0–40 cm showed good correlation with the maximum intensity.

Thresholds

The I_{max}-D threshold equations were derived empirically using the log–log scatter plot between the maximum rainfall intensity picked from the forty eight, half-hourly GPM-GIS files, for the day of the event and the total rainfall duration calculated by adding the total number of continuous rainfall days in hours. Hence the I_{max}-D threshold is identified on an I_{max}-D plot, as the minimum rainfall for which a mass movement could occur. Similarly, the SM_{max}-I_{max}D threshold equation was calculated using the maximum measured soil moisture picked from the eight, 3-hourly soil moisture files and the product of maximum intensity to total duration measured in mm. As we understand that the landslides & debris flows are characterised by different mechanism mainly in terms of the type of movement & material involved (coarse, fine and bedrock) (Varnes 1978), hence these were separately analysed to identify the triggers associated with the type of failure. For the validation of threshold equation, the 2014 database of BRO was used and the results are presented in the next section.

Results and Discussion

Rainfall Conditions and I_{max}-D Thresholds

The maximum intensity value for the mass movement database varied from as low as 0.1 mm/h to as high as 31.6 mm/h, measured only for two events, and a mean maximum intensity of 5.6 mm/h. Similarly, the total duration values varied between lows as 0.5 h to as high as 226 h with a mean of 85 h. Hence the triggering events have a wide dispersion in terms of both maximum intensity and total duration for the mass movement database. Further to analyse the differences in rainfall conditions between debris slides, rock slides and debris flows, the normal distribution

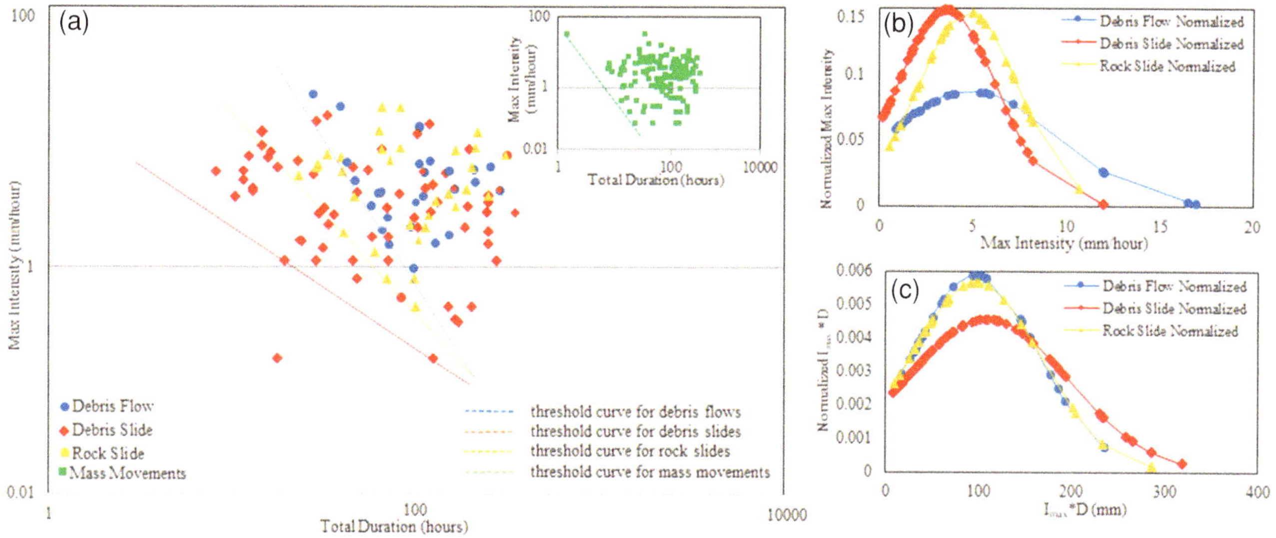

Fig. 3 **a** Imax-D thresholds for debris slide, rock slide, debris flow and mass movements, **b** normalized max intensity for debris slide, rock slide and debris flows and **c** normalized total duration for debris slide, rock slide and debris flow

curves for maximum intensity and total duration for the three mass movement typologies were analysed (Fig. 3b, c). The maximum intensity normal distribution curve showed that the highest mean max intensity of 5 mm/h was measured for debris flows followed by rock slides with a mean max intensity value of 4.8 mm/h and debris slides with a mean of 3.6 mm/h (Fig. 3b). Also, debris flows measured the highest standard deviation implying that the maximum intensity values for debris flow were highly dispersed over the database, however it was found that the dispersion was high pertaining to two events (16.5 and 16.9 mm/h) which recorded a very high maximum intensity value in comparison to the other recorded events. Based on the normal distribution of total duration, debris slides have the highest mean of 108.7 h followed by rock slides and debris flows with mean total duration of 96.4 and 96 h, respectively (Fig. 3c). This indicates that debris slide occurs at longer duration and hence lower intensity in comparison to debris flow which occurs at higher intensity, as was observed from the normalised maximum intensity. Debris slides recorded the highest standard deviation of 87 showing high dispersion in terms of total duration with values ranging from very low to high values (Fig. 3b). This is possibly because of large variation in the size and depth of debris slides.

As previously mentioned in the methodology section, the log–log scatter plots between maximum intensity to total duration were used to determine rainfall thresholds (Fig. 3a). The resultant equations obtained for different types of mass movement are as follows:

$I_{max} = 59.06D^{-1.31}$ all mass movements
$I_{max} = 21D^{-0.96}$ debris slides
$I_{max} = 11.55D^{-0.6}$ rock slides
$I_{max} = 4.45D^{-0.44}$ debris flows

The scatter plot and threshold for mass movements (all 166 events) indicated that high intensity and low duration rainfall events can cause mass movements and similarly low intensity and long duration rainfall events can result in mass movements (Fig. 3a). From the scatter plot for debris slide, rock slide and debris flows, it was observed that the debris flow thresholds lies at the highest level in the graph and has the highest slope magnitude. This implied that debris flows occur during highly intense rainfall events in comparison to rock slides and debris slides. The rock slide threshold was seen to lie above the debris slide threshold (Fig. 3a), hence it can be concluded that a higher intensity and longer duration of rainfall is required to cause a rock slide in comparison to a debris slide. Further the threshold curves were observed to become closer as the value of total duration increases and they nearly meet at 315 h. This indicates that a higher rainfall intensity is required to cause debris flow in case of short duration rainfall event compared to rock slide and debris slide, whereas a long-duration rainfall event will lead to a gradual increase in groundwater level, soil moisture, and pore water pressure (Wieczorek and Glade 2005). Therefore, similar rainfall intensity can trigger debris slide, rock slide and debris flow, given rainfall duration is sufficiently long. Also, landslides may completely or partially mobilize to form debris flows (Iverson et al. 1997), as was seen in the case of 20 recorded events wherein landslide and debris flow both occurred.

Rainfall Conditions and SM_{max}-$I_{max}D$ Thresholds

The maximum soil moisture for all mass movements varied between 27.76 mm to 205.663 mm whereas the values for the product of maximum intensity to total duration (I_{max}*D) varied from as low as 0.6 mm to as high as 31,773 mm (Fig. 4a). The normal distribution curves for maximum soil moisture and product of maximum intensity to total duration were analysed for debris slides, rock slides and debris flows and the highest mean maximum soil moisture value of 88.9 mm was recorded for debris flows, followed by rock slides with mean maximum soil moisture value of 82.44 mm and debris slides with a mean of 73.18 mm (Fig. 4b). Also, the normal distribution curves for all three mass movement typologies showed similar spread with respect to maximum soil moisture values (Fig. 4b). From the normal distribution curve for product of I_{max} and total D, debris flows showed exceptionally high mean of 8095.66 mm and a well observed largest spread i.e. highest standard deviation. It was also observed that debris flows occurred above 3000 mm of I_{max}*D value (Fig. 4c). The mean I_{max}*D value for rock slides was 472 mm which was higher than debris slides whose mean I_{max}*D was measured as 302 mm (Fig. 4c). Also, the spread for rock slides is larger than debris slides as rock slides have higher deviation because of some high values whereas debris slides have lower I_{max}*D extreme values (Fig. 4c).

The SM_{max}-$I_{max}D$ thresholds were derived using the scatter plot between maximum soil moisture and the product of max I and total D. The resultant equations obtained are given as follows:

$SM_{max} = 677.7ID^{-0.96}$ all mass movements.

$SM_{max} = 104.6ID^{-0.38}$ debris slides.

$SM_{max} = 307ID^{-0.68}$ rock slides.

$SM_{max} = 878.7ID^{-0.64}$ debris flows.

Figure 4a shows the SM_{max}-$I_{max}D$ threshold for landslides and debris flows. Here it was observed that as we move from right to left i.e. from higher value of product of I_{max} and total D to lower values of I_{max}*D, the typology of mass movements changes from debris flows to rock slides to debris slides. This shows that debris flows occur during more severe and more prolonged rainfall events in comparison to debris slides and rock slides as was previously also observed in the case of I_{max}-D thresholds. In addition, it was observed that rock slides and debris slides both occur for I_{max}*D values between 3000 to 46 mm, with only a few debris flow events. While below 46 mm of I_{max}*D, only Debris slides were recorded and beyond the I_{max}*D value of 3000 mm, majorly only debris flows were recorded. Besides for max soil moisture, it was observed that above 100 mm of recorded max soil moisture value, usually debris flows and rock slides occur. While beyond 125 mm of max soil moisture, mostly debris flows were observed. Hence from SM_{max}-$I_{max}D$ threshold curves, it can be concluded that large increments in value of I_{max}*D do not cause very high increments in resultant max soil moisture, which may be due to saturation and subsequent run-off. Another observation from the curve is that no mass movements were recorded below a maximum soil moisture value of 27.76 mm (~30 mm) for a depth of 0–40 cm of soil. Hence, it can be concluded that any increment in the value of maximum soil moisture above ~30 mm is likely to cause a slope failure. In addition it was found that for long duration rainfall events (>24 h) both landslides and debris flows occurred, when the recorded soil moisture value was greater than 47 mm. This was the case for 20 events which recorded both landslides and debris flows.

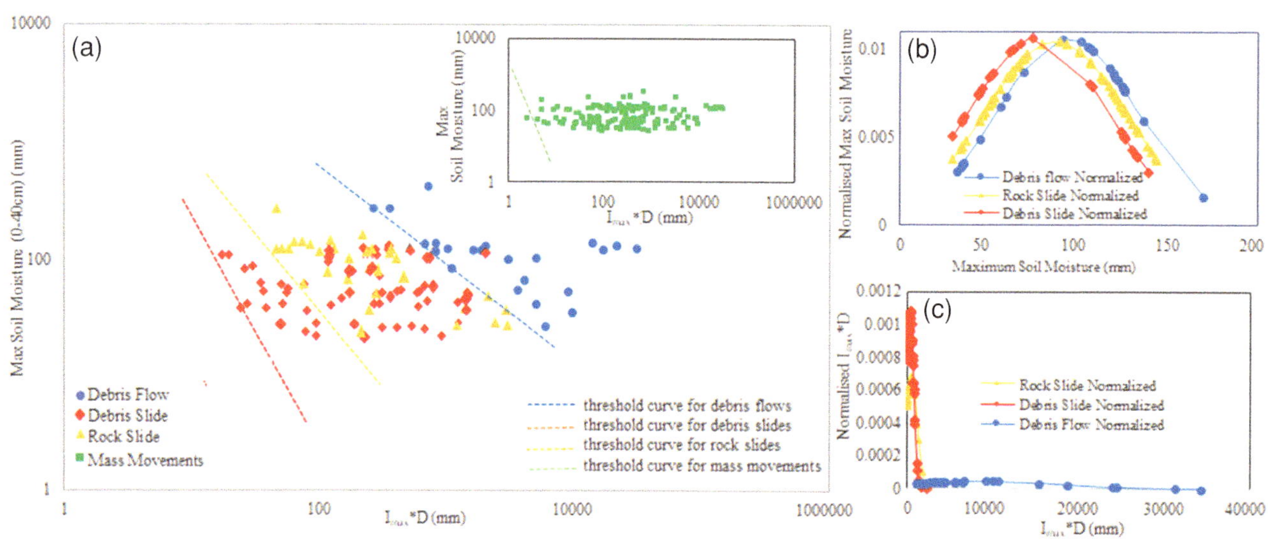

Fig. 4 **a** $I_{max}D$-SM_{max} thresholds for debris slide, rock slide, debris flow and mass movements, **b** normalized max soil moisture for debris slide, rock slide and debris flows and **c** normalized I_{max}*D for debris slide, rock slide and debris flow

Comparing $I_{max}D$ and SM_{max}-$I_{max}D$ Thresholds

The mass movement database as discussed above were dispersed in case of maximum intensity and total duration and therefore the dispersion was also high for I_{max}*D values. Whereas the observed dispersion was significantly less for max soil moisture values. From I_{max}-D thresholds it was observed that debris flows were recorded during highly intense rainfall conditions in comparison to rock slides and debris slides. The same was observed for SM_{max}-$I_{max}D$ thresholds. In addition, SM_{max}-$I_{max}D$ thresholds also indicated higher values of max soil moisture for debris flows compared to rock slides and debris slides. This indicates that debris flows are caused by highly intense rainfall events which can lead to significant increase in soil moisture. The max soil moisture thresholds also indicated that slope failure only takes place when a threshold value of ~ 30 mm is exceeded. While the rainfall thresholds suggested that similar rainfall intensity can trigger all debris slides, rock slides and debris flows, given rainfall duration is sufficiently long and landslides may completely or partially mobilize to form debris flows. On comparing the scatter plots it was observed that unlike the rainfall threshold that were more distinguishable in terms of magnitude of their slopes, the max soil moisture thresholds were distinguished by their position in the SM_{max}-$I_{max}D$ scatter plots with debris flow present to the left indicating higher $I_{max}D$ values than landslides. The overall accuracy for both $I_{max}D$ and SM_{max}-$I_{max}D$ thresholds was high for predicting slope failures in the study area.

Conclusion

The study established empirical rainfall I_{max}-D and max soil moisture thresholds for a part of NH-58 highway in the district of Chamoli, Uttarkhand, India. The derived thresholds performed well in terms of predicting mass movements along the NH-58. However, the I_{max}-D thresholds have slightly higher prediction accuracy than max soil moisture thresholds. From both the thresholds it was concluded that debris flows are triggered by highly intense rainfall events in comparison to rock slides and debris slides. Also, for a short duration rainfall event a high intensity of rainfall is required to cause debris flow while for long duration events both small and large intensities can trigger debris flow. Moreover, for a long duration rainfall event (>24 h), the same rainfall intensity can trigger both landslide and debris flow. The study also determined that large increments in the value of the product of max intensity to total duration leads to small increments in the maximum measured soil moisture and any increment in the max soil moisture value above ~ 30 mm is likely to cause a failure. Also, both landslides and debris flows can occur for a soil moisture value greater than 47 mm, for long duration rainfall event. Thus these thresholds provide a lower cut off of values below which there is lower probability of occurrence of a mass movement. Although rainfall is the main triggering factor for slope failure in the region a heavy or long duration rainfall event may not always cause a failure. Hence the use of soil moisture as another variable coupled with rainfall intensity and duration gives a more reliable threshold curve which incorporates in itself soil properties, terrain properties, geology and hydro-meteorology to an extent. Thus the soil moisture threshold reduces the probability of false alarms that may result in case of rainfall thresholds and improves the reliability of threshold to forewarn slope failures. Therefore together these thresholds can be integrated as the bases of a multilevel warning system for mass movements.

References

Aleotti P (2004) A warning system for rainfall-induced shallow failures. Eng Geol 73:247–265. https://doi.org/10.1016/j.enggeo.2004.01.007

Aleotti P, Chowdhury R (1999) Landslide hazard assessment: summary review and new perspectives. Bull Eng Geol Env 58:21–44. https://doi.org/10.1007/s100640050066

Chien-Yuan C, Tien-Chien C, Fan-Chieh Y, Wen-Hui Y, Chun-Chieh T (2005) Rainfall duration and debris-flow initiated studies for real-time monitoring. Environ Geol 47:715–724. https://doi.org/10.1007/s00254-004-1203-0

Crozier MJ (1999) Prediction of rainfall triggered landslides: a test of the antecedent water status model. Earth Surf Proc Land 24:825–833

Guzzetti F, Peruccacci S, Rossi M, Stark C (2008) The rainfall intensity–duration control of shallow landslides and debris flows: an update. Landslides 5:3–17. https://doi.org/10.1007/s10346-007-0112-1

Iverson RM, Reid ME, LaHusen RG (1997) Debris-flow mobilization from landslides. Annu Rev Earth Pl Sc 25:85–138

Irawan AM, Virgianto RH, Safril A, Gustono ST, Putranto ND (2019) Rainfall thresholds and soil moisture indexes for initiation of landslide in Banjarmangu, sub-district, Central Java, Indonesia. In: IOP conference series: earth and environmental science, vol 243, n. 1, p 012028.

Krøgli IK, Devoli G, Colleuille H, Boje S, Sund M, Engen IK (2018) The Norwegian forecasting and warning service for rainfall- and snowmelt-induced landslides. Nat Hazards Earth Syst Sci 18:1427–1450. https://doi.org/10.5194/nhess-18-1427-2018

Kuthari S (2007) Establishing precipitation thresholds for landslide initiation along with slope characterisation using GIS-based modeling. MSc dissertation, ITC, Enschede, the Netherlands

Mathew J, Babu G, Kundu S, Kumar VK, Pant CC (2013) Integrating intensity-duration based rainfall threshold and antecedent rainfall–based probability estimate towards generating early warning for rainfall-induced landslides in parts of Garhwal Himalaya. Springer Verlag Berlin Heidelberg, India

Posner AJ, Georgakakos KP (2015) Soil moisture and precipitation thresholds for real-time landslide prediction in El Salvador. Landslides 2015(12):1179

Ray R, Jacobs J (2006) Relationships among remotely sensed soil moisture, precipitation and landslide events. Nat Hazards 43:211–222

Segoni S, Piciullo L, Gaiano SL (2018a) A review of recent literature on rainfall thresholds for landslide occurrence. Landslides 1–9

Segoni S, Rosi A, Lagomarsino D, Franti R, Casagli N (2018b) Brief communication: using averaged soil moisture estimates to improve the performances of regional-scale landslide early warning system. Nat Hazards Earth Syst Sci 18(3):807–812

Validiya KS (2010) The Making of India Geodynamic Evolution. Macmillan Publishers India Ltd., Delhi. (ISBN 10: 0230-32833-4)

Varnes DJ (1978) Slope movement types and processes. In: Schuster RL, Krizek RJ (eds) Landslides, analysis and control, special report 176: transportational research board. National Academy of Sciences, Washington, DC., pp 11–33

Wieczorek G, Glade T (2005) Climatic factors influencing occurrence of debris flows. In: Jakob M, Hunger O (eds) Debris-flow hazards and related phenomena. Springer, Berlin, pp 325–362

2020 Kyoto Japan

The Efficient Early Warning with South East-Asia Oceania Flash Flood Guidance System (SAOFFGS)

Agie Wandala Putra, Nn. Ummul Choir Os, and Imaduddin Salma Faalih

Abstract

The Southeastern Asia-Oceania Flash Flood Guidance (SAOFFG) system is a part of Global Flash Flood Guidance System (FFGS). Acting as regional centre of Southeastern Asia-Oceania Flash Flood Guidance, Indonesia has the responsibility to provide regional and national verification of SAOFFGS flash flood forecasts and warnings. This paper was conducted in order to explain the implementation of SAOFFG in Indonesia and to evaluate one of the FFGS threat products used by BMKG for monitoring and forecasting floods. The selected study time period and area are based on the availability of flash flood threat forecast data i.e. IFFT 1-, 3- and 6-h. Flood event data was obtained from flood report database of BMKG Weather Early-warning subdivision. In order to acquire the statistical results, contingency tables were constructed. Results indicate that approximately one-half of the flood events were correctly detected by positive values of IFFT (POD 0.66 for 6-h IFFT). The best skills, as indicated with a CSI 0.34 occur in the verification of FFG threat product (IFFT) 1-h and 3-h.

Keywords

Flash flood guidance • Flash flood forecast • Early warning • Evaluation • Indonesia

A. W. Putra (✉) · Nn. U. C. Os · I. S. Faalih
Agency for Meteorology, Climatology, and Geophysics, Centre for Public Meteorology, Jl. Angkasa I No.2 Kemayoran, Jakarta, 10720, Indonesia
e-mail: agie.wandala@bmkg.go.id

Nn. U. C. Os
e-mail: ummul.choir@bmkg.go.id

I. S. Faalih
e-mail: imaduddin.faalih@bmkg.go.id

Introduction

The Southeastern Asia-Oceania Flash Flood Guidance (SAOFFG) system is a part of Global Flash Flood Guidance System (FFGS). In the Southeastern Asia-Oceania, flash floods play a role in causing significants number of casualties, material loss and infrastructure damage and require special attention. Referring to Indonesia National Agency for Disaster Management (BNPB) information, 385 flooding events occurred in 2019 resulting in the fatalities of 296, 2853 houses devastated, and 257 of damaged educational facilities. Consequently, relevant authorities need to improve the early warning system as a part of efforts to reduce vulnerability of regions to hydrometeorological hazards.

The planned establishment of SAOFFG regional centre has been started since February 2016 in the initial planning meeting held in Indonesia. SAOFFG is the result of collaboration between World Meteorological Organization (WMO), Hydrologic Research Centre (HRC), National Oceanic and Atmospheric Administration (NOAA), United States Agency for International Development (USAID), and Agency for Meteorological, Climatological and Geophysics (BMKG) aimed at enhancing National Meteorological Hydrological Services (NMHSs) capacities to issue timely and accurate flash flood warnings.

This paper was conducted in order to explain the implementation of SAOFFG in Indonesia and to evaluate one of the FFGS threat products used by BMKG for monitoring and forecasting floods even flash floods. Data sources used for product evaluation include a database of flood or flash flood reports collected by BMKG Weather Early-warning subdivision.

© Springer Nature Switzerland AG 2021

N. Casagli et al. (eds.), *Understanding and Reducing Landslide Disaster Risk*,
ICL Contribution to Landslide Disaster Risk Reduction,
https://doi.org/10.1007/978-3-030-60311-3_28

SAOFFG Operational

Mandate and Roles of BMKG

Member States of SAOFFGS are Indonesia which is acting as regional centre (RC), Malaysia, Brunei Darussalam, Timor Leste, Philippines and Papua New Guinea. According to WMO regulations, the RC has the responsibility to assist with tasks during the regional FFGS development and implementation phases, including: being focal point for the collection of the required historical hydro-meteorological and spatial data from the participating countries establishing data transfer tools via secure ftp, hosting high performance servers to run SAOFFG model and allow access to participating countries for SAOFFGS products, and to receive real-time data from various sources, and maintaining servers, operating systems and FFGS application software with the support of HRC (Fig. 1). In addition, the RC is asked to provide regional and national verification of SAOFFGS flash flood forecasts and warnings, and some others unstated responsibilities in this paper.

Operational Status

Implementation of SAOFFG at BMKG is divided into 3 stages i.e. the trial period beginning from May to October 2019, the operational period at National level in November 2019, and Operational period in Regional SAO level in March 2020. Daily analysis of SAOFFG report in BMKG is issued twice a day respectively at 00 and 12 UTC.

Flashflood Guidance System

Meteorological Organization (WMO)/UNESCO in International Glossary of Hydrology (WMO, N. 385 2012) define flood as follows: Rise, usually brief, in the water level of a stream or water body to a peak from which the water level recedes at a slower rate; Relatively high flow as measured by stage height or discharge. A flash flood is a rapid flooding of water over land caused by heavy rain or a sudden release of impounded water (e.g., dam or levee break) in a short period of time, generally within minutes up to several hours, a time scale that distinguishes it from fluvial floods (Hong et al.

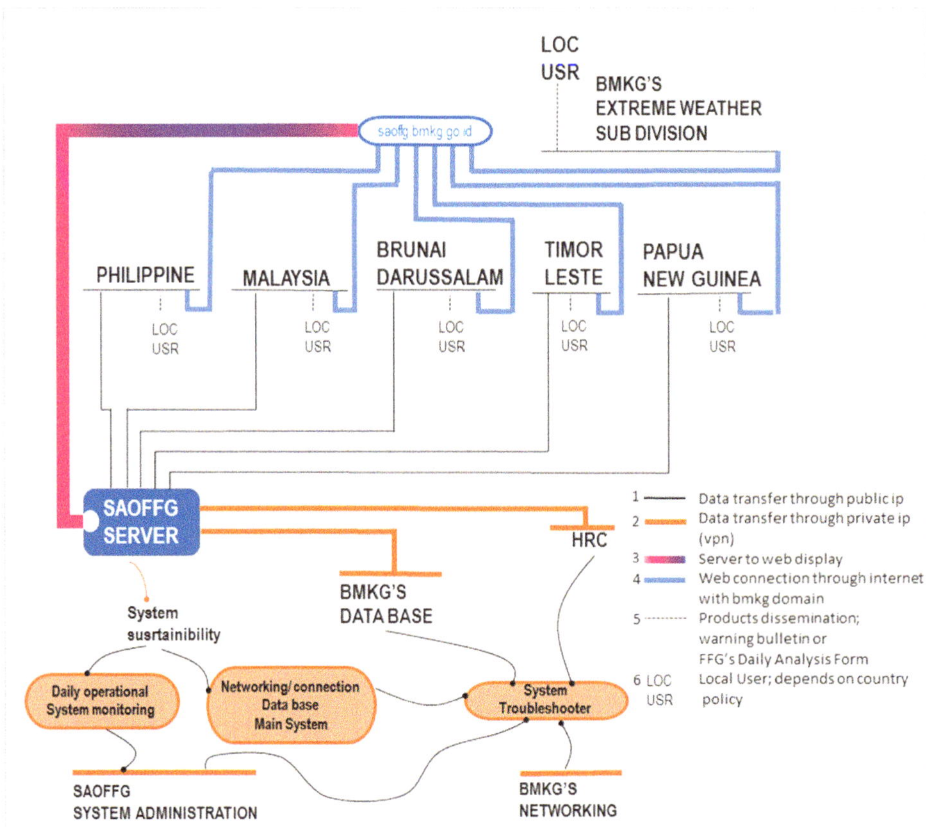

Fig. 1 SAOFFG system flow

2013). In Indonesia, floods commonly occur by cause of the rain, overflowing water in rivers, lack of water absorption and high rainfall in the upstream areas.

Flash floods need to be treated as a hydrometeorological event requiring an integration of meteorology and hydrology in real time with an infusion of local information and expertise to deliver reliable flashflood warnings. The FFGS is designed in order to facilitate this requirement. The system products are made available to forecasters as a diagnostic tool to analyse weather-related events that can initiate flash floods and then to make a rapid evaluation for a flash flood occurence at a location.

Analysis and flood forecast utilizing SAOFFG are carried out in accordance with the workflow in Fig. 2. The preliminary evaluation is conducted by condering the analysis and forecast of rainfall, soil mosture above 50%, low FFG values, and areas of concern. Afterwards, referring to the meteorological and hydrological conditions of the area of concern is necessary. To assess the threat of a local flash flood, the FFGS is designed to allow product adjustments based on the forecaster's experience with local conditions, incorporation of other information and last minute local observations, or local observer report. The FFG is normally considered as only as guidance i.e. if the forecaster believes that a warning should be issued before the rainfall rate exceeds the FFG value a warning should be issued. In some parts of the country, quantitative precipitation estimation (QPE) exceeding FFG is used as a strict threshold for issuing a warning. In other areas, forecasters may wait for QPE to reach 125% or 150% of FFG before doing so (Hong and Gourley 2015). To conduct a flood early warning, there are several steps to effectively achieve the results. These stages (Werner et al. 2005) are detection, forecasting, warning and dissemination, and response. BMKG plays a predominant role for forecasting rainfall in forecasting stage. Utilization of SAOFFG providing hydrological data enables forecaster to prepare a flood forecast by previously discussing with several relevant agencies. Thus flood warnings could be issued and be disseminated to users which allowing action or response could be taken immediately.

In this study, an evaluation of one flash flood threat products was investigated. FFG refers generally to the volume of rain of a given duration necessary to cause minor flooding on small streams (Carpenter et al. 1999). FFG value indicating the total volume of rainfall over the given duration which is adequate to cause bankfull flow at the outlet of the draining stream. A flash flood threat index is the difference between the actual, persisted or forecast precipitation and the corresponding FFG diagnostic value for the basin and duration of interest (1, 3 and 6 h). Threat products generated in FFG are Imminent Flash Flood Threat (IFFT), Persistence Flash Flood Threat (PFFT), and Forecast Flash Flood Threat (FFFT).

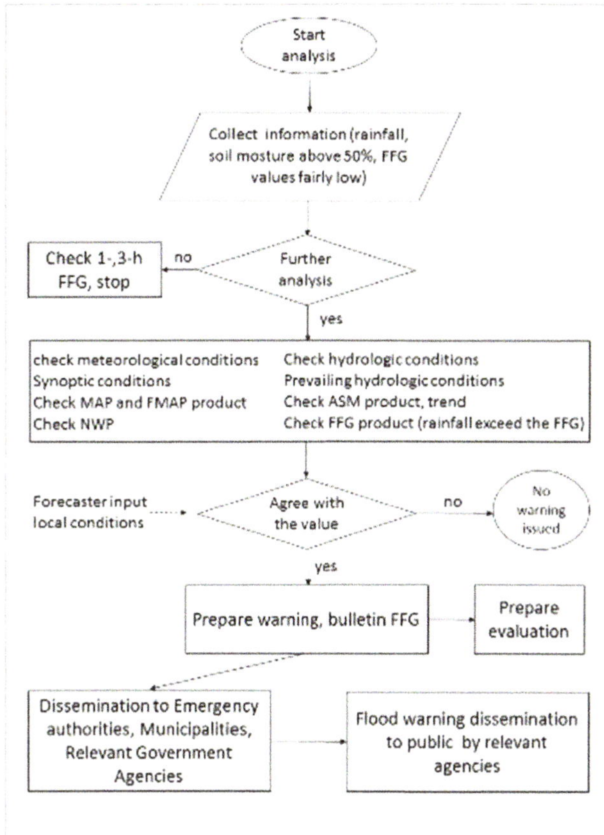

Fig. 2 Workflow of SAOFFG daily guidance at BMKG

Imminent Flash Flood Threat (IFFT)

The imminent flash flood threat index is a diagnostic index, which only includes uncertainties from estimated precipitation and land-surface model parameters (Georgakakos et al. 2019). Therefore, the uncertainty in these IFFT indices depends on the available data. IFFT products are available in image and text formats for 1, 3, and 6 h (mm) for respective basin. The IFFT value indicates the difference between Merged Mean Areal Precipitation (MAP) in the given duration and corresponds to the previous FFG model data with the same duration in each sub-basin. From this perspective, IFFT products can be considered as the current status of observation. Previous FFG products in periods 1, 3, and 6 h were carried out in conjunction with the Merged MAP in the calculation of IFFT.

a. IFFT 01-h: difference from 01-h FFG from the previous navigation hour model calculation and 01-h Merged MAP observed over following 1 h (mm/1 h).
b. IFFT 03-h: difference from 03-h FFG from previous navigation hour model calculations and 03-h Merged MAP observed over following 3 h (mm/3 h).

c. IFFT 06-h: difference from 06-h FFG from previous navigation hour model calculations and 06-h Merged MAP observed over following 6 h (mm/6 h).

Data and Methodology

The trial period of SAOFFG in BMKG has several constraints related to input data used in the running process. Numerical Weather Prediction (NWP) model resolution needs to be improved to be able to capture the meteorological events in local scale because this scale notably affects weather conditions in Indonesia region. Furthermore, the insufficient number of meteorological stations in Indonesia results in the less proportional observation data.

Considering the availability of SAOFFG data in the trial period at BMKG, this study used data from January to February 2020. Most regions in Indonesia, especially Java, experience rainy seasons in January and February. Asian cold monsoon transfers a mass of moisture air into Indonesia and the peak season of the tropical cyclone in South Indian Ocean (SIO) is within these months. The occurrence of tropical cyclones in the SIO is one of the factors that influences the pattern of weather formation in Indonesia. Heavy to extreme rain occur in these months which might induce floods.

Figure 3 shows the monthly rainfall amount in the time periode of study. The amount of January 2020 monthly rainfall is in moderate to very high category, especially in Jakarta and Central Java. The next month, February 2020, increasing amount of monthly rainfall was indicated almost uniformly in the Java region with the category of high and very high monthly rainfall. Jakarta as the Indonesia's capital experienced flooding exceeding of five times within January–February 2020 period.

The selected study time period and area area based on the availability of flash threat food forecast data i.e. IFFT 1-, 3- and 6-h. Flood event data was obtained from flood report database of BMKG Weather Early-warning subdivision which routinely collects report on the basis of information sources from the BMKG stations and mass media. The flood data is in the form of location and time data including the hour of the incident.

In order to acquire the statistical results, contingency tables were constructed (Table 1). To verify the 1-, 3-, and 6-h-duration IFFT, it was assumed that the forecast was "yes" when the threat index was greater than zero. This verification process is carried out in 2 schemes, first calculating hits and misses obtained from the determination of flood event data in the study area. Based on the verification guidelines for FFGS product arranged by HRC, IFFT data used is only at or near the FFGS delineated basin outlets. In this scheme, IFFT indexes were matched with flood data to get hits and missed. In the next scheme, false alarms are calculated by collecting positive IFFT data in study period, subsequently eliminating data that have the similar time to flood events in the first scheme.

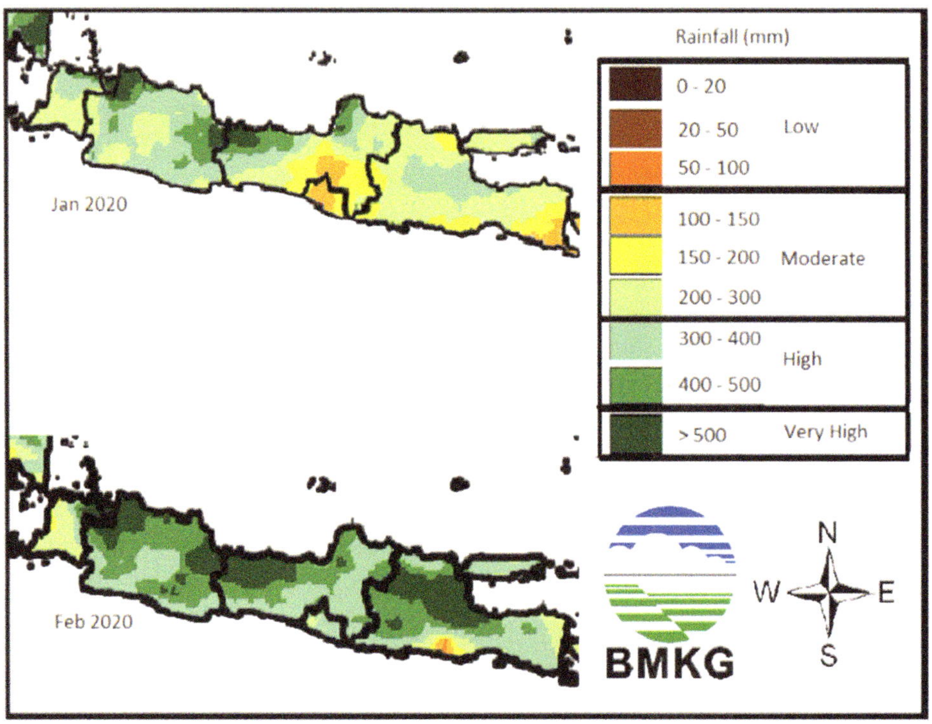

Fig. 3 Monthly rainfall analysis in study domain (Java) in Januari and Februari 2020 (BMKG)

Table 1 A 2 × 2 contingency table

		Observation (event occured)	
		Yes	No
Forecasts	Yes	Hits	False alarms
	No	Misses	Correct negative

Table 2 Statistical Indices for IFFT

IFFT	POD	FAR	CSI
1 h	0.36	1	0.34
3 h	0.53	0.62	0.34
6 h	0.66	0.55	0.28

The scores provide the most meaningful information if they are computed from large enough samples of cases. However, severe weather occurrences are rare events, thus the number of forecasts and observations of severe weather may be small, which makes the task of verification not only more important but also more challenging (WMO-No. 1132, 2014). There were 47 flood events being verified throughout Java during the study period. In the second scheme, the basin involved for the data retrieval is adjusted to the basin monitored in the first scheme.

After processing table contingency respectively, hits, misses, and false alarm were included in statistical computation presenting:

$$PoD = \frac{Hits}{Hits + Misses} \quad (1)$$

The Probability of Detection (POD) describes the fraction of the observed flood events detected correctly by the IFFT forecast. The POD ranges from 0 indicates no skill to 1 indicates a perfect scores.

$$FAR = \frac{FalseAlarms}{Hits + FalseAlarms} \quad (2)$$

The False Alarm Ratio (FAR) corresponds to the fraction of flood events forecasted by IFFT but not matched with flood event observation. The FAR ranges from 0 to 1 indicates a perfect score.

$$CSI = \frac{Hits}{Hits + Misses + FalseAlarms} \quad (3)$$

Critical Success Index (CSI) describes skill of the IFFT to flood events observed. The CSI ranges from 0 to 1 indicates a perfect skill.

Results

Table 2 presents the results of statistical computation for IFFT. POD ranges from 0.36 with 1-h IFFT to 0.66 for 6-h IFFT, which indicates that approximately one-half of the flood events were correctly detected by positive values of IFFT. Referring to high FAR values which range from 0.55 to 1, there is an indication of over-forecasting, absence of flood event data as not reported or recorded in the database of BMKG Weather Early-warning subdivision, or combination of both. A high FAR value causes a low CSI value. The highest CSI values of 0.34 are associated with the 1-h and 3-h IFFT. The lowest CSI value of 0.28 for 6-h IFFT can be caused by false alarm values which is higher than those for 1-h and 3-h IFFT.

February 2020 Jakarta Flood

On 25 February 2020, Jakarta witnessed another flood worse than the January 2020 flood. Flood in urban areas can occur for two reasons; first, urban areas are flooded due to overflowing rivers crossing the city. Second, urban flooding can occur as a special case of flash floods caused by the inability of drainage to accommodate and drain rainfall.

Referring to the FFG products, the soil moisture fraction of the upper soil was higher than 0.50 from 16 to 24 February (Fig. 4). This condition had to be monitored for possible flash flood occurrence especially if the high rainfall was possible to occur after 24 February. IFFT represents an "nowcast" weather situation or indicates that a flash flood is occuring now or is about to occur immediately. IFFT 6-h at 24 February 2020 18 UTC and IFFT 3-h at 21 UTC were higher than 10 indicating the flash flood occurence is most likely (Fig. 5). Moreover, the other FFG products and condition of Jakarta as an urban area were also taken into consideration in decision making of flood warning. Based on report collected by BMKG Weather Early-warning subdivision, the flood occured starting around 25 February 2020 06 Local Time (24 February 2020 23 UTC).

This study conducts only one method of evaluating the flash flood threat product used in operational for monitoring and predicting floods in Indonesia. Derived results show the best skills, as indicated with a CSI 0.34 occurring in the verification of flash flood threat product (IFFT) 1-h and 3-h. FFG threat product is applicable in BMKG to be used by forecasters as an indicator of flood warning. Definitely forecasters need to have other consideration before issuing flood warning which is indicated from the FFG product. Disaster mitigation efforts require coordination and training from various stakeholders, such as BNPB, experts from universities or research institutions for hydrometeorological analysis and prediction, as well as relevant sector agencies for law enforcement according to spatial planning.

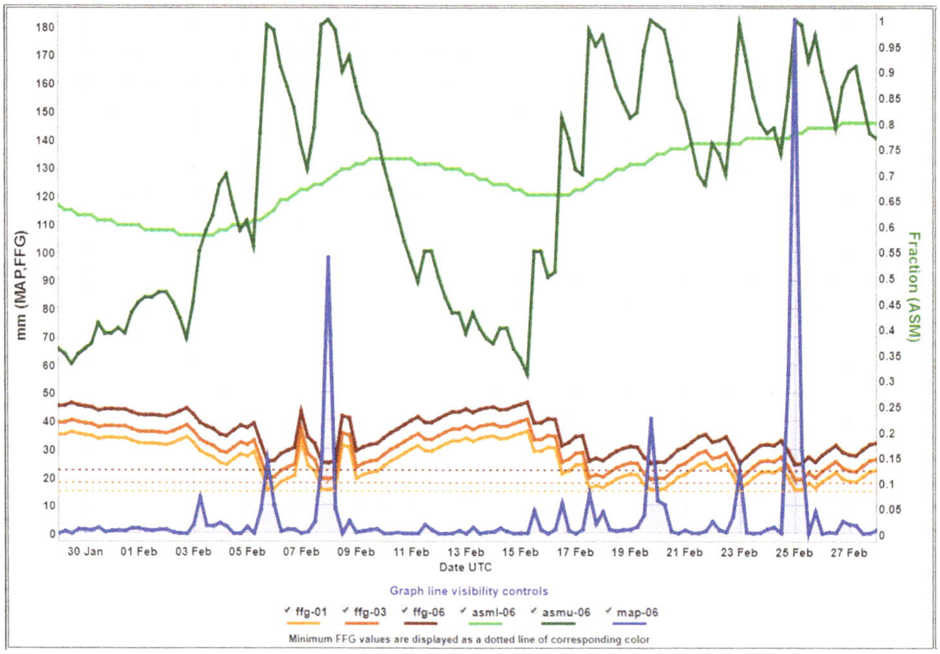

Fig. 4 FFG values indicates February 2020 flood events in Jakarta

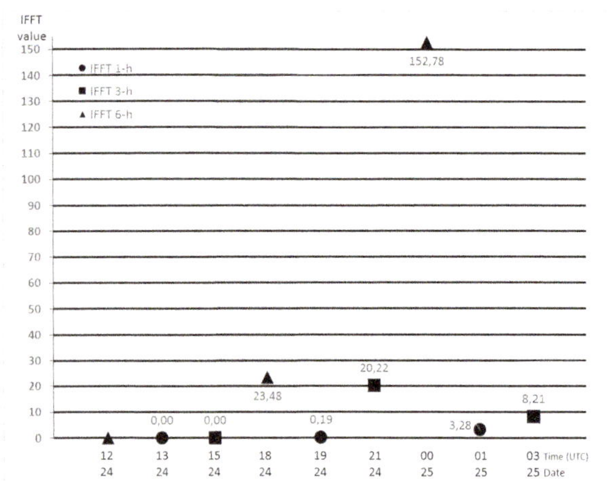

Fig. 5 IFFT values in basin 2042100129 from 24 February 2020 12 UTC to 25 February 2020 03 UTC

Communicating with user agencies is necessary for effective disaster risk reduction.

Acknowledgements This paper was supported by BMKG. We thank our General Director of BMKG who provided writers with opportunity and motivation to be involved in WLF forum.

References

Carpenter TM, Sperfslage JA, Georgakakos KP, Sweeney T, Fread DL (1999) National threshold runoff estimation utilizing GIS in support of operational flash flood warning systems. J Hydrol 224:21–44

Georgakakos KP, Theresa MMH, Shamir E, Cheng Z (2019) Verification guidelines for the flash flood guidance system component products and derivative warnings. Hydrologic Research Center San Diego USA, 120 p

Gourley JJ, Jessica ME, Yang H, Ernest BW (2012) Evaluation of tools used for monitoring and forecasting flash floods in the United States. Am Meteorol Soc J Wea Forecasting 27:158–173

Hong Y, Gourley JJ (2015) Radar hydrology: principles, models, and applications. 1st. CRC Press Taylor & Francis Group, Florida, USA, 176 p (ISBN-13:978–1–4665–1461–4)

Hong Y, Adhikari P, Gourley JJ, (2013) Flash flood. In: Bobrowsky PT (eds) Encyclopedia of natural hazards. Encyclopedia of Earth Sciences Series, Springer, Dordrecht.

Werner MGF, Schellekens J, Kwadijk JCJ (2005) Flood early warning systems for hydrological (sub) catchments. In: Anderson MG, McDonnell JJ (eds) Encyclopedia of hydrological sciences, vol 1. Wiley

WMO, UNESCO (2014) Forecast verification for the African severe weather forecasting demonstration projects. World meteorological organization Geneva Switzerland, 38 p (ISBN_978-92-63-11132-6)

WMO, UNESCO (2012) International glossary of hydrology. World meteorological organization Geneva Switzerland, 471 p (ISBN_978-92-63-03385-8)

2020 Kyoto Japan

Regional Approaches in Forecasting Rainfall-Induced Landslides

Maria Teresa Brunetti, Massimo Melillo, Stefano Luigi Gariano, Luca Ciabatta, Luca Brocca, and Silvia Peruccacci

Abstract

Hydrogeological hazards now exacerbated by the ongoing climate change pose serious challenges for the safety of the population worldwide. Among the others, the landslide risk can be mitigated by setting up efficient and reliable early warning systems. To date, rainfall thresholds are one of the most used tools to forecast the possible occurrence of rainfall-induced failures in large regions. In Italy a dense rain gauge network with hourly or sub-hourly temporal resolution is available. However, in some developing countries, where ground measurements are still absent or are available at coarser (daily) temporal resolution, satellite-based rainfall estimates could be a vital alternative. For this purpose, the reliability of rainfall thresholds defined using both satellite (SB) and ground-based (GB) data and with hourly or daily temporal resolution is assessed in a study area comprising the Abruzzo, Marche and Umbria regions (AMU), central Italy. The comparison between the performance of the different products allows to test their capability in eventually can GB rainfall measurements are gathered at hourly time steps (OBS-H) from a national network and aggregated on a daily scale (OBS-D); SB rainfall estimates are retrieved from the Climate Prediction Center Morphing Technique (CMORPH, hourly resolution), and from the SM2RASC product, based on the application of SM2RAIN algorithm to ASCAT (Advanced SCATterometer) soil moisture product (daily resolution). Results show that thresholds defined with GB rainfall data perform better than those obtained using SB estimates regardless of the temporal resolution. CMORPH and SM2RASC thresholds are still able to predict landslide occurrence although with a high number of false predictions.

Keywords

Landslides • Rainfall thresholds • Rain gauges • Satellites

Introduction

Everywhere in the world is by far evident that climate change has exacerbated hydrogeological hazards (Seneviratne et al. 2012). Since the early twenty-first century, the global temperature growth is considered to be related to ever-increasing rainfall intensities (e.g., Trenberth et al. 2003; Watterson and Dix 2003; Hegerl et al. 2004). Such peaks of precipitation are able to enhance the triggering of rapid and very rapid landslides (e.g. debris flows, soil slips) that usually do not leave people time to get rescued.

The global rise of temperature poses serious challenges in forecasting a likely rising occurrence of rainfall-induced landslides. Indeed, July 2019 was the month with the highest number of fatal landslides in the world, claiming the life of 358 people (Petley 2019). Alongside the need to review the land use policies, there is an increasing need for set up efficient and reliable landslide early warning systems. The challenge is global but the means to defend oneself are not always up to it everywhere. Most of the early warning

M. T. Brunetti (✉) · M. Melillo · S. L. Gariano · L. Ciabatta · L. Brocca · S. Peruccacci
Italian National Research Council, Research Institute for the Geo-Hydrological Protection, Via della Madonna Alta 126, 06128 Perugia, Italy
e-mail: maria.teresa.brunetti@irpi.cnr.it

M. Melillo
e-mail: massimo.melillo@irpi.cnr.it

S. L. Gariano
e-mail: stefano.luigi.gariano@irpi.cnr.it

L. Ciabatta
e-mail: luca.ciabatta@irpi.cnr.it

L. Brocca
e-mail: luca.brocca@irpi.cnr.it

S. Peruccacci
e-mail: silvia.peruccacci@irpi.cnr.it

© Springer Nature Switzerland AG 2021

N. Casagli et al. (eds.), *Understanding and Reducing Landslide Disaster Risk*, ICL Contribution to Landslide Disaster Risk Reduction, https://doi.org/10.1007/978-3-030-60311-3_29

systems for rainfall-induced landslides in the world are based on rainfall thresholds (Piciullo et al. 2018; Guzzetti et al. 2020), which are in turn mostly based on rain gauge measurements (Guzzetti et al. 2008; Segoni et al. 2018). Most of the areas which face with problems related to rainfall-triggered landslides do not have dense rain gauge networks or even they do not have any. In this case, the use of satellite-based (SB) rainfall data could be the only way to get prompt information on the (estimated) cumulated rainfall. It has been found that satellite rainfall products underestimate the precipitation responsible for landslides if compared to ground-based (GB) measurements (Rossi et al. 2017; Brunetti et al. 2018), thus resulting in lower cumulated event rainfall–rainfall duration (ED) threshold curves. However, such underestimation does not affect the product performance in terms of capability of detecting rainfall events resulting in landslides.

Another issue is that in many countries the temporal resolution of the rainfall measurements is daily thus hampering the up-to-date monitoring of the event. In this case, rainfall thresholds must be represented by equations where the duration D is in days (d) instead of hours (h), since the thresholds are valid only for D values multiple of days (Gariano et al. 2020).

The reliability of preventing landslide hazard using SB or GB data and with hourly or daily temporal resolution is assessed in a regional area of Central Italy (Fig. 1), which is used as a benchmark case study. The selected area includes three administrative regions, Abruzzo, Marche and Umbria (AMU). In this study area, ED rainfall thresholds are defined and validated both from GB and SB rainfall data at hourly and daily temporal resolutions.

Data and Methods

GB rainfall measurements are obtained at hourly time steps (OBS-H) from a rain gauge network managed by the Italian national Civil Protection Department. GB data are even aggregated on a daily scale (OBS-D) to simulate how performance changes when the landslide occurrence is predicted with lower temporal resolution data. SB rainfall estimates available in the area include the Climate Prediction Center Morphing Technique (CMORPH, Joyce et al. 2004),

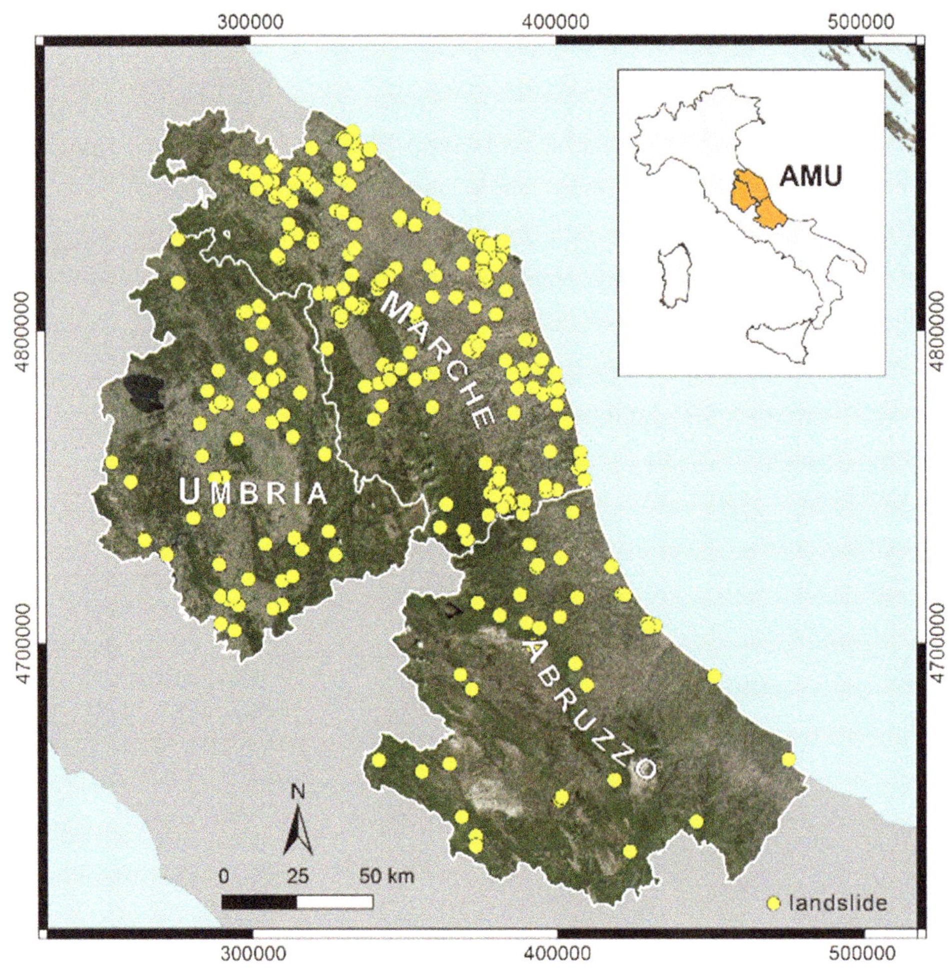

Fig. 1 Landslide distribution in AMU (Abruzzo-Marche-Umbria) study area

and the SM2RASC product that is based on the application of SM2RAIN algorithm (Brocca et al. 2014) to ASCAT (Advanced SCATterometer) soil moisture product (Wagner et al. 2013). Soil moisture data come from the Metop-A and -B satellite, and have a native spatial resolution of 25 km and a daily temporal resolution. The SM2RASC product specifically developed for Italy for the period 2008–2015 has been enhanced to 12.5 km after observation resampling. CMORPH rainfall estimates are obtained from the Climate Prediction Center of the National Oceanic and Atmospheric Administration (NOAA). Here the high-resolution product (8 km at the equator every 30 min) is used at hourly steps.

Most of the landslides used here (231) are from a published catalogue (Peruccacci et al. 2017) whereas 33 are found online for a total of 264 rainfall-induced slope failures in the 8-year period 2008–2015. The main sources of landslide information are digital and printed newspapers, blogs, technical documents, and landslide event reports (mostly coming from Fire Brigade archives). The location of the rainfall-induced landslides is shown in Fig. 1 (yellow dots).

The occurrence time of the failure is determined more or less accurately based on the information available (Peruccacci et al. 2017). For daily rainfall measurements (SM2RASC and OBS-D) the landslide is arbitrarily set at the end of the day.

Landslide information is combined with both GB and SB rainfall data using a tool that automatically reconstructs the (D, E) rainfall conditions responsible for the landslides and calculates ED rainfall thresholds (Melillo et al. 2018). The thresholds are calculated adopting a frequentist method proposed by Brunetti et al. (2010), and modified by Peruccacci et al. (2012). The threshold is a power law curve

$$E = (\alpha \pm \Delta\alpha) \times D^{(\gamma \pm \Delta\gamma)} \quad (1)$$

where E is the cumulated event rainfall (in mm), D is the duration of the rainfall event (in hours or days), α is a scaling parameter (the intercept), γ is the slope (the scaling exponent), and $\Delta\alpha$ and $\Delta\gamma$ are the uncertainties associated with α and γ, respectively. The method allows calculating thresholds at any non-exceedance probability (NEP). As an example, 5% ED thresholds are expected to leave 5% of (D, E) rainfall conditions below the threshold line.

The validation of the thresholds is used (i) to compare the performance of the GB and SB rainfall products, and (ii) to obtain the NEP value at which the threshold performs the best for each product. For the purpose, 80% of all the (D,E) rainfall conditions with landslides are selected randomly (100 times), and are used to define the curves at increasing NEPs (0.005, 0.5, 1, 1.5, 2, 3, 5, 10, 15, 20, 25, 30, 35, 40, 45, 50). Then, the remaining 20% is used to validate the thresholds using them as binary classifiers of rainfall events that triggered or did not trigger landslides. This allows building a contingency table where a (D, E) pair with landslide above the given threshold is a true positive (TP), and below is a false negative (FN). Analogously, a rainfall event without landslides above the threshold is a false positive (FP), and below is a true negative (TN). Finally, TPR (true positive rate or hit rate) and the FPR (false positive rate or false alarm rate) skill scores are calculated at the selected NEP values:

$$TPR = \frac{TP}{TP + FN} \quad (2)$$

$$FPR = \frac{FP}{FP + TN} \quad (3)$$

and are used to perform the receiver operating characteristic (ROC) analysis (Fawcett 2006) for the four data sets. The performance of the GB and SB rainfall products is then assesses based on the minimum distance δ of their ROC curve from the perfect classification point PCP (Gariano et al. 2015).

Results

In the following, the capability of the SB rainfall products at hourly and daily temporal resolution to forecast rainfall-induced landslides using the GB rainfall product as a reference is assessed. Based on the methods and approaches described above, the rainfall thresholds for the hourly-resolution (CMORPH and OBS-H), and the daily-resolution products (SM2RASC and OBS-D) are calculated and validated.

Rainfall Thresholds

Figure 2 portrays the thresholds at 5% NEP for the OBS-H ($T_{5,OBS\text{-}H}$) and CMORPH ($T_{5,CMORPH}$) data sets, which have both a temporal resolution of one hour. Note that the adopted tool does not reconstruct the rainfall conditions for all the 264 landslides as in some cases the rain is absent or it is negligible. This happens more frequently when using SB estimates. The two curves are somewhat parallel, but the threshold resulting from the CMORPH estimates is lower than that calculated with rain gauge hourly measurements (OBS-H). The threshold equations at 5% NEP are listed in Table 1.

Figure 3 shows the thresholds at 5% NEP for the OBS-D and SM2RASC data sets, which have both a temporal resolution of one day. As for the case of hourly data sets, the threshold defined using SB rainfall estimates ($T_{5,SM2RASC}$) is lower and steeper than the curve calculated with rain gauge measurements ($T_{5,OBS\text{-}D}$).

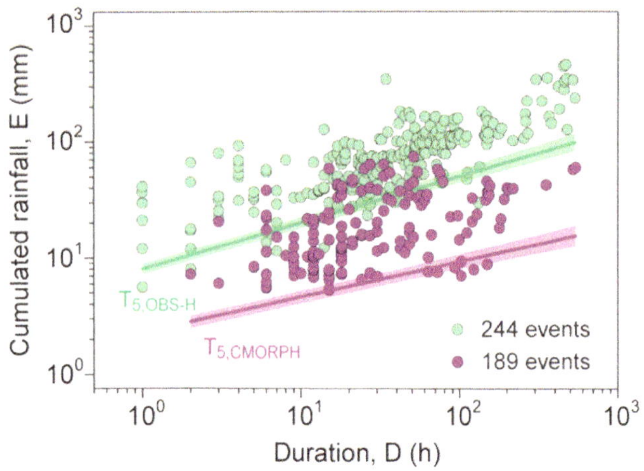

Fig. 2 Rainfall duration D (h) versus cumulated event rainfall E (mm) conditions that have resulted in landslides for the OBS-H (green) and CMORPH (violet) data sets and the corresponding ED thresholds at 5% NEPs with associated uncertainty (shaded areas)

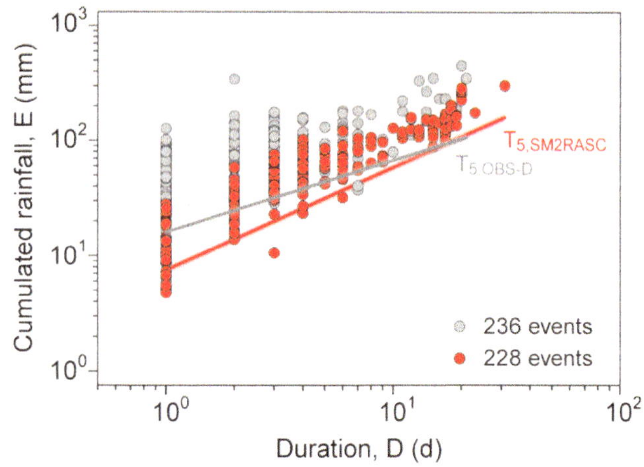

Fig. 3 Rainfall duration D (d) versus cumulated event rainfall E (mm) conditions that have resulted in landslides for the OBS-D (gray) and SM2RASC (red) data sets and the corresponding ED thresholds at 5% NEPs with associated uncertainty (shaded areas)

Validation of Thresholds

The validation procedure calculates the TPR and FPR skill scores (Eqs. 2 and 3) for each data set at the previously selected NEP values from 0.05 to 50%. The ROC curves for the four data sets are shown in Fig. 4. The OBS-H ROC curve exhibits the shortest minimum distance $\delta = 0.22$ from PCP, i.e. this data set performs the best (see the inset table in Fig. 4). The OBS-D rainfall product performs better than CMORPH and SM2RASC, the last being the worst in terms of distance from PCP. The NEP values of the thresholds that minimize δ for each data set vary from 15 to 35%.

Discussion and Conclusions

The aim of this study is assessing the reliability of rainfall thresholds obtained using different rainfall data types, as a function of their detection method (GB or SB) and their temporal resolution (hourly or daily).

Firstly, note that rainfall thresholds in Figs. 2 and 3 are calculated with hourly and daily temporal resolution, respectively, and cannot be directly compared being defined using different units of time. Nevertheless, it is worthwhile

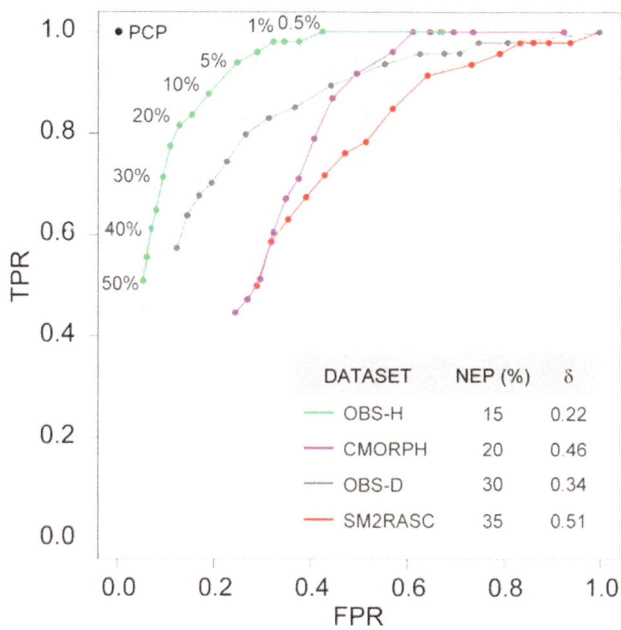

Fig. 4 ROC curves for OBS-H (green), CMORPH (magenta), OBS-D (gray), and SM2RASC (red) data sets. Horizontal and vertical bars represent variation ranges of TPR and FPR for the 100 runs in which the rainfall conditions are randomly selected. The minimum distance δ between PCP and the nearest NEP value for each curve are shown in the inset

Table 1 ED rainfall threshold equations at 5% NEP calculated with GB and SB data sets for hourly and daily temporal resolution

Data set	Rainfall threshold at 5% NEP	Best NEP %	Rainfall threshold at best NEP	Units of D	Units of E
OBS-H	$E = (7.5 \pm 0.9)\ D^{(0.41 \pm 0.03)}$	15	$E = (9.9 \pm 1.1)\ D^{(0.41 \pm 0.03)}$	h	mm
CMORPH	$E = (2.4 \pm 0.3)\ D^{(0.31 \pm 0.04)}$	20	$E = (3.8 \pm 0.5)\ D^{(0.31 \pm 0.04)}$	h	mm
OBS-D	$E = (15.4 \pm 1.9)\ D^{(0.64 \pm 0.05)}$	30	$E = (29.2 \pm 2.6)\ D^{(0.64 \pm 0.05)}$	d	mm
SM2RASC	$E = (7.3 \pm 0.6)\ D^{(0.90 \pm 0.03)}$	35	$E = (12.1 \pm 0.8)\ D^{(0.90 \pm 0.03)}$	d	mm

to compare the cumulated rainfall (E) at D = 24 h obtained from the four thresholds. The threshold value for E is 28.5 mm using $T_{5,OBS\text{-}H}$, whereas is only 16.1 mm at D = 1 d for $T_{5,OBS\text{-}D}$. The lower value obtained for the OBS-D data set is likely due to the lower accuracy of a coarser (daily) temporal resolution (also observed by Gariano et al. 2020). Conversely, for CMORPH and SM2RASC the mean cumulated rainfall at 24 h is E = 5.9 mm and 7.5 mm, respectively. the one between the two satellite products that provides a higher daily rainfall estimate is SM2RASC, despite the lower temporal resolution.

Figure 4 shows that thresholds defined with GB rainfall data (OBS-H and OBS-D) allow to better discriminate between events that have or have not triggered landslides. Indeed, for a hit rate TPR = 0.8 (80%), the number of expected false positive rate is 12% for OBS-H and 26% for OBS-D. The performance of CMORPH and SM2RASC at the same hit rate is even worse giving a probability of false alarms of 40% and 52%, respectively. For each data set, the minimum distance δ between PCP and the ROC curve identifies the NEP value (inset in Fig. 4), which optimizes the prediction performance.

Overall, the use of daily data has the global effect of degrading the reliability of the thresholds, both for GB and SB rainfall data.

The lower performance of SB with respect to GB rainfall products is easily ascribed to the high-quality of OBS-H and OBS-D data sets in AMU, which is based on ~150 rain gauges.

The use of SB rainfall products at different time resolutions shows that CMORPH performs better than SM2RASC at every NEPs. Generally, SB daily rainfall products (as SM2RASC) could be used being aware that high TPR values likely lead to have high FPR values, and therefore unwanted false alarms.

References

Brocca L, Ciabatta L, Massari C, Moramarco T, Hahn S, Hasenauer S, Kidd R, Dorigo W, Wagner W, Levizzani V (2014) Soil as a natural rain gauge: estimating global rainfall from satellite soil moisture data. J Geophys Res 119(9):5128–5141. https://doi.org/10.1002/2014JD021489

Brunetti MT, Peruccacci S, Rossi M, Luciani S, Valigi D, Guzzetti F (2010) Rainfall thresholds for the possible occurrence of landslides in Italy. Nat Hazards Earth Syst Sci 10:447–458. https://doi.org/10.5194/nhess-10-447-2010

Brunetti MT, Melillo M, Peruccacci S, Ciabatta L, Brocca L (2018) How far are we from the use of satellite rainfall products in landslide forecasting? Remote Sens Environ 210:65–75. https://doi.org/10.1016/j.rse.2018.03.016

Fawcett T (2006) An introduction to ROC analysis. Pattern Recogn Lett 27:861–874. https://doi.org/10.1016/j.patrec.2005.10.010

Gariano SL, Brunetti MT, Iovine G, Melillo M, Peruccacci S, Terranova O, Vennari C, Guzzetti F (2015) Calibration and validation of rainfall thresholds for shallow landslide forecasting in Sicily, southern Italy. Geomorphology 228:653–665. https://doi.org/10.1016/j.geomorph.2014.10.019

Gariano SL, Melillo M, Peruccacci S, Brunetti MT (2020) How much does the rainfall temporal resolution affect rainfall thresholds for landslide triggering? Nat Hazards 100:655–670. https://doi.org/10.1007/s11069-019-03830-x

Guzzetti F, Gariano SL, Peruccacci S, Brunetti MT, Marchesini I, Rossi M, Melillo M (2020) Geographical landslide early warning systems. Earth-Sci Rev 200:102973. https://doi.org/10.1016/j.earscirev.2019.10297

Guzzetti F, Peruccacci S, Rossi M, Stark CP (2008) The rainfall intensity–duration control of shallow landslides and debris flows: an update. Landslides 5(1):3–17. https://doi.org/10.1007/s10346-007-0112-1

Hegerl G, Zwiers F, Stott P, Kharin V (2004) Detectability of anthropogenic changes in annual temperature and precipitation extremes. J Clim 17:3683–3700. https://doi.org/10.1175/1520-0442(2004)017%3C3683:DOACIA%3E2.0.CO;2

Joyce RJ, Janowiak JE, Arkin PA, Xie P (2004) CMORPH: a method that produces global precipitation estimates from passive microwave and infrared data at high spatial and temporal resolution. J Hydrometeorol 5:487–503. https://doi.org/10.1175/1525-7541(2004)005%3C0487:CAMTPG%3E2.0.CO;2

Melillo M, Brunetti MT, Peruccacci S, Gariano SL, A Roccati, Guzzetti F (2018) A tool for the automatic calculation of rainfall thresholds for landslide occurrence. Environ Model Softw 105:230–243. https://doi.org/10.1016/j.envsoft.2018.03.024

Peruccacci S, Brunetti MT, Gariano SL, Melillo M, Rossi M, Guzzetti F (2017) Rainfall thresholds for possible landslide occurrence in Italy. Geomorphology 290:39–57. https://doi.org/10.1016/j.geomorph.2017.03.031

Peruccacci S, Brunetti MT, Luciani S, Vennari C, Guzzetti F (2012) Lithological and seasonal control on rainfall thresholds for the possible initiation of landslides in central Italy. Geomorphology 139:79–90. https://doi.org/10.1016/j.geomorph.2011.10.005

Petley D (2019) July 2019: a record-breaking landslide month for fatal landslides. https://blogs.agu.org/landslideblog/2019/08/06/july-2019-fatal-landslides/. Accessed 13 Dec 2019

Piciullo L, Calvello M, Cepeda JM (2018) Territorial early warning systems for rainfall-induced landslides. Earth-Sci Rev 179:228–247. https://doi.org/10.1016/j.earscirev.2018.02.013

Rossi M, Luciani S, Valigi D, Kirschbaum D, Brunetti MT, Peruccacci S, Guzzetti F (2017) Statistical approaches for the definition of landslide rainfall thresholds and their uncertainty using rain gauge and satellite data. Geomorphology 285:16–27. https://doi.org/10.1016/j.geomorph.2017.02.001

Segoni S, Piciullo L, Gariano SL (2018) A review of the recent literature on rainfall thresholds for landslide occurrence. Landslides 15:1483–1501. https://doi.org/10.1007/s10346-018-0966-4

Seneviratne SI, Nicholls N, Easterling D, Goodess CM, Kanae S, Kossin J, Luo Y, Marengo J, McInnes K, Rahimi M, Reichstein M, Sorteberg A, Vera C, Zhang X (2012) Changes in climate extremes and their impacts on the natural physical environment. In: Field CB, Barros V, Stocker TF, Qin D, Dokken DJ, Ebi KL, Mastrandrea MD, Mach KJ, Plattner G-K, Allen SK, Tignor M, Midgley PM (eds) Managing the risks of extreme events and disasters to advance climate change adaptation. A special report of working groups I and

II of the Intergovernmental Panel on Climate Change (IPCC). Cambridge University Press, Cambridge, UK, and New York, NY, USA. pp 109–230

Trenberth KE, Dai A, Rasmussen RM, Parsons DB (2003) The changing character of precipitation. Bull Am Meteorol Soc 84:1205–1217. https://doi.org/10.1175/BAMS-84-9-1205

Wagner W, Hahn S, Kidd R, Melzer T, Bartalis Z, Hasenauer S, Figa J, de Rosnay P, Jann A, Schneider S, Komma J, Kubu G, Brugger K, Aubrecht C, Zuger J, Gangkofner U, Kienberger S, Brocca L, Wang Y, Bloeschl G, Eitzinger J, Steinnocher K, Zeil P, Rubel F (2013) The ASCAT soil moisture product: a review of its specifications, validation results, and emerging applications. Meteorol Z 22(1):5–33. https://doi.org/10.1127/0941-2948/2013/0399

Watterson IG, Dix MR (2003) Simulated changes due to global warming in daily precipitation means and extremes and their interpretation using the gamma distribution. J Geophys Res 108:4379. https://doi.org/10.1029/2002JD002928

Seven Years of Landslide Forecasting in Norway—Strengths and Limitations

Graziella Devoli, Hervé Colleuille, Monica Sund, and Jaran Wasrud

Abstract

The experiences acquired by the Norwegian Landslide Forecasting and Warning Service during the first 7 years of operation are herein presented. We summarize the warnings sent in the period 2013–2019 and we present the evaluation of the warning performance and discuss some of the main strengths and limitations of the service. In our opinion, of imperative importance to the success is: A national political will, the assignation of the landslide service to an existing well consolidated flood warning service and a strong collaboration across public agencies and a multidisciplinary approach. The existence of a national landslide database and of an operational distributed hydrological model, was essential for the rapid establishment of relationships between landslides events and hydro-meteorological conditions. A strong development of IT-tools and expansion of the meteorological and hydrological network was also crucial. Yet there are still several challenges and limitations, such as an insufficient process-understanding of rainfall- and snowmelt-induced landslides. The verification of landslide occurrence is also a difficult and tedious task. Finally, another challenging task is the prediction of landslides triggered by local intense rainshowers during summer, and rapid snowmelt events during winter, due to the limitations that exist in the models and thresholds currently in use.

G. Devoli (✉) · H. Colleuille · M. Sund · J. Wasrud
Norwegian Water Resources and Energy Directorate, Middelthuns gate 29, Oslo, 0368, Norway
e-mail: gde@nve.no

H. Colleuille
e-mail: hec@nve.no

M. Sund
e-mail: mosu@nve.no

J. Wasrud
e-mail: jaw@nve.no

Keywords

Rainfall- and snowmelt-induced landslides • Forecasting and warning services • Early warning systems • Debris avalanches • Debris flows

Introduction

Early warning systems are useful mitigation options for the authorities in charge of risk management and governance. With warning messages, the authorities should invite people to implement emergency plans, take local actions and trigger contingency and emergency management in order to reduce risk of life and damages.

For an effective and successful early warning system, many efforts from the different sectors of the society are required at different steps. Politicians may have interest in establishing such of systems to prevent landslides. Researchers need to assess landslide hazard and risk, and design warning models. Forecasters have to run forecasting and warning services issuing messages, when the landslide danger increases. Finally, local authorities and population must take actions and implement emergency plans upon receipt of warning messages. Coordination and cooperation among the different sectors are essential.

The organization and the maintenance of a Landslide Early Warning Systems (LEWS) is complex and require many key components and steps, as recognized by other authors (UNISDR 2006; Di Biagio and Kjekstad 2007; Intrieri et al. 2013; Calvello 2017; Fathani et al 2016; Piciullo et al. 2018).

Two types of LEWS are found worldwide: the one that address the prevention of single landslides at slope scale, also called *local*, and the others that covers a large area predicting the occurrence of multiple landslide at regional scale, called *territorial/regional* (Piciullo et al. 2018; Pecoraro et al. 2018). The majority of the regional LEWS were established

© Springer Nature Switzerland AG 2021

N. Casagli et al. (eds.), *Understanding and Reducing Landslide Disaster Risk*, ICL Contribution to Landslide Disaster Risk Reduction,
https://doi.org/10.1007/978-3-030-60311-3_30

after 2005 and are managed often by governmental institutions. They cover regions in South-East Asia, USA, Europe and South America.

The Norwegian Landslide Forecasting and Warning Service (known as "Jordskredvarslingen" in Norwegian), has been operational since 2013 and is described in Krøgli et al. (2018) and Devoli et al. (2018). It is managed by a governmental institution, the Norwegian Water Resources and Energy Directorate (NVE), to forecast the level of danger of rainfall- and snowmelt-induced landslides, specifically shallow soil slides, debris avalanches, debris flows and slushflows (herein referred to as landslides). The service is operative 24/7 and covers the entire country. The daily management covers: the forecasters assessment of the danger level with a bulletin twice a day at www.varsom.no, improvements of organizational tasks, models, and public information. The main goal of the service is to issue correct warning levels at regional scale which trigger actions and implement emergency plans. Four awareness levels are used: green, yellow, orange and red (see Krøgli et al. 2018 for more details).

To maintain a system operational over a long time, periodic evaluations are mandatory to identify strengths or to detect problems in the system, and then propose changes and improvements (Segoni et al. 2018).

In this analysis, we aim to summarize the experience acquired between 2013 and 2019 and to evaluate some of the work done. We discuss how local and regional authorities react to warnings and how their response has changed through the years. Our experience can benefit the start-up of a similar service in other countries.

Warnings, Landslides and Warning Performance (2013–2019)

Norway is predominantly a mountainous country, with high relief and steep topography, product of repeated glaciations. Because of its elongated shape the country is exposed to a varied climate all year around. The complex geological conditions make the country also prone to different types of landslides (mainly rock falls, rock avalanches, rock slides, debris flows, debris slides, debris avalanches, clays slides and quick clay slides) but also slushflows and snow avalanches. The country is divided in 5 major physiographic regions: Northern Norway (divided in this work in two sectors Nordland and Troms/Finnmark), Central Norway (Trøndelag), Western Norway (Vestlandet), Southern Norway (Sørlandet) and South-Eastern Norway (Østlandet).

Warnings issued in the period 2013–2019

The number of days with landslide warnings varies from a minimum of about 25 days in 2016, to a maximum of 66 days, in 2013 (Fig. 1).

The figure shows all warnings sent for the different physiographic and climatic regions in Norway (from those with an annual rainfall amount of 200–300 mm to regions with rainfall amount of 3000–4000 mm). The figure do not differentiate the warnings based on the different triggering conditions. Some of these warnings were sent when high amount of rainfall episodes were forecasted, while other were sent because of snowmelt episodes or because of a combination of both. High soil moisture condition previous weather events was also an important variable in some of these cases. The fluctuations in number of warnings is due to daily fluctuation of both rainfall and snowmelt patterns in the different regions each year. In some years we sent clearly more warnings on springs as in 2013 and 2018 because these years had more snow than normal in South-East Norway and experienced fast snow melt.

On average, the yellow level is issued 40 days/year, orange 4 days/year, and red level 0,1 day/year.

Landslide warnings have been sent all year around in all regions. The average number of warnings per season is rather similar: 9 days, in summer, 10 in winter, 11 in autumn and 13 in spring (Fig. 2).

The number of days with landslides warnings during summer may be lower the last 4 years, since we have started to better differentiate the warnings sent for those days when convective clouds were expected. Convective clouds are responsible of short duration, and mostly intense rainfall, across localized areas, especially on summer (called herein heavy rainshowers). It is MET-Norway, in agreement with

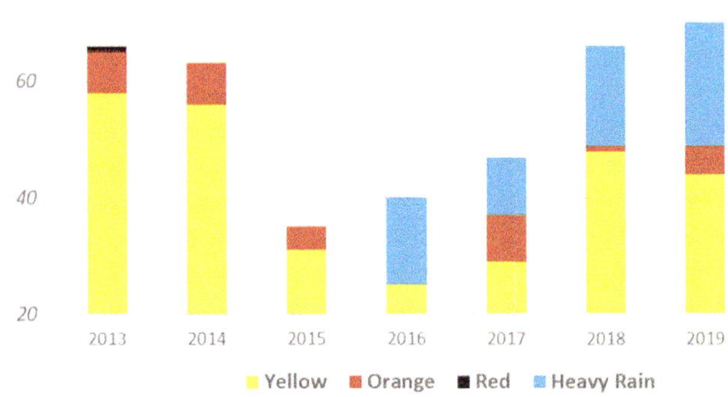

Fig. 1 Number of days with warnings issued in the period 2013–2019 with their respective warning levels, and number of days with warnings for heavy rain showers on summer (issued by MET-Norway)

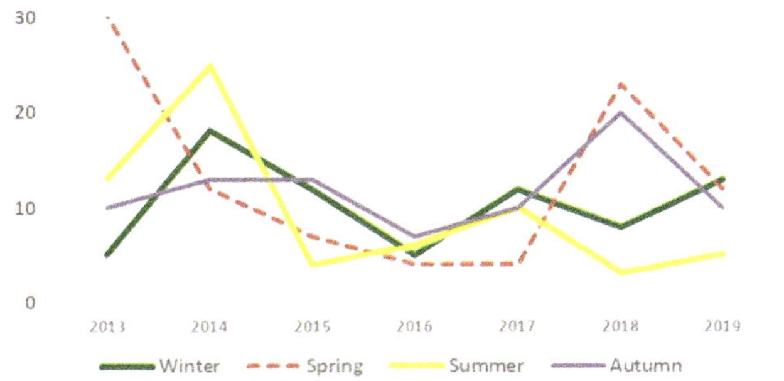

Fig. 2 Number of days with warnings by seasons in the period 2013–2019. The winter season is from December to February, week 49 to week 9, Spring from Mars to May, week 10 to 22, Summer from June to August, week 23–35, Autumn from September to November, week 36–48

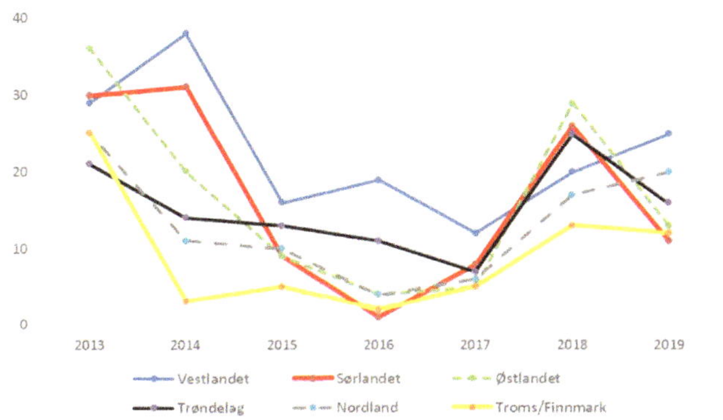

Fig. 3 Number of days with warnings divided by regions in the period 2013–2019

NVE, that issues warnings for heavy rainshowers. These warnings (indicated with a blue colour in Fig. 1) highlight the risk of surface runoff, storm water in urban areas, local flooding, local flash floods with erosional damage, debris flows and debris avalanches at locations impacted by the heavy showers.

Most of the warnings were issued in Western Norway (Vestlandet) (22 days in average) (Fig. 3), while the region that in average received the least landslide warnings (9 days) was Troms/Finnmark, part of the Northern Norway region.

Landslide events and warning performance

The performance of the service is evaluated by controlling the number, type, size and consequences of landslides, respect to the time/area and level of the warning and according to the definition of the awareness levels. The analysis is done based on a dataset of landslide events verified by the forecasters in the aftermaths of a weather event. The verified dataset is composed by 1052 landslides in the period from 2013 to 2019 (Fig. 4).

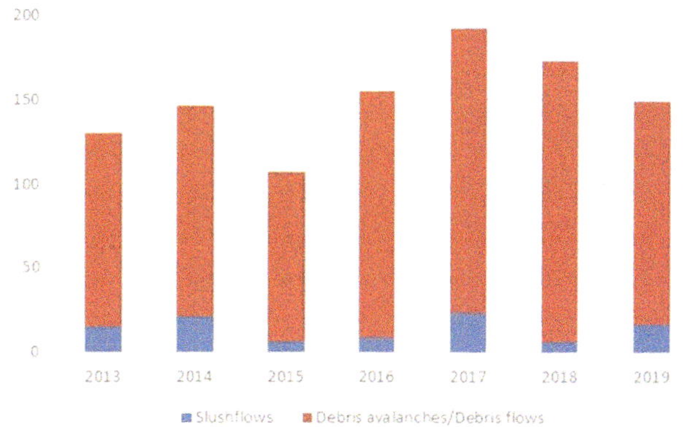

Fig. 4 Yearly distribution of landslides, from the verified dataset that forms the basis for the evaluation of the warning performance

Respect to the previous analysis presented in Krøgli et al. 2018, the results presented herein are updated with the last 2 years of observations. The performance at regional scale is about 98%. Table 1 shows the warning performance in the analyzed period. The column "correct warning" includes both true negatives (days with green level and no landslides) and true positives (days with yellow, orange or red level, and with a certain number of landslides expected for that level). The numbers of landslides expected for each level are presented in Piciullo et al. (2017). The true positives are more difficult to verify, because of the difficulty to verify the real number of landslides. In this analysis we consider only the landslide events verified by the forecasters in the aftermaths of a weather event to define the true positives. We observe also that the number of false alarms is clearly reduced from 2013 to 2015 (Table 1) due to an adjustment of the threshold in Southern and South-Eastern Norway (Krøgli et al. 2018).

User's response to the warnings

An important user of our warning is the NVE staff working in the five regional offices (Northern, Central, Western, Southern and Eastern). NVE is not a primary emergency agency, but NVE regional staff is often called upon by municipalities (and the police) for advice during emergencies and crises and in the aftermaths of landslides and floods affecting settlements. If a landslide warning level is given (yellow, orange or red), NVE declares a kind of internal emergency response divided in two levels: "Emergency" and "High emergency". "Emergency" is declared on e.g. a major single landslide involving people and/or in case of landslide orange or red warning. "High emergency" is declared if round-the-clock effort is needed from NVE regional staff, due to e.g. large geographical extent of the hazardous event (many landslides widespread in the region) or because a

Table 1 Warning performance in percentage (%) for the period 2013–2019. "Correct Warning" (C), "False Alarm" (FA), "Missing Event" (ME), "Wrong level (between yellow and orange)" (WL)

Year	C	FA	ME	WL
2019	97.7	1.0	0.8	0.3
2018	98.1	0.5	0.5	0.5
2017	96.3	1.9	1.1	0.4
2016	98.4	0.8	1.1	0.3
2015	97.9	1.4	0.3	0.4
2014	92.9	5.2	1.2	0.7
2013	94.2	3.3	2.2	0.3

large number of people is affected. These two emergency levels always involve the head chief of the regional office, responsible for the region in which the forecast has been sent. A few key persons or more may be put in emergency or standby based on the extent and the level of the warning. In cases of a lower level forecast (yellow), a NVE regional office may be set in a mode of "increased vigilance", meaning that NVE is following the situation closely, but it is not implementing any further action at the moment.

Experiences acquired at NVEs Eastern regional office indicated that the perception of landslide hazards has increased in the region among responders, authorities and the population through the years and that the forecast service has become an important tool in the emergency response phase. Moreover, the NVE regional staff consider that the value of the landslide warning is strengthened in those areas where landslide hazard has been mapped. Municipalities are then given a possibility to initiate actions for specific areas/buildings depending on the type(s) of landslide forecasted. The available hazard maps can be visualized at (https://temakart.nve.no/link/?link=Skredfaresone).

The free subscription service by e-mail and/or SMS to the warning portal varsom.no is also particularly useful for the regional offices as it provides valuable information to the expertise and engineers working in NVE throughout Norway.

A user survey conducted in autumn 2019 shows that the awareness towards landslide hazards has increase in the last 10 years among emergency authorities. 428 people answered the survey: 2/3 of them are working at responder institutions like road authorities or municipalities and 1/3 were population. 59% of the local emergency authorities answered that they know about damaging landslides and the required actions, while in 2016 only 42% of them answered to the same question and back in 2009 only the 37%. This is not related to an increased number of landslides nationwide, but to a better hazard knowledge. The increased knowledge can be attributed to the establishment of the national forecasting and warning service and all others NVE systematic efforts to better prevent landslides, like the landslide hazard mapping program conducted nationally since 2011. In the 2019 survey, more than 80% of the emergency authorities said that they have made a local assessment after receiving a warning. About 70% said that they performed actions. Over 86% of the interviewed emergency authorities expressed that the warning service is very useful, and they consider it highly reliable.

Evaluation of the Norwegian LEWS

The organization, operation and maintenance of a LEWS is complex (Table 2). Periodic evaluation of LEWS can therefore be a difficult task, especially if all components should be evaluated at the same time. In this analysis we evaluate only some of these steps, indicated by underlined text in Table 2.

Strengths and limitations of the service

Figure 5 shows the organizational history of the Norwegian LEWS. Despite a sporadic attempt of thresholds development in the late 1990s (Sandersen et al. 1996), it is only after 2005 that a common national interest grows towards the mitigation of damages caused by these types of landslides. The service was operational after a test period of two years, with research and development (models, thresholds), warning tests, and organization building (guidelines, recruitment and training of forecasters).

In Fig. 6 we have assembled the main strengths and the main reasons of success, together with the most important short- and long-term benefits.

Among the most important challenges are:

(1) A poor understanding and a limited knowledge of rainfall- and snowmelt-induced landslides and their conditioning and triggering mechanisms (i.e. weather and ground conditions). There are few studies and past investigations of these landslide types and the quality of the national landslide database is still too poor. Consequently, this has important negative effects in the

Table 2 Requirements for operational LEWS. Main key components and steps (K.C.—Key component, S.—Step and R.—Requirement). Underlined text is further evaluated here

Requirements for LEWS
K.C. Risk knowledge and setting of the system (national and institutional involvement)
S: Identify landslide risks and needs to establish a LEWS. Identify national expertise, institutions, financial support and legal statements
R. Landslides must be a risk (reliable and accurate hazard and risk analyses). LEWS is often the best and cheapest mitigation option for the society, (prefeasibility study, containing cost and benefits analyses). Political understanding and interest in the organization of EWS. Available financial support to start. Collaboration among scientific community and politicians. Scientific community with landslide expertise. Stability and long term politic
K.C. Monitoring, forecasting and warning service (institutional, researcher and forecasters involvement)
S: Implementation of a warning model and a warning service
R. Establishment of the service (internal organization). Legal mandate. Available landslide expertise. Training. Collaboration (multidisciplinary team). Guidelines and daily procedures
R. Monitoring and modelling. Effective monitoring systems in appropriate locations. Supporting tools (software, hardware, web platform for sharing data) and daily maintenance. Functioning network for receiving data and forecasts. Reliable historical data. Reliable hydro-meteorological forecasting models and thresholds
R. Daily operation and hazard assessment. Analysis and daily forecasts and model outputs. Functioning forecasts reception. Functioning supporting tools, models. Available forecasting expertise. Understanding the forecasts, model output, uncertainties. Objective interpretation of forecasts. Objective assignment of warning level. Weekly meeting and exchange experience. Freedom to do the daily hazard assessment without social pressure
K.C. Dissemination and communication (institutional, researcher and forecasters involvement)
S: Implementation of a warning model and a warning service
R. Warning service. Warning tools and platforms for communication available and functioning. Definition of warning criteria, warning areas and levels. Standards for warning text, symbols. Preparation of warning messages (text, map and level). Communication of uncertainties. Maintain contact with users and communicate risks, preparing learning material, videos. Use of social media
K.C. Response capability (local users, forecasters, researcher, institutional, national involvement)
S. Evaluate the capability response of the system
R. Emergency plan. Reception and understanding of the warning. Applying emergency plans, take actions
R. Evaluation of performance. Analysis of what happened and performance analyses. Verification of damages. Verification of emergency plans application and if mitigation actions have been undertaken. Verification of landslide occurrence: control and registrations. Field campaigns after a specific event, close contact between forecaster and users. Evaluation criteria. Training and education, also of end-users to a correct interpretation of warning messages. Periodically evaluations of the entire system, propose changes to the organization or improvements to the system. Identify needs for improve scientific development, landslide hazard education at university level, review research strategy and balance between research and operation. Building up the credibility of the scientific institution

daily landslide hazard assessment because (a) we lack of reliable thresholds for the entire country; (b) the thresholds are more reliable in those regions where past landslide records have been controlled or are more complete; (c) the lack of reliable historical records and thresholds may produce subjective assessments that depend on the forecaster's experience.

(2) The prediction of landslides triggered by local and short-intense rainfalls, product of convective clouds during summer, is a challenging task to perform with the available models and thresholds. The number of heavy rainfall events in summer is expected to increase in the future due to climate change (Hanssen-Bauer et al. 2017); the estimation of the expected rainfall amount and location is very uncertain and prediction models need to be tested; local short-intense rainfall (1–3 h) is seldom recorded by official rain gauges and the observed grid-data of rain on summer at 1 km^2, used as input in our hydrological model, is therefore often mislead. The NVEs hydrological models are running at 24 h basis. The impact of short-intense rainfall is often weakly related to pre-existing hydrological conditions (i.e. groundwater, soil water and river discharge). Due to the lack of reliable data (time of events, rainfall records, hydrological modelling), we have not yet calculated landslide thresholds for short-intense rainfall events.

(3) The prediction of landslides caused by rapid snowmelt in winter is also challenging because changes in temperature are not taken into account in our hydrological models.

(4) An important task for forecasters after sending a warning, is to verify the occurrence and extension of landslides. Therefore, an overview of spatial and temporal distribution, as well as number, type, and dimension of all occurred landslides events is strongly required. An event inventory (that ideally should register all landslides occurred during a specific weather event, Guzzetti el al. 2012) is necessary after a warning is sent, to evaluate if the warning level and the warning area were correct and, on the long term, to be used in the improvement of the landslide thresholds. NVE runs

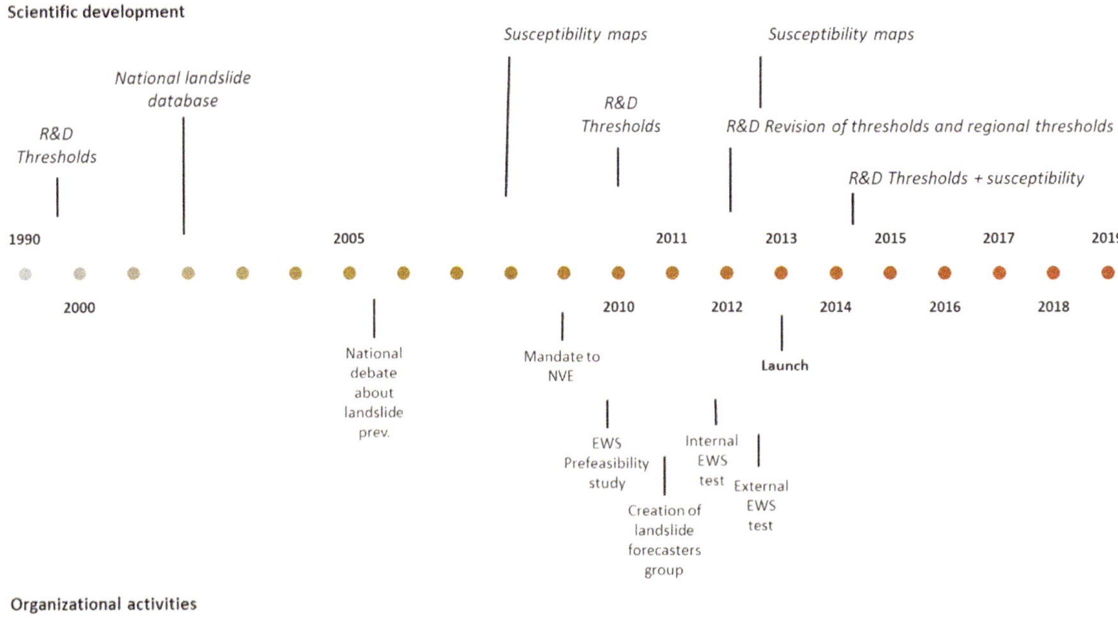

Fig. 5 Timeline of the organizational history of the Norwegian LEW. R&D stands for Research and Development

Fig. 6 Main strengths of the Norwegian LEWS

a national mass movement database in cooperation with others national institutes, where everybody can register landslide events (www.skredregistrering.no). Beside this database, NVE has developed other crowdsourcing tools to gather real-time observations from field (Krøgli et al 2018 and references therein). During a weather event many users, with variate landslide expertise, can register landslide events. Forecasters at NVE track the occurrence of landslide events under a specific weather event through media (radio, newspapers, TV) and register them in any of these tools. Beside these efforts there is not a systematic and not coordinated follow-up after landslide events, neither the preparation of event inventories after an event. Consequently, the landslide events are not systematically registered, and not systematically controlled and verified. Forecasters use a "verified landslide dataset" for warning performance analysis, but this verified information does not always match with the information in the database.

Conclusions

The landslide forecasting and warning service is in its eight year. The system is robust in terms of organization, human resources, financial supports, decision tools and forecasting models.

The service works well, predicting the main hydrometeorological conditions that can trigger landslides. The experience so far, indicates that several yellow levels were issued, when an orange level should have been sent instead. Only one day (the 22th of May 2013 in South-Eastern Norway) had red level during the period 2013–2019. It is too early to evaluate red levels after only 7 years of operation, because in general red level should occur very rarely (50 years return period in analogy to the national flood warning system). The tendency for improved performance may be explained both by more experienced forecasters, by better meteorological forecasts provided by MET Norway and by a better understanding of the uncertainty in the hydro-meteorological forecasts.

NVE is continuously working to rationalize and consolidate the service, running research projects to improve the precision and accuracy of the warnings. NVE is also improving the communication and build up users understanding. The success of such systems is like a feedback-loop. First the EWS need to be reliable enough to be taken into operation. Thereafter the users must be trained to use the available information. To ensure a proper response to a challenging situation, it is an advantage if the latter is a mutual process where the users needs are taken into account in the further development of the EWS. Sufficient coverage of landslide hazard maps is thus a key to extract the full advantage of the forecast service. NVE is continuously producing such maps for selected areas and municipalities in Norway.

We suggest to wait some years before evaluating the performance of the system. After 8 years it will be possible to see the results of the LEWS, but only if there is continuity of the service and annual events. We reccomend to start with a reliable landslide dataset and models of sufficient resolution (6, 3 h). Subsequently developing reliable regional thresholds and thresholds for short intense rainfalls. Heavy rainshowers in summer are a quite new phenomenon in Norway. There are still very few long series of rain records at hourly basis. To handle the challenges with short-intense rainfall and rapid changes in temperature NVE is currently developing a hydrological model running at 3 h steps. Future developments include landslides thresholds for 3, 6 and 12 h.

The service scores a high performance, but some steps are still challenging. The most time consuming and subjective task is the quality control of recorded landslides to be used for performance evaluation and threshold adjustment. The daily monitoring and systematic registration of landslide events has contributed to a better understanding of their physical characteristics and their spatial and temporal triggering conditions. In recent years, new technologies as use of drones and satellite images offer a possibility to register data more efficiently. Experience with these technologies for landslides in Norway, is however limited for the time being.

References

Calvello M (2017) Early warning strategies to cope with landslide risk. Riv It Geotecnica 2:63–91. https://doi.org/10.19199/2017.2.0557-1405.063

Devoli G, Tiranti D, Cremonini R, Sund M, Bøje S (2018) Comparison of landslide forecasting services in Piemonte (Italy) and in Norway, illustrated by events in late spring 2013. Nat Hazards Earth Syst Sci 18:1351–1372. https://doi.org/10.5194/nhess18-1351-2018

Di Biagio E, Kjekstad O (2007) Early warning, instrumentation and monitoring landslides. In: Proceedings of the 2nd regional training course, RECLAIM II. Phulet, Thailand

Fathani TF, Karnawati D, Wilopo W (2016) An integrated methodology to develop a standard for landslide early warning systems. Nat Hazards Earth Syst Sci 16(9):2123–2135. https://doi.org/10.5194/nhess-16-2123-2016

Guzzetti F, Mondini AC, Cardinali M, Fiorucci F, Santangelo M, Chang K-T (2012) Landslide inventory maps: new tools for and old problem. Earth Sci Rev 112:42–66

Hanssen-Bauer I, Førland EJ, Haddeland I, Hisdal H, Mayer S, Nesje A, Nilsen JEØ, Sandven S, Sandø AB, Sorteberg A, Ådlandsvik B (2017) Climate in Norway 2100–a knowledge base for climate adaptation. NCCS report no.1/2017

Intrieri E, Gigli G, Casagli N, Nadim F (2013) Landslide early warning system: toolbox and general concepts. Nat Hazards Earth Syst Sci 13:85–90. https://doi.org/10.5194/nhess13-85-2013

Krøgli IK, Devoli G, Colleuille H, Boje S, Sund M, Engen IK (2018) The Norwegian forecasting and warning service for rainfall- and snowmelt-induced landslides. Nat Hazards Earth Syst Sci 18:1427–1450. https://doi.org/10.5194/nhess-18-1427-2018

Pecoraro G, Calvello M, Piciullo L (2018) Monitoring strategies for local landslide early warning systems. Landslides 16:213–231. https://doi.org/10.1007/s10346-018-1068-z

Piciullo L, Dahl MP, Devoli G, Colleuille H, Calvello M (2017) Adapting the EDuMaP method to test the performance of the Norwegian early warning system for weather-induced landslides. Nat Hazards Earth Syst Sci 17:817–831, 2017 https://doi.org/10.5194/nhess-17-817-2017

Piciullo L, Calvello M, Cepeda JM (2018) Territorial early warning systems for rainfall induced landslides. Earth Sci Rev 179:228–247. https://doi.org/10.1016/j.earscirev.2018.02.013

Sandersen F, Bakkehøi S, Hestnes E, Lied K (1996) The influence of meteorological factors on the initiation of debris flows, rockfalls, rockslides and rock mass stability. In: Senneset K (ed) Landslides. Balkema, Rotterdam, pp 97–114

Segoni S, Rosi A, Fanti R, Gallucci A, Monni A, Casagli N (2018) A regional-scale landslide warning system based on 20 years of operational experience. Water 10(10):1297

UNISDR (2006) Available at: https://www.unisdr.org/2006/ppew/info-resources/ewc3/Global-Survey-of-Early-Warning-Systems.pdf. Accessed 11 Dec 2019

Characterization of Hillslope Deposits for Physically-Based Landslide Forecasting Models

WLF 2020 Kyoto Japan

Veronica Tofani, Gabriele Bicocchi, Elena Benedetta Masi, Carlo Tacconi Stefanelli, Guglielmo Rossi, and Filippo Catani

Abstract

Physically-based models employed for landslide forecasting are extremely sensitive to the use of geological information and a standard, universally accepted method to input maps containing information of geological interest into the models still has never been established. In this study, we used the information contained in a geo-database aimed to characterize the geotechnical and hydrological parameters of the hillslopes deposits in Tuscany, to find out how to organize and group the measurements to spatially create classes that mirror the distribution of the various types of bedrock lithology. Despite the deposits analysed are mainly consisting of well sorted silty sands, statistical analyses carried out on geotechnical and hydrological parameters highlighted that it is not possible to define a typical range of values with relation to the main mapped lithologies, because soil characteristics are not simply dependent on the bedrock typology from which the deposits originated. Instead, the analysis of the relationship of soil parameters with morphometric parameters (slope angle, profile curvature, planar curvature) shows that the highest correlation between the soil grain size class type (USCS classification) and morphometric attributes is with slope curvature, both profile and planar.

Keywords

Soil geotechnics • Physically based modelling • Landslides • Tuscany

Introduction

Many kinds of physically-based landslide prediction models for rainfall-triggered shallow landslides have been presented in the literature so far (Pack et al. 2001; Baum et al. 2002, 2010; Rosso et al. 2006; Simoni et al. 2008; Ren et al. 2010; Arnone et al. 2011; Mercogliano et al. 2013; Rossi et al. 2013; Alvioli and Baum 2016; Salciarini et al. 2017).

One of the most important factors that influences the prediction accuracy and the sensitivity of the physically-based model is the availability of detailed databases of physical and mechanical properties of rocks and soils in the selected study areas. Geotechnical and hydrological variables are often difficult to manage, and their measurement is difficult, time-consuming and expensive, especially when working on large, geologically complex areas (e.g. Baroni et al. 2010; Park et al. 2013; Tofani et al. 2017).

In addition, a poor understanding in the of the geotechnical and hydrological input parameters with respect to their spatial organization may endanger the potential application of numerical models over large areas (e.g. Tofani et al. 2017; Salvatici et al. 2018).

Data to be inputted and to feed the physically-based models can be prepared by using different strategies:

V. Tofani (✉) · F. Catani
Department of Earth Sciences, University of Florence, Via La Pira 4, 50121 Florence, Italy
e-mail: veronica.tofani@unifi.it

F. Catani
e-mail: filippo.catani@unifi.it

G. Bicocchi
Department of Earth Sciences, University of Florence and Institute of Geosciences and Earth Resources, CNR – National Research Council of Italy, Via G. La Pira 4, 50121 Florence, Italy
e-mail: gabriele.bicocchi@unifi.it

E. B. Masi · C. T. Stefanelli
Department of Earth Sciences, University of Florence, Largo Enrico Fermi 2, 50125 Florence, Italy
e-mail: elenabenedetta.masi@unifi.it

C. T. Stefanelli
e-mail: carlo.tacconistefanelli@unifi.it

G. Rossi
Civil Protection Centre, University of Florence, Largo Enrico Fermi 2, 50125 Florence, Italy
e-mail: guglielmo.rossi@unifi.it

© Springer Nature Switzerland AG 2021

N. Casagli et al. (eds.), *Understanding and Reducing Landslide Disaster Risk*, ICL Contribution to Landslide Disaster Risk Reduction,
https://doi.org/10.1007/978-3-030-60311-3_31

(1) the adoption, for each parameter, of a unique constant value for the whole study area as retrieved from experimental data or derived from literature data (e.g. Jia et al. 2012; Peres and Cancelliere 2014), (2) the use of constant values for the parameters for distinct geological, lithological or lithotecnical units, as derived from direct measurements (Segoni et al. 2009; Baum et al. 2010; Montrasio et al. 2011; Zizioli et al. 2013; Bicocchi et al. 2016; Tofani et al. 2017) or from existing databases and published data (Ren et al. 2014; Tao and Barros 2014), or (3) the definition of selected parameters, such as the cohesion and friction angle values, as random variables using a probabilistic or stochastic approach (e.g. Park et al. 2013; Chen and Zhang 2014; Raia et al. 2014; Fanelli et al. 2016; Salciarini et al. 2017).

In this work we discuss how to deal with the geotechnical and hydrological input data in regional physically-based models. In particular, we want to find an optimal way (1) to determine the ranges of variation and the characteristics of frequency distributions of the geotechnical and hydrogeological parameters that control shallow landslide triggering mechanisms, and (2) to describe the spatial variation in the geotechnical and hydrological data in relation to the information contained in the geological maps and to the physical factors such as morphometric attributes. The area of application of this approach is Tuscany region, where more than one hundred survey points of the geotechnical and hydrological parameters measurements are available (Bicocchi et al. 2019).

Geographical and Geological Description of the Study Area

Tuscany region (Fig. 1) is a topographically complex region located in central Italy strongly affected by shallow landslides occurring after major meteorological events (Giannecchini et al. 2007; Mercogliano et al. 2013; Tofani et al. 2017). The highest ridges are in the northern and eastern portion of the region. The northwestern part is characterized by mountains comprised of metamorphic rocks (i.e. Apuan Alps) and by steep valleys with thick colluvial and alluvial deposits, while the eastern part is characterized by mountains mainly formed by sedimentary rocks and by intermountain basins filled with alluvial deposits. These mountains belong to the Northern Apennine, a NE-verging fold-and-thrust orogenic belt originated from the closure of the Jurassic "Ligure-Piemontese" Ocean and the subsequent Oligocene–Miocene collision between the continental Corso-Sardinian block and the Adria microplate (e.g., Boccaletti and Guazzone 1974). The central and southern parts are characterized by hilly morphology with an isolated volcanic relief and flat plains or wide valley floors where the main rivers flow.

A lithological map of the bedrock for Tuscany was prepared (Fig. 1) by customizing that lithological map previously derived from the geological map of Italy, 1:500,000 by ISPRA (Italian National Institute for Environmental Protection and Research). The bedrock, relative to its areal distribution in the hilly and mountainous part of the region, is mainly represented by of arenaceous, calcareous and pelitic flysch units.

Geotechnical and Hydrological Measurements

The data analysed are represented by samples collected at 102 different sites (Bicocchi et al. 2019; Fig. 1). The geotechnical and hydrological parameters for characterizing the soils were determined as described in Tofani et al. 2017 and Bicocchi et al. 2019. In particular, the Borehole Shear Test (BST; Luttenegger and Hallberg 1981) for measuring the soil shear strength parameters, a constant head permeameter test performed with the Amoozemeter instrument (Amoozegar 1989) and matric suction measurements with a tensiometer, were used for the in-situ determinations. In addition, laboratory tests were conducted at the Department of Earth Sciences, University of Florence, to determinate grain size distributions, Atterberg limits, soil phase relationships (bulk porosity n; saturated, natural and dry unit weight, γ_{sat}, γ and γ_d, respectively) and the soil organic matter contents (SOM; for the latter refer to Masi et al. 2020 for further information about the analysis methods adopted).

Results

The analysed deposits are mostly classified as well-sorted silty—clayey sands, i.e. SW, SM, SC and SM-SC classes by using the Unified Soil Classification System (USCS; Wagner 1957). Nevertheless, a non-negligible part of the samples is characterized by higher contents of silt and clay (ML and MH class in the USCS), whilst an isolated sample is classified as GW (Fig. 2).

Descriptive statistics concerning dry unit weight, bulk porosity, internal effective friction angle and saturated hydraulic conductivity values are reported in Table 1. The dry unit weight (γ_d) ranges between 10.7 and 20.8 kN m^{-3}, with a mean value of 15.5 kN m^{-3}. Internal effective friction angle values (φ') vary from 15° to 45° with an average value of 32°, but much part of the value lies in a narrower interval (±5° from the arithmetic mean). The bulk porosity

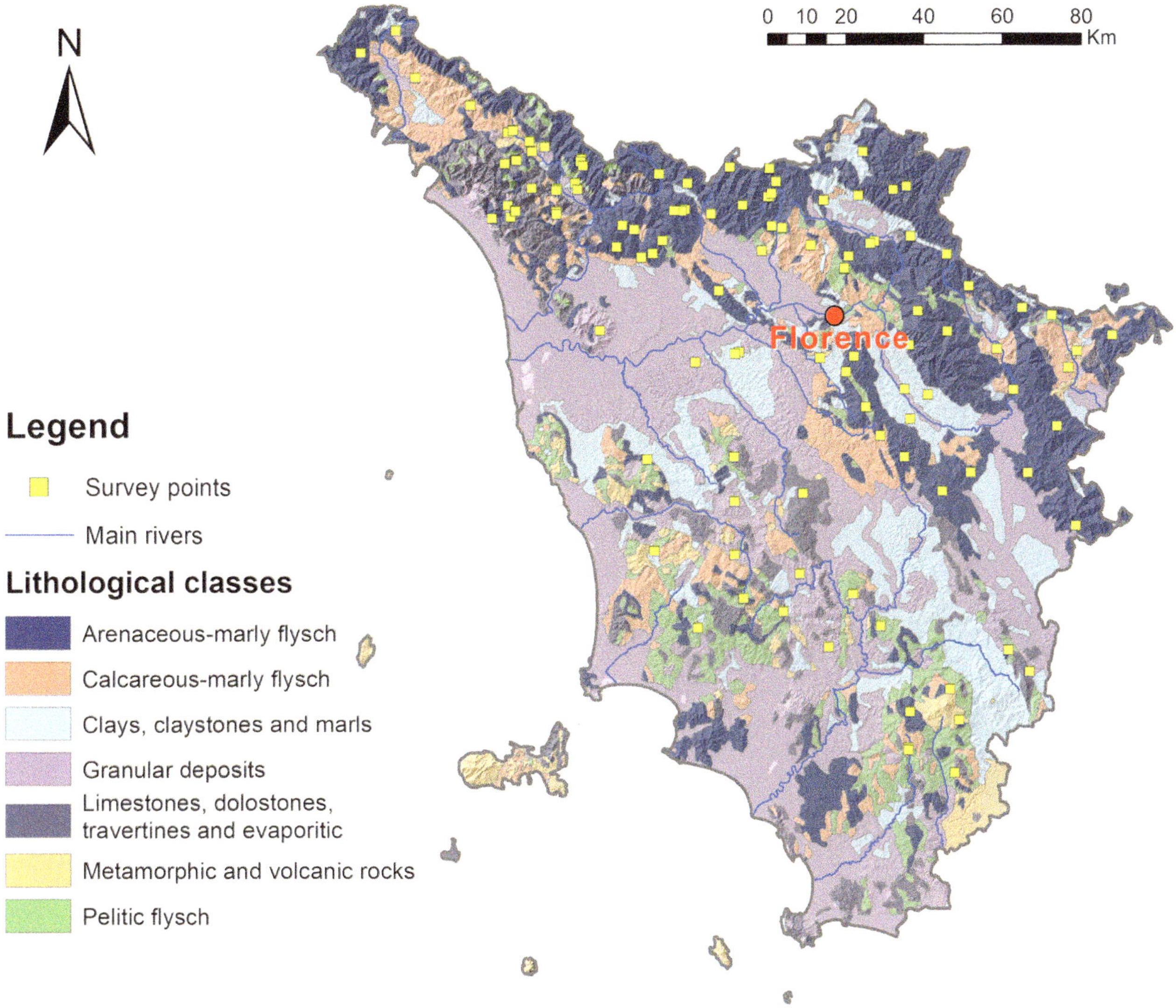

Fig. 1 Lithological map of Tuscany (bedrock) and location of the survey points (from Bicocchi et al. 2019)

(n) values span a wide interval: from 19.9 to 58.8% with a median value of 38.8%. Also, the values of k_s range in a wide interval from 4 10^{-8} to 8 10^{-5} m s^{-1}. The standard deviation (σ) values stress out this aspect since σ is ~15% of the arithmetic mean value for φ' and γ_d, while is much higher for n and k_s.

The main mineral phases detected in the samples of the 27 selected sites, representative of the different USCS soil types and bedrock lithology, are mica, quartz, non-swelling clay minerals, plagioclases, k-feldspar and calcite. The SOM normalized to the bulk samples ranges between 1.8 and 8.9% by weight, the highest values of the SOM content being associated with forest and woodlands without shrubs. The SOM values distribution showed close relationships with the abundance of the inorganic finer fractions (silt and clay) of the soil samples (Masi et al. 2020).

Discussion

In order to define a proper way to spatialize the parameters to be used for physically-based forecasting models some further analyses are carried out to examine the relationship between (1) soil parameters and bedrock lithology and (2) soil parameters and morphometric attributes.

Aggregated Data Statistics: Grouping by Bedrock Lithology Versus by USCS Classification

First of all, we studied the distribution of the soils (Table 2), classified according to USCS, with respect to the underlying bedrock types recognized in the map (Fig. 1). Arenaceous marly flysch (AMF), calcareous marly flysch (CMF),

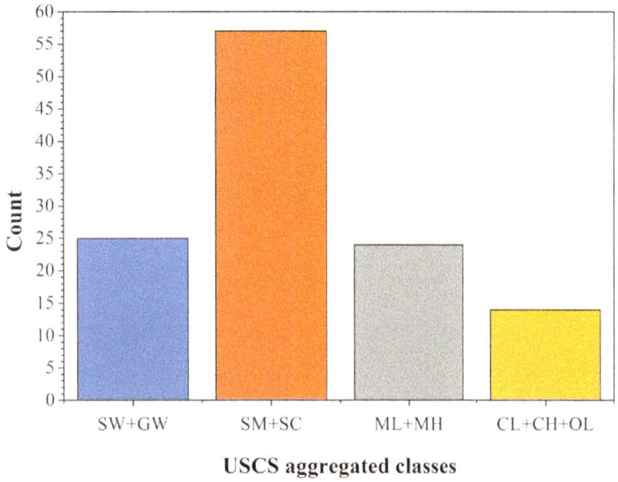

Fig. 2 Distribution of USCS aggregated classes for the samples from Bicocchi et al. (2019)

Table 1 Descriptive statistics of selected geotechnical parameters

	γ_d (kN m^{-3})	ϕ' (°)	n (%)	k_s (m s^{-1})
N° of measurements	81	109	81	119
Min	10.7	15	19.9	4.E−08
Max	20.8	45	58.8	8.E−05
Arithmetic mean	15.5	32	40.2	3.E−06
Median	15.9	32	38.8	1.E−06
Std. dev. (σ)	2.2	5	9.5	9.E−06

Table 2 Soil classification of the bedrock lithologies according to USCS classification. N°: number of survey points. AMF: arenaceous marly flysch, CMF: calcareous marly flysch, LDTE: limestones, dolomites, travertines and evaporitic deposits, PF: pelitic flysch, CCM: Clay, claystone and marls, GD: granular deposits, MVR: Metamorphic and volcanic rocks

Bed-rock	N°	SW + GW	SM + SC	ML + MH	CL + CH + OL
AMF	57	18	27	9	3
CMF	21	0	11	5	5
CCM	9	0	3	4	2
GD	11	1	3	4	3
LDTE	11	2	6	2	1
MVR	3	2	1	0	0
PF	8	1	5	2	0

limestones, dolomites, travertines and evaporitic deposits (LDTE), and pelitic flysch (PF) have mainly silty sands and clayey sands soil deposits. Clay, claystone and marls (CCM) and granular deposits (GD) show silty soils with low and high plasticity. As MVR class consists only of, three observations, every consideration is statistically poorly significant.

The variability of the main parameters (φ', γ_d and k_s), which play key roles in triggering slopes instability, was further investigated aggregating these parameter values according to the pertaining bedrock lithological and the USCS classes (Fig. 3).

Aggregation based on USCS classes show friction angle values distribution quite symmetric, especially for SM-SC and SW classes, although the range of the values covers over 20°. Conversely, the ML + MH box plot is asymmetric, because of the short distances between the box upper limit and the maximum values, while the distance between the lower limit and minimum values is quite high (over 10°). Dry unit weight box plots are symmetric in their shape, apart from SW and secondarily for CL + CH + OL, but the values are extremely dispersed. Eventually, k_s values, as for the data aggregated by bedrock class, were log-transformed prior to making up the box plots. The conductivity values are asymmetrically distributed with respect to the arithmetic mean values, which are located far above the median and often above the 3rd quartile (i.e. the upper box limits).

The aggregation based on bedrock lithology shows boxes of effective friction angle quite symmetric, as the median is very similar to the arithmetic mean. The dry unit weight box plots are symmetric in their shape, apart for the granular deposits (GD) and limestones, dolomites, travertines and evaporitic deposits (LDTE) classes, with median and arithmetic mean values very close and the space between the quartiles homogeneously distributed. Once again, hydraulic conductivity values show some distinctive asymmetric distribution with respect to the arithmetic mean values, which are located far above the median and often above the 3rd quartile.

An interesting fact to note is that, in the study area, a reliable extrapolation of soil parameters is quite difficult to achieve based on the simple observation of the underneath bedrock lithology (Bicocchi et al. 2019). Despite most of the samples in this study being classified as arenaceous marly flysch (AMF), important differences have been found concerning their grain size distribution (Table 2). The main reasons for such decoupling between the bedrock type and the deposit granulometry could be that (Bicocchi et al. 2019): (1) the deposits may have originated from a different bedrock with respect to what they overlie at present, and especially

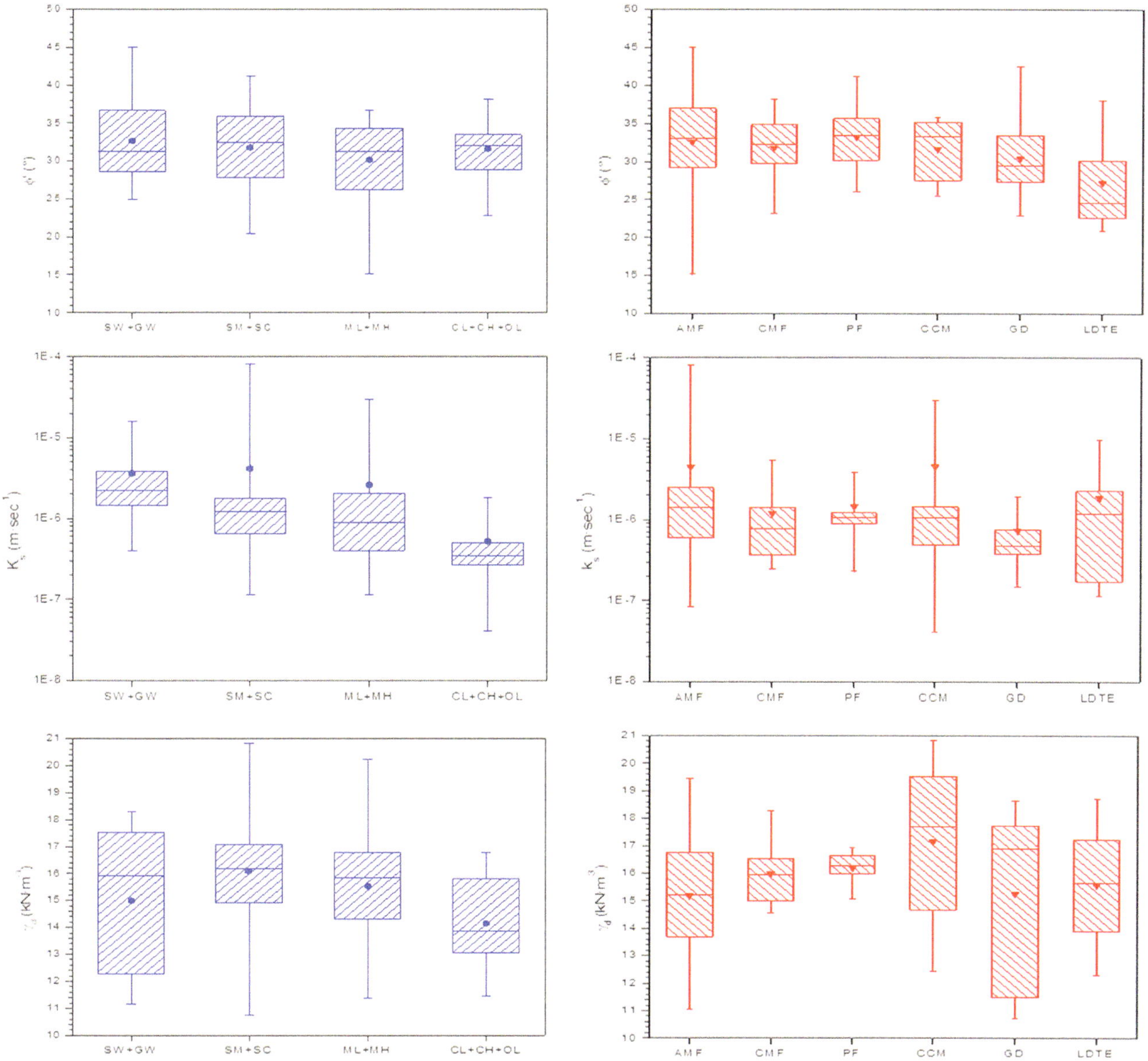

Fig. 3 Box plots of φ' (internal effective friction angle), γ_d (dry unit weight), KS (saturated hydraulic conductivity, scale is logarithmic) for different bedrock lithologies (red coloured, on the right; the dots represents the arithmetic mean value) and for USCS classes (blue coloured, on the left; the dots represents the arithmetic mean value); refer to the text for the bedrock acronyms (modified after Bicocchi et al. 2019)

(2) most of the geological units of the Northern Apennine are quite heterogeneous and intrinsically complex flysch (e.g., Martini and Vai 2001), often characterized by repeated lithological changes (e.g. sandstone to claystone and/or to limestone) in a few tens of meters, so that the characteristics of the regolith, from which the deposits formed, may vary as the bedrock lithological changes occurs.

Compared to the analysis performed by aggregating the values by bedrock lithology, the use of USCS classes, especially looking at SW + GW and CL + CH + OL for the ϕ' and k_s, appears to be more suitable for producing a symmetric distribution and a homogeneous division of the values, while both approaches substantially fail in finding an appropriate way to describe the distribution of dry unit weight box plots.

USCS Soil Type Occurrence Versus Morphometric Attributes

We have investigated the relationship between the soil type, in terms of USCS classification, and morphometric attributes with reference to slope gradient, profile curvature and planar curvature (Fig. 4).

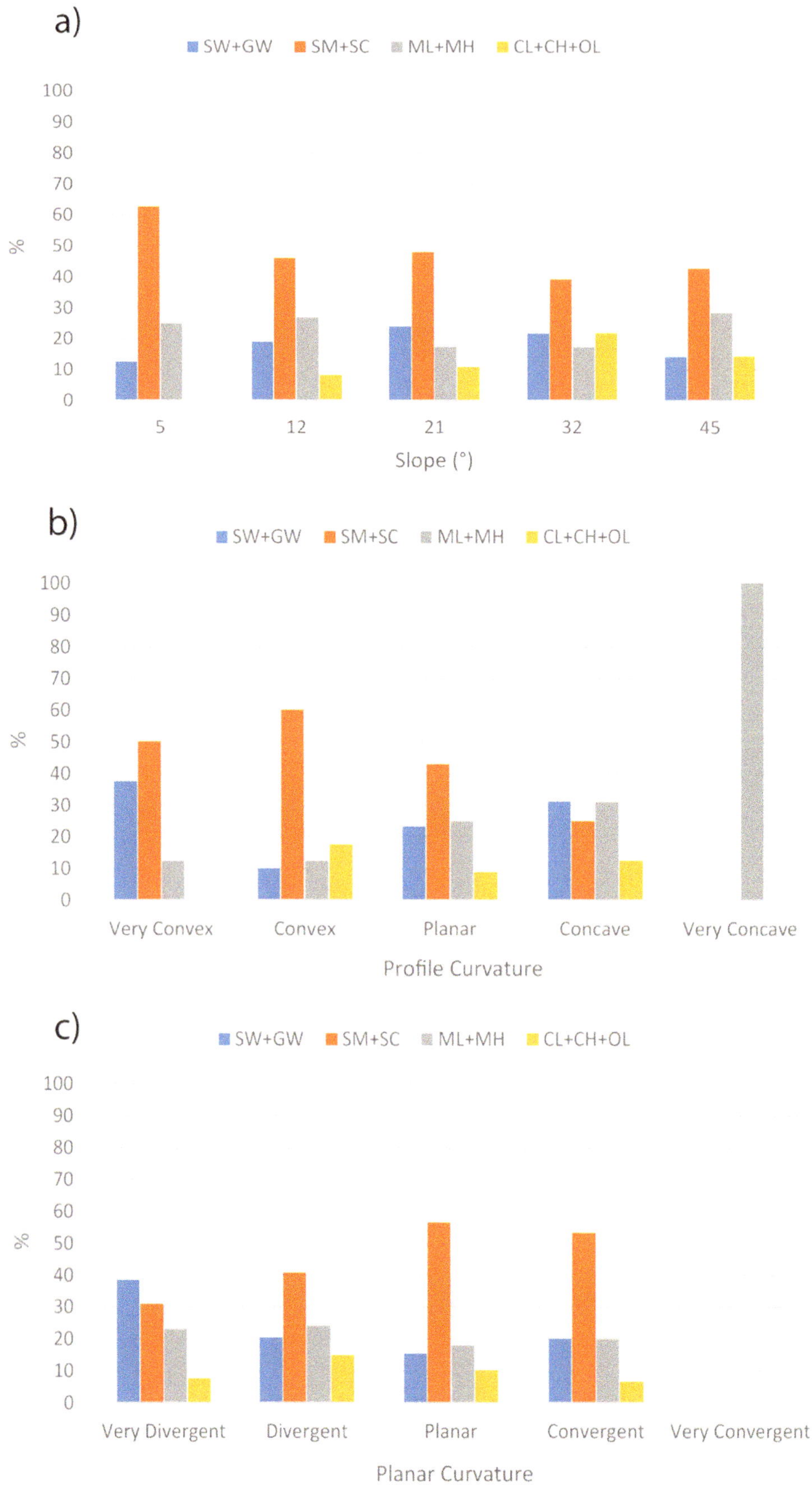

Fig. 4 Occurrence of USCS aggregated soil types with respect to the **a** slope, **b** profile curvature, **c** planar curvature in the hillslopes surveyed (modified after Bicocchi et al. 2019)

In general, at low slope angles, granular soils (SW + GW, SM + SC) are predomisnant, while with the increase of slope angle, the presence of cohesive soils (ML + MH, CL + CH + OL) increases proportionally. This behaviour can be related to the predominance of cohesive forces with respect to frictional ones from low to high slope gradients.

Owing to the profile curvature, it is worth mentioning that in convex areas granular soils (SW + GW, SM + SC) are prevalent while in concave areas the distribution of soil classes is more heterogeneous, and all the soil classes are about evenly represented. This result can be explained considering that in convex areas fine materials are more easily remodeled and transported due to various processes of surface runoff, such as rainwash and sheetwash.

The distribution of soil classes for planar curvature shows that in very divergent areas (crests) coarse granular soils (SW + GW) prevail over fine granular (SM + SC) and cohesive ones (ML + MH, CL + CH + OL). This is in line with the results coming from the profile curvature: in convex and divergent areas, rainwash and sheetwash processes produce residual soils composed of mainly coarse material. In the other classes of planar curvature, silty sands prevail. Nothing can be said about very convergent areas, where no samples have been collected in the analysed database since they usually represent incised channel bottoms or stream thalwegs.

Conclusion

In this work selected information contained in the database of geotechnical (internal effective friction angle, dry unit weight, porosity) and hydrological (saturated hydraulic conductivity) parameters for soil cover in the hillslope deposits in Tuscany (Italy) has been interpreted in order to improve the preparation of parameters for regional physically-based landslide prediction models.

An important finding, while examining the database, is that grouping the geotechnical parameters measurements with respect soil types (USCS classification) and bedrock lithology, substantially fails in giving back clearly distinguishable range of values for the different types of soils or bedrock. Indeed, in most cases the grain size distribution of soils is controlled by the intensities and the type of the acting slope processes regardless of from what bedrock typology they originated. For the same reason shear strength and hydraulic conductivity are difficult to predict on the basis of the geo-lithological maps only.

However, we have found that, instead, linkages between the different USCS soil types with morphometric parameters such as the profile of curvature of the hillslopes exist. This finding could be a starting point to develop alternative strategies to spatially organize and group in classes the geotechnical parameters.

Acknowledgements This work was financially supported by a collaboration between the Department of Earth Sciences, University of Florence and the Regional Administration of Tuscany.

References

Alvioli M, Baum RL (2016) Parallelization of the TRIGRS model for rainfall-induced landslides using the message passing interface. Environ Model Softw 81:122–135

Amoozegar A (1989) Compact constant head permeameter for measuring saturated hydraulic conductivity of the vadose zone. Soil Sci Soc Am J 53:1356–1361

Arnone E, Noto LV, Lepore C, Bras RL (2011) Physically—based and distributed approach to analyse rainfall-triggered land- slides at 320 watershed scale. Geomorphology 133(3):121–131

Baroni G, Facchi A, Gandolfi C, Ortuani B, Horeschi D, van Dam JC (2010) Uncertainty in the determination of soil hydraulic parameters and its influence on the performance of two hydrological models of different complexity. Hydrol Earth Syst Sci 14:251–270

Baum R, Savage WZ, Godt JW (2002) Trigrs: a FORTRAN program for transient rainfall infiltration and grid-based regional slope-stability 322 analysis. Open-file report, US Geol. Survey

Baum RL, Godt JW, Savage WZ (2010) Estimating the timing and location of shallow rainfall-induced landslides using a model for transient unsaturated infiltration. J. Geophys Res 115:F03013

Bicocchi G, D'Ambrosio M, Rossi G, Rosi A, Tacconi-Stefanelli C, Segoni S, Nocentini, M, Vannocci, P, Tofani V, Casagli N, Catani F (2016) Geotechnical in situ measures to improve landslides forecasting models: a case study in Tuscany (Central Italy). In: Proceedings ISL 2016 congress, pp 419–424. ISBN 9781138029880

Bicocchi G, Tofani V, D'Ambrosio M, Tacconi-Stefanelli C, Vannocci P, Casagli N, Lavorini G, Trevisani M, Catani F (2019) Geotechnical and hydrological characterization of hillslope deposits for regional landslide prediction modeling. Bull Eng Geol Env 78:4875–4891

Boccaletti M, Guazzone G (1974) Remnant arcs and marginal basins in the Cenozoic development of the Mediterranean. Nature 252:18–21. https://doi.org/10.1038/252018a0

Chen HX, Zhang LM (2014) A physically-based distributed cell model for predicting regional rainfall-induced shallow slope failures. Eng Geol 176:79–92. https://doi.org/10.1016/j.enggeo.2014.04.011

Fanelli G, Salciarini D, Tamagnini C (2016) Reliable soil property maps over large areas: a case study in central Italy. Environ Eng Geosci 22:37–52

Giannecchini R, Naldini D, D'Amato Avanzi G, Puccinelli A (2007) Modelling of the initiation of rainfall-induced debris flows in the Cardoso basin (Apuan Alps, Italy). Quat Int 171–172:108–117

Jia N, Mitani Y, Xie M, Djamaluddin I (2012) Shallow landslide hazard assessment using a three dimensional deterministic model in a mountainous area. Comput Geotech 45:1–10

Luttenegger JA, Hallberg BR (1981) Borehole shear test in geotechnical investigations. Am Soc Test Mater Spec Publ 740:566–578

Martini IP, Vai GB (2001) Anatomy of an Orogen: the Apennines and adjacent mediterranean basins. Kluwer, Dordrecht. 633 pp. https://doi.org/10.1007/978-94-015-9829-3

Masi EB, Bicocchi G, Catani F (2020) Soil organic matter relationships with geotechnical-hydrological parameters, mineralogy and vegetation cover of hillslope deposits in Tuscany (Italy). Bull Eng Geol Environ. Accepted

Mercogliano P, Segoni S, Rossi G, Sikorsky B, Tofani V, Schiano P, Catani F, Casagli N (2013) Brief communication: a prototype forecasting chain for rainfall induced shallow landslides. Nat Hazards Earth Syst Sci 13:771–777

Montrasio L, Valentino R, Losi GL (2011) Towards a real-time susceptibility assessment of rainfall-induced shallow landslides on a regional scale. Nat Hazards Earth Syst Sci 11:1927–1947

Pack R, Tarboton D, Goodwin C (2001) Assessing terrain stability in a GIS using SINMAP. In: 15th annual GIS conference, GIS

Park HJ, Lee JH, Woo I (2013) Assessment of rainfall-induced shallow landslide susceptibility using a GIS-based probabilistic approach. Eng Geol 161:1–15

Peres DJ, Cancelliere A (2014) Derivation and evaluation of landslide triggering thresholds by a Monte Carlo approach. Hydrol Earth Syst Sci 18:4913–4931. https://doi.org/10.5194/hess-18-4913-2014

Raia S, Alvioli M, Rossi M, Baum RL, Godt JW, Guzzetti F (2014) Improving predictive power of physically based rainfall-induced shallow landslide models: a probabilistic approach. Geosci Model Dev 7:495–514. https://doi.org/10.5194/gmd-7-495-2014

Ren D, Fu R, Leslie LM, Dickinson R, Xin X (2010) A storm-triggered landslide monitoring and prediction system: formulation and case study. Earth Interact 14:1–24

Ren D, Leslie L, Lynch M (2014) Trends in storm-triggered landslides over Southern California. J Appl Meteor Climatol 53:217–233

Rossi G, Catani F, Leoni L, Segoni S, Tofani V (2013) HIRESSS: a physically based slope stability simulator for HPC applications. Nat. Hazards Earth Syst Sci 13:151–166

Rosso R, Rulli MC, Vannucchi G (2006) A physically based model for the hydrologic control on shallow landsliding. Water Resour Res 42:W06410. https://doi.org/10.1029/2005WR004369

Salciarini D, Fanelli G, Tamagnini C (2017) A probabilistic model for rainfall-induced shallow landslide prediction at the regional scale. Landslides 14:1731–1746

Salvatici T, Tofani V, Rossi G, D'Ambrosio M, Tacconi Stefanelli C, Masi EB, Rosi A, Pazzi V, Vannocci P, Petrolo M, Catani F, Ratto S, Stevenin H, Casagli N (2018) Regional physically based landslide early warning modelling: soil parameterisation and validation of the results. Nat Hazards Earth Sys Sci 18:1919–1935. https://doi.org/10.5194/nhess-18-1919-2018

Segoni S, Leoni L, Benedetti AI, Catani F, Righini G, Falorni G, Gabellani S, Rudari R, Silvestro F, Rebora N (2009) Towards a definition of a real-time forecasting network for rainfall induced shallow landslides. Nat Hazards Earth Syst Sci 9:2119–2133

Simoni S, Zanotti F, Bertoldi G, Rigon R (2008) Modelling the probability of occurrence of shallow landslides and channelized debris flows using geotop-fs. Hydrol Process 22:532–545

Tao J, Barros AP (2014) Coupled prediction of flood response and debris flow initiation during warm- and cold-season events in the southern Appalachians USA. Hydrol Earth Syst Sci 18:367–388

Tofani V, Bicocchi G, Rossi G, Segoni S, D'Ambrosio M, Casagli N, Catani F (2017) Soil characterization for shallow landslides modeling: a case study in the Northern Apennines (Central Italy). Landslides 14:755–770. https://doi.org/10.1007/s10346-017-0809-8

Wagner AA (1957) The use of the unified soil classification system by the Bureau of Reclamation. In: Proceedings of the 4th international conference on soil mechanics and foundation engineering, London vol I, 125 pp

Zizioli D, Meisina C, Valentino R, Montrasio L (2013) Comparison between different approaches to modeling shallow landslide susceptibility: a case history in Oltrepo Pavese. Northern Italy. Nat Hazards Earth Syst Sci 13:559–573

WLF 2020 Kyoto Japan

Development of a Rainfall-Induced Landslide Forecast Tool for New Zealand

Brenda Rosser, Chris Massey, Biljana Lukovic, Sally Dellow, and Matt Hill

Abstract

Landslides kill 2–3 people per annum in New Zealand and cost the country on average NZ$200–300 million dollars per annum. The majority of landslides (90%) in New Zealand are triggered by rainfall and often involve thousands to tens of thousands of landslides being triggered by a single event that can extend over areas up to 20,000 km^2. Steep hillslopes (>26°) occupy over 60% of the New Zealand landmass, and much of this (5%) is classified as highly erodible land at risk of severe mass-movement erosion. To reduce the risk associated with landslides it is important to be able to predict where and when they might occur. To this end we are developing a landslide forecast tool for the National GeoHazards Monitoring Centre that will be used to forecast and warn the public of possible damaging rainfall-induced landslide events. We used logistic regression to investigate the influence of landslide triggering variables on landslide occurrence on a dataset of 20 recent and historic landslide-triggering rainfall events. From this we developed relationships to predict the probable spatial distribution of landslides triggered from a given forecast rainfall event.

Keywords

Rainfall induced landslide • Forecast tool • New Zealand • Threshold • Landslide modelling

Introduction

Landslides kill 2–3 people per annum in New Zealand and cost the country on average NZ$200–300 million dollars per annum (Page 2015). In New Zealand, landslides triggered by storm rainfall are the most common type of mass-movement erosion (Crozier 2005). Steep hillslopes (>26°) occupy over 60% of the New Zealand landmass, and much of this (5%) is classified as highly erodible land at risk of severe mass-movement erosion (Dymond et al. 2006). Rainfall-induced landslides most commonly occur as multiple-occurrence landslide events, that can cover areas ranging from a few to several thousand km^2, and usually involve first-time occurrences (Crozier 2005). They are predominantly small (<1000 m^2), rapid, shallow (<2 m deep) earth or debris slides and flows. Although individually small, cumulatively they can cause significant damage in a widespread rainfall event (Fig. 1). Regional scale landslide triggering storm events occur somewhere in New Zealand on average 2–3 times per year (Crozier 2005, 2017).

Landslide occurrence in New Zealand is highly correlated with rock type. The rock types most susceptible to rainfall induced landslides are young (Quaternary and Tertiary), poorly consolidated fine-grained sedimentary rocks or highly weathered, fractured or sheared older rocks. Land cover also plays an important role in determining landslide susceptibility.

In order to reduce the risk posed by landslide hazards to society, knowledge of when and where landslides occur is essential. In this paper we present the methodology we have used to develop a rainfall-induced landslide forecast tool for New Zealand.

B. Rosser (✉) · C. Massey · B. Lukovic · S. Dellow · M. Hill
GNS Science, 1 Fairway Drive, Avalon, Lower Hutt, 5010, New Zealand
e-mail: b.rosser@gns.cri.nz

C. Massey
e-mail: c.massey@gns.cri.nz

B. Lukovic
e-mail: b.lukovic@gns.cri.nz

S. Dellow
e-mail: s.dellow@gns.cri.nz

M. Hill
e-mail: m.hill@gns.cri.nz

© Springer Nature Switzerland AG 2021
N. Casagli et al. (eds.), *Understanding and Reducing Landslide Disaster Risk*, ICL Contribution to Landslide Disaster Risk Reduction,
https://doi.org/10.1007/978-3-030-60311-3_32

Fig. 1 Extensive shallow landslides triggered by the 2004 Manawatu storm (*Photo* Graham Hancox, GNS Science)

Fig. 2 Regions used in this study and the location of landslide-triggering rainfall events. The point marker symbolises only the location of one or more event, not the event rainfall or extent of landslide damage

Methodology

Development of Rainfall Intensity-Duration Triggering Thresholds

Rainfall intensity-duration thresholds for triggering landslides were developed for different physiographic regions of New Zealand. A database of 1029 landslide triggering rainfall events was compiled from various sources that covered the period from 1875 to 2019. The data covered all regions of New Zealand (Fig. 2), however, some regions had more landslide data than others which reflects both the reporting of landslides (or lack of) and the variation in general landslide susceptibility between different regions (Glade 1998). For each landslide triggering event, the location of the landslide or landsliding event (multiple landslides) was recorded, along with the date and the rainfall conditions that triggered the landslide. The regions reflect broad differences in physiographic features/variables that influence landslide occurrence such as topography, lithology and meteorology, and generally conform to administrative boundaries of local or regional government or agencies responsible for responding to or managing the impact of storm events (e.g. Civil defence and emergency management groups in regional councils).

Rainfall intensity-duration thresholds were developed for each of the regions by fitting a curve of the form:

$$I = \alpha D^{-\beta}$$

where I = rainfall intensity (mm/h) and D = duration (h), developed by Caine (1980), and secondly, using a probability approach to find the scale (intercept) α and the shape (slope) β of the power law curve representing the minimum and maximum thresholds (Guzzetti et al. 2008). The rainfall ID curves for all New Zealand are presented in Fig. 3.

The slope of the rainfall ID threshold curve indicates the relative importance of rainfall intensity versus rainfall duration for the initiation of shallow soil failures (Guzzetti et al. 2008). A steeper threshold curve, indicates that (short) rainfall duration is more significant for landslide generation than for a less steep curves (Guzzetti et al. 2008). This indicates that in general, short duration (<24 h) rainfall is an important control for landslide initiation in New Zealand, however there is considerable variation between regions. Guzzetti et al. (2008) attributed differences in the ID thresholds to the typical meteorological conditions and characteristic rainfall patterns which trigger shallow landslides in different climatic regions, as well as to local physiographic characteristics such as geology, soil types, vegetation and slope morphology that govern landslide susceptibility.

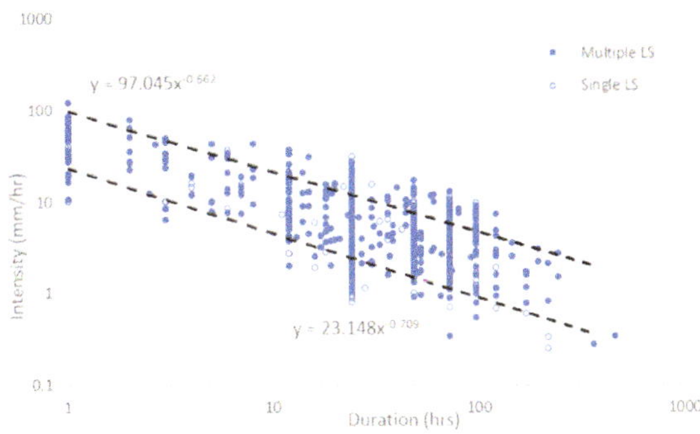

Fig. 3 Rainfall intensity-duration threshold curve for all New Zealand. The upper curve is the power fit to the 90th percentile and represents the upper ID threshold above which landslides will always occur. The lower curve is the power fit to the 2nd percentile and represents the rainfall ID threshold above which landslides will likely occur (as per Guzzetti et al. 2008)

Compilation of Storm Landslide Inventories

Existing storm landslide inventories were compiled for 20 storm events during the period 1938–2019 (Table 1). There were 16 existing storm datasets, and an additional 4 datasets compiled for the project (or as part of ongoing Geonet Landslide responses). The existing landslide datasets included a variety of data types and formats. A new GIS inventory using the data format outlined by Massey et al. (2018) was developed for the project and the data converted to a single consistent format. Both points (representing the top of the landslide source area) and polygons (representing the landslide source area) were used in the analysis. The spatial distribution of the datasets is shown in Fig. 4.

Table 1 Storm events with mapped landslide distributions used in this study

Storm, year	Storm rainfall
Kaikoura, 1975	561 mm/48 h
Coromandel, 1981	500 mm/72 h
Tasman, 2018	235 mm/18 h
Hawke's Bay, 2011	646 mm/96 h
Hunua, 2017	200 mm/12 h
Ligar Bay, 2011	674 mm/48 h
Manawatu, 2004	215 mm/52 h
Nelson, 2011	423 mm/48 h
Northland, 2007	400 mm/36 h
Orongorongo, 2002	194 mm/24 h
Orongorongo, 2004	316 mm/72 h
Orongorongo, 2006	157 mm/24 h
Orongorongo, 2008	185 mm/24 h
Paekakariki, 2000	92 mm/72 h
Paekakariki, 2003	119 mm/24 h
Paekakariki, 2005	103 mm/24 h
Paekakariki, 2006	49 mm/24 h
Paekakariki, 2007	40 mm/24 h
Uawa, 2018	252 mm/24 h
West Coast, 2019	1087 mm/48 h

Fig. 4 Spatial distribution of rainfall-induced landslide datasets used in the analysis

Compilation of Landslide Susceptibility and Rainfall Data

National GIS datasets representing landslide susceptibility factors (dominant rock type, slope, local slope height, aspect, slope curvature, vegetation) were compiled into a GIS database. The variables that contributed to landslide occurrence were determined in a pilot study carried out in Wellington City, using a database of 16,000 landslides mapped for 11 storm events over the period 1939–2016. National GIS datasets were compiled at a grid size of 30 m by 30 m for the whole country.

Rainfall data for each storm event was provided by NIWA, New Zealand's National Institute of Water and Atmospheric research from the National Climate Database (www.cliflo.niwa.co.nz). Rainfall data included 24 h maximum and storm rainfall totals derived from the national rain gauge network. Soil moisture data, representing the modelled soil moisture deficit for the day before each storm event was provided by NIWA from the Virtual Climate Station Network (https://niwa.co.nz/climate/our-services/virtual-climate-stations). The Virtual Climate Station Network estimates climatic parameters, including soil moisture, on a regular (~5 km) grid covering the whole of New Zealand. The estimates are produced every day, based on the spatial interpolation of actual data observations made at climate stations located around the country.

Data from each of the landslide susceptibility and rainfall GIS layers was extracted for each landslide in the storm landslide inventory (n = ~300,000). Each landslide point therefore had its own landslide susceptibility factors (dominant rock type, slope, local slope height, aspect, vegetation) and triggering conditions (rainfall, soil moisture) associated with it.

Logistic Regression Modelling

We used the mapped landslide distributions from the 20 storms to explore the relationships between landslide occurrence and the variables that may control its occurrence. The variables included the landslide triggering variables (rainfall, soil moisture) and landslide susceptibility variables (dominant rock type, slope, local slope relief, aspect, vegetation and anthropogenic modification). The variables that contribute to landslide occurrence must have a physical influence on landslide occurrence and contribute statistically to the fit of the model (discussed below).

We used logistic regression to investigate the influence that the triggering and susceptibility variables have on the spatial distribution of 11,000 rainfall-induced landslides attributed to 11 storm events in Wellington (Massey et al. 2019). The results indicated that the variables, which had the most influence of predicting landslide susceptibility, in rank order were: 24-hour rain; slope angle; slope aspect; local slope relief; land cover (vegetation); soil moisture; and material type. The model developed for Wellington was adapted for the entire country by including landslides from other storms (Fig. 4) along with the variables listed previously as the training data sets. The revised model-coefficients allow us to forecast landslide probability in each 32 × 32 m cell (covering the whole country) for a given 24 h rainfall and soil moisture (e.g., Fig. 5).

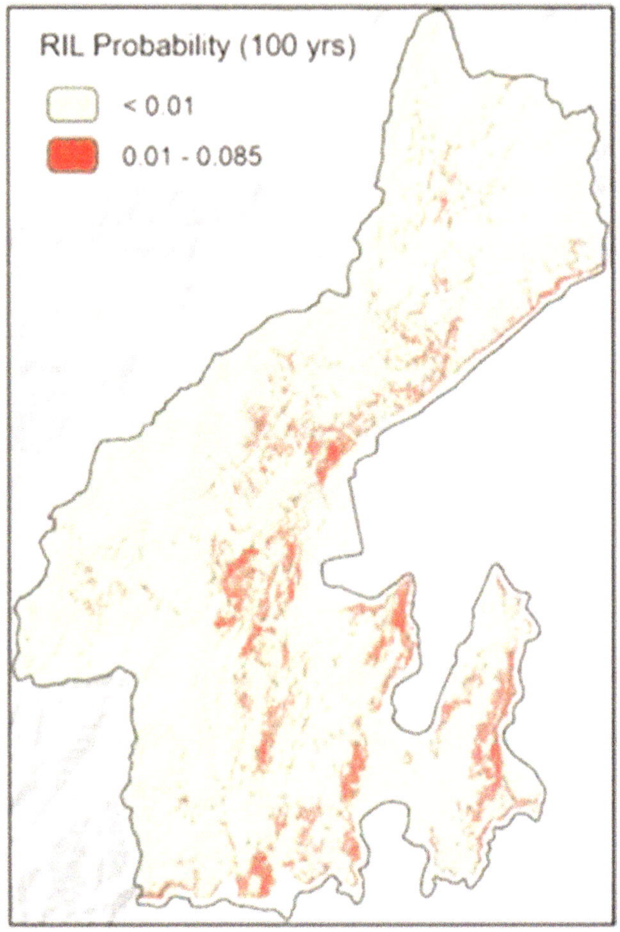

Fig. 5 Probability of a landslide occurring at a given location in Wellington if subjected to 24-hour rain amounts of 100-year return period

Rainfall-Induced Landslide Forecast Tool Development

Logistic regression model/s were converted to a forecast tool by incorporation of forecast and actual rainfall amounts and intensities supplied by MetService, New Zealand's national weather authority. The application of the Rainfall-induced Landslide (RIL) forecast tool requires two sets of data; (a) static data comprising the landslide susceptibility variables and the regional rainfall triggering thresholds and (b) dynamic event data (rainfall and soil moisture). Initially the RIL tool will run with forecast rainfall data at 72, 48, and 24 h prior to landfall, if forecast rainfall exceeds triggering thresholds, then actual rainfall data from rain radar will be used, complimented by data from the rain gauge network, and accumulated on an hourly basis. The RIL forecast tool will also be applied if the actual rainfall amounts during a storm event exceed rainfall triggering thresholds.

Conclusions

A rainfall-induced landslide forecast tool has been developed for New Zealand. Logistic regression modelling was used to determine the influence that landslide triggering and susceptibility variables had on a dataset of landslides initiated by 20 storm events across the country. The relationships developed by logistic regression modelling were applied at the national scale using national landslide susceptibility GIS datasets. The forecast tool will be run when regional rainfall ID thresholds for landslide initiation are exceeded. When combined with forecast rainfall amounts, the RIL forecast model can predict the probability of landslide occurrence, thus the spatial distribution of landslides for given forecast rainfall amounts. The RIL forecast tool will be used by the National GeoHazards Monitoring Centre to forecast and warn the public of possible damaging rainfall-induced landslide events.

Acknowledgements This work was funded by the GeoNet and National GeoHazards Monitoring Centre, funded by the Earthquake Commission (EQC).

References

Caine N (1980) The rainfall intensity-duration control of shallow landslides and debris flows. Geogr Ann 62:23–27

Crozier MJ (2005) Multiple-occurrence regional landslide events in New Zealand: hazard management issues. Landslides 2:247–256. https://doi.org/10.1007/s10346-005-0019-7

Crozier MJ (2017) A proposed cell model for multiple-occurrence regional landslide events: implications for landslide susceptibility mapping. Geomorphology 295:480–488. https://doi.org/10.1016/j.geomorph.2017.07.032

Dymond JR, Ausseil A, Shepherd JD, Buettner L (2006) Validation of a region-wide model of landslide susceptibility in the Manawatu-Wanganui region of New Zealand. Geomorphology 74:70–79

Glade T (1998) Establishing the frequency and magnitude of landslide-triggering rainstorm events in New Zealand. Environ Geol 35(2–3):160–174

Guzzetti F, Peruccacci S, Rossi M, Stark CP (2008) The rainfall intensity-duration control of shallow landslides and debris flows: an update. Landsides 5:3–17. https://doi.org/10.1007/s10346-0070112-1

Massey C, Townsend D, Rathje E, Allstadt KE, Lukovic B, Kaneko Y, Bradley B, Wartman J, Jibson RW, Petley DN, Horspool N, Hamling I, Carey J, Cox S, Davidson J, Dellow S, Godt JW, Holden C, Jones K, Kaiser A, Little M, Lyndsell B, McColl S, Morgenstern R, Rengers FK, Rhoades D, Rosser B, Strong D, Singeisen C, Villeneuve M (2018) Landslides triggered by the 14 November 2016 Mw 7.8 Kaikōura Earthquake, New Zealand. Bull Seismol Soc Am 108(3B):1630–1648. https://doi.org/10.1785/0120170305

Massey et al (2019) SLIDE (Wellington): rainfall induced and earthquake induced landslide hazard models. GNS Science Report SR2019/37

Page MJ (2015) Estimating the economic cost of landslides in New Zealand: an assessment using selected event case studies, and public utility and insurance cost data sets. GNS Science Internal Report 2014/13, 40 pp

Presenting Some Successful Cases of Regional Landslides Early Warning Systems in China

Qiang Xu, Dalei Peng, Xuanmei Fan, Xing Zhu, and Chaoyang He

Abstract

The early warning of landslides is crucial for risk management and reduction. However, it is challenging due to the complex nature of landslide behaviours and failure mechanisms. Generally, landslides undergo a certain period of evolution from deformation to failure, and own obvious three-stage deformation phases of creep slope failure. Most of landslides own obvious sudden characteristics, such as rockslide, avalanche, and loess flowslide, which are object of this study. In this paper, an early warning systems for regional landslides and a comprehensive warning model are established, taking into account deformation rate threshold and the improved tangent angle as the early warning parameters. A four-level early warning criterion is proposed. Once the early warning parameters exceeded default thresholds, relevant local authorities and scientists could immediately receive the warning information released by the 3D WebGIS-based platform. It is the 11 times our early warning system successfully forecasted landslides in four different regions since its implementation in 2017. Here, we present three typical cases of successful early warning from these three different regions and timely evacuation in advance. It could take a reference and applicable for other cases globally.

Keywords

Slope monitoring • Early warning system • Real-time • Successful early warning

Introduction

In many countries and areas, landslide is considered as one of the most severe natural hazard, which causes huge socioeconomic losses, especially in mountainous areas (Zaki et al. 2014; Fan et al. 2019a). The monitoring and early warning of landslides are crucial for risk management and the mitigation of their effects and the reduction of loss of human lives and assets (Fan et al. 2019b). The monitoring of landslides is a powerful tool for understanding kinematic aspects of mass movements and disruptive processes. Early warning should be implemented before a catastrophic event arrive, to allow individuals to take early action and to avoid a hazard turning into a human disaster (Sassa et al. 2009).

Prediction of slopes failure is a global challenge for geoscientists due to the complex nature of landslide failure mechanisms (Intrieri et al. 2019). To overcome this challenge, we have monitored the displacement of more than hundreds of multiple types of landslides in Western China. Based on these monitoring data from a large number of landslides, we found that landslides undergo a certain period of evolution from deformation to failure, and own obvious three-stage deformation phases of creep slope failure, such as initial deformation, constant deformation and accelerated deformation (Saito and Uezawa 1961; Xu et al. 2011). Displacement deformation of landslides will be obviously arisen when the depromation phase enters the accelerated deformation (Fig. 1a). (Fan et al. 2019b) (Fig. 1a). A deformation-time curve could describe the whole process of slope

Q. Xu (✉) · X. Fan · X. Zhu · C. He
State Key Laboratory of Geohazard Prevention and Geoenvironment Protection, Chengdu University of Technology, Chengdu, 610059, Sichuan, China
e-mail: xuqiang_68@126.com

X. Fan
e-mail: fxm_cdut@qq.com

X. Zhu
e-mail: zhuxing330@163.com

C. He
e-mail: hechaoyang2013@mail.cdut.edu.cn

D. Peng
Faculty of Science and Technology, Technological and Higher Education Institute of Hong Kong, Tsing Yi, Hong Kong, China
e-mail: pdchbsz@vtc.edu.hk

© Springer Nature Switzerland AG 2021
N. Casagli et al. (eds.), *Understanding and Reducing Landslide Disaster Risk*, ICL Contribution to Landslide Disaster Risk Reduction,
https://doi.org/10.1007/978-3-030-60311-3_33

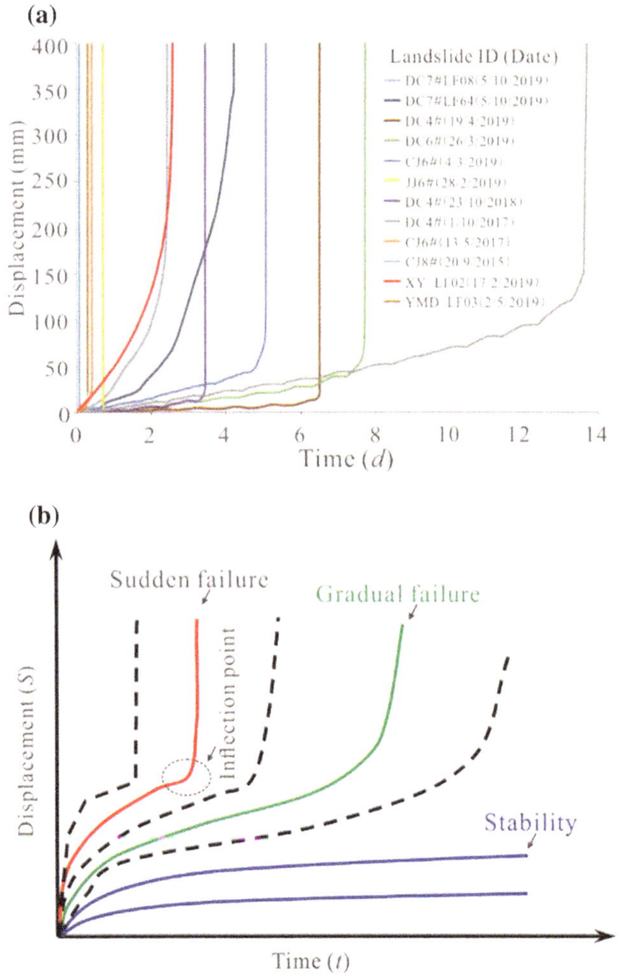

Fig. 1 Deformation characteristics of landslides. **a** Successive different landslide cumulative displacement-time curves (The difference of these landslides lies that some landslides have longer deformation period before failure (years, months, and days) than some others (hours, minutes)); **b** Movement patterns of slope displacement-time curve

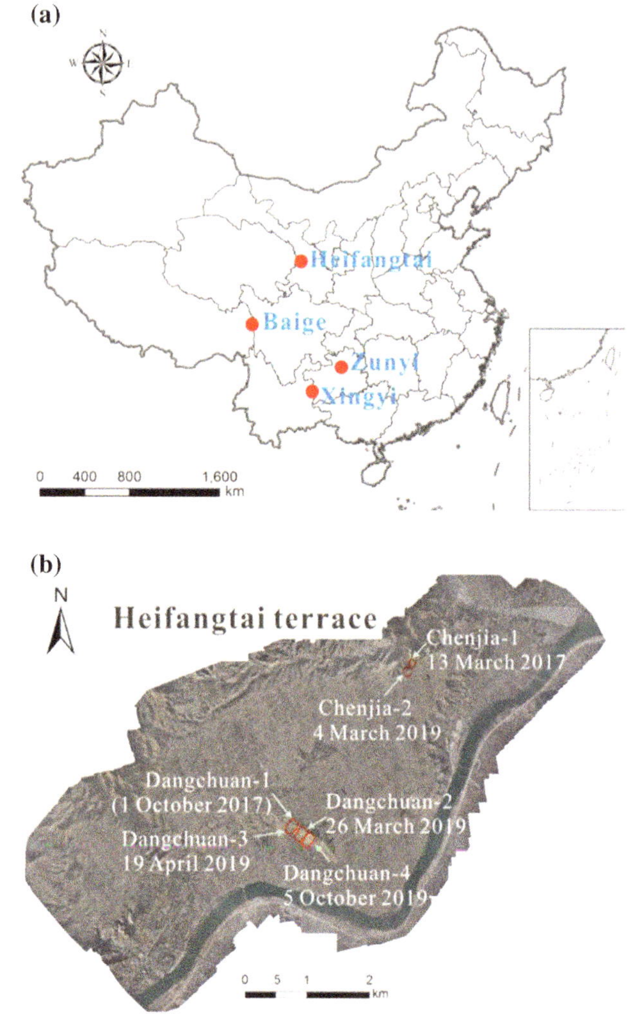

Fig. 2 Map showing locations in China where SKLGP's early warning system successfully predicted the eleven landslides in four regions (for details see Table 1)

failure. The curve is developed based on the statistical analysis of failure characteristics of landslides and on many displacement–time monitoring data (Fig. 1b) (Gance et al. 2014).

Movement patterns in different landslide have been extensively documented (Varnes 1978; Petley and Allison 1997), which could be divided into three distinct patterns being evident: stable, gradual failure, and sudden failure (Fig. 1b). (1) Stable: some landslides may undergo long periods of creep and movement will finally keep steady stable. (2) Gradual failure: there may be long-term displacement at low strain rates, frequently termed 'creep'. Creep may vary in rate on a seasonal basis but movement will rarely cease altogether, representing a gradual failure. And there are many successful warnings cases for gradual failure (Casagli et al. 2010). (3) Sudden failure: there are cases of short-term movement at very high rates of displacement (Carlà et al. 2017). With this know-how, a real-time early warning system based on the new artificial intelligence and data transmission technologies was developed by State Key Laboratory of Geohazard Prevention and Geoenvironment Protection (SKLGP). This system has successfully predicted 11 landslides since 2017 and saved thousands of lives in China (Fig. 2 and Table 1). In this study, the application of our self-developed landslide early warning systems (LEWS) to Baige rockslide-avalanche, Longjing rockslide in Xingyi, and loess flowslide at Heifangtai is described, covering monitoring strategies and early warning model.

In order to solve the key problem of "when a landslide will occur", main procedures have been undergone in this study: (1) the location and size of potential landslides were investigated based on the effective integration of Space-Sky-Ground three-dimension surveys (Xu et al. 2019). (2) a field study site was carried out and the characteristics of

Table 1 Successful and detailed landslides predictions by the LEWS (Fan et al. 2019b; Xu et al. 2020)

Landslide ID	Lead time (h)	Occurrence of landslide	Region
Chenjia-1	1.02	13/05/2017	Heifangtai
Chenjia-2	2.03	04/03/2019	Heifangtai
Dangchuan-1	11.17	01/10/2017	Heifangtai
Dangchuan-2	0.68	26/03/2019	Heifangtai
Dangchuan-3	0.30	19/04/2019	Heifangtai
Dangchuan-4	32.15	05/10/2019	Heifangtai
Baige-1	24	03/11/2018	Baige
Baige-2	0.5	11/11/2018	Baige
Baige-3	0.5	21/11/2018	Baige
XY-1	0.88	17/02/2019	Xingyi
YMD-1	0.33	02/05/2019	Zunyi

sudden-type landslides were summarized. (3) a monitoring method with self-adaptive adjusting sampling frequency is developed to monitor potential landslides. (4) a series of landslide warning thresholds with the improved tangent angle model and the deformation rate were proposed using the data processing method.

These successful experiences provide an impressive example of how a LEWS can help for preventing and mitigating landslide risks. We believe the practical implementations and the successful history of our LEWS will surely benefit other countries as well and improve global landslide resiliency.

Monitoring and Warning Methods Proposed in this Study

Monitoring Strategy

Displacement and other displacement-derived quantities are the main monitored parameters, which widely used for the early warning of landslides all over the world (Fan et al. 2019b). The cumulative displacement obtained from improved crack gauges present the most continuous and reliable data during the accelerated deformation stage (Xu et al. 2020). With the advantage of a self-adaptive data acquisition technique developed by (SKLGP) (Zhu et al. 2017), the crack gauge can automatically adjust the sampling frequency with respect to the displacement rate velocity (Fig. 3) (Fan et al. 2019b).

Multiple Criteria Warning Model

A general and quantitative criterion for LEWS was proposed by Xu et al. (2011), which described the improved tangential angle (α), referring to the deformation rate of the displacement-time curve at a given time (Eq. 1). This angle was used to specify different alert levels (Fan et al. 2019b).

$$a = \arctan\frac{v_i}{v_0} \tag{1}$$

where a is improved tangential angle; v_i is displacement rates; v_0 estimates of the creep rate at constant deformation state.

Catastrophic landslides often occur suddenly without any noticeable precursors, and are of a very short duration after entering the accelerated deformation and the speed is relatively fast in the imminent sliding stage (Qi et al. 2018). The early warning and prediction of the occurrence of landslides may be misjudged based on the traditional improved tangent angle method as the early warning criterion. Due to this reason, multiple alert thresholds are established in our LEWS. On the basis of the improved tangential angle criterion, the slope sudden failure also consider whether the deformation rate is larger than a critical rate value ($V_1 < V_2 < V_3$), the incremental rate (ΔV) at the three deformation stages of landslide (Fan et al. 2019b). The comprehensive warning criteria, taking into account three thresholds of improved tangential angle, three thresholds of deformation rate, and deformation rate increment are shown in Fig. 4.

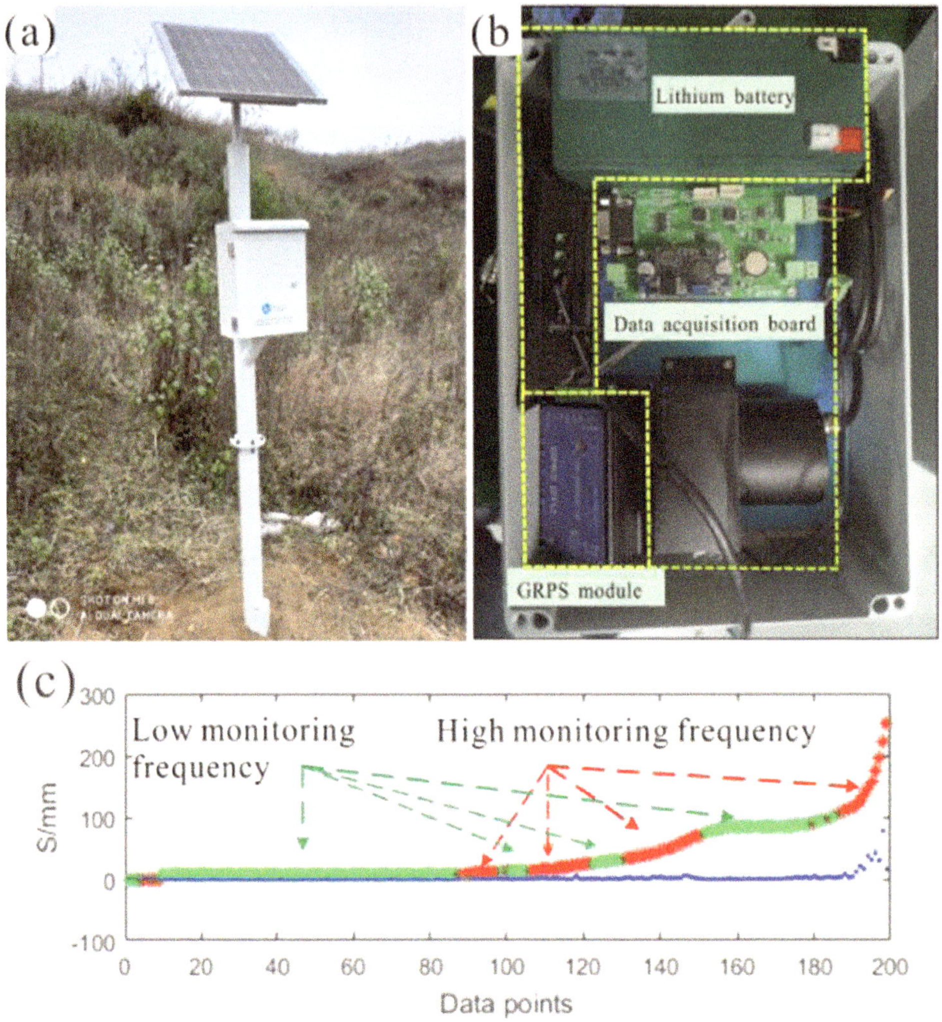

Fig. 3 Components of the improved self-adaptive crack gauge and monitoring data. **a** Intelligent crack gauge monitoring station; **b** Integrated intelligent landslide crack gauge; **c** Schematic diagram of self-adaptive acquisition frequency

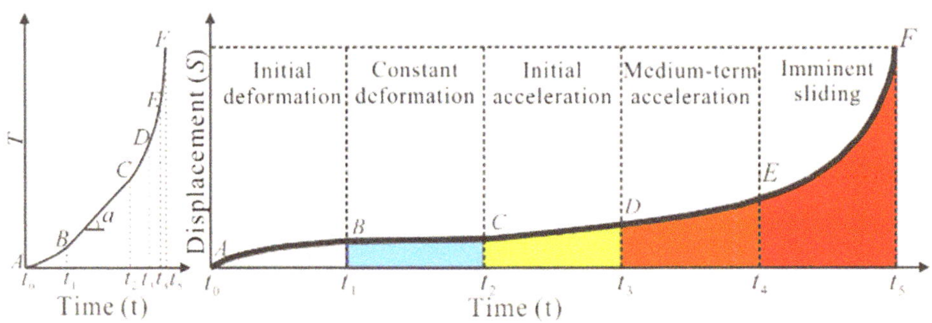

Deformation stage		Initial deformation	Constant deformation	Acceleration deformation		
				Initial acceleration	Medium-term acceleration	Imminent sliding
Early warning parameters	Deformation rate V (mm/d)	$V < V_1$	$V_1 \le V < V_2$	$V_2 \le V < V_3$	$V \ge V_3$	
	Rate increment ΔV (mm/d)	$\Delta V < 0$	$\Delta V \approx 0$	$\Delta V > 0$		
	Tangential angle α (°)	$\alpha < 45°$	$\alpha \approx 45°$	$45° < \alpha < 80°$	$80° \le \alpha < 85°$	$\alpha \ge 85°$
Early warning level		—	Attention	Caution	Vigilance	Alarm

Fig. 4 Outline of the four-level comprehensive early warning criterion based on velocity threshold and tangential angle

Application of the Method

Since the proposed the comprehensive early warning model has been established, we were able to successfully warn 11 landslides spread across China (Fig. 2 and Table 1). In this document, we present the experiences acquired in the monitoring and warning of three landslide cases named Baige landslide, loess flowslide and Xingyi rockslide. The early warning of these landslides prevented casualties and property losses, and guarantied the safety of more than 1000 people's lives and property, and achieved very good economic and social benefits.

Case 1: Baige Rockslide-Avalanche

In the early morning of October 11, 2018, a large rockslide-avalanche occurred at the border between Sichuan Province and Tibet. The landslide was located on the west bank of the Jinsha River near Baige Village, Jiangda County, Tibet. The landslide formed a dam with a water storage capacity of about 290 million m^3. It was fully drained by October 13 with the completion of partial discharge, and the danger was relieved. Then, on 3 November 2018, the landslide reactivated again in the form of a large rockslide that dammed the river for a second time and formed a new barrier lake with a volume of 5.24×10^8 m^3. These two landslides and river blocking events have aroused widespread social concern. After the October 11, 2018 event, 16 global navigation satellite system (GNSS) receivers, 16 crack gauges, and 1 rainfall gauge were placed on tensile cracks behind the source area. Sudden increases were observed three times, and they are consistent with the increase in recorded displacement time histories. First increase occurred at 18:00 local time on November 3, 2018, during the slope failure. The other two were observed at 8:00 on 11 November 2018 and at 11:00 on 21 November 2018. The early warning system performed remarkably well as predictor of the slope failure on November 3 is presented in this paper. At 21:00 on 3 November 2018, the deformation rate was 60 mm/d and the maximum tangent angle was 85°. The early warning system was able to predict the second large-scale slope failure 24 h in advance, along with minor rock falls during the spillway construction (Fan et al. 2019a). (Fig. 5).

Case 2: Heifangtai Loess Flowslide

Heifangtai is located on the fourth terrace of the Yellow River, in the semi-arid area of the Loess Plateau and at the confluence of Huangshui River and the Yellow River. A total of 77 landslides were identified on the Heifangtai terrace according to image taken on 18 January 2015 (Fig. 2). One of loess landslide named Dangchuan-1 loess flowslide is

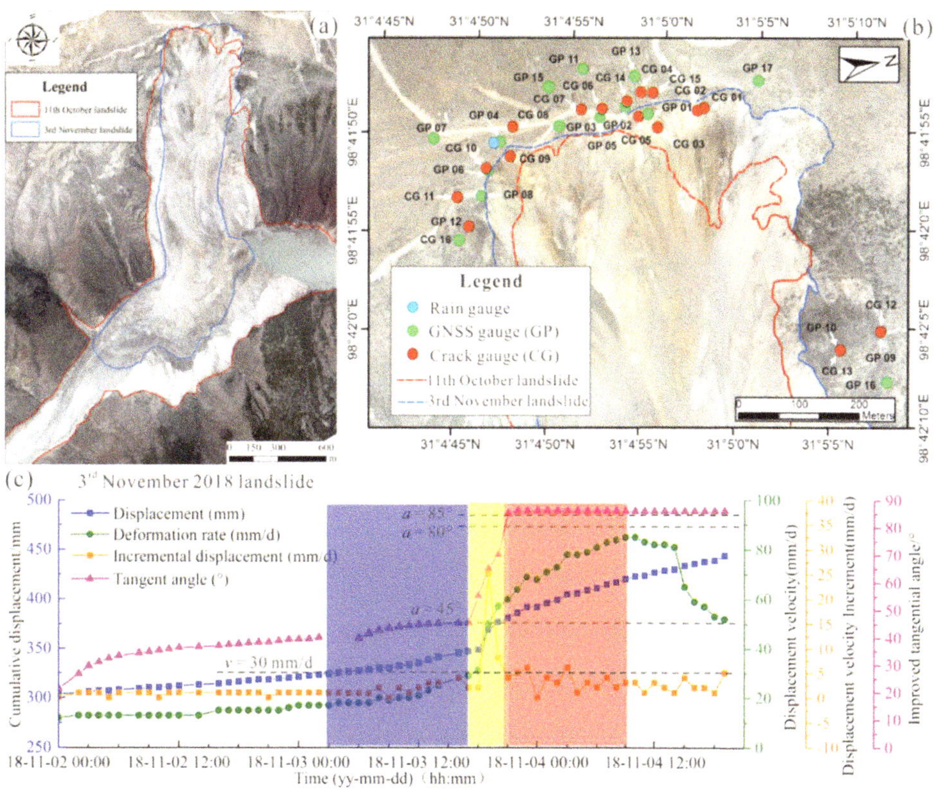

Fig. 5 Monitoring network, observed displacements and successful implementations of early warning for Baige rockslide-avalanche.
a Orthographic image of post-sliding (taken on 5 November 2018); **b** Location map of installed monitoring instruments (GNSS gauge (GP) refers to the GNSS displacement sensors and CG refers to crack gauges);
c Displacement curve before the landslide event on 3 November 2018 (using crack gauge CG02)

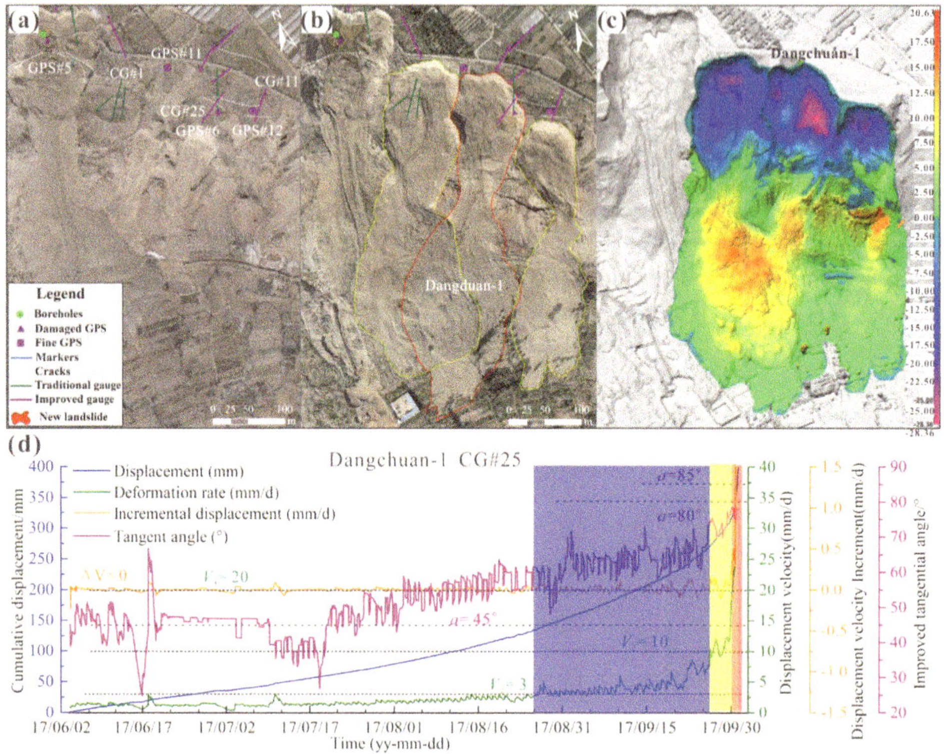

Fig. 6 Monitoring network, observed displacements and successful implementations of the loess flowslide Dangchuan-1 at Heifangtai. **a** Orthophoto image and monitoring system layout before landslide; **b** Orthophoto image and monitoring system layout after landslide; **c** Elevation change from the pre-slide to the post-slide; **d** Complete monitoring curve of cumulative displacement-time, deformation rate, increment of deformation rate and tangent angle curve

located on the southwest of the Heifangtai terrace. The source area of the landslide is 300 m in length and 20 m in width. Volume of the runout materials is about 3.4×10^5 m^3. The landslide has begun to deform and the rate of deformation has gradually increased since the end of August 2017. At 23:41 on 30 September 2017, the deformation rate was 99.45 mm/day and the maximum tangent angle was 88.27°. A successful prediction of the flowslide was achieved 8 h before the occurrence (Xu et al. 2020) (Fig. 6).

Case 3: Xingyi Rockslide

On 17 February 2019, a large rockslide occurred in Longjing village, Xingyi city, Guizhou Province. The rockslide was initially triggered in 2014 due to the removal of rock mass near the slope toe for road construction. Since then, the rockslide had become a potential threat to the local residents, pedestrians, and traffic. The deployment of the crack sensors and its distribution on the site are illustrated in Fig. 7a. To avert human and economic losses, emergency mitigation measures were implemented and a self-developed real-time LEWS. At 5 a.m. on 17 February 2019, the deformation rate was of about 251 mm/day, the cumulative displacement was of 829.2 mm, and the tangent angle was more than 85°. The real-time geological disaster monitoring and early warning system automatically send out the red early warning level to scientists, local government and residents, and then evacuate the relevant personnel by local government and residents in the site. A successful prediction of the rockslide was achieved 53 min before the occurrence on 17 February 2019. Prompt action taken by scientists and local authorities averted human and economic losses completely (Fig. 7).

Conclusions

Early warning systems are valuable tools for risk reduction, and they empower the community by providing timely alerts spanning hours, thus saving invaluable lives. Our self-developed LEWS helped to predict and mitigate eleven landslides, including two new landslides and nine reactivation landslide, eventually achieving zero casualties or injuries and almost no property losses. These cases of successful monitoring, early warning, and prediction and mitigation of landslides are very exceptional worldwide due to their complex nature. In this study, we presented three successful cases of early warning for disastrous landslides in China. The improved data monitoring and processing method were introduced. A comprehensive early warning model was established to realize the comprehensive dynamic and real-time early warning of landslide, taking into account the deformation rate threshold and improved tangent angle as the early warning parameters. Four alert levels were proposed on the basis of analysing the characteristics of displacement time

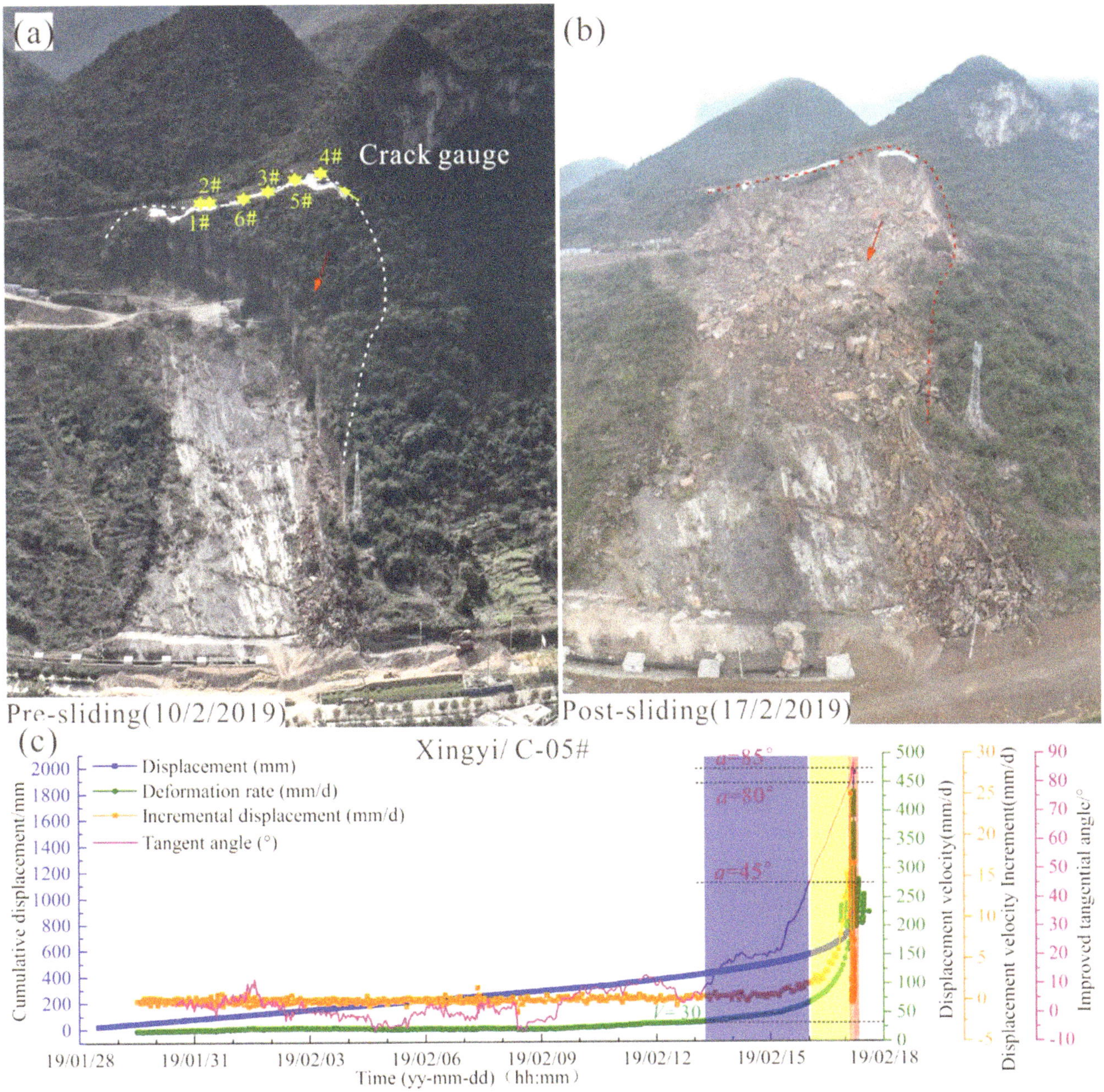

Fig. 7 Monitoring network, observed displacements and successful implementations of the Xingyi rockslide. **a** Location and distribution of installed monitoring instruments and pre-sliding image; **b** Image of post-sliding; **c** Cumulative displacement obtained by crack gauge C-05 and the derived curves of α, v, and Δv with corresponding alert levels a data from 27 January 2019 to 17 February 2019

series. The warning time in advance was from 0.3 h to 32 h. Our study also contributed to the compilation of a comprehensive database of displacement time histories of landslides that will be helpful for predicting similar landslides in the future.

Acknowledgements This research is financially supported by the National Natural Science Foundation of China (Nos. 41630640, 41790445 and 41877254), the Research Grants Council of the Hong Kong Special Administrative Region, China (No. UGC/FDS25/E11/17).

References

Carlà T, Farina P, Intrieri E, Botsialas K, Casagli N (2017) On the monitoring and early-warning of brittle slope failures in hard rock masses: Examples from an open-pit mine. Eng Geol 228:71–81. https://doi.org/10.1016/j.enggeo.2017.08.007

Casagli N, Catani F, Del Ventisette C, Luzi G (2010) Monitoring, prediction, and early warning using ground-based radar interferometry. Landslides 7(3):291–301. https://doi.org/10.1007/s10346-010-0215-y

Fan XM, Xu Q, Alonso-Rodriguez A, Subramanian SS, Li WL, Zheng G, Dong XJ, Huang RQ (2019a) Successive landsliding and damming of the jinsha river in eastern Tibet, China: Prime investigation, early warning, and emergency response. Landslides 16(5):1003–1020. https://doi.org/10.1007/s10346-019-01159-x

Fan XM, Xu Q, Liu J, Subramanian SS, He CY, Zhu X, Zhou L (2019b) Successful early warning and emergency response of a disastrous rockslide in Guizhou province, china. Landslides. https://doi.org/10.1007/s10346-019-01269-6

Gance J, Malet JP, Dewez T, Travelletti J (2014) Target detection and tracking of moving objects for characterizing landslide displacements from time-lapse terrestrial optical images. Eng Geol 172:26–40. https://doi.org/10.1016/j.enggeo.2014.01.003

Intrieri E, Carlà T, Gigli G (2019) Forecasting the time of failure of landslides at slope-scale: a literature review. Earth-Sci Rev 193:333–349. https://doi.org/10.1016/j.earscirev.2019.03.019

Petley DN, Allison RJ (1997) The mechanics of deep-seated landslides. Earth Surf Proc Land 22:747–758. https://doi.org/10.1002/(SICI)1096-9837(199708)22:8<747::AID-ESP767>3.0.CO;2-%23

Qi X, Xu Q, Liu FZ (2018) Analysis of retrogressive loess flowslides in Heifangtai, China. Eng Geol 236:119–128. https://doi.org/10.1016/j.enggeo.2017.08.028

Saito M, Uezawa H (1961) Failure of soil due to creep. In: Fifth international conference on soil mechanics and foundation engineering, Montreal, Que., Canada

Sassa K, Picarelli L, Yin YP (2009) Monitoring, prediction and early warning. In: Canuti P (ed) Landslides—disaster risk reduction. Springer, Berlin Heidelberg, pp 351–375

Varnes DJ (1978) Slope movement types and processes. In: Schuster RL, Krizek RJ (eds) Landslides: analysis and control, transportation research board special report, Washington, D.C., pp 11–33

Xu Q, Dong XJ, Li WL (2019) Integrated space-air-ground early detection, monitoring and warning system for potential catastrophic geohazards. Geomatics Inform Sci Wuhan Univ 44(7):957–966 (Chinese). https://doi.org/10.13203/j.whugis20190088

Xu Q, Peng DL, He CY, Qi X, Zhao KY, Xiu DH (2020) Theory and method of monitoring and early warning for sudden loess landslide: a case study on the heifangtai terrace. J Eng Geol 28(1):111–121 (Chinese). https://doi.org/10.13544/j.cnki.jeg.2019-038

Xu Q, Yuan Y, Zeng YP, Hack R (2011) Some new pre-warning criteria for creep slope failure. Sci China Technol Sci 54(S1):210–220. https://doi.org/10.1007/s11431-011-4640-5

Zaki A, Chai HK, Razak HA, Shiotani T (2014) Monitoring and evaluating the stability of soil slopes: a review on various available methods and feasibility of acoustic emission technique. Cr Geosci 346(9–10):223–232. https://doi.org/10.1016/j.crte.2014.01.003

Zhu X, Xu Q, Qi X, Liu HX (2017) A self-adaptive data acquisition technique and its application in landslide monitoring. In: Mikoš M, Arbanas Ž, Yin YP, Sassa K (eds) Advancing culture of living with landslides. Springer International Publishing, Switzerland, pp 71–78

Towards an Early Warning System for Instable Slopes in Georgia: The Large Tskneti-Akhaldaba-Landslide

Klaus-Peter Keilig, Markus Bauer, Peter Neumann, and Kurosch Thuro

Abstract

In June 2015 a flash flood caused by a failure of a natural dam originated by a hazardous debris flow in the Vere valley hit Georgias capital Tbilisi. 23 persons lost their lives and property damages were USD 24 mio. Along with the planning of the reconstruction of two destroyed roads an early warning system shall be developed an implemented for the safe use of the new roads. Therefore, some detailed geological investigations were carried out or are still in progress. This includes detailed engineering-geological mapping, hazard and risk mapping, geophysical measurements as well as planning and execution of exploration boreholes. Furthermore, a setup of a multi-sensor network was designed and is already installed in large parts. First data is providing some evidence of geological, hydrogeological and geotechnical setting in the Tskneti region as well as the occurring deformation. Further monitoring combined with numerical modelling will ultimately lead to the implementation of an early warning system, which is the main goal of this research. This paper tries to give an idea of the general geomorphological and geological setting as well as the occurring processes. It shows, the setup of the monitoring system and how it already delivers safety relevant data.

Keywords

Early warning system • Rock slide • Geomorphology • Monitoring network

K.-P. Keilig (✉) · K. Thuro
Technical University of Munich, 80333 Munich, Germany
e-mail: kp.keilig@tum.de

K.-P. Keilig · M. Bauer · P. Neumann
Baugeologisches Büro Bauer GmbH, 80807 Munich, Germany

Introduction

On the night of 13–14 June 2015 a very large landslide of reportedly 1 mio m^3 occurred in the Vere valley west of Georgias capital Tbilisi (Gaprindashvili et al. 2016). The landslide mass temporarily blocked the Vere river and a flash flood impacted in Tbilisi after failing of the dam (UNDP 2015a; Gaprindashvili et al. 2016). More than 700 citizens were directly affected, direct physical damage was estimated to be at least USD 24 mio and most tragically it caused 23 fatalities (UNDP 2015a). The catchment area is a region of high landslide susceptibility with a range of active and expectable processes with differing intensities and volumes. The event of 2015 must be seen as megaevent with a recurrence period of several 1000 s of years or even more. However, the landslide has created even more unstable conditions and weakened an already semi-stable system. As conclusion, the likelihood for medium to large subsequent events has risen significantly.

As part of the reconstruction of two damaged roads in the landslide area an early warning system (EWS) is to be developed based on detailed geological and geotechnical investigations. While some EWS for dams, mainly of hydropower plants, already exist in the country (e.g. CAE S. p.A. 2018) and the UNDP is currently developing a climate-based regional multi-hazard EWS for the whole country (UNDP 2015b), this is the first local EWS in development for a single landslide in Georgia.

A joint-venture of the local construction company Caucasus Road Projects (CRP), the Austrian manufacturer of retention structures Trumer Schutzbauten and the geological-geotechnical consultants of Baugeologisches Büro Bauer (Munich, Germany) is working on this project and developing the EWS in cooperation with Technical University of Munich, Engineering Geology.

Being located between the two main mountain ranges of the Caucasus, the Greater Caucasus in the north and the Lesser Caucasus in the south, about 65% of Georgias

© Springer Nature Switzerland AG 2021

N. Casagli et al. (eds.), *Understanding and Reducing Landslide Disaster Risk*, ICL Contribution to Landslide Disaster Risk Reduction,
https://doi.org/10.1007/978-3-030-60311-3_34

landmass is mountainous. Therefore, it is not surprising that the country has a broad landslide inventory with all major landslide processes occurring. Yet, a concept for landslide mitigation is only just developing in the country, partly as a response to the 2015 disaster (UNDP 2015b). Even though the potential for landslide processes becomes clear just by looking at the country's geomorphology the first ever rock fall protection fence in Georgia has been installed as part of this project.

The goal of the research presented in this article is to develop and implement a local early warning system for the mountain range west of Tskneti. It shall provide the highest safety possible for the newly reconstructed roads.

Study Area

Geological Setting

The large landslide is located about 10 km west of Tbilisi between Tskneti and Akhaldaba on the northern edge of the so-called Lesser Caucasus (Fig. 1). The Lesser Caucasus is part of the Alpine-Himalayan mountain belt and is characterized by complex thrust tectonics and strong volcanic activity (Sosson et al. 2010; Danelian et al. 2011). After Sosson et al. (2010) and Danelian et al. (2011) it can be divided into four major units:

The South Armenian Block (1) in the south of the mountain range is a remnant of the former Gondwana continent and is known for its Middle to Upper Paleozoic sedimentary rocks. The active Eurasian continental margin (2) is mainly characterized by Jurassic and Cretaceous volcanic deposits and follows to the northeast. In between lies the ophiolite complex of the Sevan-Akera zone (3), which reflects the area of the subduction zone during the continental collision. In addition, some 1,000 m thick molasse deposits (4) can be found. They were formed by the erosion of the young mountain range from the Paleocene to the Miocene and they are accompanied by a variety of volcanic deposits (Adamia et al. 2010; Sosson et al. 2010).

Two sedimentary rock formations of tertiary age occur in the study area, which belong to the Molasse deposits (4) (Gudjabidze 2003; Sosson et al. 2010). On the one hand, very thin-bedded alternating Oligocene sandstones and claystones appear in the lower part of the slope. The portion of sandstones within this formation increases from the bottom to the top (Gaprindashvili et al. 2016). On the other hand, there are flyschoid Eocene alternating deposits of thick-bedded to massive conglomerates and sandstones with very thin-bedded clay- and siltstones (Gaprindashvili et al. 2016). Both formations are, partly strongly, folded and tectonically stressed. In general, the rock layers are dipping very unfavorably with about 25–50° parallel to the slope. See Fig. 3 for an impression of the thick bedded formation.

The upper scarp of the large landslide is located at an altitude of about 1410 m.a.s.l. directly at the top of the mountain range west of Tskneti, which forms the southern valley flank of the Vere valley (see also Fig. 2). The average slope angle is 29° and therefore very steep between the main scarp (1410 m) and the Akhaldaba road (910 m). Below the Akhaldaba road, the terrain is significantly shallower at an average of 6.5° until the Vere river.

The Tskneti-Akhaldaba Landslide

The landslide leading to the blockage of the Vere river occurred between Tskneti and Akhaldaba south of the Vere river west of Tbilisi (Fig. 1) and was a highly complex process of different types of landslides, such as rock slides, debris slides, earth slides and debris flows (UNDP 2015a; Gaprindashvili et al. 2016). In the landslide area two important roads were completely destroyed by rock slide (upper Samadlo road) and debris flow (lower Akhaldaba road), isolating Akhaldaba from Tskneti, which is the villages main source for supplies, water and food (UNDP 2015a; Gaprindashvili et al. 2016).

Both landslide and flash flood were caused by exceptionally long and heavy rainfalls in the previous ten days to the event resulting in an already high discharge of the Vere river. After the landslide mass temporarily blocking the river and ultimately causing the flash flood after failing of the dam peak discharge during the event has been estimated to be 468 m^3/s (UNDP 2015a; Gaprindashvili et al. 2016). This almost doubles the discharge during the catastrophic flood in 1960 (259 m^3/s discharge; UNDP 2015a). Following a flood recorded on 4 June 2015 (155 m^3/s discharge) this were the highest consecutive floods ever recorded in the Vere river (UNDP 2015a). Vere river flows into Mtkvari river in Tbilisi.

Field Investigation

The development and implementation of an early warning system is an interdisciplinary and complex task. For a successful execution, extensive investigations are necessary in order to finally obtain a coherent geological and geotechnical model of the slope and to install the appropriate monitoring devices at the right locations.

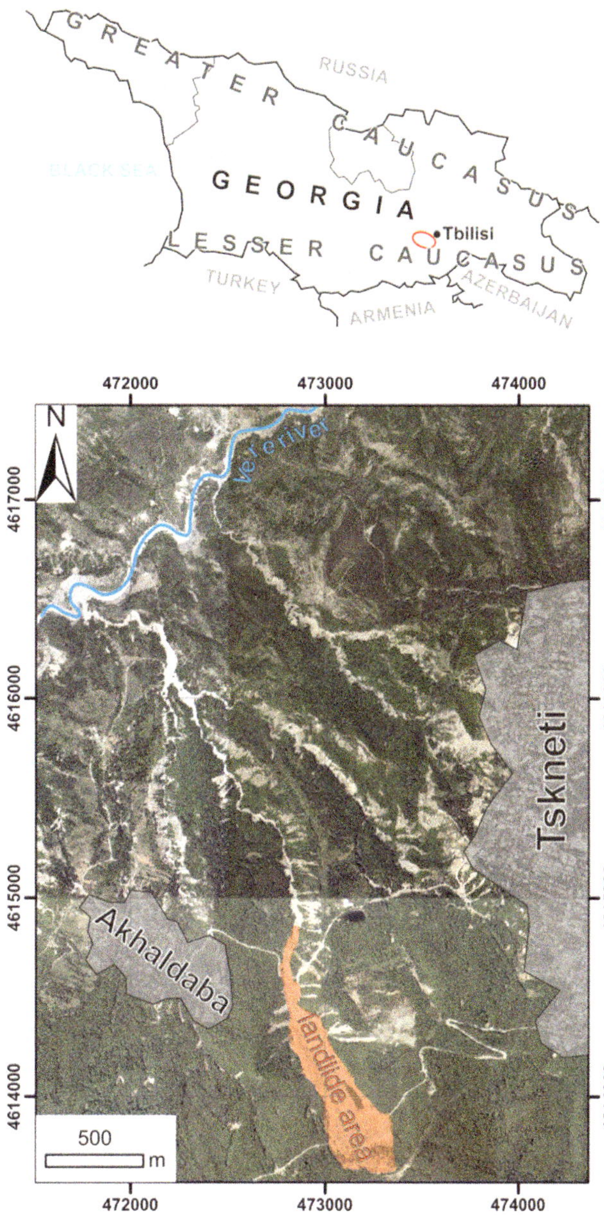

Fig. 1 Geographical overview of the study site

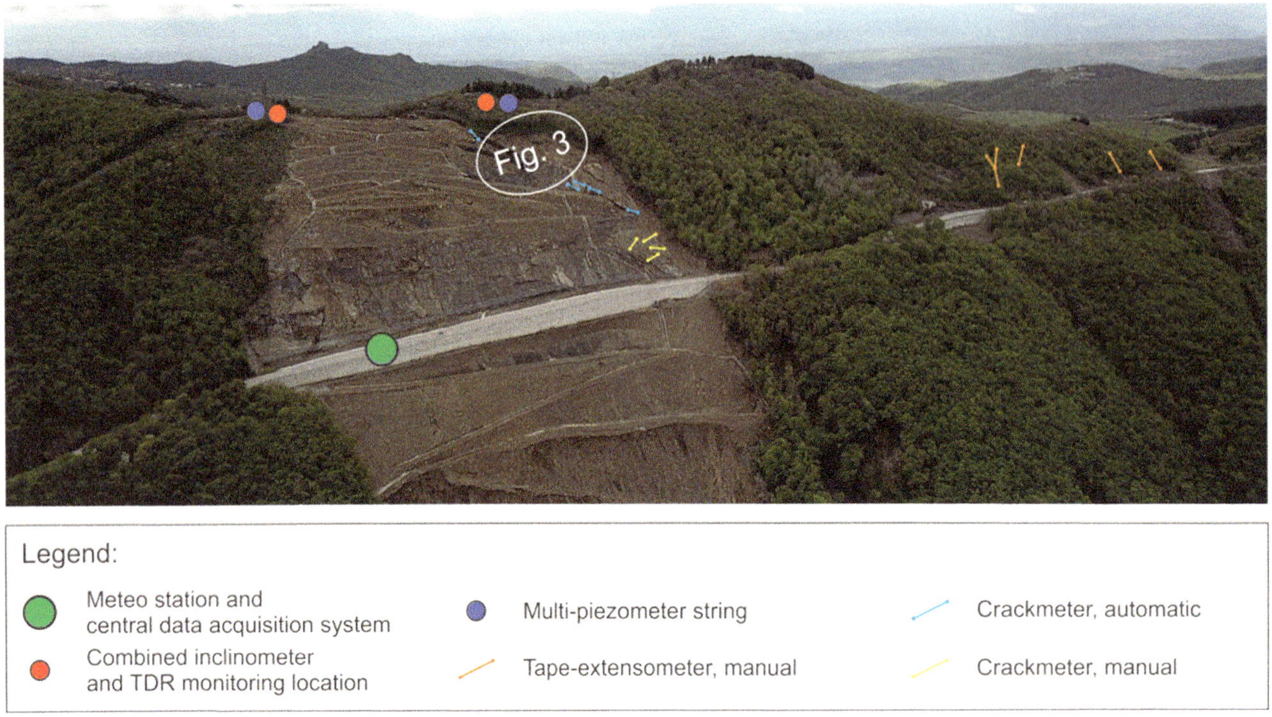

Fig. 2 Overview on the sensor network installed in the upper part of the landslide area. Extent of Fig. 3 is indicated

Fig. 3 Large cliff in the upper part of the landslide that is being monitored with an inclinometer

Geological-Geotechnical Investigation

The area of the large landslide, in particular the upper scarp area, as well as the closer vicinity have already been mapped in large parts from a geological-geotechnical point of view. In addition to the spatial distribution and the geotechnical properties of the rocks, geomorphological features of slope movements were recorded in order to draw conclusions about the different movement processes. The mapping has been carried out on a scale of 1:1.000 and forms the basis for the development of a geological model and the construction of a sensor network. It will also provide important information for further investigations.

In addition to the exploratory drillings along the roads, which are to be rebuilt as part of the construction work, further drillings were carried out on the basis of the first mapping results. The aim was, on one hand, geological exploration in other areas of the work area and, on the other hand, to install monitoring equipment into the boreholes.

Representative rock samples were taken from both boreholes and the surface to determine geotechnical parameters such as uniaxial compressive strength, tensile strength and durability in the laboratory.

Additionally, a very detailed characterization of discontinuities based on ISRM (1978) was carried out. This was done with special attention to spacing, persistence and roughness of sets of different discontinuities in order to create statistically relevant input data for future numerical stability calculations.

Optical Investigations

Optical methods are used in order to observe the development of the slope. UAV surveys are used to create digital elevation models (DEMs) by the photogrammetric reconstruction of the terrain surface. This is done using the commonly used "structure from motion—multi-view stereo" (SfM MVS) workflow (Carrivick et al 2016). Multiple DEMs can be compared following the approach of Wheaton et al. (2010) in order to detect areas with geomorphic changes.

Terrestrial laser scanning (TLS) is used to monitor particularly susceptible areas such as the 12 m high reinforced earth construction (as seen in Fig. 2) or the main cliff in the upper part of the study site (Fig. 3).

Monitoring Network

After the geological-geotechnical mapping was largely completed, numerous monitoring devices were installed in the project area in the summer of 2018. An overview of the installed sensors and monitoring location is shown in Fig. 2.

Ground Deformation

In order to observe movements in the unstable rock mass or between larger blocks, several crackmeters were installed in the project area. With these devices the aperture of discontinuities can be measured. In order to keep the costs for the instrumentation as low as possible, a very simple construction was chosen. This construction consists of two common rock bolts, which are installed on both sides of a discontinuity or crack. The distance between these bolts can be manually measured with a caliper. If the aperture of a discontinuity changes, so does the distance between the screws. Advantages of this easy to apply method are the extremely low costs and the robustness against damage. Additionally, a total of five automatic vibrating wire crackmeters were installed in particularly dangerous areas which were also very difficult to access. Just as the simple rock bolt construction, this monitoring device is fixed on both sides of a discontinuity and by detecting frequency changes of the vibrating wire, deformations can be detected at sub-mm range.

Furthermore, several monitoring locations with tape-extensometers were equipped to detect deformations at longer distances e.g. between large blocks and/or trees and a main slide scarp. This also represents a cost-effective and simple device, but subsequent manual readings are necessary. The measurement accuracy also is in the sub-mm range and can provide important insights into the movement patterns.

At the moment, preparations are under way for the installation of reflectors, which are to be regularly surveyed geodetically with a tachymeter.

Deep Deformation

Two boreholes (42 m and 50 m) above the main scarp of the landslide were equipped in order to perform inclinometer measurements. This allows to detect and quantify movements perpendicular to the borehole axis. In order to be able to carry out continuous deformation monitoring in case of movement acceleration without the high investment costs of a chain inclinometer, a coaxial cable was installed additionally in the boreholes. Time domain reflectometry (TDR) uses a transceiver to transmit electrical pulses into the cable and measure the reflection. If the cable is kinked by shearing movement, this can be detected in the reflected signal (Singer et al. 2006, 2009). In addition to the continuous measurement, another advantage is that also with large deformations, measurements can be made even if the inclinometer tube has already been squeezed and the inclinometer is not able to be lowered in the borehole (Singer et al. 2006, 2009; Thuro et al. 2010).

Hydrological Measurements

In five boreholes with 30–50 m depth, a multi-sensor chain of five piezometers was installed to measure water pressure at different elevations. The alternation of impermeable clay layers with more permeable sandstones suggests that non-correspondence aquifers exist at different depths. On the one hand, this hypothesis should be checked with the chosen installation. On the other hand, possible connections between pore water pressures and deformation measurements can be recognized.

Meteorological Measurements

In order to be able to correlate measured deformations and pore water pressures with intensity, duration and magnitude of precipitation events, a meteorological station was installed in the project area. This station measures precipitation, temperatures and humidity. It is equipped with a solar cell and thus self-contained.

First Results and Outlook

First measurement results show active movements in multiple sensors. Figure 4 shows parts of the inclinometer measurements monitoring the large cliff at the main scarp (see Fig. 3). A very defined translational shear movement in about 30 m depth can be detected and quantified. The results prove the ongoing deformation of the slope and provide vital information about the activity, velocity and intensity of the rock slide at the cliff. The results will also play an important role for future numerical modelling by indication the depth and activity of the movement.

Tape extensometer measurements have detected a high acceleration of a shallow slide directly next to the road west of the main landslide. Based on this data the slide was stabilized by a simple wall construction and damage to the road was avoided successfully. Monitoring of this slide continues and movement rates have decreased dramatically after the stabilization.

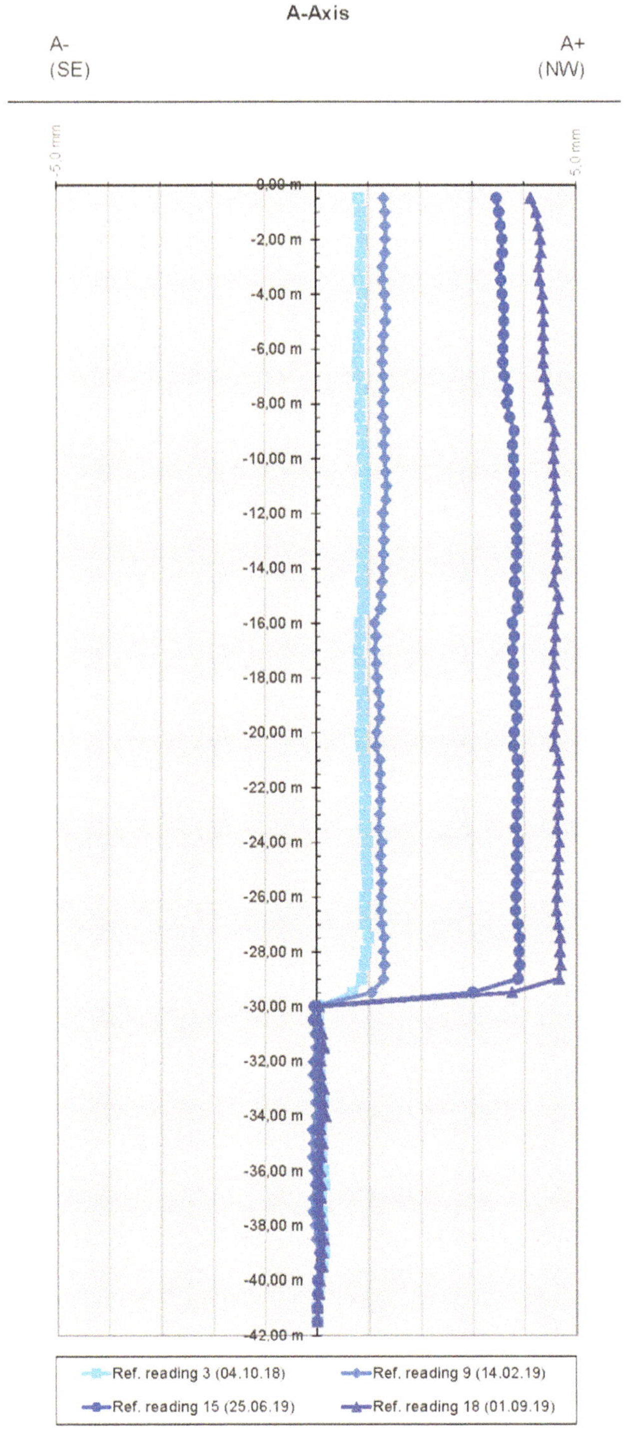

Fig. 4 Exemplary inclinometer measurements below the main cliff shown in Fig. 3 near the main scarp at the top

Overall, the installed monitoring system already proves to be very valuable for the safety of the road and will continue to provide important data for the completion of the EWS.

In the further progression of the project the data acquisition will be optimized with improved radio antennas and updated power supply. Ultimately, data collection and analysis should be carried out simultaneously through an integrated database system, as developed in the alpEWAS project (Thuro et al. 2009, 2010).

To develop a true local EWS, thresholds must be defined that are able to trigger the closure of the affected roads. Such thresholds can be large amounts of precipitation, critical pore water pressures or high deformation rates. To avoid false alarms, a great deal of effort must be made in deriving and evaluating these alarm thresholds correctly. Although there are many different ways in which such thresholds can be determined, the determination of such a value is still a challenge and a constant adaptation to new information is necessary. In most cases, time series analysis and/or numerical modeling are used to identify the thresholds (Nadim et al. 2009; Festl and Thuro 2016). In this project we aim for a multi-level threshold consisting of a threshold for precipitation, pore water pressure and deformation rate. In theory, significant precipitation causes a rising water table and therefore rising pore water pressure, which can result in higher deformation rates. Between each step is a certain time lag, which provides the possibility to both review the data and have a timely and early warning in its true meaning.

The most important task for the near future will be to complement and refine the geological model. These include e.g. the ongoing laboratory tests of geotechnical parameters, further drilling and their equipment to new measurement locations or the implementation and evaluation of planned geophysical measurements like ERT electrical resistivity tomography.

In addition, an application process is currently underway with Georgian colleagues for the promotion of a project aimed at developing a regional EWS for the whole Vere catchment.

References

Adamia S, Alania V, Chabukiani A, Chichua G, Enukidze O, Sadradze N (2010) Evolution of Late Cenozoic basins of Georgia (SW Caucasus): a review, pp 239–259. In: Sosson M, Kaymakci N, Stephenson RA, Bergerat F, Starostenko V (eds) Geological Society, London, Special Publications, p 340: Sedimentary basin tectonics from the Black Sea and Caucasus to the Arabian platform. The Geological Society, London (ISBN 9781862393080), 509 pp

CAE S.p.A. (2018) Dams: emergency plans, early warning and public safety systems. The case of Zhinvali in Georgia. https://www.cae.it/eng/news/dams-emergency-plans-early-warning-and-public-safety-systems.-the-case-of-zhinvali-in-georgia-nw-1181.html. Last accessed 24 Jan 2020

Carrivick J, Quincey D, Smith M (2016) Structure from motion in the geosciences: new analytical methods in earth and environmental science. Wiley Blackwell, Chichester (ISBN 9781118895849), 197 pp

Danelian T, Sosson M, Avagyan A, Galoyan G, Asatryan G, Rolland Y, Sahakyan L, Müller C, Grigoryan A, Person A, Corsini M, Jrbashyan R, Melkonyan R (2011) A brief geological outline of the lesser caucasus: new insights on its tethyan-alpine evolution based on recent results of a french-armenian collaboration. Bull Soc Geol Fr 18(2):65–75

Festl J, Thuro K (2016) Determination of thresholds at the Aggenalm landslide (Bayrischzell, Germany) by time series analysis and numerical modeling. In: Landslides and engineered slopes. Experience, theory and practice: proceedings of the 12th international symposium on landslides, 12–19 June 2016. Napoli, Italy, pp 909–916

Gaprindashvili G, Gaprindashvili M, Tsereteli E (2016) Natural disaster in Tbilisi City (Riv. Vere Basin) in the Year 2015. Int J Geosci 7 (9):1074–1087

Gudjabidze GE (2003) Geological map of Georgia 1:500.000. Georgian State Department of Geology and National Oil Company "Saqnavtobi", Tbilisi

ISRM—International Society for Rock Mechanics (1978) Suggested methods for the quantitative description of discontinuities in rock masses. Int J Rock Mech Mining Sci Geomech 15(6):319–368

Nadim F, Cepeda J, Sandersen F, Jaedicke C, Heyerdahl H (2009) Prediction of rainfall-induced landslides through empirical and numerical models. In: Proceedings of the 1st Italian workshop on landslides, 8–10 June 2009. Napoli, Italy, pp 206–215

Singer J, Schubäck S, Wasmeier P, Thuro K, Heunecke O, Wunderlich T, Glabsch J, Festl J (2009) Monitoring the Aggenalm landslide using economic deformation measurement techniques. Aus J Earth Sci 102(2):20–34

Singer J, Thuro K, Sambeth U (2006) Development of a continuous 3D-monitoring system for unstable slopes using TDR. Felsbau 24 (3):16–23

Sosson M, Rolland Y, Müller C, Danelian T, Melkonyan R, Kekelia S, Adamia S, Babazadeh V, Kangarli T, Avagyan A, Galoyan G, Mosar J (2010) Subductions, obduction and collision in the Lesser Caucasus (Armenia, Azerbaijan, Georgia), new insights, pp 329–352. In: Sosson M, Kaymakci N, Stephenson RA, Bergerat F, Starostenko V (eds) Geological Society, London, Special Publications, p 340: Sedimentary basin tectonics from the Black Sea and Caucasus to the Arabian platform. The Geological Society, London (ISBN 9781862393080), 509 pp

Thuro K, Singer J, Festl J, Wunderlich T, Wasmeier P, Reith CH, Heunecke O, Glabsch J, Schubäck S (2010) New landslide monitoring techniques—developments and experiences of the alpEWAS project. J Appl Geodesy 4(2):69–90

Thuro K, Wunderlich T, Heunecke O, Singer J, Schuhbäck S, Wasmeier P, Glabsch J, Festl J (2009) Low cost 3D early warning system for alpine instable slopes—the Aggenalm Landslide monitoring system. Geomech Tunnel 2(3):221–237

UNDP—United Nations Development Programme (2015a) Tbilisi disaster needs assessment 2015. Part 1 (Final draft), 59 pp

UNDP—United Nations Development Programme (2015b) Towards a multi-hazard early warning system for Georgia, 25 pp

Wheaton JM, Brasington J, Darby SE, Sear DA (2010) Accounting for uncertainty in DEMs from repeat topographic surveys: improved sediment budgets. Earth Surf Proc Land 35(2):136–156

An EWS of Landslide and Slope Failure by MEMS Tilting Sensor Array

Lin Wang, Makoto Fukuhara, Taro Uchimura, Gallage Chaminda, and Tharindu Abeykoon

Abstract

A low-cost and simple method of monitoring rainfall-induced landslides is proposed, with the intention of developing an early-warning system (Uchimura et al. 2015). Surface tilt angles of a slope are monitored using this method, which incorporates a Micro Electro Mechanical Systems (MEMS) tilt sensor and a volumetric water content sensor. In several case studies, the system detected distinct tilt behaviour in the slope in pre-failure stages. Based on these behaviours and a conservative approach, it is proposed that a precaution for slope failure be issued at a tilting rate of 0.01°/h, and warning of slope failure issued at a rate of 0.1°/h. The development of this system can occur at a significantly reduced cost compared with current and comparable monitoring methods, which such as extensometer or borehole inclinometers. Increasing the number of installed sensors, thus increasing the accuracy of the early warning thresholds and predictions, so that given the cost reduction, slopes can be monitored at many points, resulting in detailed observation of slope behaviours, but the potentially large number of monitoring points for each slope does induce a financial restriction. Therefore, the selection of sensor positions needs to be carefully considered for an effective early warning system. These case studies will henceforth be helpful in determining the installation of the sensor array of early warning system.

Keywords

Landslide • Slope failure • Early warning • MEMS tilting sensor

L. Wang (✉) · M. Fukuhara
Chuo Kaihatsu Corporation, Technology Center, Tokyo, 169-8612, Japan
e-mail: wang@ckcnet.co.jp

M. Fukuhara
e-mail: fukuhara@ckcnet.co.jp

T. Uchimura
Geotechnical and Geosphere Research Group, Saitama University, Saitama, 338-8570, Japan
e-mail: uchimurataro@mail.saitama-u.ac.jp

G. Chaminda · T. Abeykoon
Science and Engineering Faculty, Queensland University of Technology, Brisbane, QLD 4001, Australia
e-mail: chaminda.gallage@qut.edu.au

T. Abeykoon
e-mail: a.abeykoon@qut.edu.au

Introduction

There is a long history of prevention and mitigation of rainfall and/or scouring-induced landslides. Mechanical countermeasures to prevent slope failure have been widely used, including retaining walls and ground anchors. However, these methods can be expensive and are not always realistically applicable for all slopes of varying scale and potential risk factors. Therefore, careful monitoring of slope behaviour and consequent early warning of failure provides a reasonable and slope-specific alternative.

In this paper, an early warning system for slope failure is proposed and its development is described (Fig. 1) (Uchimura et al. 2015). The system consists of a minimum number of low-cost sensors strategically placed on a slope, with monitoring data that are collected being transmitted via a wireless network. It is anticipated that this low-cost and simple system will provide at risk residents with access to accurate and timely precautions or warnings of slope failure.

Uchimura et al. (2015) summarized case studies of slope tilting rates during pre-failure stages obtained on several natural slope sites under natural or artificial heavy rainfall. Figure 2 presents an example of the typical monitoring data obtained, in which the tilting rate (X-axis) can be related with the time elapsed until slope failure or slope stabilization (Y-axis). Figure 3 shows the definition of the tilting rate and

© Springer Nature Switzerland AG 2021
N. Casagli et al. (eds.), *Understanding and Reducing Landslide Disaster Risk*, ICL Contribution to Landslide Disaster Risk Reduction, https://doi.org/10.1007/978-3-030-60311-3_35

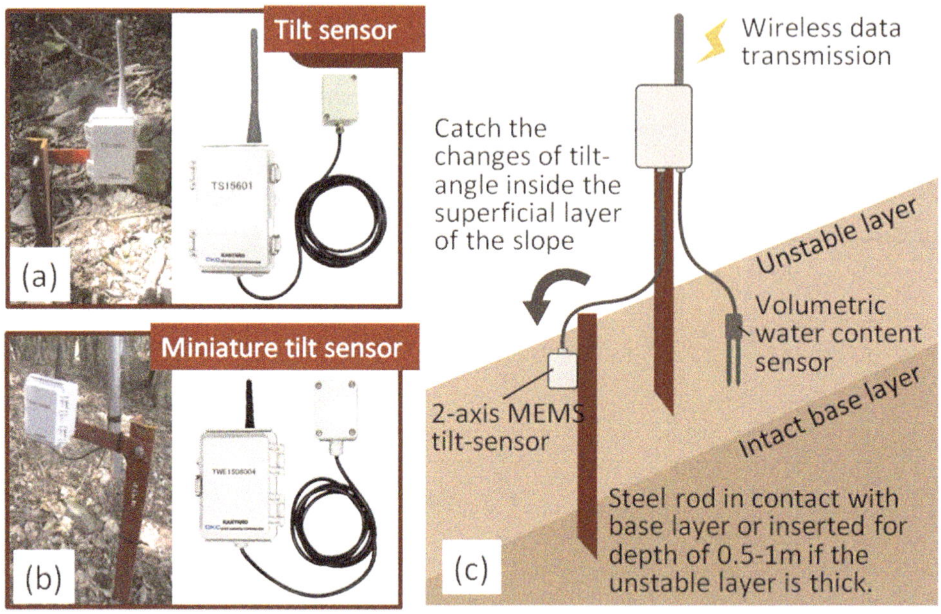

Fig. 1 Schematic illustration of MEMS tiltmeter sensor for early warning

the time in Fig. 2, in which Ti is the time until failure or stabilization, and Ri is tilting rate.

In cases where the slope failed at the position of the tilt sensor, the elapsed time is measured from the time when tilting accelerated to the time of failure. In cases where the slope did not fail but instead stabilized, the time is measured from when tilting decelerated to the time when the slope stabilized.

According to Fig. 2, the order of tilting rate observed with slope deformation varied widely, from 0.0001 to 10°/h depending on a number of factors. The tilting rate tends to increase towards failure with a relatively short time until failure, when a higher tilting rate is observed. The observed tilting rate was >0.01°/h for all the cases in which the slope failed or nearly h was observed before failure for a tilting rate of 0.1°/h.

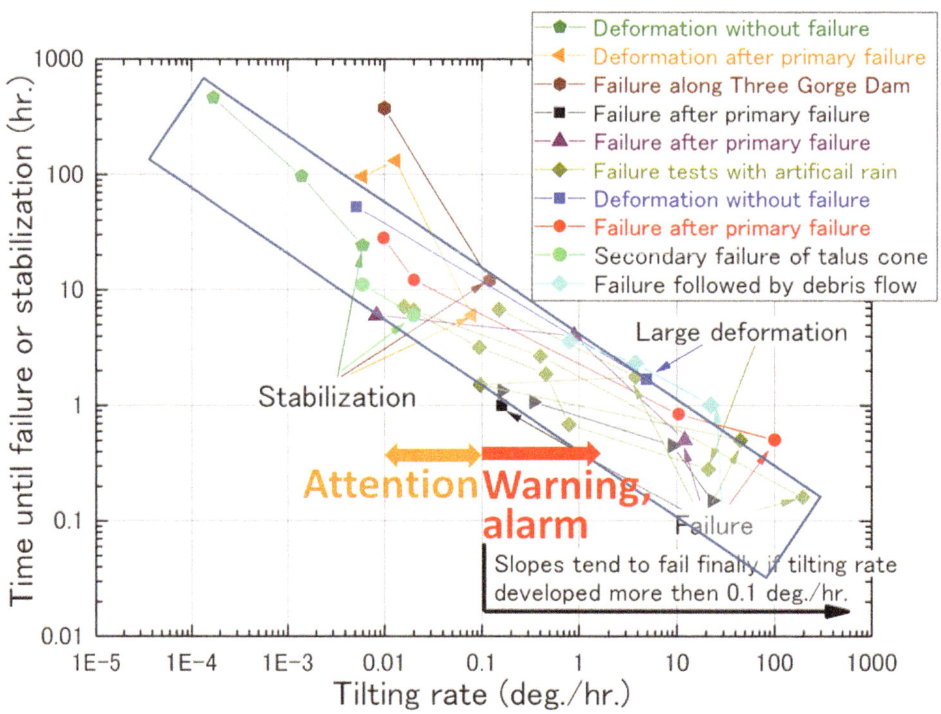

Fig. 2 Graphic illustration of the tilting rate as a function of time before slope failure (or stabilization) for several case studies

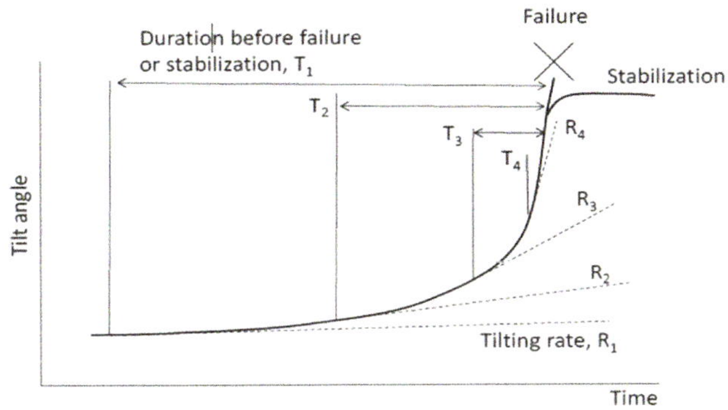

Fig. 3 Definition of the tilting rate and the durations

Based on the past case studies, it is proposed that when the tilting rate exceeds 0.1°/h a warning of slope failure should be issued, and a precaution issued at a tilting rate of 0.01°/h, taking safety into account. Additionally, this paper explores efforts by the current authors to improve the applicability of the monitoring and early warning system. The miniature tilt sensors are modified from that currently available to be more cost-effective, smaller in size and weight, and simpler to install, maintain and operate. As a result, it is possible to install a larger number of sensors on a given slope, thereby providing greater coverage and higher data density.

Figure 4 illustrates the typical arrangement of two types of proposed sensors, with data transfer pathways also shown. Despite the advantages described above, the new type miniature tilt sensors have relatively short radio transmission distances (~30 m in non-ideal conditions). They are arranged densely on high-risk areas of a slope, with one conventional tilt sensor unit collecting all the data of each area. The data are transmitted over greater distances (300–600 m), and uploaded to an internet server. If the data transfer is interrupted for other reasons, the data is reacquired and an alert is issued when the signal is restored.

Field Validation in Japan and Australia

A Case of Detection of Rain Induced Landslides in Critical Slopes the Lake Baroon Catchment, Maleny Plateau, Brisbane, Australia

This case study investigated the applicability of real-time monitoring and wireless data transmission in predicting rain-induced slope instability in critical slopes. The ground inclinometers equipped with MEMS tilt sensors, volumetric water content sensors, a rain gauge and a wireless data transmission unit (DTU) for real-time slope monitoring. The study employed a wide range of data collected in the period from 10th May 2016 to now, this year of 2020, for the prediction of the slope failure under rainfall infiltration.

Study area is Lake Baroon catchment, Maleny (Fig. 5) is located approximately 100 km north of Brisbane (26.76 0S 152.85 0E). Mapleton—Maleny plateau, which contains Lake Baroon catchment have been documented and discussed since the mid-1950s as a highly susceptible area for rainfall-induced slope failure. Slope failure and mass movement of sediment into the waterways within the Lake Baroon catchment are recognized as a significant risk to water quality and the water storage capacity of Lake Baroon, which is used to supply water to South East Queensland. Approximately 170 mass movement landforms have been identified within the Baroon catchment, and the study area is one such high-risk slope. This landslide site hosted a voluminous, single-failure rotational landslide in 2008 following heavy rainfall (Abeykoon et al. 2018).

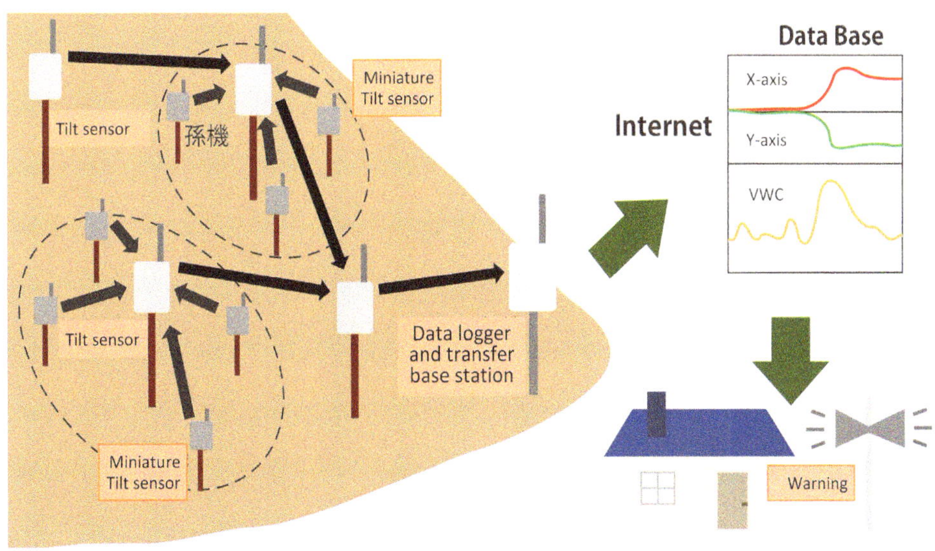

Fig. 4 An early warning system of slope failure by multi-point tilt and volumetric water content

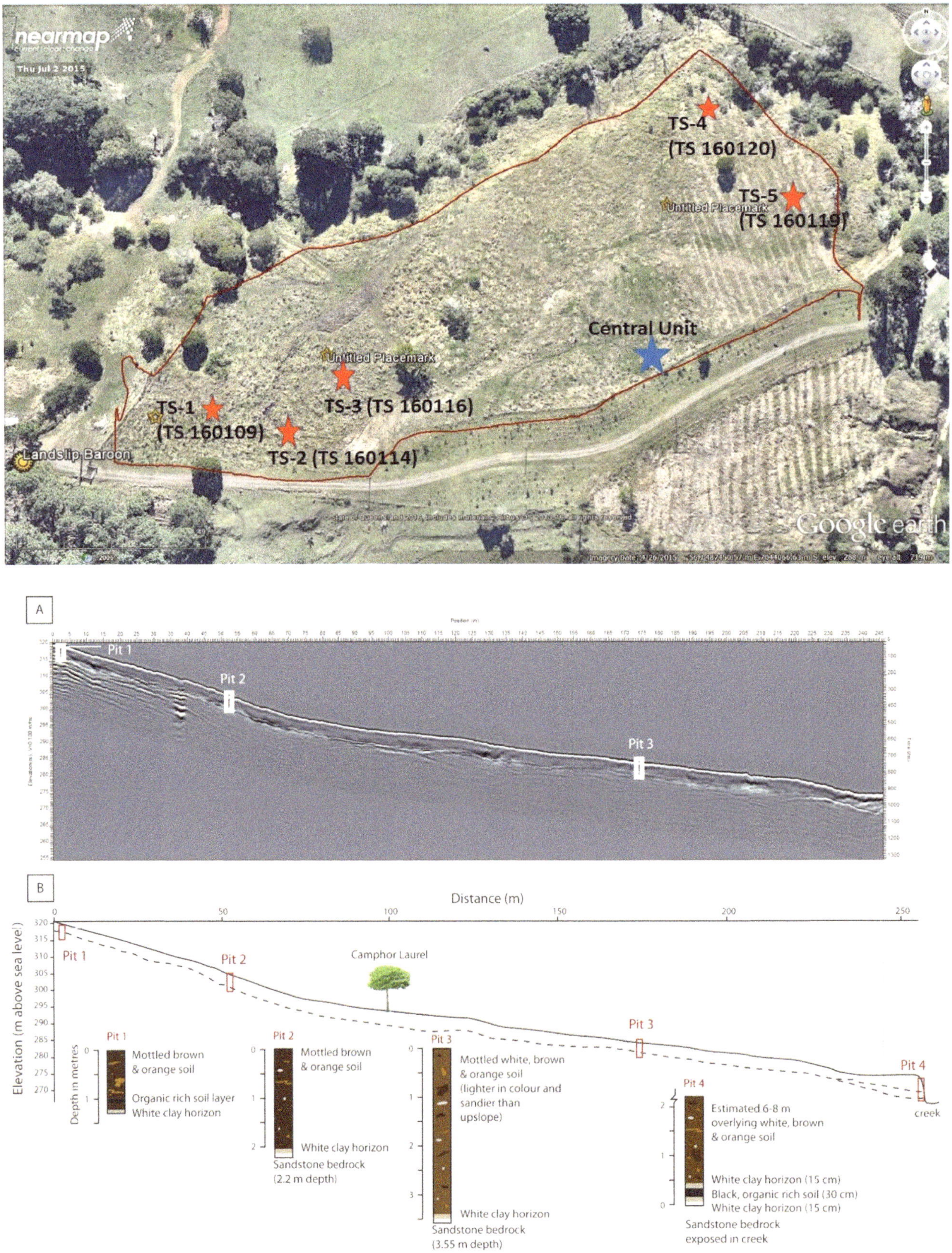

Fig. 5 **a** Locations of the sensors. X denotes the local downslope direction, whereas Y denotes the direction perpendicular to downslope. **b** Cross-section from the above GPR profile, showing the position of the white clay/bedrock reflector (dashed line)

Table 1 Soil index properties

Classification test	Results
Grain size	% finer than 75 μm > 79%
Distribution	Clay % = 41.0%
Atterburg limits	LL = 67.2% PI = 28.2%
Linear shrinkage	LS = 13.4%

Soil Properties

The soil extracted from the monitoring site was subjected to laboratory tests to determine the required soil properties for the numerical analysis. Table 1 summarises the results of the laboratory tests conducted to determine the index properties of the soil according to Australian standards.

Strength Properties

Strength properties of the soil were determined by the direct shear test, in which all the samples were prepared to achieve a dry density of 1.25 g/cm^3 to replicate the in situ dry density of the soil. Direct shear tests were conducted for four different water contents. Figure 6 shows the variation of apparent cohesion with gravimetric water content. However, the soil friction angle did not significantly vary with the gravimetric water content, which was determined as 15.90.

The pre-2008 landslide topography was subsequently reset by pushing failed soil and colluvium back onto the original slope. Vegetation (planting and growing trees) was suggested as an effective slope stabilization method for this area. Additionally, the five inclinometers slope monitoring

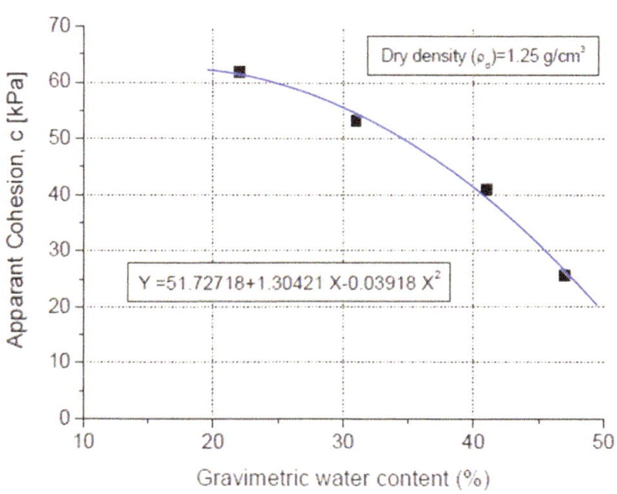

Fig. 6 The variation of apparent cohesion with gravimetric water content

experiment were installed. The real-time slope monitoring system aimed to measure the efficacy of revegetation as a slope stabilization method for this slope (Abeykoon et al. 2018).

The real-time monitoring system that consists of five sensor units (TS1, TS2, TS3, TS4, and TS5) and a central logging station was installed in the slope as shown in Fig. 5. Each sensor unit consists of a logging and transmission unit, MEMS tilt sensor, volumetric soil moisture sensor, and temperature sensor. The central unit comprises a central data logger, power supply unit (solar panel and back-up battery), data receiving unit (from sensor units), rain gauge as shown in Fig. 5a.

After characterizing the soil profile by determining the interface between soil and underlying bedrock by ground penetrating radar (GPR) survey, four locations were selected to excavate pits for determining the composition of soil layers, soil layer thicknesses and verification of GPR survey results. Figure 5b illustrate the longitudinal GPR profile and the GPR survey transect line and a cross-section of soil profile along the transect line with the locations of excavation pits, respectively.

The tilt angles accumulated distribution due to each rain are summarized as Fig. 7, which included the accumulated distribution results from 15/Jun/2016 to 01/Nov/2016. Red means the inclinometer is tilted in the direction of the landslide slope, and blue means the inclinometer is tilted in the opposite direction of the slope. It was found that the slope was clearly deformed as the time was increased. This slow movement is considered a typical landslide deformation.

TS1, which is located at top of the slope failed area, tilted (rotated) more than 2 degree in slope-direction during this period. TS2 which is located outside the failed area did not respond to the failure of the slope. However, TS2 started showing minor rotation with the reactivation of the failure above its location, which could be due to overloading the area of TS2 by the failed soil mass above its location. TS3, which is located at center of the slope, tilted (rotated) more than 2 degree in opposite direction of landslide slope. TS5, which is located at bottom of the landslide slope, is rotating in the direction of the slope, was pushed by the top mass of slope failure.

These results clearly show the movement of the entire slope shown in Fig. 7 as time order. It can be seen that the head and the bottom of the slope are inclined in the direction of landslide slope inclination, and the middle of the slope is a circular arc slide. By arranging five sensor arrays, it is thought that false alarms can be prevented by issuing an alert based on the movement of the entire slope, instead of local fluctuations based on a single sensor result.

Figure 8 shows the aerial view of the landslide area that the photos were taken in 28/Oct/2017, and Fig. 9 shows the aerial view of the landslide deformation area that the photos

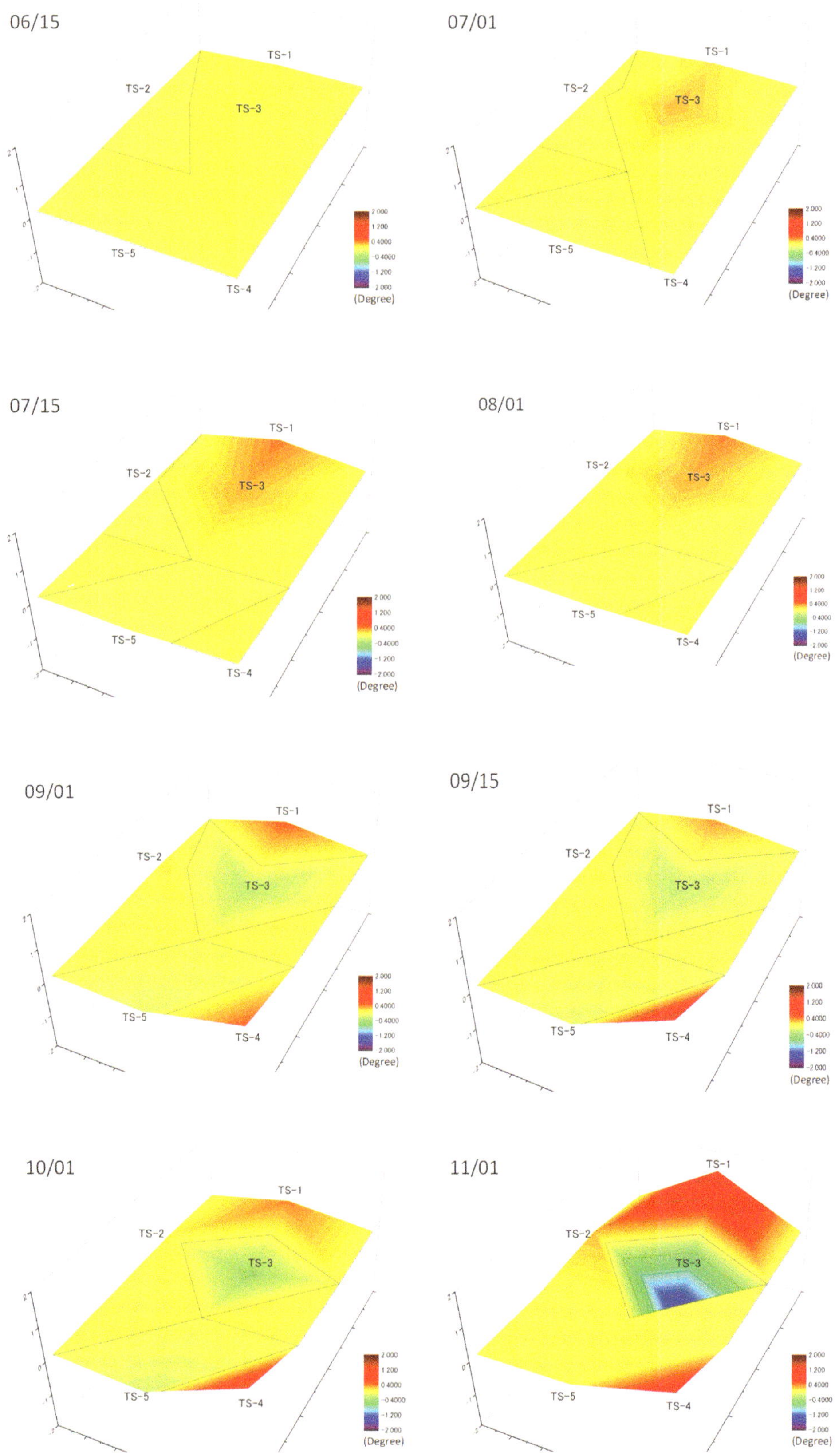

Fig. 7 Distribution of accumulated tilt angle of year 2016

Fig. 8 Areal view of the landslide area (28/Oct/2017)

Fig. 9 View of the landslide area (20/Feb/2018)

taken in 20/Feb/2018. These photographs show that the results correspond to the landslide collapse measured by the sensor array.

Monitoring Slope Failure at Manzawa, Yamanashi, Japan

The Manzawa area in the Yamanashi Prefecture of Japan contains a large-scale reactivation of old slope failures featuring rockfalls that involve the detachment and rapid downward movement of rock.

Because most traditional slope monitoring methods are expensive, difficult to control and may not be suitable for application in this civilian area, the simple and low-cost monitoring system was deployed on a test slope to validate field performance. It should be noted that the research is supported by the Japanese Government, and the following result that is reported in this paper is intermediate.

Figure 10 shows the scale of Manzawa slope failure site, and Fig. 11 shows the arrangement of the multi-point tilt sensors and locations, where two types of tilt sensor were used. The arrangement interval of the sensor is designed to 5 m. A total of 66 sets of sensors were deployed.

The system proposed in this study implemented wireless sensors consisting of MEMS accelerometers to measure tilt from angular movements. This orientation change data from the MEMS accelerometers were transmitted wirelessly to a remote monitoring facility. A real-time monitoring system would be an effective tool for the transmission of alerts and immediate activation of emergency procedures, thus providing ample time to save lives and property.

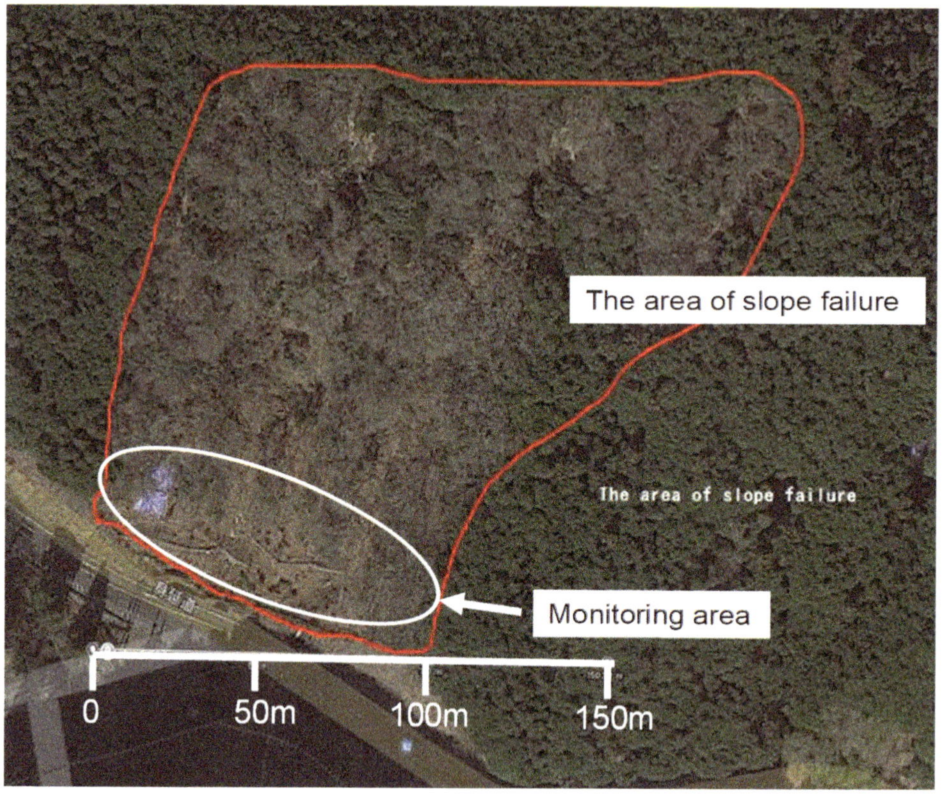

Fig. 10 Area of slope failure at Manzawa site, Japan

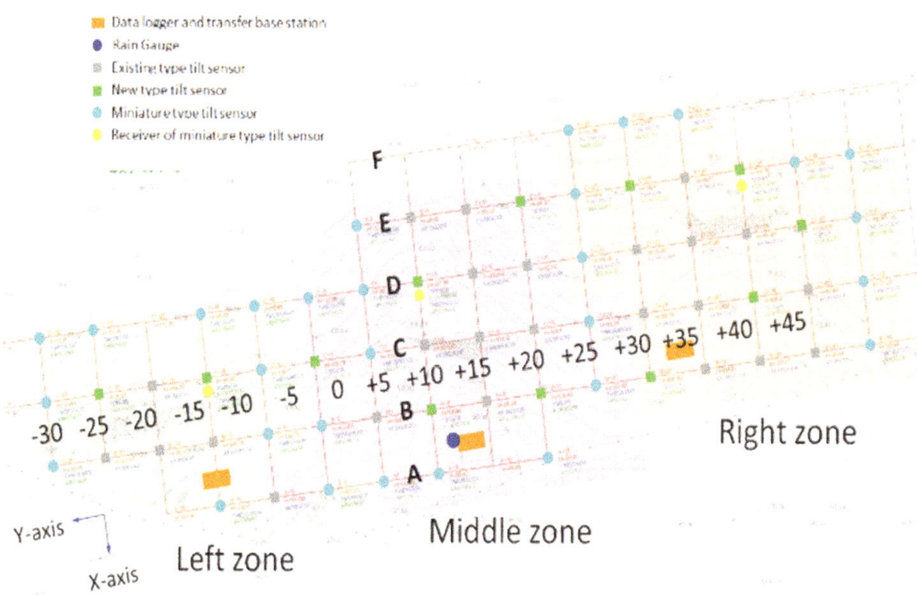

Fig. 11 Arrangement of the multi-point tilt sensors

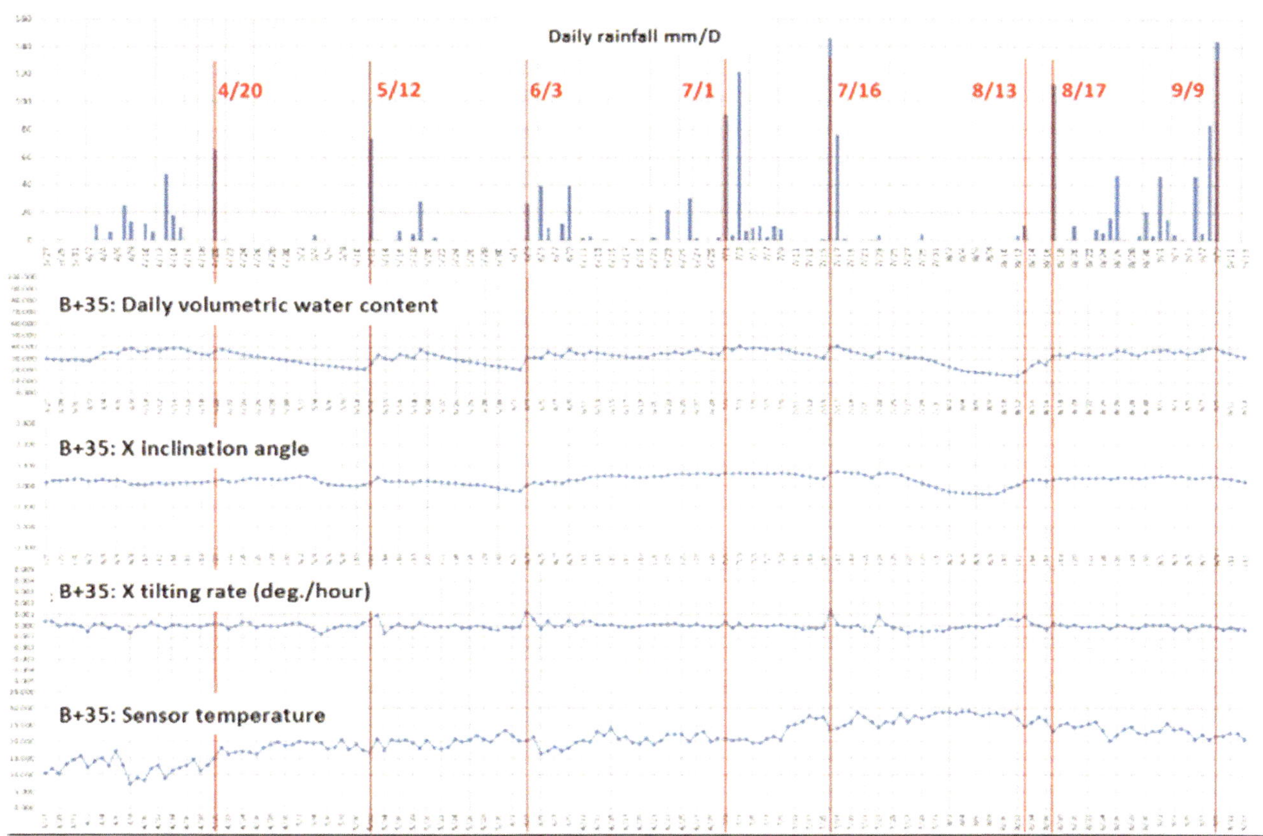

Fig. 12 Time histories of movements in rainy days

Necessary components of the system include sensors with the required resolution and software with the capacity for signal interpretation and failure alert algorithms. The challenges exist in identifying methods to minimize energy consumption of the units (i.e. improving battery life), keeping the appropriate number of devices for deployment and recognizing patterns of movement so that incipient sliding can be distinguished from random movements and environmental effects. The requirement for battery lifetime should ideally be longer than one year to reliably monitor the most critical time period without interruption and multiple year lifetimes should be achievable given the progress being made in battery technology.

Algorithms can then be developed to account for these movements and the sensitivity of these to varying threshold values can be evaluated. Finally, an effective early warning system can be developed.

The 66 sensor units are divided into three groups, left/middle/right zone, and one data receiver unit and one logger/gateway unit for internet collect all the data from respective group as shown in Fig. 11. There were eight heavy rainfall events during summer of 2015 shown in Fig. 12, and the tilt angles accumulated distribution due to each rain are summarized as Fig. 13. The tilting rate averaged during each rainfall event is shown in Fig. 14. Distribution of tilting behaviours is figured out by multi-point monitoring.

For practice, criteria for issuing early warning have to be defined based on data from the large number of sensors. One of very simple index for the criteria is simple sum of tilting rate from the sensors:

$$V_{alarm} = \sum_{n=1}^{n} \left(|V_n| * \frac{A_n}{A_0} * \partial_n \right) \tag{1}$$

Here, n is serial number of tilt sensors, V_n is tilting rate of slope sliding direction at the n-th sensor, An is the area of installation of the n-th sensor, A_0 is the total area of monitored slope, and ∂_n is a constant weight for the n-th sensor decided considering geology, geography, vegetation, and other factors. As the simplest example, values calculated with n = 1 for all the sensors are indicated in Fig. 14. The rain on 4/20, 6/3, and 8/13 caused relatively higher value of V_{alarm} in this case, but did not exceed precaution threshold of 0.01°/h.

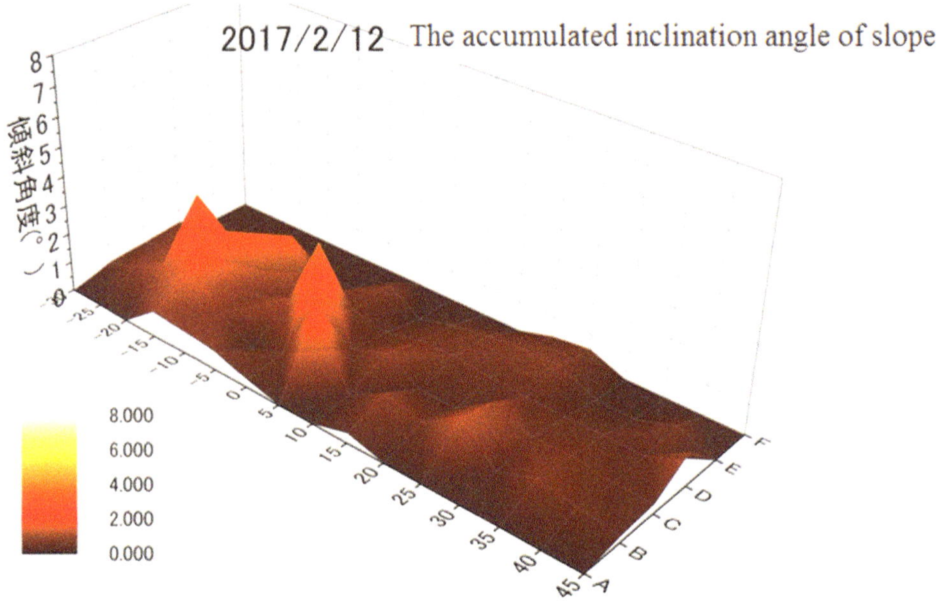

Fig. 13 Distribution of accumulated inclination angle

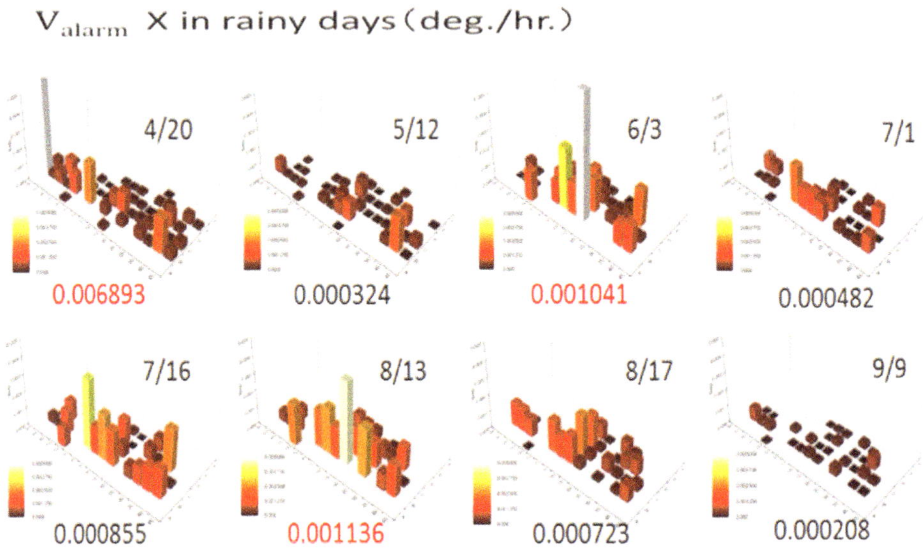

Fig. 14 Distribution of tilting rates during each rain day

Conclusion

A low-cost and simple monitoring method for an early warning system of rainfall-induced landslides has been proposed. Tilting angles in the surface layer of the slope are mainly monitored using this method and, in several case studies, distinct behaviours in the tilting angles in the pre-failure stages were detected. From this behaviour it is recommended that, from a regulatory perspective, a precaution is issued when the tilting rate of a slope is 0.01°/h, and a warning issued when the tilting rate is 0.1°/h.

Improvement in the applicability and development of the monitoring and early warning system has been made by modifying the equipment to be lower in cost, smaller in size and weight, and simpler to operate. It is estimated that the total cost for the monitoring system is reduced by one third, compared to regular systems, and thus a larger number of sensors can be deployed at the same cost (if desired). This will assist in improving data density and real-time feedback on slope behavior. These case studies will henceforth be helpful in determining the installation of sensor array of early warning system.

Acknowledgements This research is supported by Council for Science, Technology and Innovation, "Cross-ministerial Strategic Innovation Promotion Program (SIP), Infrastructure Maintenance, Renovation, and Management". (funding agency: NEDO), Grants-in-Aid for Scientific Research of Japan Society for the Promotion of Science (JSPS), and Core-to-Core Program "B. Asia-Africa Science Platforms" (JSPS).

References

Abeykoon AGTBB, Gallage C, Dareeju (2018) Real-time monitoring and wireless data transmission to predict rain induced landslides in critical slopes. Aust Geomech J 53(3):61–76

Uchimura T, Towhata I, Wang L, Nishie S, Yamaguchi H, Seko I, Qiao J-P (2015) Precaution and early warning of surface failure of slopes by using tilt sensors. Soils Found 55(5):1086–1099

WLF 2020 Kyoto Japan

Influence of Intervals Measuring Surface Displacement on Time Prediction of Slope Failure Using Fukuzono Method

Naoki Iwata and Katsuo Sasahara

Abstract

Time-prediction methods based on monitoring the displacement of a slope are effective for the prevention of sediment-related disasters. Several models have been proposed to predict the failure time of a slope based on the creep theory of soil, which describes the accelerating surface displacements that precede slope failure. Fukuzono's method has been widely adopted in practice. This method can only be applied to the period when the surface displacement accelerates. However, the observed surface displacement appears to increase monotonically, slightly repeating the increase and decrease. These results decrease the accuracy of the predicted failure time. Thinning out the observed data is effective for minimising the influence of fluctuations. In this study, we predicted the failure time of a sandy model slope under artificial rainfall using four methods based on Fukuzono's model, compared the prediction accuracy of each method and examined the influence of measurement intervals on the predicted failure time using extracted data at different measurement intervals. The results showed that the variation of the extracted data group decreases and the prediction accuracy of the failure time improves if the measurement interval increases. Moreover, when the failure time of a slope is predicted using statistical methods, the accuracy of the prediction is further improved.

Keywords

Monitoring • Slope failure • Surface displacement • Time prediction • Fukuzono's model • Measurement interval • Prediction accuracy

Introduction

Disasters of natural and artificial slopes around roads, railways and residential area are caused by heavy rains. Furthermore, in construction sites, slope failures occur by the destabilisation of the ground owing to change in stress caused by the embankment and cut earth. Time prediction methods based on monitoring the displacement of a slope using sensors are effective at preventing sediment-related disasters.

Several methods have been proposed to predict slope failure using the surface displacement of a slope. The formulae proposed by Fukuzono (1985) have been widely adopted in practice because of their simplicity. These models were proposed to predict the failure time of a slope based on the creep theory of soil, which is divided into three stages: primary creep (decreasing velocity), secondary creep (constant velocity) and tertiary creep (increasing velocity).

Fukuzono's model formulates the relationship between the velocity of the surface displacement and the acceleration in the tertiary creep stage. Fukuzono found that the logarithm of the acceleration of the surface displacement is proportional to the logarithm of the velocity in model slope experiments. Time integration of this relationship leads to the typical trends of the time variation in the inverse-velocity of the surface displacement before failure. The failure time of a slope can be predicted when the extrapolation curves approach zero. Fukuzono's method has been widely applied because of its simplicity and convenience of use.

N. Iwata (✉)
Chuden Engineering Consultants Co. Ltd., Nuclear Project Department, Hiroshima, 7348510, Japan
e-mail: n.iwata@cecnet.co.jp

K. Sasahara
Kochi University, Natural Science Cluster, Kochi, 7808520, Japan
e-mail: sasahara@kochi-u.ac.jp

© Springer Nature Switzerland AG 2021

N. Casagli et al. (eds.), *Understanding and Reducing Landslide Disaster Risk*, ICL Contribution to Landslide Disaster Risk Reduction,
https://doi.org/10.1007/978-3-030-60311-3_36

His method can only be applied to the period when the surface displacement accelerates. However, the actual displacement of the slope is complicated owing to variations in the rainfall intensity and the inhomogeneity of the surface layer, and it is not easy to specify the period when the surface displacement accelerates. Moreover, the observed displacement appears to increase monotonically, slightly repeating the increase and decrease, the acceleration varies widely and some data points become negative. These results decrease the accuracy of the predicted failure time. Thinning out the observed data is effective for minimising the influence of fluctuations.

In this study, we predicted the failure time of a sandy model slope under artificial rainfall with constant rainfall conditions using four methods based on Fukuzono's model to extract data at difference time intervals. We compared the prediction accuracy of each method and examined the influence of measurement intervals on the predicted failure time.

Methods for Predicting the Failure Time

Fundamental Equation of Fukuzono's Model

Fukuzono (1985) proposed that the logarithm of the velocity of the surface displacement is proportional to the logarithm of the acceleration in the tertiary creep stage, which describes the accelerating surface displacement before slope failure, given as

$$\frac{d^2x}{dt^2} = a\left(\frac{dx}{dt}\right)^{\alpha} \tag{1}$$

where x is the downward surface displacement along the slope, t is the time, dx/dt is the velocity, dx^2/dt^2 is the acceleration, and a and α are constants. α is greater than 1 in the period during which the surface displacement accelerates.

After integrating Eq. (1), the inverse-velocity of the surface displacement can be written as follows:

$$\frac{dt}{dx} = \frac{1}{v} = \{-a(a-1)\}^{1/(\alpha-1)}(t_r - t)^{\frac{1}{(\alpha-1)}} \tag{2}$$

where v is the velocity of the surface displacement and t_r is the failure time. Eq. (2) shows a downward slope; further, the time approaches the time immediately prior to slope failure as the inverse-velocity of the surface displacement, $1/v$, approaches zero. The curve is linear for $\alpha = 2$, convex for $\alpha > 2$ and concave for $1 < \alpha < 2$. The value of α for actual slope failure ranges from 1.5 to 2.2.

Precise Prediction Method Using Inverse-Velocity

As Eq. (2 becomes linear for $\alpha = 2$, the failure time is calculated easily using two inverse-velocity values at different times. However, when $\alpha \neq 2$, it is difficult to predict it accurately using two values owing to the curvature of the inverse-velocity curve. Therefore, Fukuzono (1985) proposed the time-prediction method expressed in Eq. (3) by a time differential in Eq. (2).

$$(1/v)\Big/\frac{d(1/v)}{dt} = -(a-1)\times(t_r - t) \tag{3}$$

The curve of Eq. (3) is linear; the failure time, t_r, is predicted by inserting two inverse-velocity values ($1/v_{i-1}$, $1/v_i$) and two inclination values ($d(1/v_{i-1})/dt$, $d(1/v_i)/dt$) at two different times (t_{i-1}, t_i) into Eq. (4) as follows:

$$t_r = \frac{t_i \times (1/v_{i-1})/\left(\frac{d(1/v_{i-1})}{dt}\right) - t_{i-1}\times(1/v_i)/\left(\frac{d(1/v_i)}{dt}\right)}{(1/v_{i-1})/\left(\frac{d(1/v_{i-1})}{dt}\right) - (1/v_i)/\left(\frac{d(1/v_i)}{dt}\right)} \tag{4}$$

Three Data Prediction Method

Tsuchiya and Omura (1989) proposed the time-prediction method using time–velocity relationship in Eq.[2] and time–displacement relationship by time integral in Eq. (2) from three surface displacement values at equal time intervals. We improved this method to be applicable to different time intervals. The failure time, t_r, is predicted by inserting three displacement values (x_{i-2}, x_{i-1}, x_i) at three different times (t_{i-2}, t_{i-1}, t_i) as follows:

$$t_r = \frac{(t_{i-1}+t_i)\times\left(\frac{x_i - x_{i-1}}{x_{i-1}-x_{i-2}}\times\frac{t_{i-1}-t_{i-2}}{t_i - t_{i-1}}\right)^{\alpha-1} - (t_{i-1}-t_{i-2})}{2\times\left\{\left(\frac{x_i - x_{i-1}}{x_{i-1}-x_{i-2}}\times\frac{t_{i-1}-t_{i-2}}{t_i - t_{i-1}}\right)^{\alpha-1} - 1\right\}} \tag{5}$$

$$\alpha = 2 - \frac{\log\left(\frac{x_{i-2}-x_{i-1}}{x_{i-1}-x_i}\right)}{\log\left(\frac{x_{i-1}-x_{i-2}}{x_i - x_{i-1}}\cdot\frac{t_i - t_{i-1}}{t_{i-1}-t_{i-2}}\right)} \tag{6}$$

Least Squares Prediction Method

Assuming the curve of Eq. (2) to be linear ($\alpha = 2$), the relationship between the inverse-velocity and time is calculated via the least squares method using all previous inverse-velocity values from the start of the measurement.

The failure time, t_r, corresponds to the intercept of the straight line with the time axis.

Nonlinear Regression Prediction Method

The constants a and α in Eq. (1) are calculated by the least squares methods using velocity and acceleration data plotted in a double logarithmic chart. The failure time, t_r, is predicted by inserting the constants a and α into Eq. (7) as follows:

$$t_r = \frac{v^{1-\alpha}}{a(a-1)} + t \tag{7}$$

Experimental Set-up and Observed Data

In this study, we used the observed surface displacement of a slope failure experiment that was conducted using the large-scale rainfall simulator at the National Research Institute for Earth Science and Disaster Prevention (Sasahara and Ishizawa 2016). Figure 1 shows a photograph of the model slope. The model was 300 cm long and 150 cm wide in the horizontal section and 600 cm long and 150 cm wide in the slope section with an inclination of 30°. The soil layer was 50 cm thick and composed of granitic soil. The surface of the slope was parallel to the base of the slope.

The surface displacement was measured using an extensometer with a non-linearity of approximately 0.1 mm; it was fixed at the upper boundary of the flume. The surface displacement was defined as the distance between the upper boundary of the flume and the moving pole at the surface of the slope at 160 cm from the toe of the slope. The surface displacement was measured every 10 s.

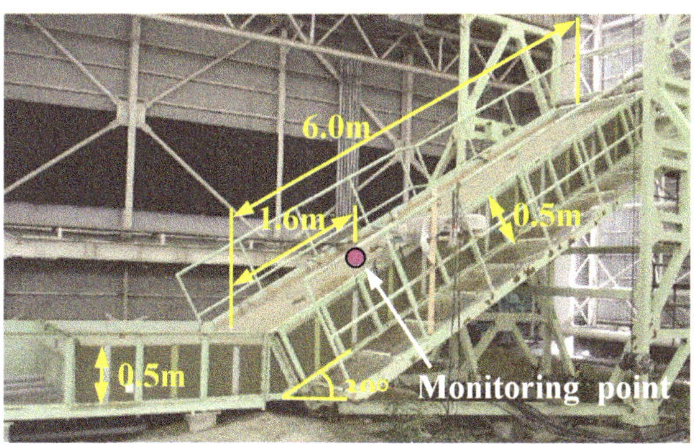

Fig. 1 Overview of the model slope

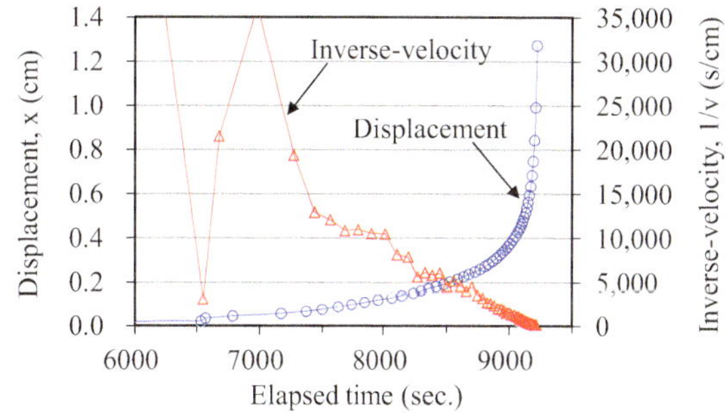

Fig. 2 Time variation of the surface displacement and inverse-velocity of the surface displacement

The rainfall had an intensity of 50 mm/h and continued until the onset of the failure of the model slope. Slope failure occurred at 9,220 s, and the surface displacement just before the slope failure was 1.27 cm.

As the extensometer has an accuracy of 0.1 mm, data were extracted to be greater than 0.1 mm between the two measurements of the surface displacement. Figure 2 shows the time variation of the surface displacement and the inverse-velocity of the surface displacement. The slope of the inverse-velocity curve before 7,500 s suddenly increases and decreases and displays a uniform downward slope afterwards.

Data for the Prediction

The time-prediction methods based on Fukuzono's model are applied to the tertiary creep stage. However, a curve of the inverse-velocity has fluctuations and it is difficult to predict the time of onset of the tertiary creep in actual practice. Therefore, in this study, all data from the start of monitoring onwards are used to examine the influence of measurement intervals of surface displacement, Δx, and predict the failure time. The data were extracted to be greater than difference Δx from previous extracted data of the surface displacement: 0.01, 0.05, 0.1 and 0.2 cm.

The velocity of the surface displacement, v_i, is calculated from $v_i = (x_i - x_{i-1})/(t_i - t_{i-1})$, where x_i and x_{i-1} are the surface displacements at times t_i and t_{i-1}. Because v_i is the average velocity between t_{i-1} and t_i, the time against v_i, t'_i, is set as the mid of t_i and t_{i-1}, specifically $t'_i = (t_i - t_{i-1})/2$. The acceleration of the surface displacement, $(dv/dt)_i$, is calculated from $(dv/dt)_i = (v_i - v_{i-1})/(t'_i - t'_{i-1})$, where v_i and v_{i-1} are the velocities at times t'_i and t'_{i-1}. The time against $(dv/dt)_i$, t''_i, is to set to the middle of t'_i and t'_{i-1}, i.e. $t''_i = (t'_i - t'_{i-1})/2$.

The failure time is inferred via four prediction methods: (1) Precise prediction method, (2) Three data prediction method, (3) Least squares prediction method and (4) Nonlinear regression prediction method, using the extracted data from the start of monitoring and onwards. We compare the prediction accuracy of each method and examine the influence of measurement intervals on the predicted failure time.

Variation of the Extracted Data

Figure 3 shows the time variation of the displacement and inverse-velocity for different Δx and the relationship between velocity and acceleration. The time variation of the displacement, shown in Fig. 3a, demonstrates behaviour similar to that of tertiary creep after approximately 6500 s. When $\Delta x = 0.1$ cm or less, the measurement value of $\Delta x = 0.01$ cm is usually reproducible. However, when Δx 0.1 cm, the data could not be extracted at the initial stage of the tertiary creep and the deviation of the data at the initial stage increased at 8000 s and earlier. The time variation of the inverse-velocity, shown in Fig. 3b, significantly varied at

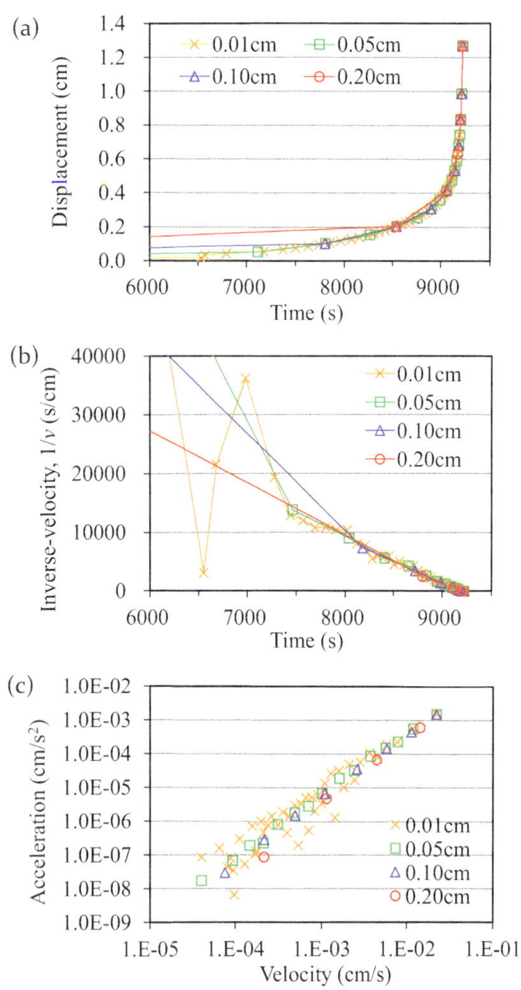

Fig. 3 Comparison of the extracted data: **a** time variation of displacement; **b** time variation of inverse-velocity; **c** velocity–acceleration relationship

7500 s and earlier for $\Delta x = 0.01$ cm; however, this variation was eliminated by increasing the Δx. When Δx increases, the number of displacement data to be extracted decreases, and there are no data at the initial stage of the tertiary creep; therefore, it is impossible to predict until immediately before the failure.

As shown in Fig. 3c, the velocity–acceleration relationship originally showed displacement behaviour similar to that of the tertiary creep stage, and the variation was small. The larger the Δx, the smaller the number of the data and the smaller the variation. When the variation decreases, it is assumed that applicability to Fukuzono's prediction formula increases, whereas the previously mentioned issues will exist.

Results of Time-Prediction of Slope Failure

Figure 4 shows the comparison of the time variation of difference between the predicted failure time and the elapsed time, t_r–t, obtained using the different prediction methods. The t_r–t implies the time interval for the slope failure. When the accuracy of prediction is high, t_r–t values are plotted in the positive domain of the graph and tend to zero as the time approaches the slope failure time (i.e. the time variation of t_r–t has a downward slope). When t_r–t values appear in the negative domain of the graph, the predicted failure time precedes the elapsed time and the slope failure time is thus unpredictable.

Using the precise prediction method, when $\Delta x = 0.01$ cm, t_r–t is nearly 0, i.e. the current time is predicted as the precise prediction time, indicating that the failure time is unpredictable. Further, when $\Delta x = 0.05$ cm, t_r–t significantly fluctuates, indicating that the prediction accuracy is poor. However, when $\Delta x \geq 0.1$ cm, the prediction result is close to the black broken line, indicating high prediction accuracy. This result is caused by the precise prediction method, which predicts using the displacement velocity data at two different times. If Δx is small, the prediction accuracy will decrease because of the increasing and decreasing of the surface displacement even if the data exhibit a behaviour resembling the tertiary creep.

Based on the three-data prediction method, t_r–t significantly increases or decreases for both $\Delta x = 0.01$ and 0.05 cm, after 8700 s at $\Delta x = 0.05$ cm, the prediction result is close to the black broken line with improved accuracy. When $\Delta x = 0.1$ cm, a table prediction result was obtained using the precise rediction method; however the accuracy before 8800 s was reduced by the three-data prediction method. This demonstrates that the predicted value can considerably vary depending on the data to be extracted.

As the least squares prediction method predicts using all data from the start of monitoring onwards, the variation is

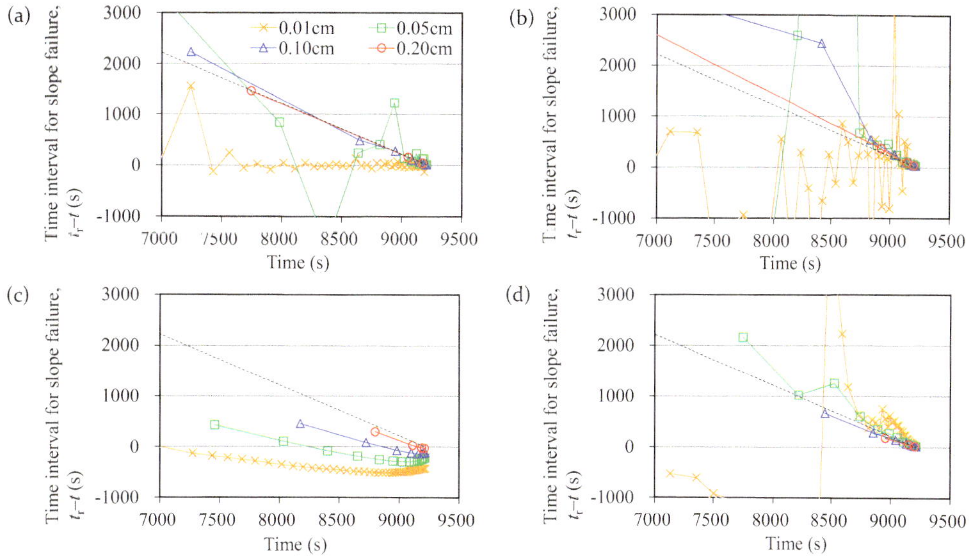

Fig. 4 Comparison of the time variation of the predicted failure time according to the different prediction methods using the extracted data at different measurement intervals: **a** precise prediction method; **b** three data prediction method; **c** least squares prediction method; **d** nonlinear regression prediction method

limited and shows a right-downward tendency. The predicted value is plotted below the black dashed line and is negative near the failure time, resulting in low prediction accuracy. As Δx increases, the results approach the black broken line with improved prediction accuracy. This is because of data in the initial stage of measurement, showing that the inverse-velocity is very large. As Δx increases, the number of data and the value of inverse-velocity in the initial stage decrease, thus improving the prediction accuracy.

The non-linear regression prediction method is close to the black dashed line after 8700 s, even when $\Delta x = 0.01$ cm, and comparatively high prediction accuracy results are obtained. When $\Delta x \geq 0.05$ cm, good prediction accuracy is obtained because the variation in velocity–acceleration is small, as seen in Fig. 3c. When $\Delta x = 0.01$ cm, the constants are $a = 1.32$ and $\alpha = 1.79$ and the correlation coefficient is 0.94. When $\Delta x \geq 0.05$ cm, the constant a increases from 2.0 to 5.0 and α increase from 1.8 to 2.0. However, the prediction accuracy is high because the correlation coefficient is around 0.99 and the curve between inverse-velocity and time is almost linear (i.e. α is around 2.0).

Discussion

As shown in Fig. 3a, as Δx increases, the discrepancy between measured and extracted data increases. This discrepancy is evaluated using the root mean square error (*RMSE*), which is a method for evaluating the variation in data given by Eq.[8].

$$\mathrm{RMSE} = \sqrt{\frac{1}{\mathrm{N}}\sum_{i=1}^{\mathrm{N}}(\mathrm{F}_i - A_i)^2} \tag{8}$$

where N is number of target data, F_i is the measured displacement at time t_i, A_i is the extracted displacement at time t_i and F_i-A_i indicates an error. If there are no data at the time t_i among the extracted data, the datum for t_i is projected by a proportional distribution of the extracted data before and after time t_i. In the *RMSE*, the smaller the value, the smaller the variation of the extracted data group, i.e. the greater the reproducibility.

Figure 5 shows a comparison between the *RMSE* and the correlation coefficient of the nonlinear regression equation for the velocity–acceleration relationship when Δx is 1/5000, 1/2500, 1/1670, 1/1000, 1/500 and 1/250 of soil layer thickness T (0.01, 0.02, 0.03, 0.05, 0.1 and 0.2 cm, respectively). When $\Delta x/T$ exceeds 1/1670–1/1000, the *RMSE* rapidly increases and the reproducibility of measurement data decreases; however, the correlation coefficient exceeds 0.98, indicating a very high correlation. Moreover, the correlation decreases when the $\Delta x/T$ is less than 1/1670–1/1000. These results imply that the appropriate interval for achieving high prediction accuracy in this study is 1/1670–1/1000.

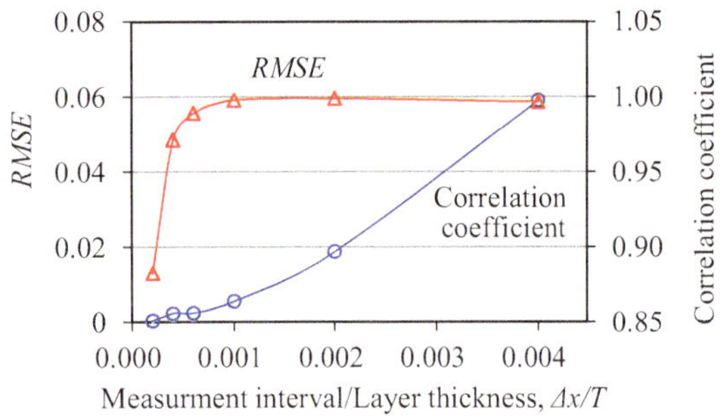

Fig. 5 Comparison between the *RMSE* and correlation coefficient for the velocity–displacement relationship with the change in measurement interval

In the least squares prediction method and nonlinear regression prediction method, it was found that the prediction accuracy tended to improve upon increasing Δx. This suggests that additional improvement in the accuracy of prediction is possible by rejecting the data at the initial stage when the velocity is low (i.e. $1/v$ is very large). Therefore, the method based on the moving average acceleration (i.e. extracting the period where the average acceleration of the five preceding steps is continuously positive) performed by Iwata et al. (2017) was used to extract the time after 7600 s at the tertiary creep stage. The failure time was predicted using the least squares prediction method and nonlinear regression prediction method for this period.

Figure 6 shows a comparison of the time variation of t_r-t with the tertiary creep stage. The prediction accuracy of the least squares prediction method is significantly improved in comparison with Fig. 4c. Consequently, it can be seen that to improve the prediction accuracy, rejecting data at the initial stage is very effective. Furthermore, the influence of the difference of Δx is not significant because the variation in data was small, as seen in Fig. 3c.

In comparison with Fig. 4d, the non-linear regression prediction method shows a slight improvement in the prediction accuracy above 9000 s at $\Delta x = 0.01$ cm but no improvement at 9000 s or less. However, when $\Delta x = 0.05$ cm, the prediction accuracy for 8700 s or less declines because the quantity of the data used in the regression analysis is small, and the coefficient obtained in the regression analysis significantly differs owing to the slight changes in the data.

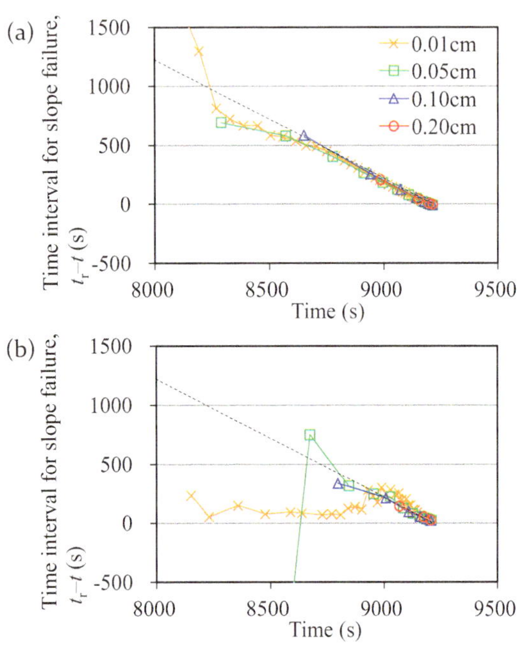

Fig. 6 Comparison of the time variation of the predicted failure time in the tertiary creep stage at different measurement intervals: **a** least squares prediction method; **b** nonlinear regression prediction method

Conclusion

We predicted the failure time of a sandy model slope using four methods based on Fukuzono's model and examined the influence of measurement intervals on the predicted failure time. The findings obtained from this study can be summarised as follows:

1. The prediction accuracy decreases owing to the fluctuations of the surface displacement even if the data exhibit a behaviour resembling the tertiary creep.
2. As the measurement intervals of displacement increase, the number of the extracted data decreases, the variation of the extracted data group decreases, thus improving the prediction accuracy.
3. The prediction accuracy using the precise prediction method and three-data prediction method is inferior to other methods because a small number of data are used to predict the failure time and these methods are susceptible to fluctuations.
4. The least squares prediction method and non-linear regression prediction method give relatively stable prediction results. However, when the data in the initial stage where the velocity is low are included in the used data, the prediction accuracy decreases.
5. The appropriate measurement interval to achieve high prediction accuracy in this study is 1/1670–1/1000 of soil layer thickness.
6. If the surface displacement data in the tertiary creep stage can be extracted, the prediction accuracy of the least squares prediction method and non-linear regression can be significantly improved.

In this study, one problem was that the quantity of the extracted data decreases as Δx increases because the measurement data are for a small model slope. However, as the displacement just before the slope failure is large at the actual slope, it is assumed that even if Δx is increased, there is no problem with the number of acquired data being reduced.

References

Fukuzono T (1985) A new method for predicting the failure time of a slope. In: *Proceedings of 4th international conference and field workshop on landslides*. Tokyo, pp 145–150

Iwata N, Sasahara K, Watanabe S (2017) Improvement of fukuzono's model for time prediction of an onset of a rainfall-induced landslide. In: Mikos M et al (eds) Advancing culture of living with landslides. Springer, Berlin, pp 103–110

Sasahara K, Ishizawa T (2016) Time prediction of an onset of shallow landslides based on the monitoring of the groundwater level and the surface displacement at different locations on a sandy model slope. Jpn Geotech J 11(1):69–83 ((in Japanese with English abstract))

Tuchiya S, Omura H (1989) Forecast methods of slope failure time and their case studies. Landslides 26(1):1–8 ((in Japanese with English abstract))

Velocity and Acceleration of Surface Displacement in Sandy Model Slope with Various Slope Conditions

Katsuo Sasahara

Abstract

Measurement of surface displacement was implemented in three model slope test cases under constant intensity of artificial rainfall with different initial water content or slope inclination. The aim was to examine the relationship between velocity and acceleration of the increase in the surface displacement, which is the basis for predicting the failure time of a slope (Fukuzono T (1985) A New Method for Predicting the Failure Time of a Slope. In Proceedings of the IVth International Conference and Field Workshop on Landslides, Tokyo, Japan, pp 145–150.). The velocity and acceleration were derived from actual measured surface displacements. The relationship between the velocity and acceleration of the increase in the surface displacement was unique at different locations on the slope and with different pore pressure loading mechanisms (under unsaturated conditions or increased groundwater levels). The relationship was also unique under different slope inclination. This suggests the possibility of deriving the relationship by indoor shear tests with the same soil of actual slopes before monitoring of displacement on the slope, for predicting the failure time of the slope.

Keywords

Monitoring • Surface displacement • Model slope • Slope angle • Water content

Introduction

Measurement of slope displacement or deformation has often been adopted for early warning against landslides. This has also been implemented in many experiments involving model or natural slopes to examine the mechanism of landslides.

Acceleration of slope displacement or deformation have been observed prior to failure in many reports such as Moriwaki et al. (2004) for a model slope, and Ochiai et al. (2004) and Askarinejad et al. (2018) for natural slopes. Kromer et al. (2015) observed variation in the topography of a rock wall surface before and after collapse and identified accelerating increase in displacement by analysing the displacement of targets on the wall. Accelerative increase in displacement has been recognised as a precursor of slope instability. Many studies (Saito 1965; Saito and Yamada 1973; Varns 1982; Fukuzono 1985; Voight 1988, 1989; Xiao et al. 2009; Bozzano and Mazzanti, 2012) helped establish the method for better prediction of time of an onset of slope failure, based on the measurement of an accelerative increase in displacement.

Fukuzono (1985) reported that there is a linear relationship between velocity and acceleration of an increase in surface displacement (hereafter the displacement velocity and acceleration) on a logarithmic scale prior to slope failure of the model, and showed that any prediction method of slope failure time is preferable to be based on the relationship. This idea proposed by Fukuzono has been widely adopted around the world. Further, he insisted that the relationship could only be established during the tertiary creep stage, when there is an accelerating increase in displacement. It is for now not clear whether unique relationship can be derived for slopes with different geometry or initial conditions even though they are made of the same soil. An attempt is made herein to examine, through a series of indoor experiments, if we can derive a unique relationship for various slope models of the same soil, expecting that the

K. Sasahara (✉)
Kochi University, 2-5-1, Akebonocho, Kochi, 780-8520, Japan
e-mail: sasahara@kochi-u.ac.jp

© Springer Nature Switzerland AG 2021
N. Casagli et al. (eds.), *Understanding and Reducing Landslide Disaster Risk*, ICL Contribution to Landslide Disaster Risk Reduction,
https://doi.org/10.1007/978-3-030-60311-3_37

result would help develop a rational protocol for early warnings.

Displacements and groundwater levels in model slopes with different initial conditions were measured and those data were analysed to clarify whether the relationship between the displacement velocity and acceleration is unique under different initial conditions of the slope model in this study.

Methodology

Experimental Apparatus

Figure 1 shows a lateral view of the model slope and arrangement of monitoring devices. The model slope was in a steel flume with a lateral wall of glass and consisted of granite soil, with the grain size distribution shown in Fig. 2. The density of soil particles was 2.489 (g/cm^3), and the maximum and minimum densities were 1.307 (g/cm^3) and 1.073 (g/cm^3), respectively. The upper slope was 110 cm long, 60 cm wide, 12 cm thick, and with a inclination of 40°. The lower slope was 50 cm long, 60 cm wide, 12 cm thick at the boundary of both slopes, with 10° of surface slope. The shape of the upper slope was rectangular, and that of the lower slope was triangular. Steel blades (1 cm high) were set at every 50 cm in a downward direction, and coarse sand was glued on the base plate of the flume to prevent slippage between the soil and the base plate.

Downward and vertical displacement at the surface and pore pressure at the bottom of the upper slope were measured at distances of 25, 55, and 85 cm from the upper boundary of the flume. The downward and vertical displacements were measured by the system shown in Fig. 3. Downward or vertical movement of the moving plate on the surface of the slope pulled wires connected to displacement gauges, which were set lateral to the flume. Dimension of moving plate is 10 by 10 cm with four steel blades of 1 cm height behind it to prevent slippage on the surface. Weight of the moving plate was 700 g to resist reaction force of displacement gauge. The direction of the wire was changed from downward to vertical and from vertical to lateral by two pulleys for the measurement of downward displacement, while it was changed from vertical to lateral by a pulley for the measurement of vertical measurement. Pullies were made of acrylic resin to prevent friction between pullies and wires. Only downward displacement was adopted as the surface displacement in this study, because vertical displacement could not be accurately measured. Many noises and scattering were found in the vertical displacement measurements. The accuracy of the displacement gauge was 0.2 mm, and the accuracy of the pore pressure gauge was 50 Pa, which corresponded to 0.5 mm of water level. The measured pore pressure was converted into a water level in this study. Trial in small scale model revealed the accuracy of the measurement in groundwater level was around 1 cm.

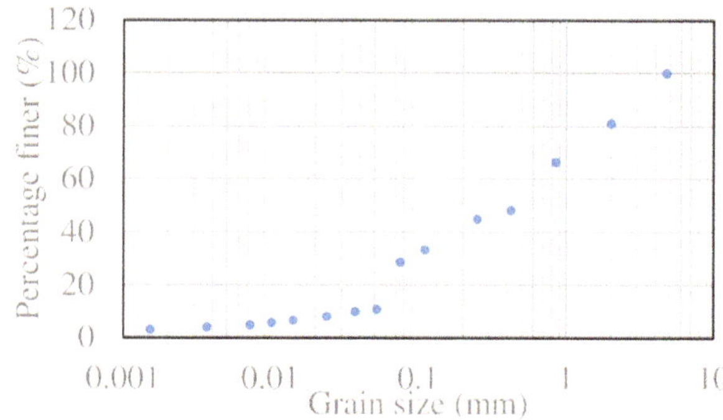

Fig. 2 Grain size distribution of the soil of the model

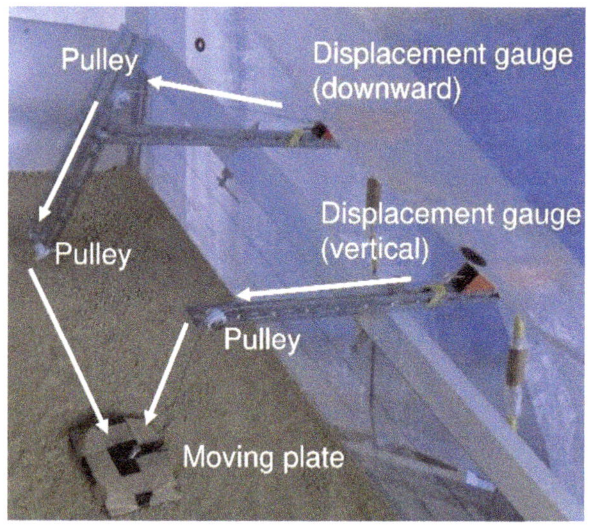

Fig. 3 Measurement of downward and vertical displacements

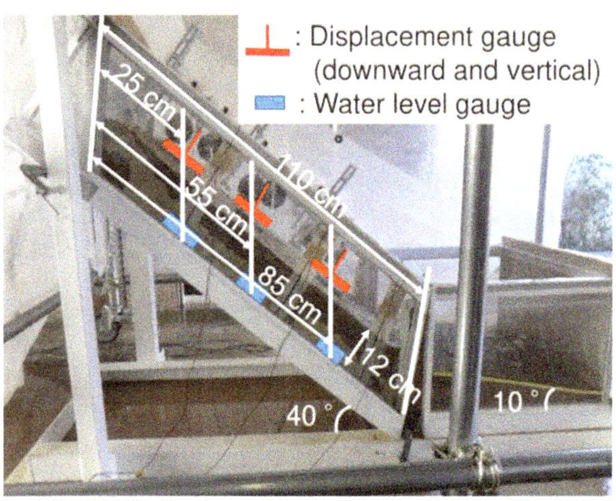

Fig. 1 Model slope and arrangement of monitoring devices in Case 1

Table 1. Conditions of the model slope

	Case1	Case2	Case3
Rainfall intensity (mm/h)	46	46	46
Dry unit weight (g/cm^3)	1.32	1.21	1.32
Water content (%)	0	15.1	0
Upper slope angle (deg.)	40	40	35

Water was sprinkled at a constant rate on the model slope from the rainfall simulator above the flume. The quantity of water sprinkled was expressed in rainfall intensity (mm/h) in these experiments, as shown in Table 1.

Experimental Conditions

Three cases of the experiment with different model slope conditions (Table 1) were implemented to examine the relationship between surface displacement velocity and acceleration for different slope conditions. Water content was different between Case 1 and Case 2. It was 0 and 15.1% in Case 1 and Case 2, respectively. Soil was dried in oven before the construction of the model and water content was measured by soil moisture gauge of METER EC-5 with 2–3% resolution. The wet unit weight in Case 2 was 1.32 (g/cm^3), which was the same as the dry unit weight of Case 1. Oven dry soil with this dry unit weight was adopted after many trial using small model to make non-uniformity of soil quantities in model slope least. The upper slope angle was 40° in both cases. The upper slope angle was 40° and 35° in Case 1 and Case 3, respectively. The slope of the model was almost same with shallow landslides in Japan. The dry unit weight and water content was the same in both cases, and the rainfall intensity was 46 mm/h in all cases. This was also decided by actual rainfall intensity at landslide disaster in Japan. Two times of tests with the condition of Case 1 were conducted and results of both cases were compared to ensure the repeatability of the experiment. The surface displacement was 22 cm at 8,400 s in first case while it was 25 cm at the same time in second case.

The surface displacement velocity and acceleration were derived from measured data of surface displacement, by the process shown in Fig. 4. Quantitative variation of the vector of the surface displacement was adopted here. Surface displacement velocity was defined as the increase in surface displacement divided by the difference in time. Surface displacement acceleration was derived as the increase in surface velocity divided by the difference in time.

Video images were recorded from lateral side of the model and no slippage was observed from the images.

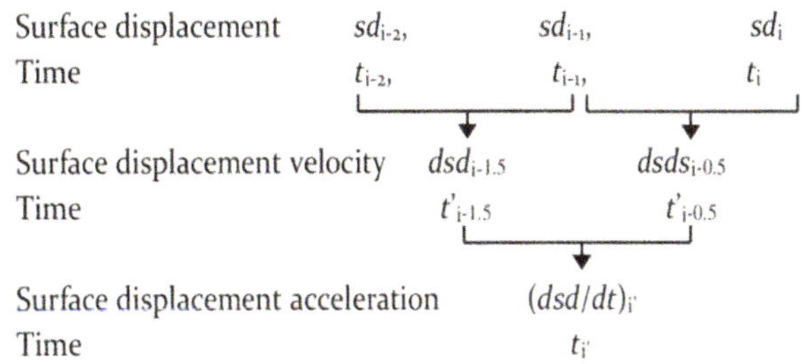

Fig. 4 Definition of surface displacement velocity and acceleration

Results of the Experiments

Figure 5a and b show the time variation in the surface displacement and the groundwater levels, respectively, at different locations on the model slope for Case 1. The surface displacement gradually increased from the start of the water

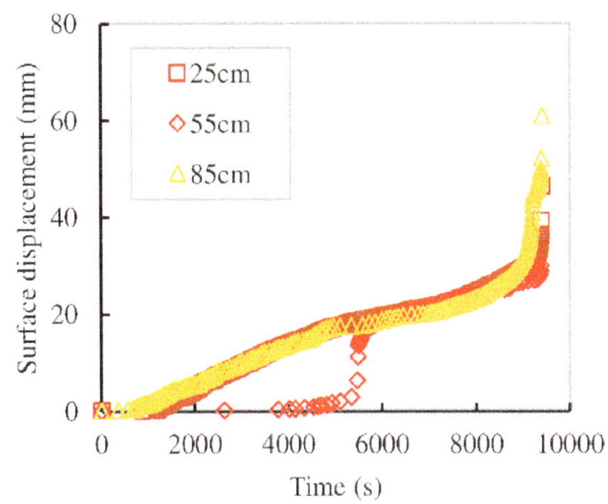

(a) The surface displacement

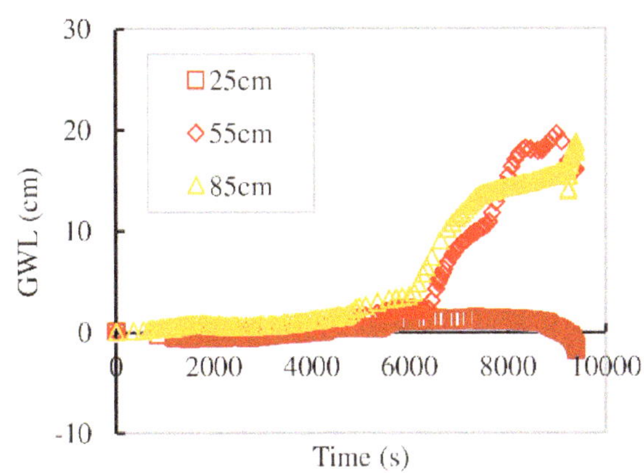

(b) The groundwater level

Fig. 5 Time variation of the surface displacement and the groundwater level at Case 1. GWL: groundwater level sprinkling, and exhibited an accelerative increase just prior to failure of the slope at 25 and 85 cm, while it remained at almost zero until 5,500 s and then significantly increased until 20 at 55 cm. It subsequently increased gradually, and increased again acceleratively up to failure at 55 cm. Groundwater levels remained almost zero until 6,000 s, and increased up to 15–20 cm until the failure at 55 and 85 cm, while it remained almost zero until the failure at 25 cm. It is noteworthy that surface displacement increased even before the generation of the groundwater level. Shear deformation of the slope might have been generated even under unsaturated conditions at this stage. The surface displacement increase accelerated with the increase in groundwater level. Time variations in the surface displacement and the groundwater level also had the same tendency in Case 3, which showed a different tendency to Case 2.

Figure 6a and b show comparisons between the surface displacement and the groundwater level at 85 cm in each case. The surface displacement significantly increased after the generation of the groundwater level, and exhibited accelerative increase with the increase in groundwater level just prior to failure in Case 1 and Case 3. It increased linearly with time without the generation of the groundwater level in Case 2. Failed soil mass moved into earth flow and deposited on the surface of the lower slope in Case 1 and Case 3. It contained a considerable quantity of water because the groundwater level was high at slope failure; thus, it could flow into the earth flow. While the failed soil mass was relatively dry in Case2. This was because the groundwater level was almost zero at failure, even if the initial water content was higher in Case 2 than in the other cases. It was observed by video image that the dry soil mass moved almost in one-piece, and it was pushed back by the soil layer of the lower slope in Case 2. Movement and deformation of the soil mass of the upper slope might have been restrained in this way in Case 2.

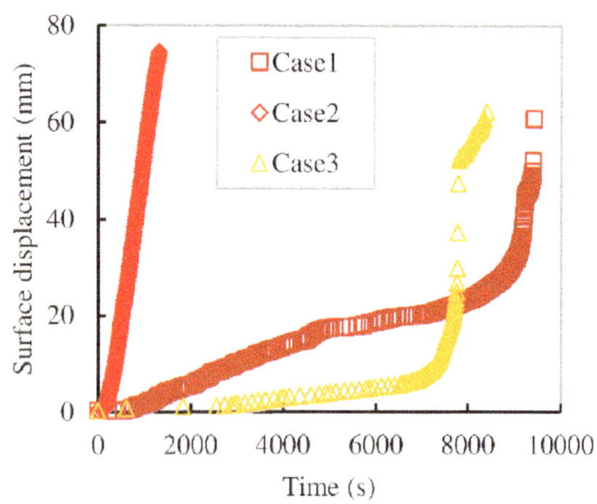

(a) The surface displacement

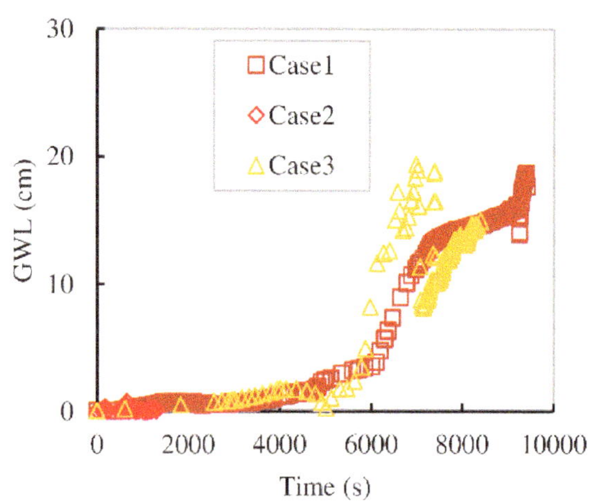

(b) The Groundwater level

Fig. 6 Comparison of the surface displacement and the groundwater level at 85 cm from upper boundary in each case. GWL: the groundwater level

Discussion

Figure 7 shows the relationship between the surface displacement velocity and acceleration at different distances from the upper boundary of the slope in Case 1 on a logarithmic scale. It was recognised that the relationship may have been linear on a logarithmic scale with some scatter, while the time variation of the surface displacement was different in each case. The relationship may have been almost unique, even though there was some scatter at different locations on the model slope.

Figure 8 shows the relationship between the surface displacement velocity and acceleration before and after the generation of the groundwater level at 85 cm from the upper

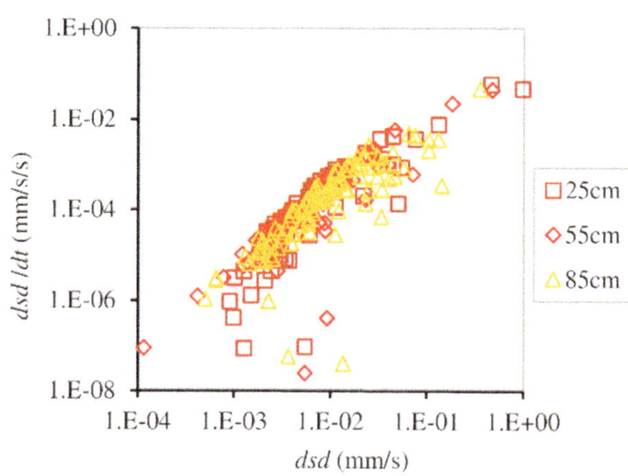

Fig. 7 Relationship between the surface displacement velocity and acceleration at different locations in Case 1. *dsd*: surface displacement velocity (mm/s). *dsd/dt*: surface displacement acceleration (mm/s/s)

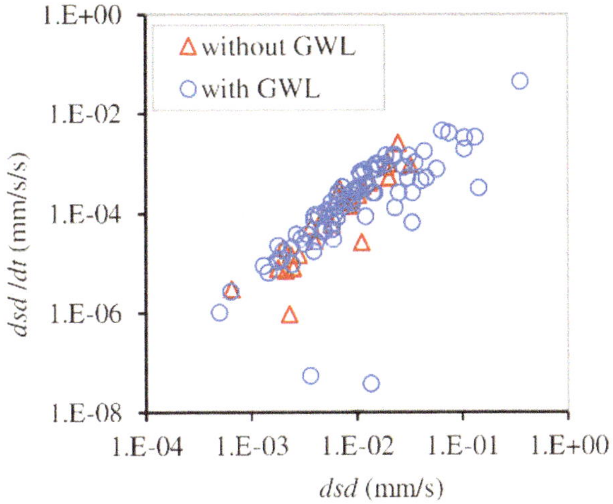

Fig. 8 Relationship between the surface displacement velocity the acceleration before and after the generation of the groundwater level at 85 cm from the upper boundary of the slope in Case 1. *dsd*: surface displacement velocity (mm/s). *dsd*/*dt*: surface displacement acceleration (mm/s/s). GWL: groundwater level

boundary of the model in Case 1. Red triangles symbolise before the generation of the groundwater level, while blue circles represent after the generation of the groundwater level. The relationship before the generation of the groundwater level was linear on a logarithmic scale, and almost the same as after the generation of the groundwater levels. The surface displacement was generated by the decrease in the suction of the slope (in unsaturated conditions) before the generation of the groundwater level, while it was generated by the decrease of effective stress in the slope due to the increase of static pore pressure (groundwater level) after the generation of the groundwater level. The relationship between the surface displacement velocity and acceleration was the same in both cases.

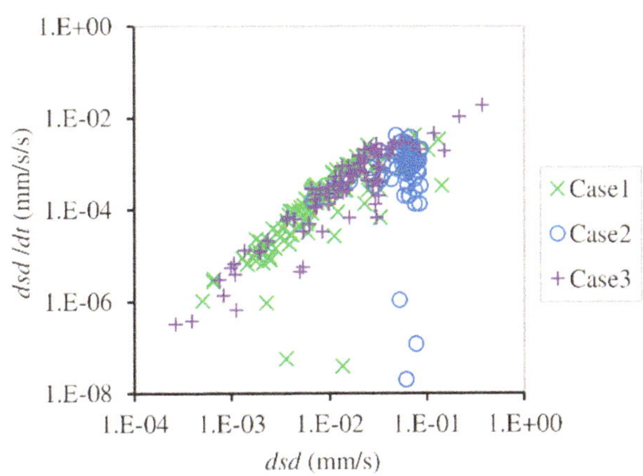

Fig. 9 Comparison of the relationship between the surface displacement velocity and acceleration at 85 cm from the upper boundary of the slope in each case. *dsd*: surface displacement velocity (mm/s). *dsd*/*dt*: surface displacement acceleration (mm/s/s)

Figure 9 shows a comparison of the relationship between the surface displacement velocity and acceleration at the same distance (85 cm) from the upper boundary of the slope in each case on a logarithmic scale. The relationship in Case3 was also linear on a logarithmic scale and almost the same as that in Case1 in the range of the surface displacement velocity from 1E-03 to 1E-01 mm/s. While the relationship in Case 2 gathered around the linear trend in Case1 and Case3 from 1E-02 to 1E-01 mm/s of the surface displacement velocity and did not indicate a clear trend. The range of the surface displacement velocity in Case2 was from 1E-02 to 1E-01, much smaller than the range (1E-03 to 1E-01 m/s) in Case1 and Case3. These might have been due to the restraint of the movement of soil mass in Case2. It was revealed that the relationship between the surface displacement velocity and acceleration was unique under different inclination of model slope.

Conclusion

Measurement of surface displacement was implemented in three cases of model slope test under artificial rainfall conditions. Initial water content and slope inclination were different in the three cases. The surface displacement velocity and acceleration were derived from the measured surface displacement and the relationship between those in each case was compared. Following are the results of the examination.

(1) The surface displacement showed accelerative increase just prior to the failure of the slope with the increase in the groundwater level in Case 1 and Case 2 with different slope inclinations, while it increased monotonically with time in Case 2 without the generation of the groundwater level. Movement of the soil mass in upper slope was restrained by the lower slope in a relatively dry condition for Case 2.
(2) The relationships between the surface displacement velocity and acceleration at different locations were unique in Case 1 and Case 3.
(3) The relationship between the surface displacement velocity and acceleration before the generation of the groundwater level was similar to that during the increase of the groundwater level. The relationship in unsaturated conditions was same with that after the generation of the groundwater level.
(4) The relationship between the surface displacement velocity and acceleration was unique under different slope inclinations. However, the influence of initial water content to the relationship was not clear in this

study. Accordingly, further examination is necessary to reveal the influence of the water content to the relationship.

It was recognised that the relationship between the surface displacement velocity and acceleration was unique under different slope inclinations with the same soil. This fact suggests the possibility of deriving the relationship by indoor shear test before monitoring the displacement of actual slopes to predict the failure time of the slope. While much more examination are necessary to verify the relationship in a different condition and scale in a slope.

Part of this research is supported by the Grant-in-Aid for Scientific Research B -KAKENHI-, 18H01674, JSPS.

References

Askarinejad A, Akca D, Springman SM (2018) Precursors of instability in a natural slope due to rainfall: a full-scale experiment. Landslides 15:1745–1759

Bozzano F, Mazzanti P (2012) Assessing of failure prediction methods for slope affected by human activities. In: Eberhardt E, Froese C, Turner AK, Leroueil S (eds) Landslides and engineered slopes: protecting society through improved understanding (Proceedings of 11th international and 2nd North American symposium on landslides and engineered slopes, Banff, Canada, 3–8 June 2012). CRC Press/Balkema, Leiden, pp 1465–1471

Fukuzono T (1985) A new method for predicting the failure time of a slope. In: Proceedings of the IVth international conference and field workshop on landslides. Tokyo, Japan, pp 145–150

Kromer RA, Huchinson DJ, Lato MJ, Gautheir D, Edwards T (2015) Identifying rock slope failure precursors using LiDAR for transportation corridor hazard management. Eng Geol 195:93–103

Moriwaki H, Inokuchi T, Hattanji T, Sassa K, Ochiai H, Wang G (2004) Failure processes in a full-scale landslide experiment using a rainfall simulator. Landslides 1:277–288

Ochiai H, Okada Y, Furuya G, Okura Y, Matsui T, Sammori T, Terajima T, Sassa K (2004) A fluidized landslide on a natural slope by artificial rainfall. Landslides 1:211–219

Saito M (1965) Forecasting the time of occurrence of a slope failure. In: Proceedings of 6th international conference on soil mechanics and foundation engineering. Montreal, Canada, 2, pp 537–541

Saito M, Yamada G (1973) Forecasting and result in case of landslide at Takabayama. In: Proceedings of 8th international conference on soil mechanics and foundation engineering, vol 4, issue 3. Moscow, USSR, pp 325–327

Varns DJ (1982) Time-deformation relations in creep to failure of earth materials. In: Proceedings of 7th southeast asian geotechnical conference, vol 2, pp 107–130

Voight B (1988) A relation to describe rate-dependent material failure. Science 243:200–203

Voight B (1989) Materials science law applies to time forecasts of slope failure. In: Landslide news (Japan Landslide Society (Eds)), vol 3. Tokyo, pp 8–10

Xiao J-Q, Ding D-X, Xu G, Jiang F-L (2009) Inverted S-shaped model for nonlinear fatigue in rock. Int J Rock Mech Min Sci 46:643–548

![WLF logo]

2020 Kyoto Japan

Comparison of Moving-Average, Lazy, and Information Gain Methods for Predicting Weekly Slope-Movements: A Case-Study in Chamoli, India

Praveen Kumar, Priyanka Sihag, Ankush Pathania, Pratik Chaturvedi, K. V. Uday, and Varun Dutt

Abstract

Landslide incidence is common in hilly areas. In particular, Tangni in Uttrakhand state between Pipalkoti and Joshimath has experienced a number of landslide incidents in the recent past. Thus, it is important to forecast slope-movements and associated landslide events in advance of their occurrence to avoid the associated risk. A recent approach to predicting slope-movements is by using machine-learning techniques. In machine-learning literature, moving-average methods (Seasonal Autoregressive Integrated Moving Average (SARIMA) model and Autoregressive (AR) model), Lazy methods (Instance-based-k (IBk) and Locally Weighted Learning (LWL)) and information-gain methods (REPTree and M5P) have been proposed. However, a comparison of these methods for real-world slope-movements has not been explored. The primary objective of this paper is to compare SARIMA, AR, LWL, IBk, REPTree and M5P methods in their ability to predict soil-movements recorded at the Tangni landslide in Chamoli, India. Time-series data about slope-movements from five-sensors placed on the Tangni landslide hill were collected daily over a 78-week period from July 2012 to July 2014. Different model parameters were calibrated to the training data (first 62-weeks) and then made to forecast the test data (the last 16-weeks). Results revealed that the moving-average models (SARIMA and AR) performed better compared to the lazy and information-gain methods during both training and test. Specifically, the SARIMA model possessed the smallest error compared to other models in test data. We discuss the implications of using moving-average methods in predicting slope-movements at real-world landslide locations.

Keywords

SARIMA • Instance-based • Locally weighted learning • M5P • REPTree • Tangni landslide

Introduction

Landslides and associated slope-movements are caused due to the rain, mostly in the monsoon season in hilly areas (Pande et al. 2006). These landslides cause massive damage to life and property (Parkash et al. 2011). In fact, landslides are a major problem in India with a shocking 11,000 deaths over the past 12 years in the country (Akter 2018). Thus, one needs to monitor, forecast, and timely alert people about slope-movements on hills prone to landslides (Chaturvedi et al. 2016). A way of predicting slope-movements is via machine-learning (ML) algorithms (Korup and Stolle 2014). Here, ML algorithms could learn patterns in data collected by sensor systems that are installed in different landslide-prone sites (Mali et al. 2019).

There is a large class of ML algorithms that have been proposed for making predictions about slope-movements. For example, the Seasonal Autoregressive Integrated

P. Kumar (✉) · P. Sihag · A. Pathania · V. Dutt
Applied Cognitive Science Laboratory, Indian Institute of Technology Mandi, Mandi-175075 Suran, India
e-mail: bluecodeindia@gmail.com

P. Sihag
e-mail: priyankasihag8993@gmail.com

A. Pathania
e-mail: ankushpathania.ap@gmail.com

V. Dutt
e-mail: varun@iitmandi.ac.in

P. Chaturvedi
Defence Research and Development Organization (DRDO), Defence Terrain Research Laboratory, 110054 New Delhi, India
e-mail: prateek@dtrl.drdo.in

K. V. Uday
Geohazard Studies Laboratory, Indian Institute of Technology Mandi, Mandi-175075 Medford, India
e-mail: uday@iitmandi.ac.in

© Springer Nature Switzerland AG 2021

N. Casagli et al. (eds.), *Understanding and Reducing Landslide Disaster Risk*, ICL Contribution to Landslide Disaster Risk Reduction,
https://doi.org/10.1007/978-3-030-60311-3_38

Moving Average (SARIMA) algorithm (Duan and Niu 2013; Qiang and Duan-you 2005) and the Autoregressive (AR) algorithm (Pu et al 2015; Li et al. 2018) are popular moving-average methods, where the previous values in a time-series are used to predict upcoming values (Duan and Niu 2013; Qiang and Duan-you 2005). Similarly, algorithms like the Instance-based-k (IBk) and Locally Weighted Learning (LWL) are lazy algorithms, where the most similar training data are used to predict the test data (Cheng and Hoang 2015,2016). There have also been algorithms like the REPTree and M5P that maximize the information gain in data to predict test data (Chen et al. 2017; Pham et al. 2019).

Although several ML algorithms have been used in literature to predict slope-movements in prior research, the application and comparison of moving-average methods, information-gain methods, and lazy methods has not been explored. The primary objective of this paper is to compare certain moving- average methods (SARIMA and AR) with lazy methods (IBk and LWL) and information-gain method (M5P and REPTree) in their ability to predict slope-movements in an active landslide site in the Himalayan mountains. Specifically, we use weekly slope-movement movement data collected via sensors at an active landslide (the Tangni landslide between Pipalkoti and Joshimath) between 2013 and 2014 and use it to predict slope-movements one-week ahead of time. Tangni is a major landslide site in Uttrakhand, India, which has seen a number of landslides in the recent past (Pathak 2016).

In what follows, we first detail the background literature concerning the use of ML algorithms for slope-movements predictions. Next, we detail the working of the SARIMA, AR, IBk LWL, REPTree and M5P algorithms and the method of calibrating these algorithms to data from the Tangni landslide. Finally, we present our results from different algorithms and discuss the implication of using moving-average, information-gain, and lazy methods for slope-movement predictions.

Background

Several prior research studies have used tree-based models for predicting slope movements (Krkac et al. 2017). For example, reference (Krkac et al. 2017) presented a methodology for prediction of landslide movements using random forests, a machine learning algorithm based on regression trees. According to reference (Krkac et al. 2017), the random forest method was established based on time-series for landslide movement, groundwater level, and precipitation gathered from the Kostanjek landslide monitoring system and nearby meteorological stations in Zagreb (Croatia) over a 2-year period. The validation results showed the capability of the information-gain model to predict the evolution of daily displacements over a period up to 30 days.

Some researchers have found the moving-average models to perform accurately in predicting slope movements (Duan and Niu 2013; Qiang and Duan-you 2005). For example, reference (Duan and Niu 2013) used the Autoregressive Integrated Moving Average (ARIMA) model was employed to forecast the cumulative displacement of the Bazimen landslide. Results indicated that the ARIMA method improved the mining result of traditional static data. Similarly, reference (Qiang and Duan-you 2005), have compared the ARIMA model and the considerable auto regressive (CAR) model and found both these methods to yield good results.

Furthermore, reference (Pu et al. 2015) have used Autoregression (AR) model and a detrended fluctuation analysis (DFA) method to model debris-flow. The coefficient and variance of AR and scaling exponent of DFA were estimated using a sliding window. Results showed that the DFA scaling exponent could indicate the abrupt change in debris-flows. Similarly, reference (Li et al. 2018) used autoregressive moving-average time-series models to analyze the autocorrelation of landslide triggering factors.

Research has also used lazy methods (like IBk) to predict slope movements (Cheng and Hoang 2015). For example, reference (Cheng and Hoang 2015; Bui et al. 2017; Chen et al. 2017) proposed a Swarm-Optimized Fuzzy Instance-based Learning (SOFIL) model for predicting slope collapses. The proposed model utilized the Fuzzy k-Nearest Neighbor (FKNN) algorithm as an instance-based learning method to predict slope collapse events. Similarly, reference (Cheng and Hoang 2016) proposed a novel approach for slope collapse assessment using the IBk algorithm, where the IBk approach provided a probabilistic slope stability estimation. A database containing 211 slope evaluation samples was collected in the Taiwan Provincial Highway Nos. 18 and 21, where the IBk achieved a roughly 8% improvement in accuracy rate compared with other benchmark methods. In the same way, reference (Bui et al. 2017) proposed combined a fuzzy IBk algorithm and a differential evolution (DE) optimization for spatial prediction of rainfall-induced shallow landslides at a tropical hilly area of Quy Hop, Vietnam. Overall, the fuzzy IBk model performed better compared to other support vector machines and tree-based models.

Although literature has proposed moving-average, information-gain, and lazy methods for slope movement predictions, a comparison of these methods on real-world slope movement data is yet to be undertaken. In this paper, we address this literature gap by comparing moving-average models (SARIMA and AR) with Information-gain models (REPTree and M5P) and lazy models (IBk and LWL). We perform our investigation by considering the prediction of slope movements on the Tangni landslide in Chamoli, India. To best of authors' knowledge, this study is the first of its kind to compare moving-average methods with lazy methods and information-gain methods on the Tangni landslide in India.

Study Area

The study was performed on the Tangni landslide in Chamoli district of Uttarakhand, India. The study area covers an area of 0.72 km2. It is located in the northern Himalayan region at latitudes 30° 27′ 54.3″ N and longitudes 79° 27′ 26.3″ E, at an altitude of 1450 m (Fig. 1a and b). As seen in Fig. 1b, the Tangni landslide is located on National Highway 58, which connects Ghaziabad in Uttar Pradesh near New Delhi with Badrinath and Mana Pass in Uttarakhand. The geology of this area consists of slate and dolomite rocks (Chaturvedi et al. 2016). The natural formation slope is 30° above the road level and 42° below the road level. The nearby area is a forest of oak and pinewood trees. There have been several prior slope-failures in this area causing road-blocks and economic losses to tourism (Landslides near Badrinath in Uttarakhand 2013).

Soil movement data was collected from the inclinometers, across five different boreholes in Tangni landslide area at a daily scale between 1st July 2012 and 1st July 2014. These five boreholes are represented by different colors in Fig. 1b (red—borehole 1, green—borehole 2, yellow—borehole 3, blue—borehole 4, and pink—borehole 5). Each borehole contained five sensors at different depths (3, 6, 9, 12, and 15 m). Data from some of these five boreholes was used for evaluating different moving-average and support-vector methods.

Methodology

Data Pre-processing

Data from Tangni landslide in Chamoli, India was obtained from the Defense Terrain Research Laboratory, Defense Research and Development Organization. The monitoring system in each of the five boreholes at the Tangni landslide contains inclinometer sensors at different depths (3, 6, 9, 12, and 15 m). These sensors measure tilt in mm per m units (essentially the angle the inclinometer tilts). Each inclinometer sensor is a 0.5-m long sensor that is installed vertically at different depths in a borehole. The monitoring system at Tangni landslide has five sensors per borehole across five boreholes. Thus, in total there are 25 sensors across 5 boreholes. Figure 2 shows the inclinometer sensor installed in its casing at a certain depth.

As shown in Fig. 2, if there is a tilting movement (θ) of the inclinometer of length L, then the horizontal displacement in the tilting direction is L sin θ. For better understanding, we converted the displacement in mm per m units into a θ angle (degrees), where 1 mm/m displacement equaled 0.05729°.

First, we calculated the relative tilt angle of each sensor from its initial reading at the time of installation. Second, we chose only those sensors from each borehole that gave the maximum average tilt angle over a two-year period. Thus, the data was reduced to five time-series, where each time-series represented the relative tilt per borehole from the sensor that moved the most in the borehole across the two-year period. As the daily data was sparse, we averaged the tilt over weeks to yield 78 weeks of average tilt data per time-series. Figure 3a–e represents the average relative tilt per week from five sensors across five boreholes (one sensor

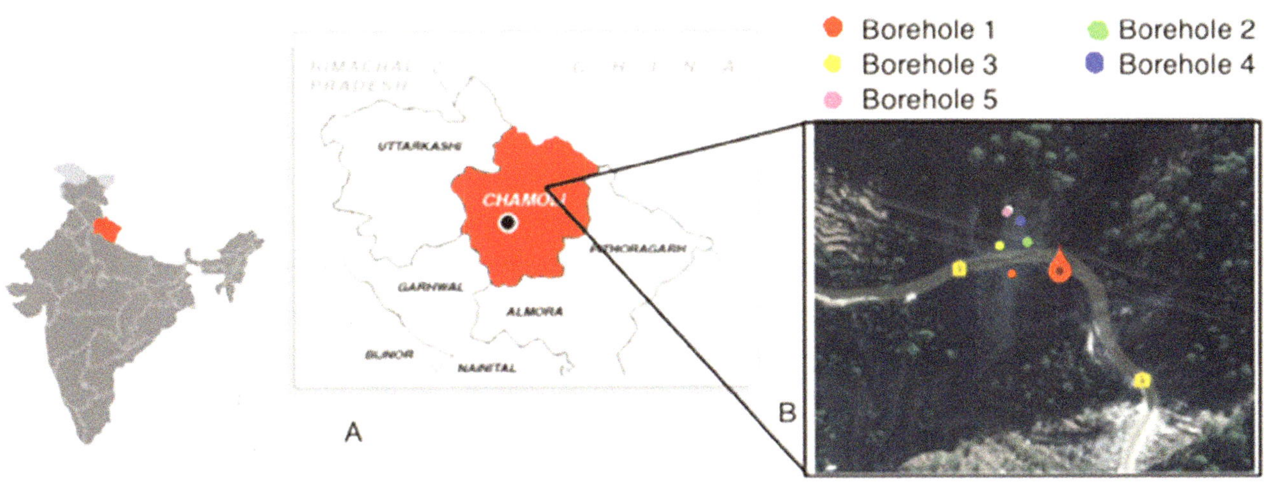

Fig. 1 **a** Location of the study area. **b** The Tangni landslide on Google Maps$'Fig1<>

$tforig 18mm, 178.3mm<>

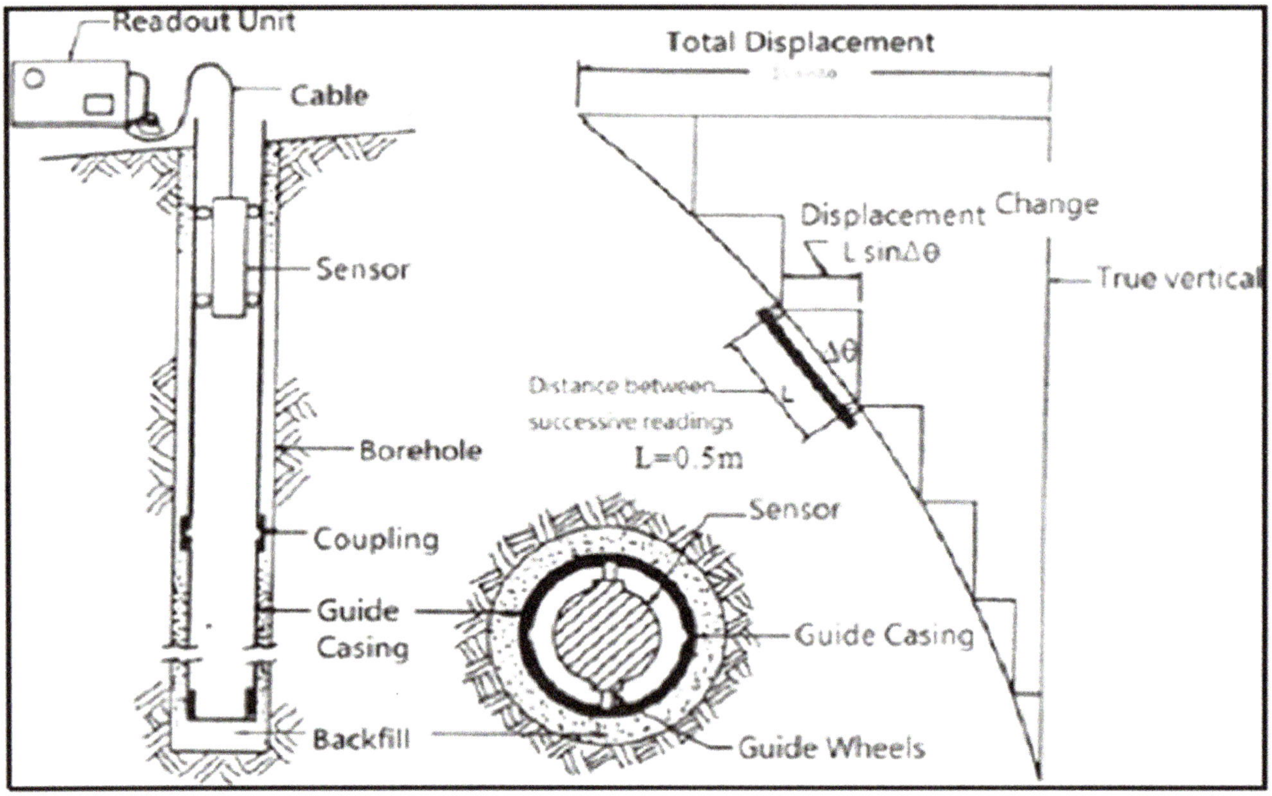

Fig. 2 Inclinometer sensor installed in its casing at a certain dept. *Source* Wilson and Mikkelsen 1977

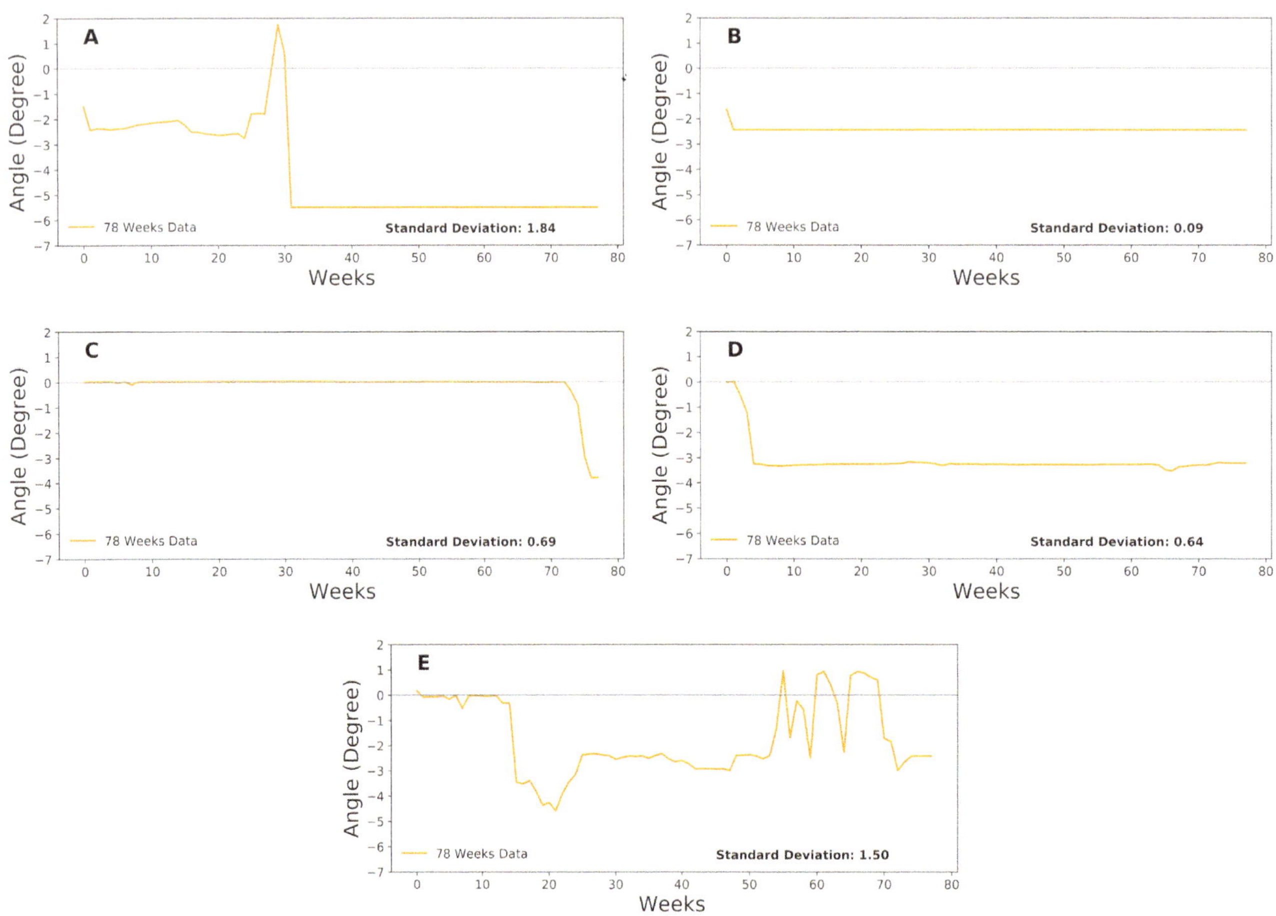

Fig. 3 Average tilt angle in degrees across five sensors (one per borehole). **a** Borehole 1 and 3 m sensor. **b** Borehole 2 and 12 m sensor. **c** Borehole 3 and 6 m sensor. **d** Borehole 4 and 15 m sensor. **e** Borehole 5 and 15 m sensor

per borehole) that caused the maximum average tilt across 78 weeks. These five time-series were used to compare the moving-average methods with the Lazy and information gain methods.

By convention, a negative tilt angle was downhill motion and a positive tilt angle was an uphill motion. As seen in Fig. 3c, a downhill motion starts from −0.11° in the 73rd week and suddenly becomes larger (−4.4°) in the last four weeks. The data was split in an 80:20 ratio (sixty-two weeks for training and the last sixteen weeks for testing) across different machine learning algorithms.

Machine Learning Models

Seasonal Auto-Regressive Integrated Moving-Average (SARIMA)

Seasonal Auto-Regressive Integrated Moving-Average (SARIMA) is a statistical forecasting method popular for univariate time-series data that may contain trend and seasonal components. It predicts the time-series by describing the autocorrelations in data (Asteriou and Hall 2011).

Stationarity of Time-Series: A time-series with constant values over time for mean, variance, and autocorrelation is stationary. Most statistical forecasting methods assume that a time-series can be made approximately stationary using mathematical transformations such as differencing (Hyndman and Athanasopoulos 2018). The first step of building a SARIMA model is stationarizing the data.

Auto Regressive (AR): An auto-regressive component in the SARIMA model predicts a variable at current state by passing the past values of the same variable to the model. Thus, an auto-regressive model is defined as:

$$y_t = c + \phi_1 y_{t-1} + \phi_2 y_{t-2} + \ldots + \phi_p y_{t-p} + \epsilon_t \tag{1}$$

where p is the auto-regressive trend parameter, ϵ_t is white noise and $y_{t-1}, y_{t-2} \ldots y_{t-p}$ denote the movement data of the previous weeks (Asteriou and Hall 2011).

Moving Average (MA): A moving-average component in the SARIMA model uses past prediction errors in a regression model, which is given in Eq. (2). A moving-average model is defined as:

$$y_t = c + \epsilon_t + \theta_1 \epsilon_{t-1} + \theta_2 \epsilon_{t-2} + \ldots + \theta_q \epsilon_{t-q} \tag{2}$$

where q is the moving-average trend parameter, ϵ_t is white noise and $\epsilon_{t-1}, \epsilon_{t-2} \ldots \epsilon_{t-q}$ are the error terms at previous weeks.

If we combine auto-regression (AR) in Eq. (1) and a moving-average (MA) in Eq. (2) model on stationary data, we obtain a non-seasonal ARIMA model, which is defined as:

$$y'_t = c + \phi_1 y'_{t-1} + \ldots + \phi_p y'_{t-p} + \theta_1 \epsilon_{t-1} + \ldots + \theta_q \epsilon_{t-q} + \epsilon_t \tag{3}$$

SARIMA builds upon an ARIMA model by adding seasonal parts to the ARIMA model (Asteriou and Hall 2011; Hyndman and Athanasopoulos 2018). The seasonal parts of an ARIMA model can have an AR factor, a MA factor, and an order of difference term. All these factors in seasonal data operate across multiples of the number of lagged periods in a season. In SARIMA, the three trend elements that require calibration are the trend AR order 'p', the trend difference order 'd' and trend MA order 'q'. Additional four seasonal elements that require calibration are the seasonal AR order 'P', the seasonal difference order 'D', the seasonal MA order 'Q' and the number of time steps 'm' for a single seasonal period. A SARIMA model performs differencing of the order D at a lag equal to the number of seasons 'm' to remove additive seasonal effects. As with lag 1 differencing to remove the trend, the lag 'm' differencing introduces a moving-average term. The SARIMA model also possess a trend parameter that captures the constant or linear trend in data.

Instance Based-K Learner (or K—Nearest Neighbour)

The IBk algorithm (Ade and Deshmukh 2014) uses a distance measure to locate k "closest" instances in the training data for each test instance. Next, the algorithm uses those selected instances to make a prediction. We calibrated the value of the parameter k to find its best value for slope-movement data.

Locally Weighted Learning (LWL)

The LWL (Englert 2017) is a general algorithm for locally weighted learning. It assigns weights using an instance-based method. In a supervised learning problem, where each input is associated to one output, the algorithm uses a similarity model to predict values for test data points.

Auto Regression (AR)

Auto regression (AR) is a time series model that uses observations from previous time steps as input to a

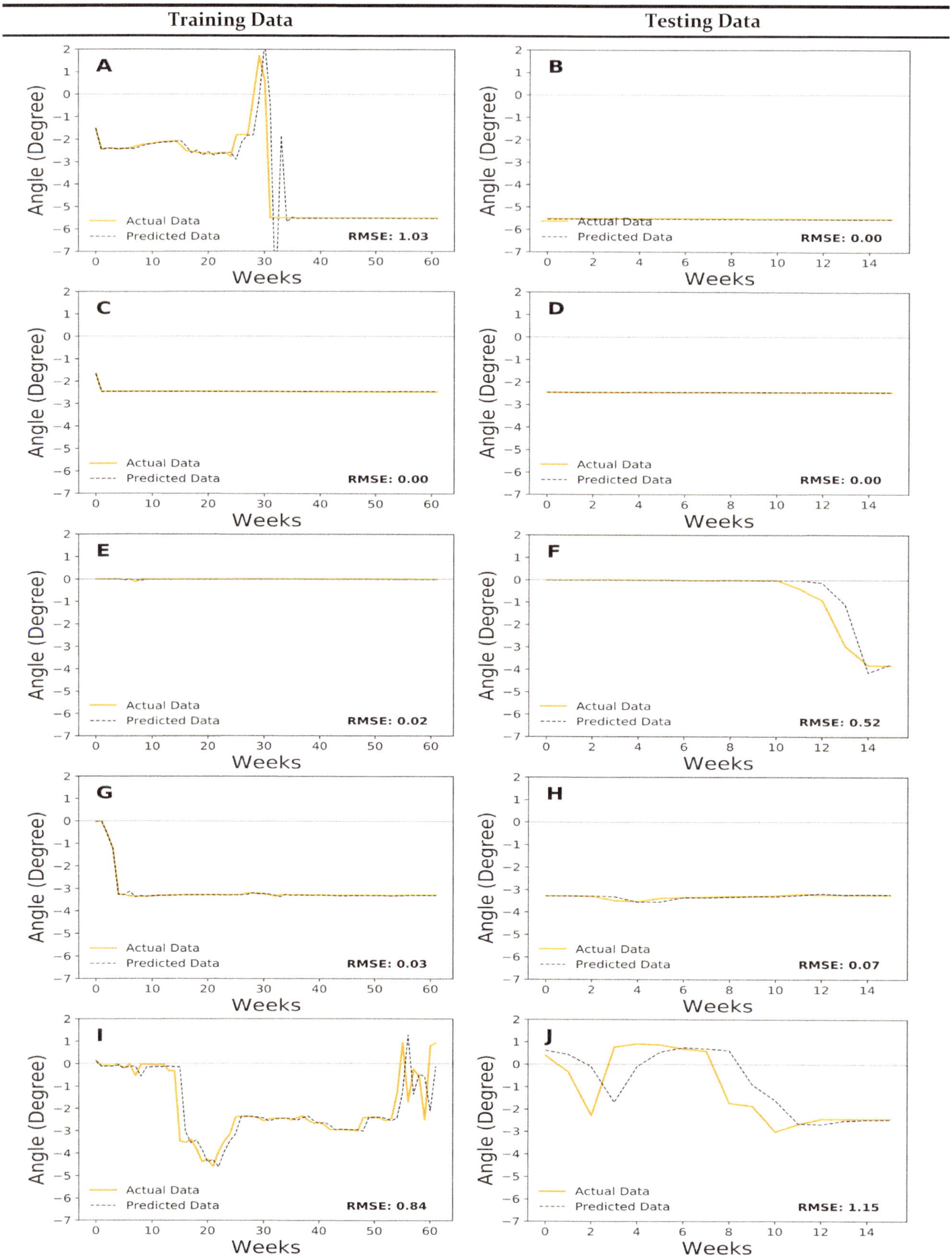

Fig. 4. Relative angle (in degree) over training data (62 weeks) and test data (16 weeks) from the best performing combined SARIMA model. **a** and **b**: borehole 1, 3m depth. **c** and **d**: borehole 2, 12m depth. **e** and **f**: borehole 3, 6m depth. **g** and **h**: borehole 4, 15m depth. **i** and **j**: borehole 5, 15m depth

Table 1 Parameters and their variation in SARIMA

Parameters	Range of values
Trend auto regressive (p)	[0, 1, 2]
Trend eifferencing (d)	[0, 1]
Trend moving-average (q)	[0, 1, 2]
Trend	[Absent, Constant, Linear, Constant with Linear Trend]
Seasonal auto-regressive (P)	[0, 1, 2]
Seasonal differencing (D)	[0, 1]
Seasonal moving-average (Q)	[0, 1, 2]
Seasonal periods (m)	[0, 1]

regression equation to predict the value at the next time step (Asteriou and Hall 2011; Hyndman and Athanasopoulos 2018). The autoregression model is identical to the AR part in the SARIMA model.

Reduced Error Pruning Tree (REPTree)

Basically, the reduced error pruning Tree (or REPTree) algorithm uses the regression tree logic and creates multiple trees in different iterations (Kalmegh 2015). After the trees are created, the algorithm selects the best tree from all generated trees as the representative. In pruning the tree, the measure used is the mean square error on the predictions made by the tree. Overall, the REPTree is a fast decision tree learning algorithm that builds a decision tree based on maximizing the information gain on attributes present in data (Al-Nabi and Ahmed 2013).

M5P

M5P tree algorithm was developed by Wang and Witten (Zhan et al. 2011). They modified the original M5 tree algorithm which was developed by Quinlan (1992). M5P tree algorithm was developed by modifying M5 tree to handle enumerated attributes and attribute missing values. In the M5P tree algorithm, before tree construction, all enumerated attributes are converted into binary variables.

M5P is powerful because it combines decision trees and linear regression for predicting a continuous variable (Braga et al. 2007a). Moreover, the M5P implement both regression tree and model tree (Braga et al. 2007b). The M5P algorithm adapts decision trees, firstly developed for classification, to regression problem.

These problems involve predicting a continuous numeric value instead of a discrete class. The M5P algorithm has three stages (Braga et al. 2007a, b): building a tree, pruning the tree and Smoothing.

Optimization of Model Parameters

Sarima

This model has eight parameters p, d, q, P, D, Q, m, and Trend. Table 1 shows the range of variations for different parameters in the SARIMA model. The trend parameter has four different values, where absent means no trend, constant means constant (horizontal) trend, linear means linear trend, and finally, the constant with linear trend means there is both a constant and linear trend. The m parameter means the number of time steps for a single seasonal period. A zero value for a parameter means that we do not include that parameter in the model. A reason for using the SARIMA model was that it allows one to account for a seasonal trend present in the time-series. We had two SARIMA models, one calibrated to each borehole (individual SARIMA model) and one calibrated across all boreholes (combined SARIMA model).

IBK

The size of the neighborhood was controlled by the k parameter. We varied the k parameter in steps of 2 from 1 to 21 instances. Also, we varied the distance (similarity) metric used as Manhattan distance or Euclidean distance.

LWL

The size of the neighbourhood was controlled by the k parameter in this model as well. We varied the k parameter in steps of 2 from 1 to 21 instances. Also, we varied the distance (similarity) metric used as Manhattan distance or Euclidean distance.

AR

This algorithm has parameters varied from 1 to 7, i.e., the model could have beta coefficients corresponding to the last 7 lag terms.

REPTree

In this algorithm, we varied the minimum number of instances per leaf between 1 and 4 and number of folds for reduced error pruning between 1 and 5.

M5P

In this algorithm, we varied the minimum number of instances per leaf between 1 and 5. Pruning of tree was used as part of the M5P algorithm.

Results

Each model was calibrated to each time-series, independently. Table 2 shows the root-mean squared error (RMSE) results of training of IBk, AR, LWL, M5P, REPTree and SARIMA models on the training data across the five boreholes. As can we see in Table 2, among all the algorithms, the LWL and M5P performed the best and second best for training data.

Tables 3, 4, and 5 shows the optimized values of different parameters in the AR, IBk, LWL, REPTree, M5P, and SARIMA algorithms. For example, in Table 3, the best value of the k parameter was 9 for a Manhattan distance function across all boreholes for the IBk algorithm. The SARIMA models showed non-zero seasonal Q parameter and non-seasonal q parameter. In certain cases, the seasonal P, the non-seasonal p, and the non-seasonal d parameters possessed non-zero values. The combined SARIMA model only possessed non-zero seasonal parameters. Overall, the combined SARIMA model produced the best result with 0.38° RMSE in the training dataset. Whereas, the SARIMA model calibrated to each borehole yielded a poorer RMSE of 0.71° in the training dataset.

Table 6 shows the RMSEs from different models across different boreholes in the last 16-weeks of test data. As can be seen in the Table 6, the individual and combined

Table 2 The RMSE of different algorithms in the training dataset

Algorithm	Root-mean squared error (RMSE) in degree of angle					
	BH[a] 1 03 m	BH 2 12 m	BH 3 06 m	BH 4 15 m	BH 5 15 m	Avg[b]
LWL	0.17	0.00	0.00	0.00	0.17	0.07
M5P	0.48	0.00	0.00	0.03	0.61	0.22
AR	0.84	0.00	0.00	0.02	0.71	0.31
REPTree	0.20	0.00	0.01	0.01	1.34	0.31
IBk	1.06	0.00	0.00	0.02	0.82	0.38
SARIMA[c]	1.03	0.00	0.02	0.03	0.84	0.38
SARIMA[d]	0.86	0.00	0.03	1.04	1.11	0.71

[a]Borehole [b]Average [c]Combined model across all borehole [d]Individual model across all boreholes

Table 3 Optimized parameters for Autoregression

Sensor borehole	Location depth	Function ($\phi_{t+1} = b_0 + b_1 * \phi_{t-1} \ldots b_n * \phi_{t-n}$)
1	03-m	$\phi_{t+1} = -3.9 - 0.57t + 1.0\phi_{t-1} - 0.34\phi_{t-2}$
2	12-m	$\phi_{t+1} = -42.96$
3	06-m	$\phi_{t+1} = 0.01 + 1.23\phi_{t-1} - 0.61\phi_{t-2} + 0.44\phi_{t-3} - 0.18\phi_{t-4}$
4	15-m	$\phi_{t+1} = -16.94 + 0.72\phi_{t-1} - 0.33\phi_{t-5} + 0.51\phi_{t-6} - 0.20\phi_{t-7}$
5	15-m	$\phi_{t+1} = -23.64 + 0.26t + 0.57\phi_{t-1} + 0.42\phi_{t-3} - 0.46\phi_{t-4} + 0.24\phi_{t-5} + 0.30\phi_{t-6} - 0.41\phi_{t-7}$

Table 4 Different ML algorithms and their calibrated parameters value

Algorithm	Parameters	Calibrated values
IBk	K	9
	Distance function	Manhattan
LWL	K	9
	Distance function	Manhattan
REPTree	Minimum number of instances	1
	Number of folds	3
M5P	Minimum number of instances	4

Table 5 Optimized parameter for SARIMA

SARIMA model	Location		Best set of parameters [(p, d, q), (P, D, Q, m), 'Trend']
	BH[a]	Depth	
Individual	1	03 m	[(0, 1, 0), (0, 0, 1, 0), 'Absent']
	2	12 m	[(0, 0, 1), (1, 0, 1, 1), 'Constant']
	3	06 m	[(0, 1, 1), (0, 0, 0, 0), 'Absent']
	4	15 m	[(2, 0, 1), (0, 0, 1, 0), 'Absent']
	5	15 m	[(1, 0, 0), (0, 0, 0, 0), 'Absent']
Combined	All		[(0, 0, 0), (0, 1,1, 1), 'Absent']

[a]Borehole

Table 6 The RMSE of different algorithms in the testing dataset

Algorithm	Root-mean squared error (RMSE) in degree of angle					
	BH[a] 1 03 m	BH 2 12 m	BH 3 06 m	BH 4 15 m	BH 5 15 m	Avg[b]
SARIMA[c]	0.00	0.00	0.52	0.06	1.05	0.33
SARIMA[d]	0.00	0.00	0.52	0.07	1.15	0.35
AR	0.02	0.00	0.57	0.07	1.42	0.53
LWL	0.00	0.00	1.55	0.09	1.12	0.55
IBk	0.00	0.00	1.55	0.09	1.35	0.60
REPTree	0.00	0.00	1.55	0.09	1.70	0.67
M5P	0.46	0.00	0.67	0.63	2.13	0.78

[a]Borehole [b]Average [c]Combined model across all borehole [d]Individual model across all boreholes

SARIMA and AR models performed best and second best compared to all other ML algorithms.

Figure 4 shows the fits of the combined SARIMA Model to the time-series data across the five boreholes in the training and test datasets. Overall, these results are reasonably good with very small RMSE values.

Discussion and Conclusion

One focus of machine-learning (ML) algorithms could be the prediction of slope-movements in advance to timely alert people about impending landslides. In this work, we applied moving-average models (SARIMA and AR), lazy models (IBk and LWL) and information-gain models on weekly slope-movement data from the Tangni landslides in Chamoli, India. Our results revealed that the moving-average models (SARIMA and AR) outperformed the lazy models (IBk and LWL) and information-gain models (REPTree and M5P) on test data. One likely reason for the SARIMA model to perform better compared to other models could be because this model has seasonal, auto-regressive, integrated, and moving-average components built into its working. Similarly, the AR model may also contain the AR component and this component helped this model to perform better. A likely reason for the combined SARIMA model to perform similar to or better compared to the individual SARIMA models could be that seasonal components fitted the dataset better compared to the non-seasonal components. It could also be that the similar geological and hydrological conditions across co-located boreholes made the combined SARIMA model to perform better compared to the individual SARIMA models. In fact, the combined SARIMA model showed similar results across both training and test data compared to individual SARIMA models, where the latter performed poorly in training data.

In this paper, we were able to show that moving-average methods to outperform lazy and information-gain algorithms for real-world slope-movement predictions. These results may not agree with findings in literature (Bui et al. 2017), where lazy algorithms have been shown to dominate other models. Thus, slope-movement predictions at different places may need different algorithms.

As part of our future research, we plan to extend these analyses to other ML algorithms including neural-network methods including the use of both artificial neural networks as well as recurrent neural networks (e.g., long short-term memory models). Some of these ideas form the immediate next steps in our program on slope-movement predictions using machine-learning techniques.

Acknowledgements The project was supported from grants (awards: IITM/NDMA/VD/184, IITM/DRDO-DTRL/VD/179, and IITM/DCo N/VD/204) to Varun Dutt. We are also grateful to Indian Institute of Technology Mandi for providing computational resources for this project.

References

Ade R, Deshmukh PR (2014) Instance-based versus batch-based incremental learning approach for students classification. Int J Comput Appl 106(3)

Akter M (2018) Why India needs to be concerned about landslides. https://weather.com/en-IN/india/science/news/2018-10-18-why-india-needs-to-be-concerned-about-landslides, 17 Dec 2019

Al-Nabi D, Ahmed SS (2013) Survey on classification algorithms for data mining: (comparison and evaluation). Comput Eng Intell Syst 4 (8):18–24

Asteriou D, Hall S (2011) ARIMA models and the Box Jenkins methodology. Appl Econ 2(2):265–286

Braga PL, Oliveira AL, Meira SR (2007a) Software effort estimation using machine learning techniques with robust confidence intervals. In: 7th international conference on hybrid intelligent systems, IEEE, 17–19 Sept 2007. Kaiserslautern, Germany, pp 352–357

Braga PL, Oliveira AL, Ribeiro GH, Meira SR (2007b) Bagging predictors for estimation of software project effort. In: International joint conference on neural networks, IEEE. 12–17 Aug 2007. Orlando, Florida, USA, pp 1595–1600

Bui DT, Nguyen QP, Hoang ND, Klempe H (2017) A novel fuzzy K-nearest neighbor inference model with differential evolution for spatial prediction of rainfall-induced shallow landslides in a tropical hilly area using GIS. Landslides 14(1):1–17

Chaturvedi P, Srivastava S, Kaur PB (2016) Landslide early warning system development using statistical analysis of sensors? data at Tangni Landslide, Uttarakhand, India. In: Proceedings of sixth international conference on soft computing for problem solving, 23–24 Dec 2016. Patiala, India, pp 259–270

Chen W, Shirzadi A, Shahabi H, Ahmad BB, Zhang S, Hong H, Zhang N (2017) A novel hybrid artificial intelligence approach based on the rotation forest ensemble and naïve Bayes tree classifiers for a landslide susceptibility assessment in Langao County, China. Geomat, Nat Hazards Risk 8(2):1955–1977

Cheng MY, Hoang ND (2015) A swarm-optimized fuzzy instance-based learning approach for predicting slope collapses in mountain roads. Knowl-Based Syst 76:256–263

Cheng MY, Hoang ND (2016) Slope collapse prediction using Bayesian framework with k-nearest neighbor density estimation: case study in Taiwan. J Comput Civ Eng 30(1):04014116

Duan GH (2013) A method of dynamic data mining for landslide monitoring data. J Yangtze River Sci Res Inst 30(5):10

Englert P (2017) Locally weighted learning. Encyclopedia of machine learning and data mining. Sammut C, Webb G (2). Boston. (978-1-4899-7685-7). 759 p

Hyndman RJ, Athanasopoulos G (2018) Forecasting: principles and practice. Hyndman R (2). OTexts, Australia. (978-0-9875071-1-2). 225 p

Kalmegh S (2015) Analysis of weka data mining algorithm REPTree, simple cart and randomtree for classification of indian news. Int J Innov Sci, Eng Technol 2(2):438–446

Korup O, Stolle A (2014) Landslide prediction from machine learning. Geol Today 30(1):26–33p

Krkac M, Spoljaric D, Bernat S, Arbanas SM (2017) Method for prediction of landslide movements based on random forests. Landslides 14(3):947–960

Landslides near Badrinath in Uttarakhand (2013) Landslides near Badrinath in Uttarakhand. https://www.Indiatvnews.com/news/india/landslides-near-badrinathin-uttarakhand-26296.html, 17 Dec 2019

Li H, Xu Q, He Y, Deng J (2018) Prediction of landslide displacement with an ensemble-based extreme learning machine and copula models. Landslides 15(10):2047–2059

Mali N, Chaturvedi P, Dutt V, Uday KV (2019) Training of sensors for early warning system of rainfall-induced landslides. Recent Adv Geo-Environ Eng, Geomech Geotech, Geohazards 1:449–452

Pande RK (2006) Landslide problems in Uttaranchal, India: issuesand challenges. Disaster Prev Manag: Int J 15(2):247–255

Parkash S (2011) Historical records of socio-economically significant landslides in India. J South Asia Disaster Stud 4(2):177–204

Pathak D (2016) Knowledge based landslide susceptibility mapping in the Himalayas. Geo Environ Disasters 3(1):8

Pham BT, Nguyen MD, Van Dao D, Prakash I, Ly HB, Le TT, Ngo HT (2019) Development of artificial intelligence models for the prediction of compression coefficient of soil: an application of Monte Carlo sensitivity analysis. Sci Total Environ 679:172–184

Pu F, Ma J, Zeng D, Xu X, Chen N (2015) Early warning of abrupt displacement change at the Yemaomian landslide of the three gorge region. China Nat Hazards Rev 16(4):04015004

Qiang LI, Duan-you LI (2005) Research of dynamic predication technique for landslide displacement monitoring. J Yangtze River Sci Res Inst 22(6)

Quinlan JR (1992) Learning with continuous classes. In 5th Australian joint conference on artificial intelligence, Hobart, Tasmania, 16–18 Nov 1992, vol 92, pp 343–348

Wilson SD, Mikkelsen PE (1977) Foundation instrumentation—inclinometers, FHWA TS-77-219. U.S. Department of Transportation, FHWA

Zhan C, Gan A, Hadi M (2011) Prediction of lane clearance time of freeway incidents using the M5P tree algorithm. IEEE Trans Intell Transp Syst 12(4):1549–1557

New Insights into the Spatiotemporal Precursory Failure Dynamics of the 2017 Xinmo Landslide and Its Surrounds

WLF 2020 Kyoto Japan

Antoinette Tordesillas, Shuo Zhou, Federico Di Traglia, and Emanuele Intrieri

Abstract

Sentinel-1 data on the kinematics of the 2017 Xinmo landslide and its surrounds are studied to understand the precursory failure dynamics of a large region with a historical predisposition to landslides. We perform a systematic spatiotemporal analysis over a period of two years to identify high-risk regions and discriminate between their precursory failure dynamics. We found the 2017 Xinmo landslide source to exhibit a unique kinematic signature which can be distinguished, almost a year in advance, from those of other sites of instabilities. Findings pave the way for the development of a new framework that exploits these differences in the dynamics of motions to accurately predict the location and size of a catastrophic landslide, and distinguish it from false alarms and/or smaller land slips early in the pre-failure regime.

Keywords

Sentinel-1 • Kinematics • Clustering • Precursory failure dynamics • Spatiotemporal

Introduction

The level of hazard and hence risk of a landslide depends crucially on its location and size. Consequently, the accurate and timely prediction of the location and geometry of an impending landslide is essential for effective risk management and decision-making (Bellugi et al. 2015; Intrieri et al. 2012). Unfortunately, despite tremendous advances in the acquisition of monitoring data for early warning systems, identifying precisely where a catastrophic slope failure will likely occur remains a challenge. In particular, one key aspect remains poorly understood: the precursory failure dynamics of landslide kinematics and its influence on the location and size of the rupture area. Landslide kinematics bears spatiotemporal structure; hence data of high spatial and temporal resolution are needed to properly characterise its salient features (Intrieri et al. 2012; Schulz et al. 2017). To this end, we study data from multi temporal interferometric synthetic aperture radar (MT-InSAR) using a method that integrates advanced analytics with state-of-the-art knowledge of granular media failure dynamics (Das and Tordesillas 2019; Singh and Tordesillas 2020; Tordesillas et al. 2018; Zhou et al. 2020; Wang et al. 2020; and references therein).

MT-InSAR techniques offer data with high spatial and temporal coverage of surface deformation. This is achieved by processing a long stack of synthetic aperture radar images, leading to measurements of millimetre-scale displacements over long time frames with reduced temporal decorrelation, atmospheric artefacts and other error sources (Ferretti et al. 2011). But a major problem with large MT-InSAR datasets is that they are difficult to interpret due to the huge number of moving points. For this reason, various techniques for mapping fast moving areas have been proposed, mainly based on clustering analysis (Lu et al. 2012). These studies, however, do not predict the area that will actually collapse. Indeed, landslide forecasting from satellites has so far proved to be feasible only in the aftermath of the event, when the precise location of the landslide is known (Carlà et al. 2019). Significant progress in MT-InSAR data usage would thus be achieved by a method that can precisely predict the site of collapse (and not simply detect areas displaying large movements) from a very large MT-InSAR dataset – without need for any other information (i.e. Landslides Inventory Maps, Digital Elevation Models, field surveys etc.). Such a method, which can potentially

A. Tordesillas (✉) · S. Zhou
School of Mathematics and Statistics, University of Melbourne, VIC 3010, Parkville, Australia
e-mail: atordesi@unimelb.edu.au

F. Di Traglia · E. Intrieri
Department of Earth Sciences, University of Florence, 50121 Florence, Italy

© Springer Nature Switzerland AG 2021

N. Casagli et al. (eds.), *Understanding and Reducing Landslide Disaster Risk*, ICL Contribution to Landslide Disaster Risk Reduction,
https://doi.org/10.1007/978-3-030-60311-3_39

identify newly formed landslides with a high probability of collapse, is proposed here.

Using data from Sentinel-1 (Intrieri et al. 2018), we develop and test our method for a large region comprising 130,000 measurement points (MPs) including the source of the 2017 Xinmo landslide. The region is in an area of active tectonics and is highly susceptible to recurring landslides in multiple locations – in tuned with earthquake and rainfall cycles (Fan et al. 2017). The study period is around 2 years and culminates at the time of catastrophic failure (ToF) on 24 June 2017. Given the history of the region, many local sites of instability prior to ToF may have led to smaller landslides while others may have never eventuated into an actual collapse – a defining feature that, for our purposes, can be defined as "false alarms".

The specific question we seek to answer is: *Did the landslide source exhibit unique precursory dynamics that distinguish it from other sites of instability in the region and, if so, how early?* The impetus of this research into this question in part stems from the "hot issues" raised in Fan et al. (2017): Did the source mass exhibit any precursors before sliding and what was its deformation history? Is it possible to perform early recognition and warning of such kind of landslides, originating in inaccessible, high-elevation and steep mountain slopes with dense vegetation, and, if so, how?

Sentinel-1 Data on Xinmo Landslide

On 24 June 2017 a rock avalanche occurred on the mountain above Xinmo village. Initially, 4.3 million m^3 of rock material detached from near the crest of the mountain and gained momentum along the slope; the volume increased up to 13 million m^3 as the rock avalanche entrained new material until it eventually reached Xinmo village with an estimated velocity of 250 km/h (Fan et al. 2017). This caused the death of 83 people and the destruction of 64 houses. Given the remoteness of the source area (located on a 55°-60° steep slope at an elevation of 3400 m a.s.l.), detecting early signs of instability through field survey was virtually impossible and, although Sentinel-1 data were regularly acquired, no data elaboration and interpretation was performed. This made it only possible to assess that, in hindsight, the displacement time series gathered by Sentinel-1 showed clear signs of tertiary creep leading to failure (Carlà et al. 2019; Intrieri et al. 2018). The data used to perform this analysis (and which are also adopted in the present paper) consisted in 45 SAR images acquired by Sentinel-1 constellation in C-band (6.5 cm wavelength), at 5 × 14 m spatial resolution, along the descending orbit (incidence angle of 40.78°) and spanning from 9 October 2014 to 19 June 2017 (i.e. five days before the failure). Data cover a 460 km^2 area, encapsulating 130,000 measurement points, elaborated using SqueeSAR algorithm (Ferretti et al. 2011).

It is apparent that making an actual prediction would have been difficult because of the small size of the initial critically unstable area (Fig. 1) with respect to the 460 km^2 area, especially because other zones were displaying movements. This highlights the necessity of a method that, starting from a wide-scale map, is able to precisely and timely pinpoint slopes where failure is approaching, even before they start displaying a tertiary creep behaviour (Intrieri et al. 2018).

The method we propose here is different from both Intrieri et al. (2018) and Carlà et al. (2019) where individual displacement time series for the MPs in the landslide source were analysed and, as such, can be vulnerable to noise effects and estimation errors in displacement. By contrast, here the motion of all 130,000 MPs are examined to uncover emergent spatiotemporal patterns. In this task, the *relative* (not the absolute) values of displacements matter the most in identifying the dynamics that differentiate the landslide source area from the other high-risk areas. Consequently, one of the advantages of our approach is that it is more robust to noise compared to traditional landslide forecasting methods of analysing a univariate displacement time series from selected locations.

Methods

Here we describe our proposed method which is designed to both characterise the spatiotemporal dynamics of motions of MPs and identify patterns that distinguish the true landslide source from the other regions of high risk of a landslide. First, we propose a procedure to identify those regions of highest risk of a landslide. Next, the actual landslide source, together with three other representative high-risk regions, are characterised. To do this, we employ a range of techniques, including knowledge of granular failure dynamics gained from new data-driven approaches, to uncover salient kinematic patterns that distinguish the region most likely to undergo catastrophic failure.

Identification of High-Risk Regions

We proceed in two steps. In step 1, we partition the whole monitoring domain into a 40 × 40 equal-sized grid partition. Each grid cell covers 0.3 km^2, which is of similar size to the landslide source area around the Xinmo village. On average, each grid cell contains around 80 MPs. Since MPs are not uniformly distributed in the monitoring space, we eliminate grid cells that have less than 100 MPs from further analysis, resulting in a total of 368 valid grid cells. In step 2, we

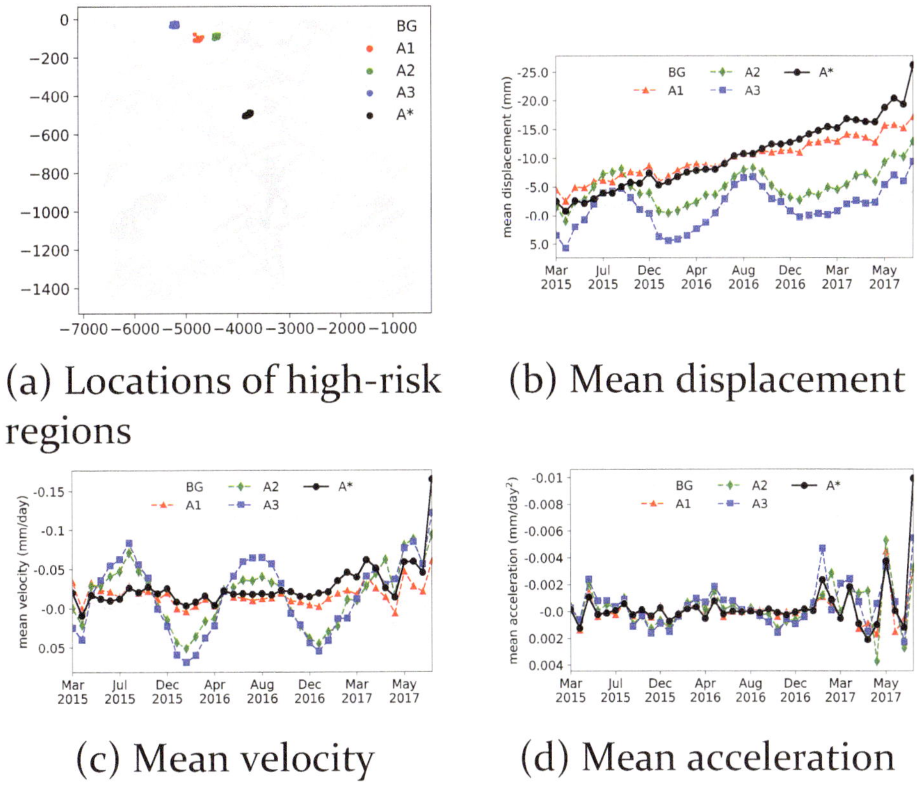

Fig. 1 Locations and temporal evolutions of the kinematic features of four high-risk regions A1, A2, A3 and A*. Region A* encapsulates the Xinmo landslide source. BG denotes all other MPs

identify the high-risk regions. To do so, we first form a candidate set of grid cells that show motions considerably larger than all others. Cumulative displacement, velocity and acceleration are used to identify three high-risk regions (denoted by A1-A3), to compare against the landslide source (A*). At each time stamp, cells are ranked by their cumulative displacement; those in the top 2.5% are added to the candidate set. The cell with the highest frequency in this displacement candidate set is treated as the high-risk region A1. This protocol is repeated for velocity and acceleration to find A2 and A3, respectively.

Characterisation of High-Risk Regions

Each of the high-risk regions, A1-A3 and A*, contains hundreds of MPs. It is difficult to directly compare these regions and discover patterns that can discriminate A* from the other high-risk areas based on the raw multivariate time series. Consequently, we propose to characterise each region using advanced analytics techniques and concepts – in addition to basic statistical properties. Specifically, we build on recent studies which examined kinematic clustering (Das and Tordesillas 2019; Tordesillas et al. 2018; Zhou et al. 2020) and explosive percolation (Singh and Tordesillas 2020) in the precursory failure regime. Such studies, which analysed radar data of a rockslide in an open pit mine and various laboratory tests, used knowledge of kinematic patterns that emerge when different sites of instabilities interact in the precursory failure regime in granular systems. For example, in laboratory samples, antecedent localised failure zones (e.g., cracks and shear bands) are known to interact, leading to structural arrest in some parts while failure amplifies in other parts, before catastrophic failure occurs. Similar phenomena at the large scale have been reported (Intrieri et al. 2018). Findings from Singh and Tordesillas (2020) suggest that the spatiotemporal dynamics of this interaction, at both small and large scales, manifest a distinct pattern of kinematic clustering and, crucially, the quality and temporal persistence of the "active cluster" for a given region determine the likely location and time of catastrophic failure. **We emphasise that the characterisation of the spatiotemporal pre-failure dynamics and the determination of the active cluster (set of MPs that form a cluster with the highest mean displacement) for each region is performed entirely in the displacement-state-space (DSS). No information on the locations of the MPs is used to inform these studies, except for visualisation and validation of the spatial location and geometry of the predicted area of collapse.**

Following Zhou et al. (2020), we perform a *k-means* clustering (Lloyd 1982) analysis in DSS for each region at each time stamp. That is, MPs are assigned into k clusters such that each MP has the minimal distance in DSS to its cluster centre (i.e., mean displacement over MPs within the cluster). Here we perform a quantitative evaluation of the clustering quality to find the optimum number of clusters: 2 for all time stamps. Accordingly, we set $k = 2$: one is the active cluster which corresponds to the unstable sliding area; the other is the stable area.

We characterise the dynamics of the kinematic clustering pattern from a number of aspects. First, since the active cluster provides an early prediction of the likely sliding area, we can apply the well-known Fukuzono's method (Fukuzono 1985) to study the velocity time series of the MPs belonging to this area. Specifically, for a given region and time stamp, a linear regression is fitted to the inverse of the mean velocity of the predicted sliding area for the last 6 time stamps to current time, in accordance with the size of the time window in Intrieri et al. (2018). The goodness of fit is used to identify an acceleration regime, by choosing the time period with the highest r-squared and a slope lower than -0.2. We refer to this time period as the tertiary creep phase.

Two metrics are used to summarise the dynamics of the clustering pattern: normalised mutual information (*NMI*) (Vinh et al. 2010) and the Silhouette score S (Rousseeuw 1987). *NMI* is a clustering similarity measurement, used here to quantify the persistence of clustering assignments between consecutive time stamps as follows:

$$NMI(Y_t, Y_{t+1}) = \frac{I(Y_t; Y_{t+1})}{\sqrt{H(Y_t)H(Y_{t+1})}}$$

where Y_t and Y_{t+1} are the clustering assignments at time t and $t+1$, $I(Y_t; Y_{t+1})$ is the mutual information between Y_t and Y_{t+1}, and $H(\cdot)$ is the entropy of the corresponding clustering partition. $NMI = 1$ means the clustering pattern at the current and previous times are exactly the same. Thus, the higher the *NMI* score, the more persistent and robust is the clustering pattern across time.

In addition, the Silhouette score S, which is the Silhouette coefficient averaged among the MPs in each region, is used to quantify the clustering quality. Specifically, the Silhouette coefficient of the i-th MP at time t, $s_t^{(i)}$, is a measure of how similar it is to its own cluster compared to the other cluster:

$$s_t^{(i)} = \frac{b_t - a_t}{\max(a_t, b_t)}$$

where a_t is the mean distance (measured in DSS) of the i-th MP to other MPs in the same cluster at time t, b_t is the averaged distance to MPs in the other cluster. The Silhouette score S ranges from -1 to 1. A value close to 1 means that MPs in different clusters are well separated in terms of their displacements, while MPs in the same cluster are very similar – *i.e., MPs move in near-rigid-body motion.*

A more powerful technique than *k-means* clustering analysis is the method proposed by Singh and Tordesillas (2020) using the concept of explosive percolation (Achlioptas et al. 2009). Distinct from *k-means*, this method does not depend on a prescribed k. Its key merit is that it provides a succinct summary of the time evolution of the number, size and member MPs of individual clusters, as well as their kinematic separation in DSS. At each time stamp, MPs within a kinematic-distance r in DSS are classified in the same cluster. The growth in the size of the largest cluster, $p(r, t) = G(r, t)/n$, is tracked as r is systematically increased, such that $G(r, t)$ is the number of MPs in the largest cluster and n is the total number of MPs in the studied high-risk region. In the absence of a clustering pattern in DSS, $p(r, t)$ should increase with r continuously until the largest cluster contains all n MPs. On the other hand, if the region is fractured, then member MPs may exhibit subdivided motions: e.g., a group of MPs may show accelerated motions and thus form a separate cluster in the high-displacement end of DSS, away from the MPs in the stable cluster in the low-displacement end of DSS. Depending on the number of clusters, this partitioning in motion results in multiple discontinuous jumps $p(r, t)$ as r increases. Each such jump or "explosive" growth in the size of the largest cluster results when a cluster is amalgamated into the largest cluster. These mergers lead to a distinctive stair-case pattern of consecutive "run-rise" cycles in the profile $p(r, t)$ vs r. The number of rises in each profile corresponds to one less than the number of clusters. The height of each "rise" corresponds to the size of the newly merged cluster, while the width of each "run" preceding the rise corresponds to the separation between the mean displacements of MPs in the largest cluster and its newly subsumed members. Readers are referred to (Singh and Tordesillas 2020) for complete details.

Results and Discussions

The positions of the three identified high-risk regions (A1, A2, A3 as described in the previous section) relative to the entire monitoring domain is shown in Fig. 1a, along with a comparison of their motions against those in the region A* containing the landslide source. Relative to A1-A3, A* exhibits the highest mean displacement (Fig. 1b) of constituent MPs in the year leading up to ToF (24 June 2017), followed by a sharp rise in velocity and acceleration in the last 9-week period of 20 April 2017 to 19 June 2017 (Fig. 1c, d). On average, spatially and temporally, the motions in A1 are similar to those in A*. Although, in the latter half of the monitoring period, motions in A* accelerate and surpass those in A1, the real difference between the two is only recorded in the last time stamp, when A* experiences a strong acceleration; apart from that, a manifestly similar pattern of behaviour can be observed in the time series for A* and A1 thus giving no hint of their vastly different outcome. As for A2 and A3, relatively large fluctuations can be observed in their mean displacement and velocity, though the mean displacement in these regions remain consistently lower than those observed in A*.

In Figs. 2–6, we show the results from recently developed analytics approaches to characterise the spatiotemporal dynamics of landslides in the pre-failure regime. Different tertiary creep phases are identified for different regions: 20 April 2017 to 19 June 2017 for A* and A3.

A3, 20 Jan 2016 to 19 May 2016 for A1, and 7 April 2015 to 5 August 2015 for A2 (Fig. 2). The corresponding clustering pattern for each region during its respective tertiary creep phase is shown in Fig. 3. The spatial pattern of the predicted sliding area in A* is coherent while those in A1, A2 and A3 are relatively fragmented. This suggests that the active sliding area in A* will likely detach as a solid whole.

In Fig. 4a, we observe a consistently high clustering persistence in time for A* with $NMI > 0.8$ from February 2016. The clustering quality (Silhouette score S in Fig. 4b) of A* is also gradually increasing with increasing time. This means that for A*, the separation among its MPs in DSS manifests way before the ToF and is steadily developing to the final failure. By contrast, this pattern is absent in the other high-risk regions. Focussing on the tertiary creep phase for each region (highlighted in Fig. 4), we observe A* shows consistently high NMI and S ($NMI > 0.8$ and $S > 0.75$). By comparison, although both A1 and A2 show relatively good quality clustering during their tertiary creep phases ($S \sim 0.7$ for A1 and $S \sim 0.6$ for A2), their clustering patterns do not persist in time ($NMI \sim 0.5$ for A1 and $NMI \sim 0.2$ for A2).

The results from the explosive percolation analysis are given in Fig. 5 during the tertiary creep phase and, more broadly, for the entire monitoring period in Fig. 6. Consistent with the recent study of a developing rockslide (Singh and Tordesillas 2020), we similarly find explosive percolation behaviour as evident in the distinctive stair-case pattern in all the high-risk areas during their respective tertiary creep phase. However, the pattern in A* (Fig. 5a) is distinct from those observed in the other regions (Fig. 5b–d). Specifically, there is one major step transition or jump for the same value of $p(r, t)$ across all time states in the tertiary creep regime for A*. This signifies the emergence of two major clusters that remain essentially invariant with respect to member MPs and whose inter-cluster motions become increasingly separated with time. By contrast, the clusters in the other regions keep changing in size, separation and number with time (Fig. 5b–d). Finally, the profiles in Fig. 6 suggest that a critical transition to the precursory failure regime for A* occurred around April 2016. Prior to this time, the profiles are relatively smooth and lacking the characteristic *prolonged run followed by a step transition* that define the so-called explosive percolation transition (Singh and Tordesillas

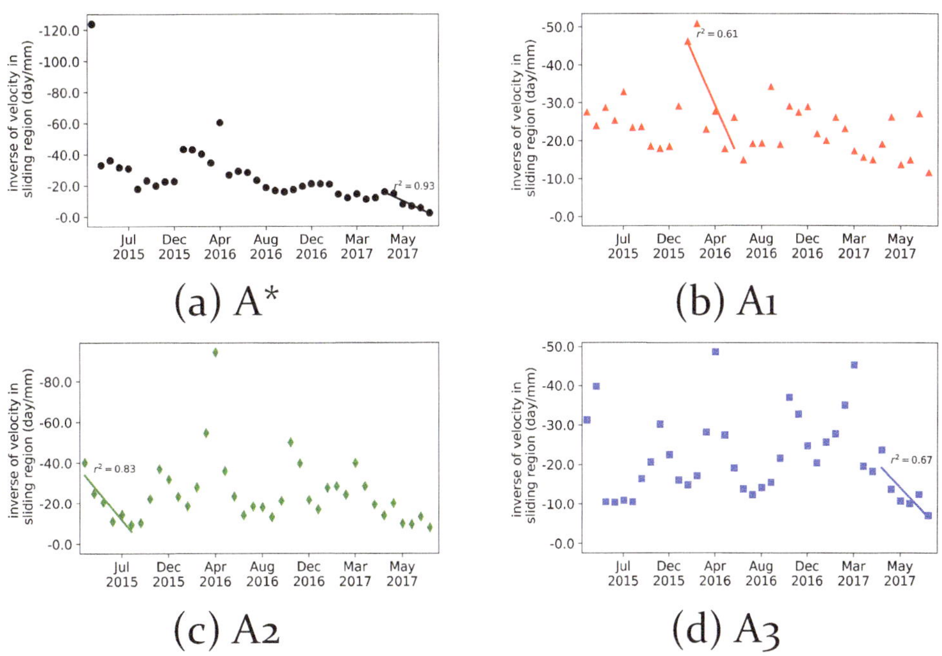

Fig. 2 Evolution with time of the inverse mean velocity of the predicted sliding zone in each high-risk region. The regression line is shown for the tertiary creep phase, that time period when the goodness of fit is maximum

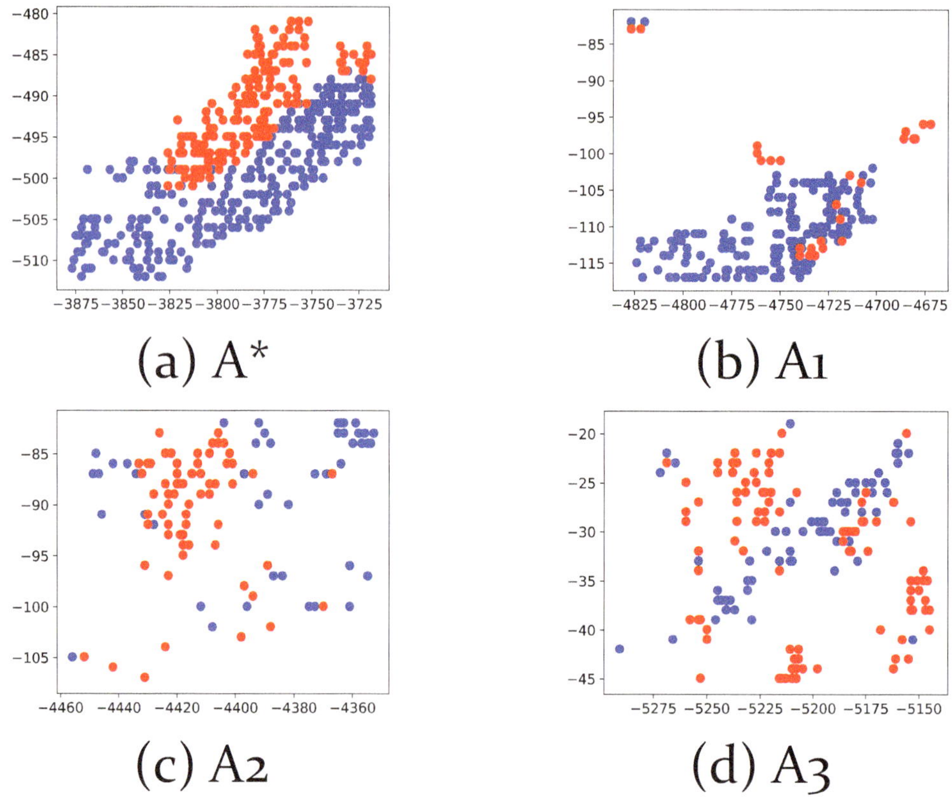

Fig. 3 The kinematic clustering pattern for $k = 2$ clusters mapped to the spatial domain during the tertiary creep phase. The active cluster (red MPs) corresponds to the predicted sliding area for each high-risk region

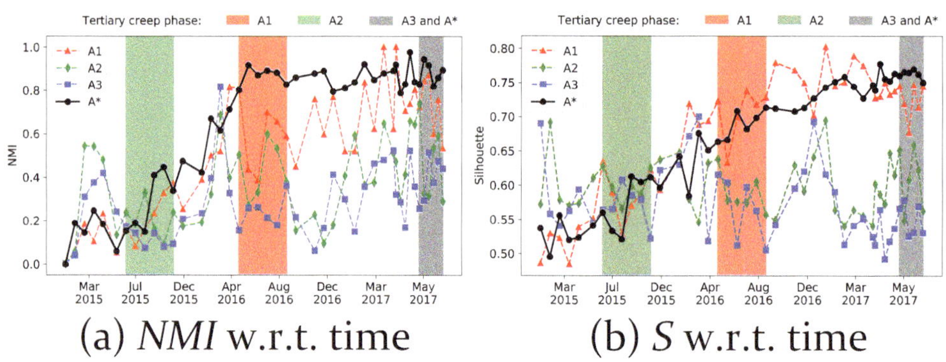

Fig. 4 Spatiotemporal dynamics of the clustering pattern in the motions of the constituent MPs

2020). Most notably, the inverse velocity method only anticipates failure from April 2017, i.e. one year later (Intrieri et al. 2018).

Conclusions

We studied cumulative displacement time series data across a large region with a long history of susceptibility to landslides from Sentinel-1 satellites. In particular, we characterised the spatiotemporal precursory failure dynamics of motions in regions of high risk of a landslide, including that encapsulating the 2017 Xinmo landslide source. The patterns uncovered effectively distinguish the true landslide region from the other high-risk regions that may constitute false alarms or smaller landslides. Specifically, our key findings are:

- In the tertiary creep phase, A* distinguishes itself from A1-A3 by its increasingly high Silhouette and consistently high *NMI* scores. This suggests that MPs in the sliding area increasingly move "in rigid-body motion" as ToF draws near.
- The distinct spatiotemporal dynamics of motions in A* during the tertiary creep phase can be observed even before the tertiary creep phase: from April 2016, more than a year before ToF.
- Results here suggest a regime change point occurred for A* on April 2016. Fan et al. (2017) reported a large quantity of pre-existing inter-connected cracks and

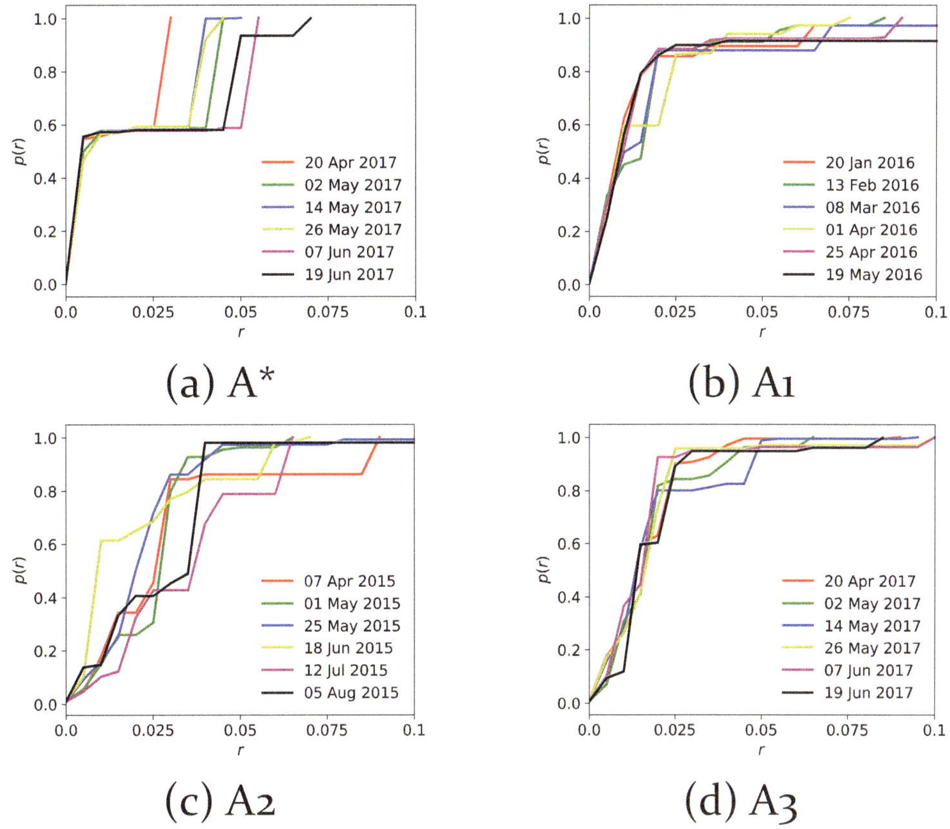

Fig. 5 Explosive percolation profiles at different time states during the tertiary creep phase for each high-risk region

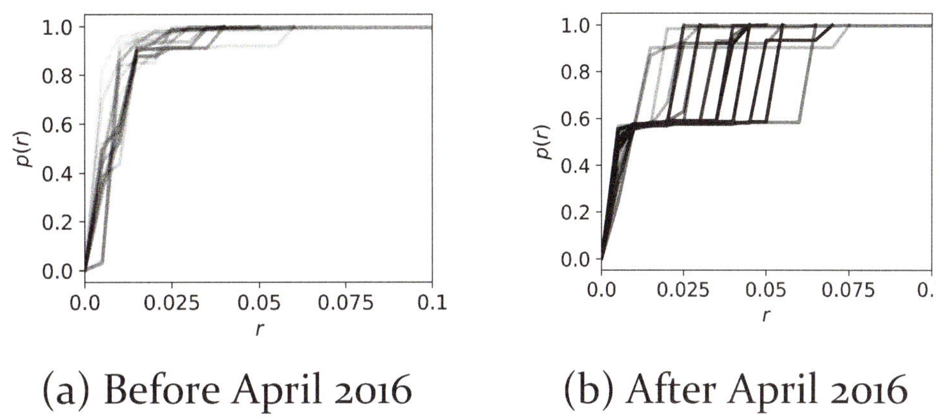

Fig. 6 Explosive percolation profiles before and after April 2016 for A*. Light grey to black corresponds to increasing time

fractures within the rock mass in this source area from prior earthquakes. As April marks the start of the rainy monsoon season, it is likely that the infiltration of water into these cracks from the ensuing heavy rainfall may have induced further growth of the old cracks, as well as created new cracks and fractures in the rock mass. The clustering pattern uncovered here suggests a significant crack growth of the sliding surface initiated as early as April 2016.

Ongoing work is now focussed on the development of a new machine learning framework that exploits these differences in pre-failure dynamics of motions to predict the location and size of a catastrophic landslide and distinguish it from false alarms and/or smaller land slips early in the precursory failure regime.

Acknowledgements A. T. and S. Z. acknowledge support from the U.S. Army International Technology Center Pacific (ITC-PAC) and US DoD High Performance Computing Modernization Program (HPCMP) Contract FA5209-18-C-0002.

References

Achlioptas D, D'Souza RM, Spencer J (2009) Explosive percolation in random networks. Science 323:1453–1455

Bellugi D, Milledge DG, Dietrich WE, McKean JA, Perron JT, Sudderth EB, Kazian B (2015) A spectral clustering search algorithm for predicting shallow landslide size and location. J Geophys Res: Earth Surf 120:300–324

Carlà T, Intrieri E, Raspini F, Bardi F, Farina P, Ferretti A, Colombo D, Novali F, Casagli N (2019) Perspectives on the prediction of catastrophic slope failures from satellite insar. Sci Rep 9:1–9

Das S, Tordesillas A (2019) Near real-time characterization of spatio-temporal precursory evolution of a rockslide from radar data: integrating statistical and machine learning with dynamics of granular failure. Remote Sensing 11:2777

Fan X, Xu Q, Scaringi G, Dai L, Li W, Dong X, Zhu X, Pei X, Dai K, Havenith H-B (2017) Failure mechanism and kinematics of the deadly june 24th 2017 xinmo landslide, maoxian, sichuan, china. Landslides 14:2129–2146

Ferretti A, Fumagalli A, Novali F, Prati C, Rocca F, Rucci A (2011) A new algorithm for processing interferometric data-stacks: squeesar. IEEE Trans Geosci Remote Sens 49:3460–3470

Fukuzono T (1985) A new method for predicting the failure time of a slope. In: Proceedings of 4th international conference and field workshop on landslide, pp 145–150

Intrieri E, Gigli G, Mugnai F, Fanti R, Casagli N (2012) Design and implementation of a landslide early warning system. Eng Geol 147:124–136

Intrieri E, Raspini F, Fumagalli A, Lu P, Del Conte S, Farina P, Allievi J, Ferretti A, Casagli N (2018) The maoxian landslide as seen from space: detecting precursors of failure with sentinel-1 data. Landslides 15:123–133

Lloyd S (1982) Least squares quantization in PCM. IEEE Trans Inf Theory 28:129–137

Lu P, Casagli N, Catani F, Tofani V (2012) Persistent scatterers interferometry hotspot and cluster analysis (psi-hca) for detection of extremely slow-moving landslides. Int J Remote Sens 33:466–489

Rousseeuw PJ (1987) Silhouettes: a graphical aid to the interpretation and validation of cluster analysis. J Comput Appl Math 20:53–65

Schulz WH, Coe JA, Ricci PP, Smoczyk GM, Shurtleff BL, Panosky J (2017) Landslide kinematics and their potential controls from hourly to decadal timescales: insights from integrating ground-based insar measurements with structural maps and long-term monitoring data. Geomorphology 285:121–136

Singh K, Tordesillas A (2020) Spatiotemporal evolution of a landslide: a transition to explosive percolation. Entropy 22:67

Tordesillas A, Zhou Z, Batterham R (2018) A data-driven complex systems approach to early prediction of landslides. Mech Res Commun 92:137–141

Vinh NX, Epps J, Bailey J (2010) Information theoretic measures for clusterings comparison: variants, properties, normalization and correction for chance. J Mach Learn Res 11:2837–2854

Wang H, Qian G, Tordesillas A (2020) Modeling big spatio-temporal geo-hazards data for forecasting by error-correction cointegration and dimension-reduction. Spatial Statistics 36:100432

Zhou S, Bondell H, Tordesillas A, Rubinstein BIP, Bailey J (2020) Early prediction of a rockslide location via a spatially-aided gaussian mixture model. Annals of Applied Statistics 14:977–992

Cutting-Edge Technologies Aiming for Better Outcomes of Landslide Disaster Mitigation

Kazuo Konagai

The International Consortium on Landslides (ICL) and The Global Promotion Committee of the International Programme on Landslides (GPC/IPL) have been responsible for organizing the World Landslide Forums (WLFs) every three years since 2008. Ever since the 1st WLF, the forums have long been the arena for landslide researchers and practitioners to exchange up-to-date information of recent devastations caused by landslides, cutting-edge technologies for landslide disaster mitigations and early warnings etc. to establish synergies among all participants worldwide.

Though the upcoming WLF5 has officially been postponed by one year to 2 to 6 November 2021 due to the global disruption caused by the coronavirus pandemic, the WLF5 will be all the more important with the Kyoto Landslide Commitment 2020 (KLC2020) to be launched as planned in the final online signatory meeting on 5 November 2020; the KLC 2020 is intended to be our action goals as the further advanced successor of the 'Sendai Landslide Partnerships 2015–2025 for Global Promotion of Understanding and Reducing Landslide Disaster Risk' in line with some of 17 Sustainable Development Goals (SDGs), particularly SDG 11, "Make cities and human settlements inclusive, safe, resilient and sustainable," of the United Nations.

For these important goals, the ICL has been inviting sponsorship from industries, businesses, and government agencies; all leading players in landslide science and technologies. They have been supporting a variety of the ICL/IPL activities such as publishing the International full-color journal "Landslides (Journal of the International Consortium on Landslides), full-color books for WLFs, exhibiting their cutting-edge technologies in WLFs, etc. Here follow short introductions of their activities with their names, addresses and contact information:

Marui & Co. Ltd.

1-9-17 Goryo, Daito City, Osaka 574-0064, Japan.
URL: https://marui-group.co.jp/en/index.html
Contact: hp-mail@marui-group.co.jp

Marui & Co. Ltd. celebrates its 100th anniversary in 2020. Marui, as one of the leading manufacturers of testing apparatuses in Japan, has been constantly striving to further improve its service since its foundation in 1920, thus contributing to the sustainable development of our nation and society. Our main products cover a wide variety of destructive and non-destructive testing apparatuses in the fields of geotechnical engineering, concrete engineering (mortar, aggregates, etc.), and ceramic engineering. Of special note is that Marui has been helping manufacture ring-shear apparatuses half-century long based on the leading-edge idea of Dr. Kyoji Sassa, Professor Emeritus at the Kyoto University. Marui has delivered total 7 ring-shear apparatuses to the Disaster Prevention Research Institute, Kyoto University, and 2 to the International Consortium on Landslides. Also the apparatuses were exported to the United States of America, China, Croatia and Vietnam.

Marui & Co. Ltd. takes great pleasure in developing, manufacturing, and providing new products of high value sharing the delight of achievement with our customers, and

K. Konagai (✉)
Secretary General, Organizing Committee of the Fifth World Landslide Forum, International Consortium On Landslides, Kyoto, 606-8226, Japan
e-mail: konagai@iclhq.org

© Springer Nature Switzerland AG 2021
N. Casagli et al. (eds.), *Understanding and Reducing Landslide Disaster Risk*, ICL Contribution to Landslide Disaster Risk Reduction, https://doi.org/10.1007/978-3-030-60311-3_40

thus contributing to the social development. The whole staff of Marui & Co. Ltd. are determined to devote ceaseless effort to keep its organization optimized for its speedy and high-quality services, by the motto "Creativity and Revolution", and strive hard to take a step further, as a leading manufacturer of testing apparatuses, to answer our customer's expectations for the twenty-second century to come.

Nippon Koei Co., Ltd.

5-4 Kojimachi, Chiyoda-ku, Tokyo 102-8539, Japan.
URL: https://www.n-koei.co.jp/english/
Contact: https://www.n-koei.co.jp/english/contact/input.

Nippon Koei Co., Ltd. and its group companies conduct many projects to support the growth of developing countries in Asia, Africa, the Middle and Near East, Latin America and other regions. Examples of their efforts include environmental measures to combat global warming, development of regional transportation infrastructure to support the rapid growth of emerging economies, and reconstruction assistance for regions affected by conflict and/or natural disasters.

OSASI Technos, Inc.

65-3 Hongu-cho, Kochi City, Kochi 780-0945, Japan.
URL: https://www.osasi.co.jp/en/
Contact: cs@osasi.co.jp

OSASI Technos, Inc. has been making its best efforts to develop its cutting-edge technologies for landslide early warning. Its unique compact and lightweight sensors making up the Landslide Early Warning System enable long-term monitoring of unstable landslide mass movements, precipitations, porewater pressure buildups, etc., in a remote mountainous area where commercial power is often unavailable. OSASI Technos, Inc. is also proud of its advanced technology to transfer observed data even in areas with poor telecom environments as proven in the successful implementations in South Asia.

All stuff members of OSASI Technos work together for mitigation of landslide disasters worldwide.

Godai Corporation

1-35 Kuroda, Kanazawa City, Ishikawa Prefecture 921-8051, Japan.
URL: https://soft.godai.co.jp/En/Soft/Product/Products/LS-RAPID/
Contact: pp-sales@godai.co.jp

Ever since its foundation in 1965, Godai Kaihatsu Co., Ltd., a civil engineering consulting firm, has long been providing a variety of software and measures particularly for natural disaster mitigation. With its rich expertise in both civil engineering and information technology (IT), the company has its primary goal to address real world needs of disaster mitigation. All the staff of Godai Kaihatsu Co., Ltd. feel it more than happy that their cutting-edge technologies help mitigate natural disasters.

Japan Conservation Engineers & Co., Ltd.

3-18-5 Toranomon, Minato-ku, Tokyo 1,050,001, Japan.
URL: https://www.jce.co.jp/en/
Contact: go_info@jce.jp

Japan Conservation Engineers & Co., Ltd. (JCE) is a general consulting firm working on landslide prevention research and consulting. JCE provides various disaster prevention technologies for debris flows, landslides, slope failures, rockfalls, etc. In addition, JCE is proud of its expertise having been conducting surveys and consulting works on coastal erosions and tsunami countermeasures for about 20 years. JCE contributes to the world through its activities in the realm of both structural and non-structural measures to build a resilient society.

OYO Corporation

7 Kanda-Mitoshiro-cho, Chiyoda-ku, Tokyo 101-8486, Japan.
URL: https://www.oyo.co.jp/english/
Contact: https://www.oyo.co.jp/english/contacts/.

OYO Corporation, the top geological survey company in Japan established in Tokyo in 1957, is well known as one of leading companies providing cutting-edge technologies and measures for natural disasters such as landslides, earthquakes, tsunamis, and floods. Not just developing and selling measuring instruments related to disaster prevention, OYO also delivers a market-leading services in 3D ground/geological modeling and 3D exploration technologies.

Kokusai Kogyo Co., Ltd.

2 Rokubancho, Chiyoda-ku, Tokyo 102-0085, Japan.
URL: https://www.kkc.co.jp/english/index.html
Contact: overseas@kk-grp.jp

Kokusai Kogyo Co., Ltd. as a leading company of geospatial information technologies, has long been providing public services with its comprehensive expertise to address real world needs and cutting-edge measurement technologies. Kokusai Kogyo Co., Ltd. helps rebuild "Green Communities," which has been of our great concern in terms of "environment and energy," "disaster risk reduction" and "asset management". Kokusai Kogyo Co., Ltd. offers the advanced and comprehensive analyses of geospatial information for developing new government policies, maintaining and operating social infrastructures safe and secure, and implementing low-carbon measures in cities.

Influenced by the recent global climate change, extreme rainfall events have become more frequent worldwide and resultant hydro-meteorological hazards are creating more deaths and devastations particularly in many developing countries where effective advanced countermeasures are not readily available. Kokusai Kogyo Co., Ltd. is proud of its achievements in establishing resilient infrastructure systems and implementing effective monitoring/early warning systems in developing countries, which have long been helping reduce the risks from natural hazards.

Geobrugg AG

Aachstrasse 11, 8590 Romanshorn, Switzerland.
URL: https://www.geobrugg.com
Contact: info@geobrugg.com

Swiss company Geobrugg is the global leader in the supply of high-tensile steel wire safety nets and meshes – with production facilities on four continents, as well as branches and partners in over 50 countries. True to the philosophy "Safety is our nature" the company develops and manufactures protection systems made of high-tensile steel wire. These systems protect against natural hazards such as rockfall, landslides, debris flow and avalanches. They ensure safety in mining and tunneling, as well as on motorsport tracks and stop other impacts from falling or flying objects. More than 65 years of experience and close collaboration with research institutes and universities make Geobrugg a pioneer in these fields.

Ellegi Srl

Via Petrarca, 55 I-22070 Rovello Porro (CO) Italy.
URL: https://www.lisalab.com/engl/?seze=1
Contact: info@lisalab.com

Ellegi srl provides worldwide monitoring services and produces Ground Based synthetic aperture radar (GBInsAR) for remote measurement of displacements and deformations on natural hazards and manmade buildings using its own designed and patented LiSALab system.

Its activities started in 2003 as a spin off project to exploit commercially the Ground Based Linear Synthetic Aperture Radars technology developed by European Commission's Ispra Joint Research Centre and based on the results of more than 10 years of research. Since then Ellegi has industrialized and developed the core technology of the LiSALab system and latest LiSAmobile system represents the 5th generation of development.

In 2003 it was the first commercial company in the world to provide GBInSAR measurements of natural hazards and structure.

Ellegi srl offers:

- Displacement fields measurement, control and monitoring of the deformation caused by natural hazards, like landslides, rockslides, sinkhole, volcanic deformation in every operative condition, including emergencies,
- Structural strain fields measurement, control, monitoring and diagnosis of the deformation affecting buildings, bridges, viaducts, dams.
- GBInSAR monitoring systems, installation, management and maintenance in order to provide information about natural hazards or anthropic activity, that can generate or cause slopes failures or buildings instabilities.
- In all the above-mentioned activities Ellegi srl uses the GBInSAR LiSALab technology that represents a real "break-through".

Chuo Kaihatsu Corporation

3-13-5 Nishi-waseda, Shinjuku-ku, Tokyo 169-8612, Japan.
URL: https://www.ckcnet.co.jp/global/
Contact: https://www.ckcnet.co.jp/contactus/.

Chuo Kaihatsu Corporation (CKC) was founded in 1946, and has been aiming to become the "Only One" consultant for our customers. We engage in the hands-on work that will "Remain with the earth, Remain in people's hearts, and Lead to a prosperous future." We focus on road, river and dam engineering to flesh out industrial infrastructures specifically by means of geophysical/geotechnical/geological investigations, civil engineering surveys and project implementations. In recent years, we make significant efforts on earthquake disaster mitigation, sediment disaster prevention/mitigation and ICT information services. Many achievements of ours have already contributed to mitigation of natural disasters such as landslides, earthquakes and slope failures in Japan, Asia and the Pacific Region.

IDS GeoRadar s.r.l.

Via Augusto Righi, 6, 6A, 8, Loc. Ospedaletto, Pisa, Italy, 56,121.
URL: https://idsgeoradar.com/
Contact: info@idsgeoradar.com

IDS GeoRadar, part of Hexagon, provides products and solutions, based on radar technology, for monitoring applications including landslides, rockfalls, complex structures, mining and civil engineering. The company is a leading provider of Ground Penetrating Radar (GPR) and Interferometric Radar solutions worldwide. IDS GeoRadar is committed to delivering best-in-class performance solutions and to the pursuit of product excellence, through the creation of application-specific, innovative and cost-efficient systems for a wide range of applications.

METER Group, Inc.

2365 NE Hopkins Court, Pullman, WA 99,163, USA.
URL: https://metergroup.com/wlf5.
Contact: bryan.wacker@metergroup.com

METER Group provides accurate, rugged, and dependable instrumentation to monitor moisture in all its phases within an unstable slope. METER specializes in instrumentation for near real-time monitoring of incoming moisture in the form of rain and weather. In addition, we provide real-time below-surface monitoring of existing moisture conditions like moisture content and soil suction which show how the soil profile is filling with water to saturation, including the transition to positive pore water pressure.

The ZL6 advanced cloud data logger works together with ZENTRA Cloud data software to simplify and speed up data collection, management, visualization, and alerting. Our well-published instrumentation is used worldwide in universities, research and testing labs, government agencies, and industrial applications.

For almost four decades, scientists and engineers have relied on our instrumentation to understand critical moisture parameters. We've even partnered with NASA to measure soil (regolith) moisture on Mars. Wherever you measure, and whatever you're measuring, rely on METER for accuracy, affordability, and simplicity that will make your job easier.

Asia Air Survey Co., Ltd.

Shinyuri 21 BLDG 3F, 1-2-2 Manpukuji, Asao-Ku, Kawasaki, Kanagawa, 215-0004, Japan.
URL: https://www.ajiko.co.jp/en/
Contact: service@ajiko.co.jp

Asia Air Survey (AAS), as one of the leading engineering and consulting companies, has long been providing disaster prevention and mitigation services for over 65 years, particularly in the fields of landslide, debris flow, erosion control, etc. AAS is proud of being the inventor of Red Relief Image Map (RRIM), which is a cutting-edge 3D terrain visualization method allowing great geomorphological details to be visualized in one glance, thus has been used in various facets of disaster prevention and mitigation.

Kiso-Jiban Consultants Co., Ltd.

Kinshicho Prime Tower 12 Floor, 1-5-7 Kameido, Koto-ku, Tokyo 36-8577, Japan.
URL: https://www.kisojiban.com/
Contact: kisojiban-contactus@kiso.co.jp

Kiso-Jiban Consultants, established in 1953, is an engineering consulting firm especially well known in the field of geotechnical engineering. The areas of its comprehensive services are listed below:

Geological and Geotechnical Survey
Geotechnical Analysis and Design
Disaster Prevention and Management
GIS (Geographic Information Systems)
Soil and Rock Laboratory Tests
Instrumentation and Monitoring
Geophysical Exploration and Logging
Distribution of Geosynthetics Products

Much-talked-about new service is Kiso-SAR System allowing accurate estimation of both extent and rate of landslide movements based upon a comprehensive interpretation of InSAR results from geotechnical and landslide engineering viewpoint (see the one-page introduction of Kiso-Jiban Consultants Co., Ltd.). With Kiso-SAR system, the following pieces of important geotechnical information can be provided:

(1) Extent of a deforming landslide mass (and the rate of its movement.
(2) Consolidation buildup in soft clay underlying a fill.
(3) Deformation buildups induced by slope cutting.

Okuyama Boring Co., Ltd.

10-39 Shimei-cho, Yokote City, Akita 013-0046, Japan.
URL: https://okuyama.co.jp/en/
Contact: info@okuyama.co.jp

Okuyama Boring Co., Ltd. is proud of its achievements in various projects to help solve many landslide problems. The company has been offering services in geological surveys and analyses, developing rational countermeasures against various geotechnical problems as well as safe workflow diagrams, and providing necessary pieces of advice for ensuring safety during landslide countermeasure works. For this purpose, Okuyama Boring Co., Ltd. works on monitoring, observations, field surveys, numerical analyses, countermeasure works, etc. of landslides.

Kawasaki Geological Engineering Co. Ltd.

Mita-Kawasaki Bldg, 2-11-15 Mita, Minato-ku, Tokyo10 8-8337, Japan.
URL: https://www.kge.co.jp/
Contact: post-master@kge.co.jp

Kawasaki Geological Engineering Co., Ltd. as one of the leading members of SAAM Research Group, has proactively been involved in developing "Sustainable Asset Anchor Maintenance (SAAM, hereafter) System," enabling easy maintenance of ground anchors. Its unique jack, weighing about half the weight of a conventional jack, together with a newly developed jig, can be applied to any type of anchor even with a short extra length, thus allowing for in-situ lift-off tests on these anchors. The SAAM system also has an optional weight meter that can be installed after performing a lift-off test.

Nissaku Co., Ltd.

4-199-3 Sakuragi-cho, Omiya-ku, Saitama 330-0854, Japan.
URL: https://www.nissaku.co.jp/
Contact: survey@nissaku.co.jp

Nissaku Co., Ltd., founded in 1912 as a well drilling company, provides services for far-flung fields of not only groundwater exploitation but also measures for landslides. Having its rich expertise in these fields, Nissaku Co., Ltd. offers general reliable one-stop technical services including designs, investigations, analyses, constructions, and maintenances.

Full-color presentations from the above seventeen exhibitors focusing on their landslide technologies are shown on the following pages. Their cutting-edge technologies have of course been instrumental in the progress that we have made in landslide risk-reduction worldwide, and we want to exert even greater effort to aim high given the KLC 2020 as our new action goals. The International Consortium on Landslides seeks volunteers willing to support our activities introducing their brand-new technologies for landslide disaster mitigation in our international journal "Landslides," full color books for WLFs, exhibitions at WLFs, etc. If you are interested in being engaged in supporting ICL activities, please contact the ICL secretariat <secretariat@iclhq.org>.

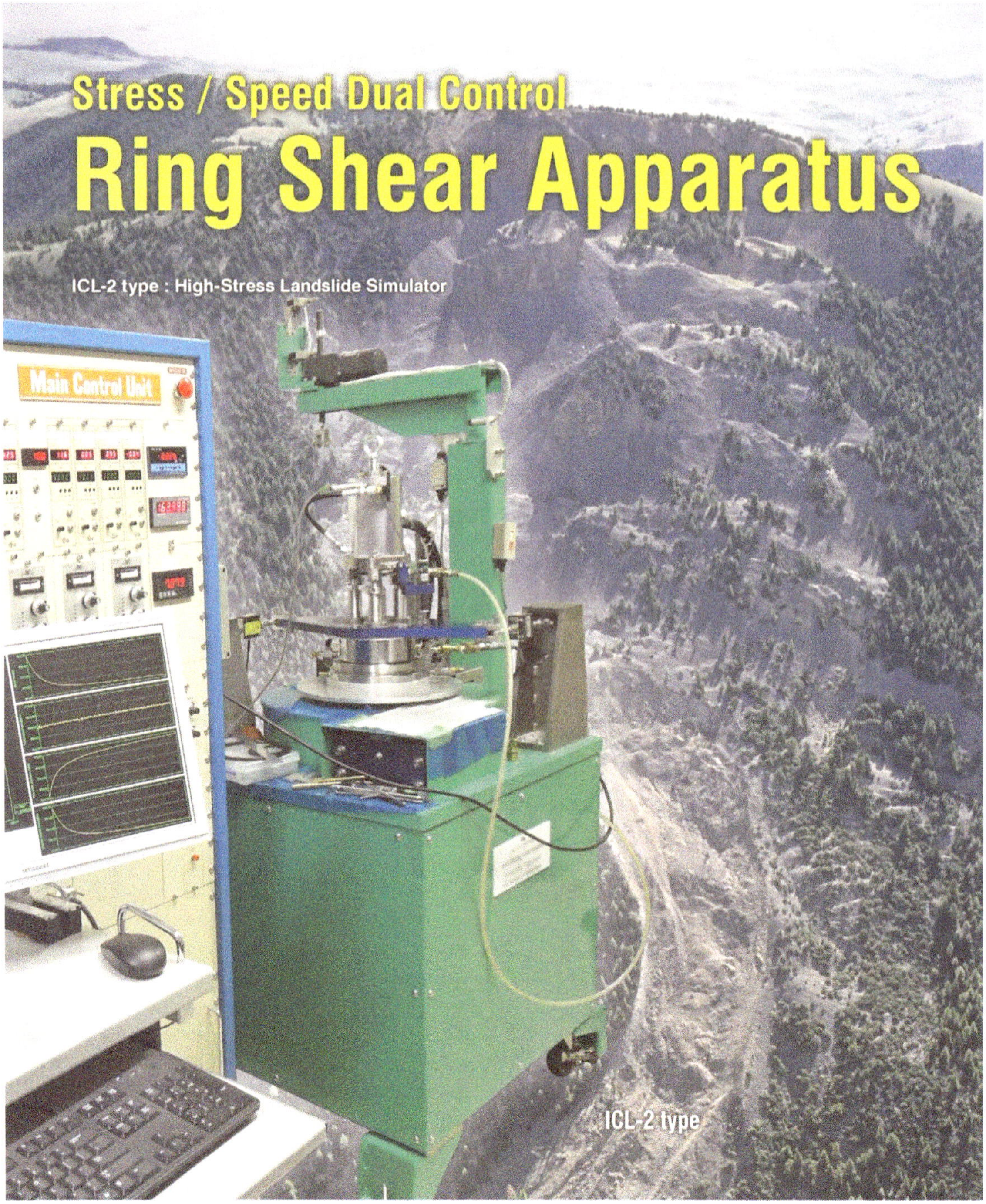

MARUI & CO., LTD.

Web site : https://www.marui-group.co.jp/en/
E-mail : hp-mail@marui-group.co.jp
Address : 1-9-17 Goryo, Daito City, Osaka Prefecture, 574-0064, Japan
Phone : 81-72-869-3201
F a x : 81-72-869-3205

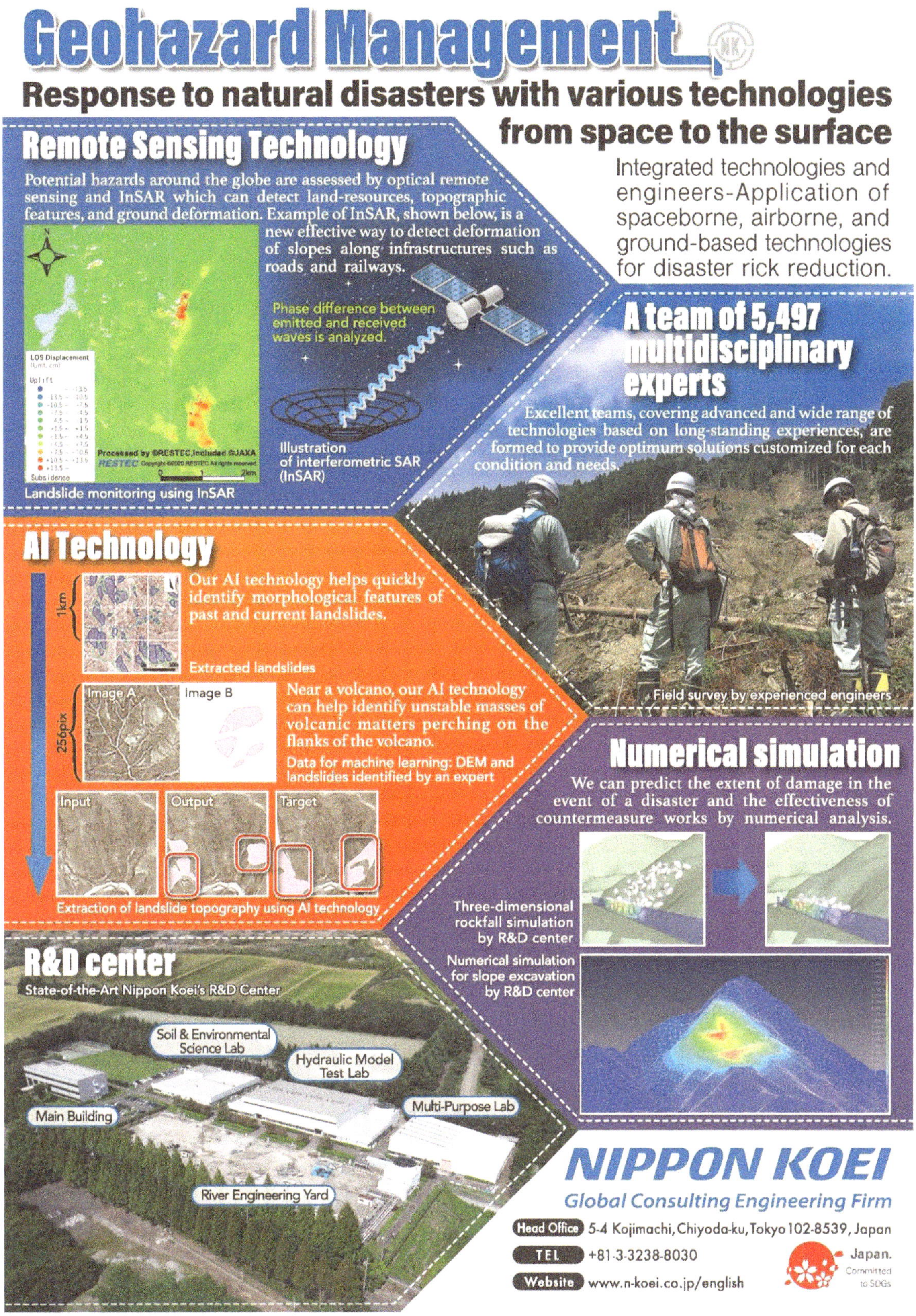

Geohazard Management
Response to natural disasters with various technologies from space to the surface
Integrated technologies and engineers-Application of spaceborne, airborne, and ground-based technologies for disaster rick reduction.
Remote Sensing Technology
Potential hazards around the globe are assessed by optical remote sensing and InSAR which can detect land-resources, topographic features, and ground deformation. Example of InSAR, shown below, is a new effective way to detect deformation of slopes along infrastructures such as roads and railways.
Phase difference between emitted and received waves is analyzed.
Illustration of interferometric SAR (InSAR)
Landslide monitoring using InSAR
A team of 5,497 multidisciplinary experts
Excellent teams, covering advanced and wide range of technologies based on long-standing experiences, are formed to provide optimum solutions customized for each condition and needs.
Field survey by experienced engineers
AI Technology
Our AI technology helps quickly identify morphological features of past and current landslides.
Extracted landslides
1km
256pix
Image A
Image B
Near a volcano, our AI technology can help identify unstable masses of volcanic matters perching on the flanks of the volcano.
Data for machine learning: DEM and landslides identified by an expert
Input
Output
Target
Extraction of landslide topography using AI technology
Numerical simulation
We can predict the extent of damage in the event of a disaster and the effectiveness of countermeasure works by numerical analysis.
Three-dimensional rockfall simulation by R&D center
Numerical simulation for slope excavation by R&D center
R&D center
State-of-the-Art Nippon Koei's R&D Center
Soil & Environmental Science Lab
Hydraulic Model Test Lab
Main Building
Multi-Purpose Lab
River Engineering Yard
NIPPON KOEI
Global Consulting Engineering Firm
Head Office 5-4 Kojimachi, Chiyoda-ku, Tokyo 102-8539, Japan
TEL +81-3-3238-8030
Website www.n-koei.co.jp/english
Japan. Committed to SDGs

We Pass on Voices of the Earth
Protection of Human Life by Measurement
Measurement technology × Transmission technology
Early Warning System
Extensometer
Communication device
Multi-point Inclinometer
Alarm device
Rain gauge
OSASI
OSASI TECHNOS INC.
Corporate Headquarters (Kochi)
65-3 Hongu-cho, Kochi-shi, Kochi 780-0945, Japan
TEL +81-88-850-0535 E-mail : cs@osasi.co.jp
http://www.osasi.co.jp/en/
Tokyo Headquarters
Sumitomo Seimei Nishi-Shimbashi Building 4F,
1-10-2, Nishi-shimbashi, Minato-ku, Tokyo, 105-0003, Japan
TEL +81-3-5510-1391 E-mail : cs@osasi.co.jp

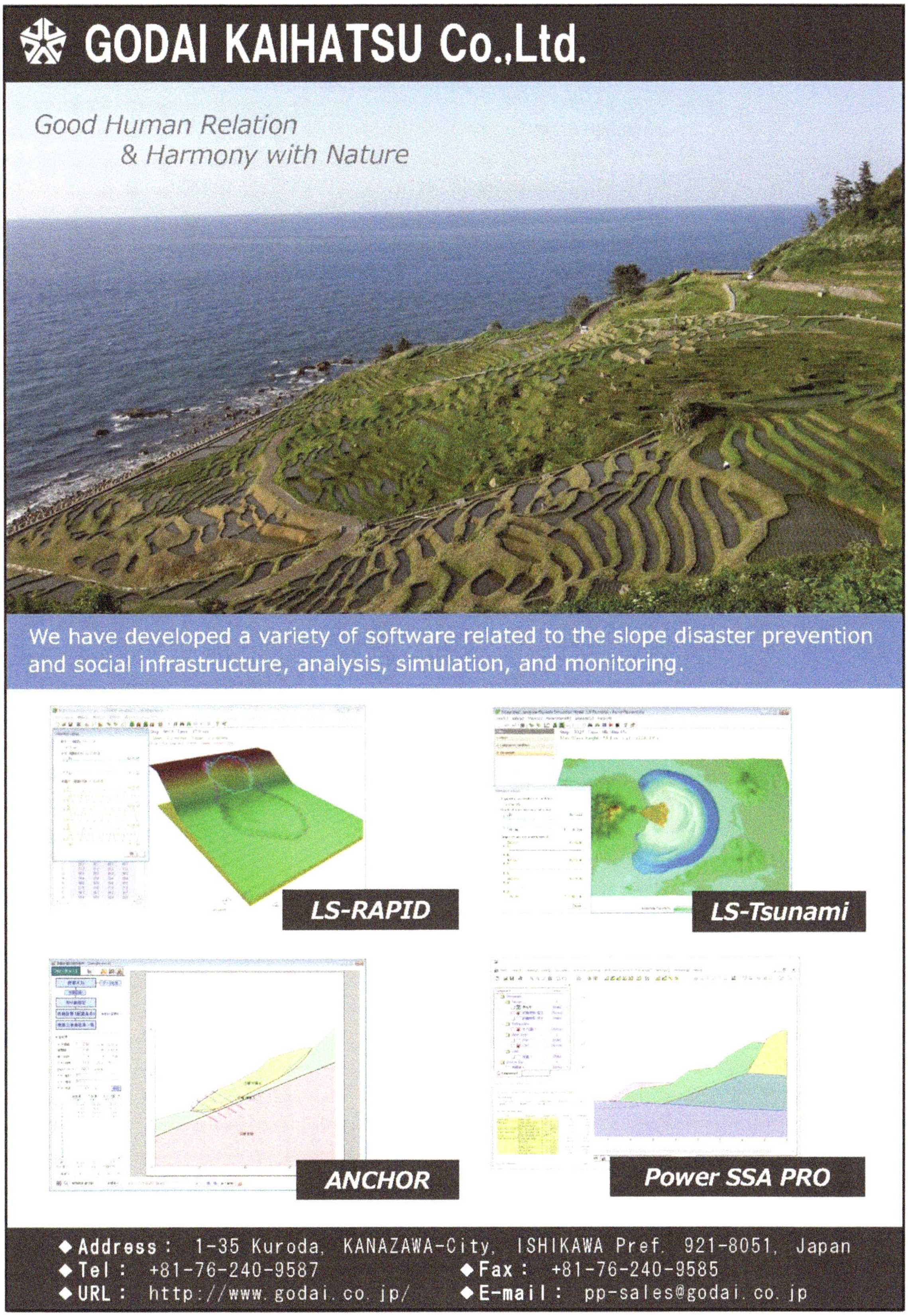
GODAI KAIHATSU Co.,Ltd.
Good Human Relation
& Harmony with Nature
We have developed a variety of software related to the slope disaster prevention and social infrastructure, analysis, simulation, and monitoring.
LS-RAPID
LS-Tsunami
ANCHOR
Power SSA PRO
◆Address : 1-35 Kuroda, KANAZAWA-City, ISHIKAWA Pref. 921-8051, Japan
◆Tel : +81-76-240-9587
◆Fax : +81-76-240-9585
◆URL : http://www.godai.co.jp/
◆E-mail : pp-sales@godai.co.jp

We aim to establish a sustainable country by various technology related to earth, water, and vegetation.

Consulting Services

Slope Disaster Management/
Forest Conservation and Afforestation/
Community Based Disaster Risk Reduction/
Disaster Risk Assessment Technology Transfer

Construction and Supervision

Construction of Landslide and Slope Disaster Prevention Measures/
Countermeasures Against Aging Infrastructure

Research and Development

Slope Protection Technology/
Afforestation Technology/
Geospatial Information Technology/
3D Simulation Technology/
Laboratory Testing of Soil and Rocks

JAPAN CONSERVATION ENGINEERS & CO., LTD.

3-18-5, Toranomon, Minato, Tokyo,Japan 105-0001
Tel: +81.3.3436.3673 E-mail: go_info@jce.jp URL https://www.jce.co.jp/en

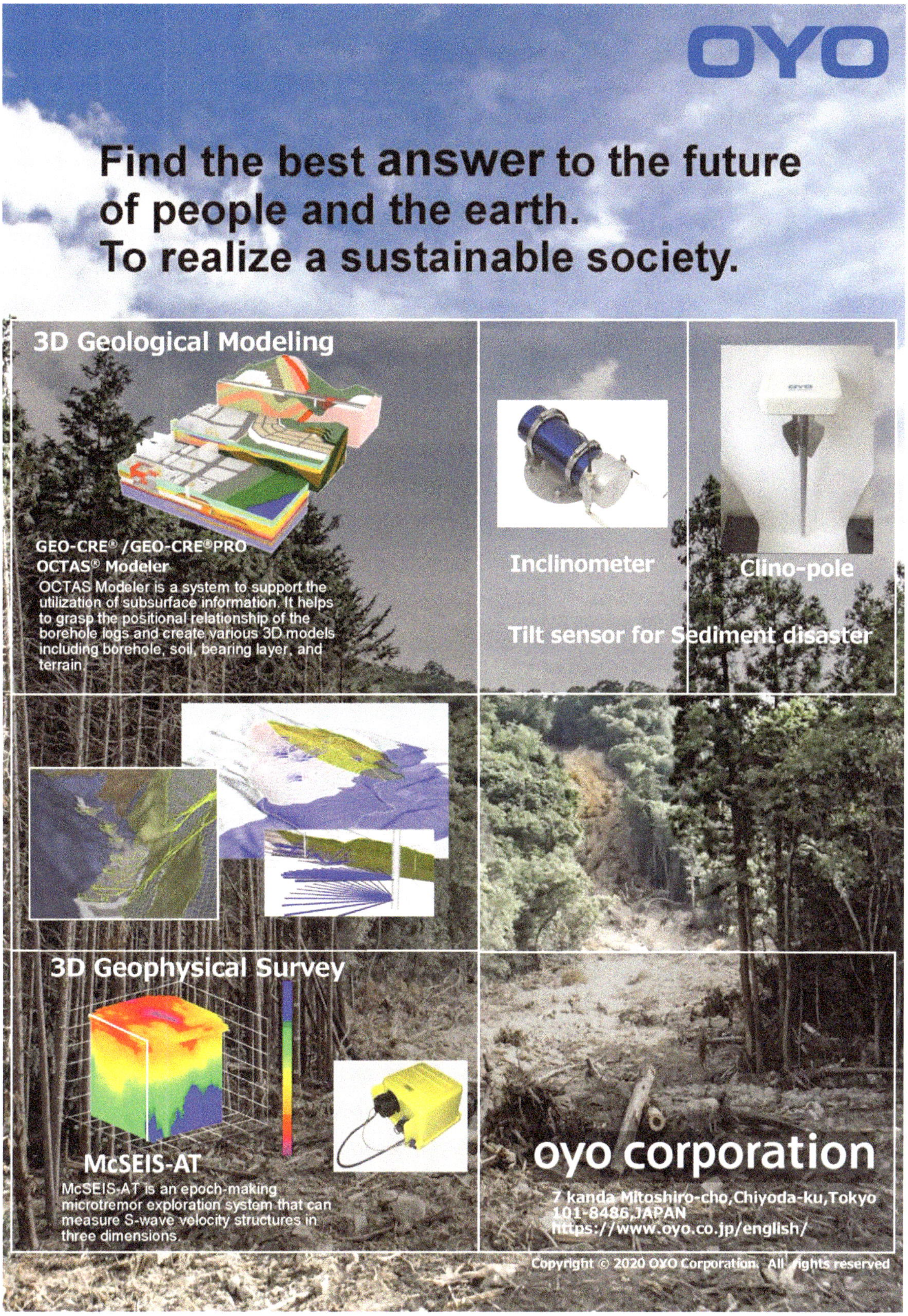
OYO
Find the best answer to the future
of people and the earth.
To realize a sustainable society.
3D Geological Modeling
GEO-CRE® /GEO-CRE®PRO
OCTAS® Modeler
OCTAS Modeler is a system to support the utilization of subsurface information. It helps to grasp the positional relationship of the borehole logs and create various 3D models including borehole, soil, bearing layer, and terrain.
Inclinometer
Clino-pole
Tilt sensor for Sediment disaster
3D Geophysical Survey
McSEIS-AT
McSEIS-AT is an epoch-making microtremor exploration system that can measure S-wave velocity structures in three dimensions.
oyo corporation
7 kanda Mitoshiro-cho,Chiyoda-ku,Tokyo
101-8486,JAPAN
https://www.oyo.co.jp/english/
Copyright © 2020 OYO Corporation. All rights reserved

Kokusai Kogyo as a leading company of geospatial information technologies, has been contributing to the improvement of public services with advanced measurement technologies and a wide range of consulting technologies. Kokusai Kogyo supports the development of "Green Communities" representing the new era public concerns on "environment and energy," "disaster risk reduction" and "asset management". Kokusai Kogyo offers the advanced analysis of geospatial information consultancy for developing new government policies, maintaining and operating social infrastructures with safe and secure city planning, and building low-carbon cities.

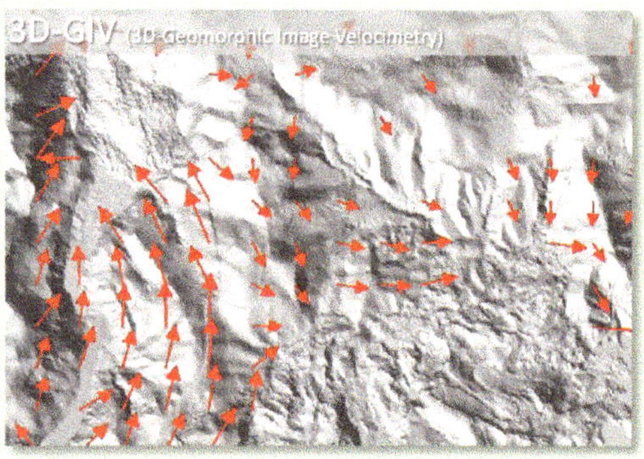

3D-GIV can help grasp the ground surface displacements caused by natural phenomena such as landslides by analyzing differences between digital geomorphic images obtained through ad hoc Airborne Laser Surveys.

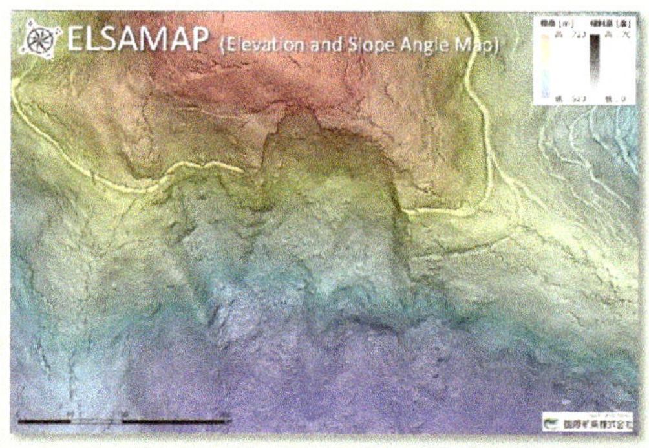

ELSAMAP is our cutting-edge 3D terrain visualization method allowing great geomorphological details to be visualized in one glance with gray-scaled slope inclinations and colored altitudes. ELSAMAP has been used to interpret micro-topographies, landslides and some other things.

Our Realtime Hazard Map reflects up-to-date information of soil natures and precipitations at landslide hazard sites, etc. that can constantly be changing, and evaluates area-wide hazard risk in real time.

"shamen-net" is a total monitoring system integrating GNSS and other monitoring devices (Measurement precision: +/-1mm, on a real time basis)

2 Rokubancho,Chiyoda-ku,Tokyo,Japan 102-0085 https://www.kkc.co.jp/english/ info_overseas@kk-grp.jp

www.geobrugg.com/debrisflow
GEOBRUGG
BRUGG
Safety is our nature
Free dimensioning tool
www.mygeobrugg.com
Ring net barriers made of high-tensile steel wire
THE ECONOMICAL SOLUTION
TO DEBRIS FLOW
Geobrugg AG | CH-8590 Romanshorn | www.geobrugg.com

LiSALab
Powered by **Ellegi srl**

... A step ahead in monitoring structures and natural hazards ....

by Ground Based Interferometric Synthetic Aperture Radar LiSALab technology

Ellegi's **provides services and products for remote sensing measurement of displacements and deformations of natural hazards and manmade buildings** using a ground based SAR system, known as **LiSALab system**, in software production and system integration, production and developments of data acquisition, visualization and data-analysis systems.

LiSALab system at present is at its 5th generation of development since 2003.

One of the biggest point of strength of Ellegi is based on its **high vertical integration**. It can internally design, produce, sell, maintains and provide services and products using GBInSAR LiSALab technology, **the customer can have all the answers**.

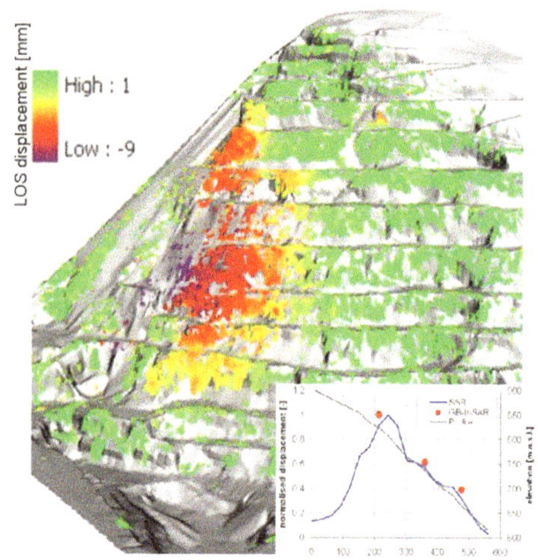

GBInSAR LiSALab technology quarry monitoring example and displacements' field comparison between the GBInSAR measurements and FEM model results.

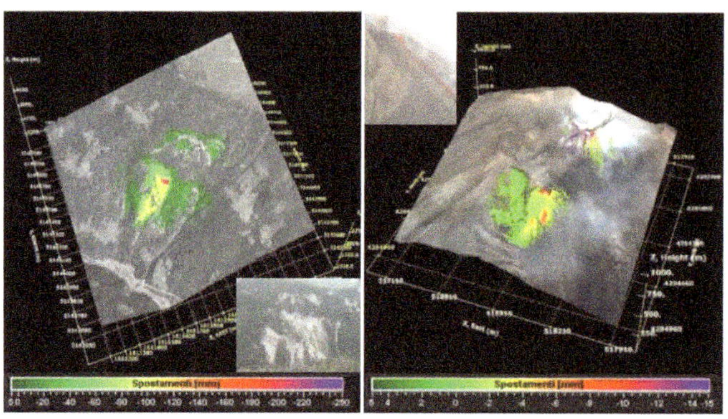

GBInSAR LiSALab technology results in monitoring a slope affected by a landslide (left) and a volcanic slope affected by deformation (right).
Landslide or moving areas mapping and boundaries identification is made easy by GBInSAR LiSALab tecnology.

Ellegi srl offers:

- **displacement fields measurement, control and monitoring** of the deformation caused by natural hazards, in every operative conditions, including emergencies;
- **structural strain fields measurement, control, monitoring** and diagnosis of the deformation affecting buildings, bridges, viaducts, dams, etc. etc.;
- **integrated monitoring systems** design, installation, management and maintenance in order to provide information about natural hazards or anthropic activity, that can generate or cause slopes or buildings instabilities.

Since 2003 Ellegi has provided services, systems and technologies in the world to the most important players in the monitoring sector.

Ellegi srl
Via Bandello, 5 I-20123 Milano Italy
Headquarters: via Petrarca, 55 I-22070 Rovello Porro CO Italy
Tel. +39 02 9443 5051 Fax +39 +39 02 9443 5052
info@lisalab.com - www.lisalab.com

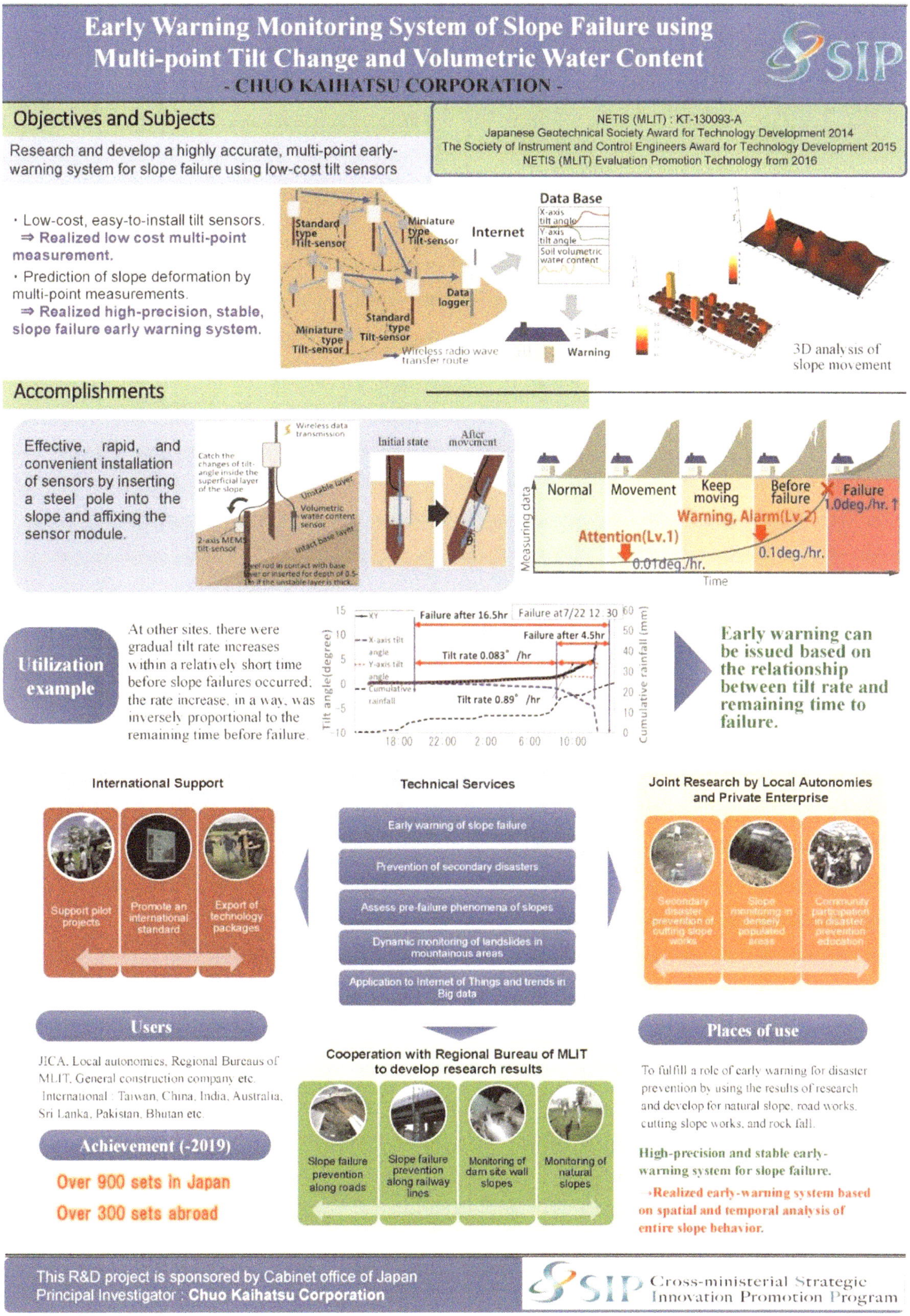

Early Warning Monitoring System of Slope Failure using Multi-point Tilt Change and Volumetric Water Content
- CHUO KAIHATSU CORPORATION -
SIP
Objectives and Subjects
Research and develop a highly accurate, multi-point early-warning system for slope failure using low-cost tilt sensors
NETIS (MLIT) : KT-130093-A
Japanese Geotechnical Society Award for Technology Development 2014
The Society of Instrument and Control Engineers Award for Technology Development 2015
NETIS (MLIT) Evaluation Promotion Technology from 2016
・Low-cost, easy-to-install tilt sensors.
⇒ Realized low cost multi-point measurement.
・Prediction of slope deformation by multi-point measurements.
⇒ Realized high-precision, stable, slope failure early warning system.
Standard type Tilt-sensor
Miniature type Tilt-sensor
Data logger
Internet
Data Base
X-axis tilt angle
Y-axis tilt angle
Soil volumetric water content
Wireless radio wave transfer route
Warning
3D analysis of slope movement
Accomplishments
Effective, rapid, and convenient installation of sensors by inserting a steel pole into the slope and affixing the sensor module.
Wireless data transmission
Unstable layer
Volumetric water content sensor
Intact base layer
2-axis MEMS tilt-sensor
Initial state
After movement
Normal
Movement
Keep moving
Before failure
Failure
Attention(Lv.1)
Warning, Alarm(Lv.2)
0.01deg./hr.
0.1deg./hr.
1.0deg./hr.
Measuring data
Time
Utilization example
At other sites, there were gradual tilt rate increases within a relatively short time before slope failures occurred; the rate increase, in a way, was inversely proportional to the remaining time before failure.
Failure after 16.5hr
Failure at7/22 12:30
Failure after 4.5hr
Tilt rate 0.083° /hr
Tilt rate 0.89° /hr
X-axis tilt angle
Y-axis tilt angle
Cumulative rainfall
Tilt angle(degree)
Cumulative rainfall (mm)
18:00 22:00 2:00 6:00 10:00
Early warning can be issued based on the relationship between tilt rate and remaining time to failure.
International Support
Support pilot projects
Promote an international standard
Export of technology packages
Technical Services
Early warning of slope failure
Prevention of secondary disasters
Assess pre-failure phenomena of slopes
Dynamic monitoring of landslides in mountainous areas
Application to Internet of Things and trends in Big data
Joint Research by Local Autonomies and Private Enterprise
Secondary disaster prevention of cutting slope works
Slope monitoring in densely populated areas
Community participation in disaster prevention education
Users
JICA, Local autonomies, Regional Bureaus of MLIT, General construction company etc.
International : Taiwan, China, India, Australia, Sri Lanka, Pakistan, Bhutan etc.
Achievement (-2019)
Over 900 sets in Japan
Over 300 sets abroad
Cooperation with Regional Bureau of MLIT to develop research results
Slope failure prevention along roads
Slope failure prevention along railway lines
Monitoring of dam site wall slopes
Monitoring of natural slopes
Places of use
To fulfill a role of early warning for disaster prevention by using the results of research and develop for natural slope, road works, cutting slope works, and rock fall.
High-precision and stable early-warning system for slope failure.
→Realized early-warning system based on spatial and temporal analysis of entire slope behavior.
This R&D project is sponsored by Cabinet office of Japan
Principal Investigator : Chuo Kaihatsu Corporation
SIP Cross-ministerial Strategic Innovation Promotion Program

IDS GeoRadar

IBIS FM - Natural Hazard Monitoring Solution

GB-InSAR System with Advanced Atmospheric Correction

IDS GeoRadar is nowadays the leader company in the Radar monitoring solutions applied to landslide monitoring, thanks to the combination of the highest performance radar technology with the most advanced data processing algorithm for the removal of atmospheric artifacts integral part of Guardian Software.

The unique IBIS-FM radar system accurately monitors multiple scales of displacements in real time, from early detection of slow movements to fast accelerations associated with slope collapse. The great operative range, up to 4500 m, allows to safely deploying the system in comfortably accessible areas, without exposing people and equipment to hazardous zones.

Fully automatic

The atmospheric correction procedure does not require any input from the user (does not need any stable area selection), it is a completely automatic software processing, not requiring advanced know-how to the user. The atmospheric compensation algorithm is based on hundreds of thousands of pixels automatically selected by the software and updated at each radar scan. It permits to achieve the best performances even in case of extreme or unstable atmospheric conditions.

Advanced 3D atmospheric effects modelling

IDS GeoRadar algorithm for atmospheric correction offers unique capability able to model and remove complex atmospheric effects in extensive areas overcoming the limits of the standard linear model algorithms, providing very clean displacement maps in the most extreme atmospheric conditions environment.

Slow and fast movement detection

IDS GeoRadar algorithm is able to distinguish very slow movement together with the faster ones providing the capability to measure in real time movements across four orders of magnitudes: from very fast movements (up to 150 mm/hour) to extremely low displacements (few mm/month)

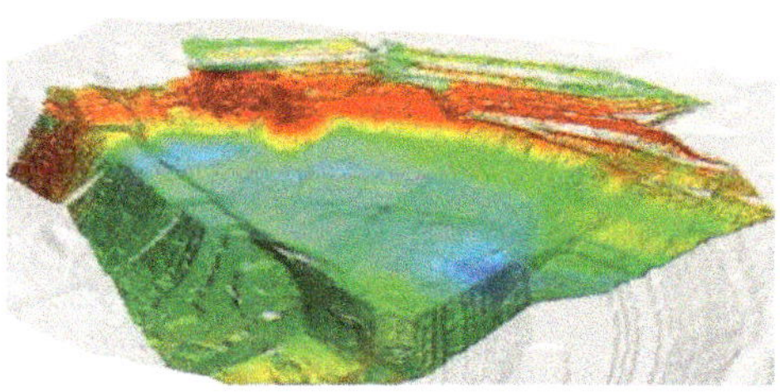

Displacement map obtained with standard processing and standard atmospheric correction

Displacement map obtained with IDS GeoRadar Guardian atmospheric correction and processing

IDS GeoRadar Srl
Part of Hexagon

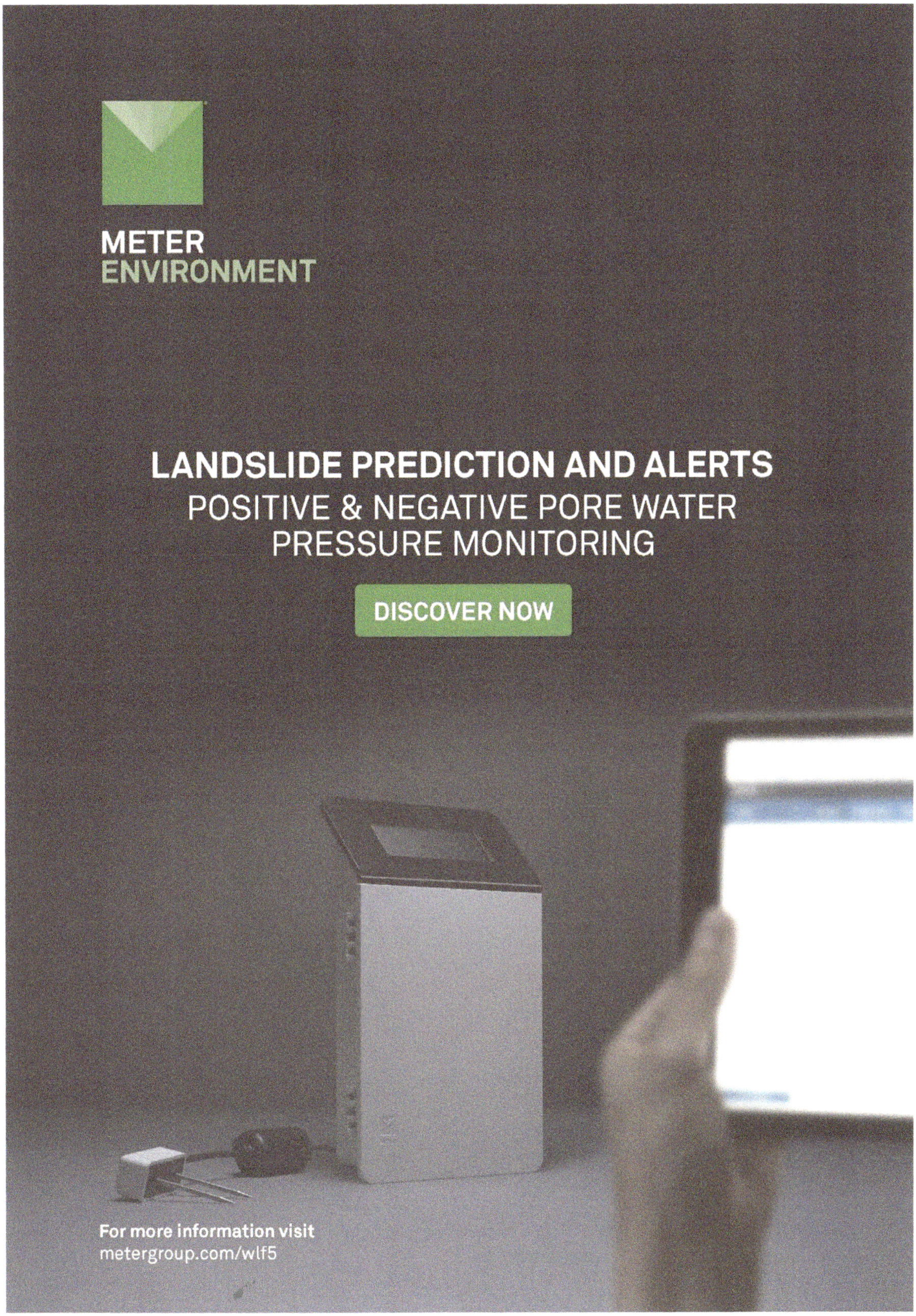
METER
ENVIRONMENT
LANDSLIDE PREDICTION AND ALERTS
POSITIVE & NEGATIVE PORE WATER
PRESSURE MONITORING
DISCOVER NOW
For more information visit
metergroup.com/wlf5

ASIA AIR SURVEY CO.,LTD.

PIONEERING THE FUTURE

A novel 3D visualization technique

Red Relief Image Map (RRIM)

Developed by AAS, RRIM is a simple and effective tool for representing and interpreting ground surface features. The RRIM allows great geomorphological details to be visualized in one glance with a single map without any device, at any map scale, from any viewing angles without shades.

AAS provides one-stop services of RRIM.

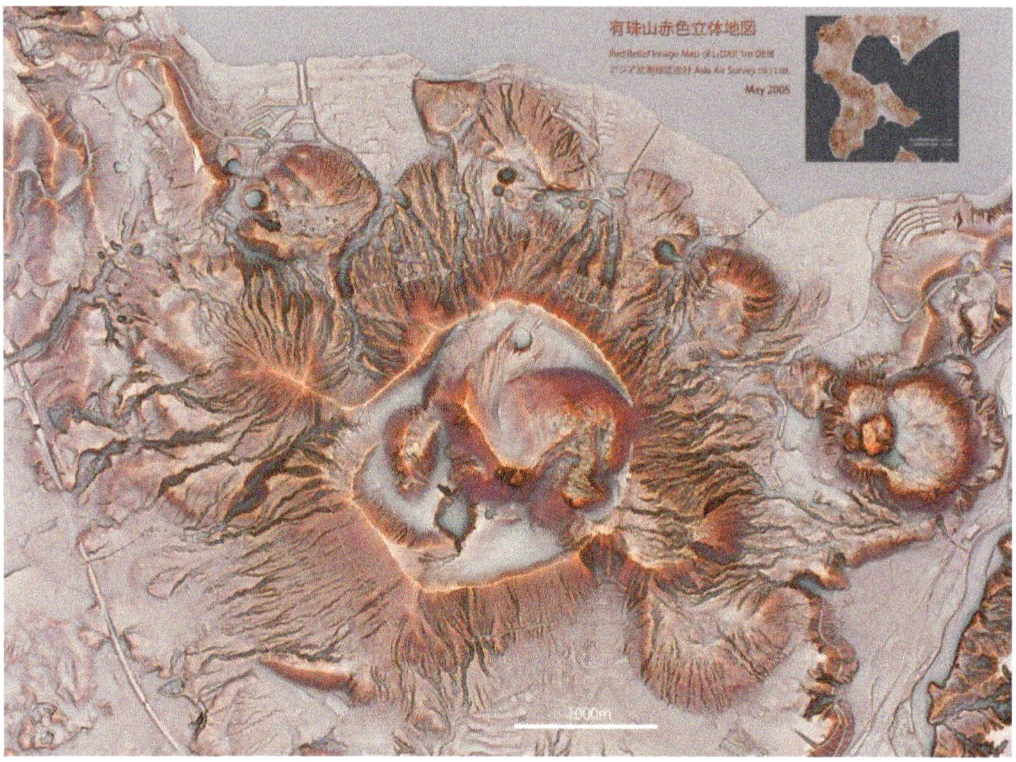

RRIM of Mt.Usu, Japan, from 1m mesh Lidar data.

Contact:

Shinyuri 21 BLDG 3F, 1-2-2 Manpukuji, Asao-Ku, Kawasaki City, Kanagawa Prefecture, 215-0004, Japan
https://www.ajiko.co.jp/en/
email:service@ajiko.co.jp
Tel. +81-44-969-7510 Fax. +81-44-965-0029

Kisojiban

Kiso-Jiban Consultants Co., Ltd.

Monitoring Service for Landslide by Kiso-SAR System

What's SAR ?

SAR (Synthetic Aperture Radar) is a technique that utilizes interference of radio waves for precise determination of distance.
InSAR (interferometric SAR), the phase of the received backscattered signal from two images of the same scene are used to measure path length differences with an accuracy of a few millimeters.

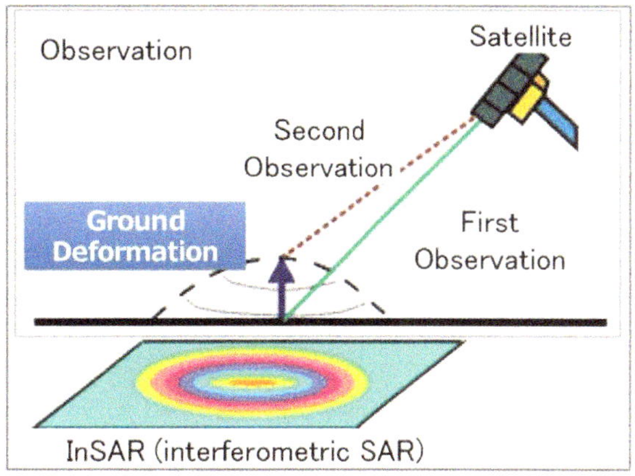

Ground Deformation Estimation in Landslides

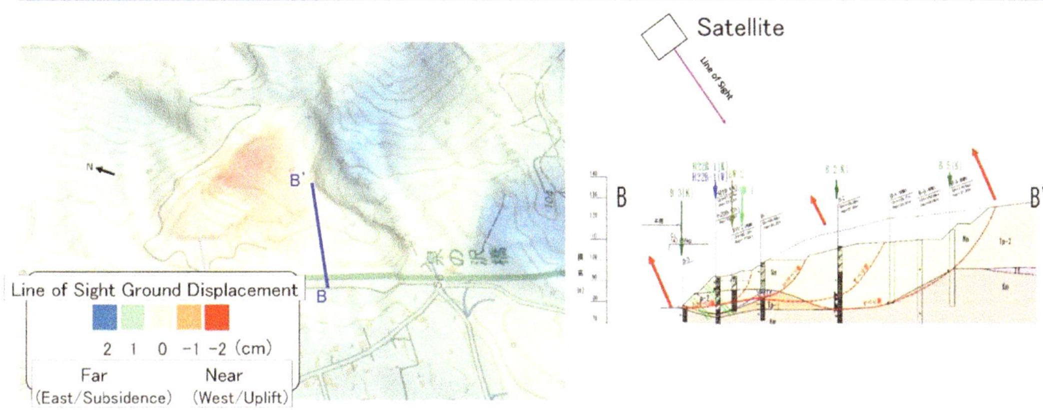

Ground Deformation of Landslide Observed by Kiso-SAR System

Kisojiban

Kiso-Jiban Consultants Co., Ltd.
Kinshicho Prime Tower 12 Floor,1-5-7
Kameido,Koto-ku,Tokyo 136-8577,Japan
Tel.:+81-3-6861-8800

Small & simple water drainage drilling system* for landslide disaster prevention

Simple & Small

- Simple structure.
- At a narrow space (0.3m×1.5m for guide rail, 0.8m×2.0m for drilling machine base).
- Lightweight equipment (Max. weight of drilling machine: 25kg).
- Easy installation of one operational well requires only 2 persons.

Quick & Short-term construction

- All pieces of equipment are man-portable.
- Preparation time for starting drilling is only 30 min.
- 50% reduction in construction time compared with conventional construction method.

Low cost & High effectivity

- Additional works (ex. scaffold or construction road) are not required.
- 50% reduction in construction cost compared with conventional construction method.
- About 30m long water drainage hole can be drilled in clayey and/or soft rock layers.

Safety & Environmentally friendly

- Small sound and low vibration.
- Construction yard is not required.

*Japanese utility model registration No.3186011

奥山ボーリング株式会社
Okuyama Boring Co.,Ltd.

https://okuyama.co.jp/en/

10-39 Shinmei-cho, Yokote-City, AKITA Pref. 013-0046, JAPAN

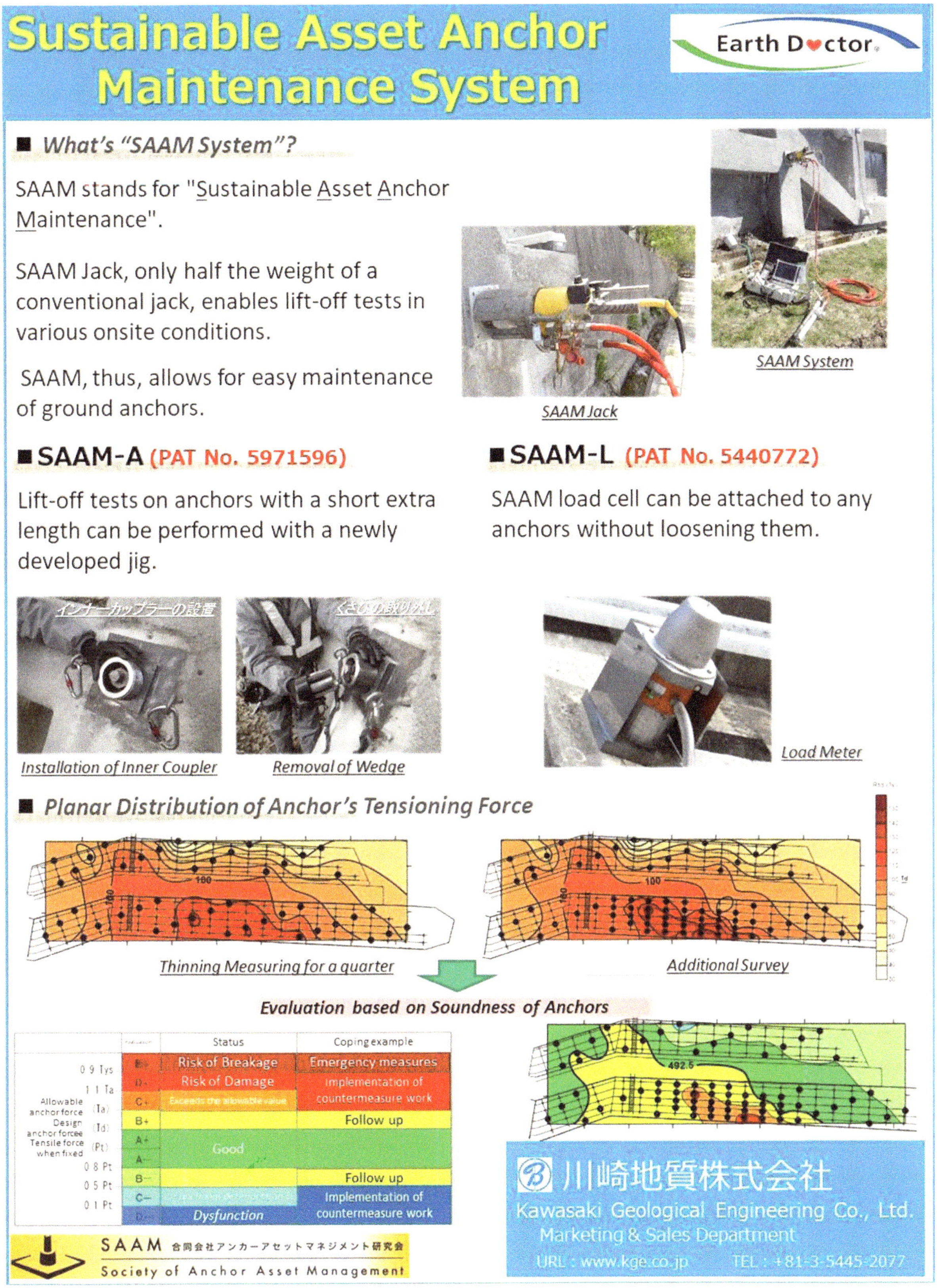
Sustainable Asset Anchor Maintenance System
Earth Doctor
■ What's "SAAM System"?
SAAM stands for "Sustainable Asset Anchor Maintenance".
SAAM Jack, only half the weight of a conventional jack, enables lift-off tests in various onsite conditions.
SAAM, thus, allows for easy maintenance of ground anchors.
SAAM System
SAAM Jack
■ SAAM-A (PAT No. 5971596)
Lift-off tests on anchors with a short extra length can be performed with a newly developed jig.
■ SAAM-L (PAT No. 5440772)
SAAM load cell can be attached to any anchors without loosening them.
インナーカップラーの設置
くさびの取り外し
Installation of Inner Coupler
Removal of Wedge
Load Meter
■ Planar Distribution of Anchor's Tensioning Force
Thinning Measuring for a quarter
Additional Survey
Evaluation based on Soundness of Anchors
Status
Coping example
0.9 Tys
1.1 Ta
Allowable anchor force (Ta)
Design anchor forces (Td)
Tensile force when fixed (Pt)
0.8 Pt
0.5 Pt
0.1 Pt
Risk of Breakage
Risk of Damage
Exceeds the allowable value
Emergency measures
implementation of countermeasure work
C+
B+
Follow up
A+
A−
Good
B−
Follow up
C−
D−
Dysfunction
Implementation of countermeasure work
492.5
川崎地質株式会社
Kawasaki Geological Engineering Co., Ltd.
Marketing & Sales Department
URL : www.kge.co.jp TEL : +81-3-5445-2077
SAAM 合同会社アンカーアセットマネジメント研究会
Society of Anchor Asset Management

Established in 1912
NISSAKU CO., LTD.
WELL DRILLINGS & CONSULTING ENGINEERS

Geological Survey, Construction Consultancy Service

Civil Construction

Well Drilling

Well Rehabilitation

Facility Construction

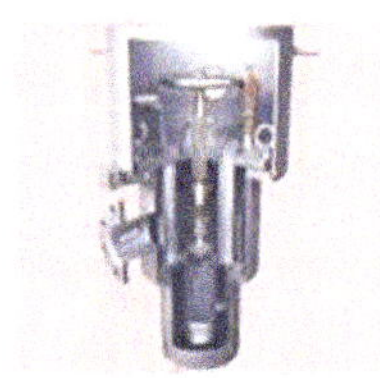
Manufacturing and Sales of Equipment

Overseas Business

Technology Development

Head Office No. 199-3, Sakuragi-cho 4-chome, Omiya-ku, Saitama-shi, Saitama 330-0854, Japan
TEL : 048-644-3911 FAX : 048-644-3958

Branch Office : Sendai, Niigata, Jyoetsu, Tokyo, Nagano, Komatsu, Nagoya, Shizuoka, Osaka, Fukuoka, Naha and Konosu (Saitama factory)

URL : http://www.nissaku.co.jp/ E-Mail : honsha@nissaku.co.jp

International Consortium on Landslides

International Consortium on Landslides

An international non-government and non-profit scientific organization promoting landslide research and capacity building for the benefit of society and the environment

President: Peter T. Bobrowsky (Geological Survey of Canada)
Vice Presidents: Matjaž Mikoš (University of Ljubljana, Slovenia), Dwikorita Karnawati (Agency for Meteorology, Climatology, and Geophysics, Indonesia), Nicola Casagli (University of Florence, Italy), Binod Tiwari (California State University, USA), Željko Arbanas (University of Rijeka, Croatia)
Executive Director: Kaoru Takara (Kyoto University, Japan), Treasurer: Kyoji Sassa (Prof. Emeritus, Kyoto University, Japan)

ICL Full Members:

Geotechnical Engineering Office, Hong Kong Special Administrative Region, China
UNESCO Chair for the Prevention and the Sustainable Management of Geo-hydrological Hazards - University of Florence, Italy
Korea Institute of Geoscience and Mineral Resources (KIGAM)
University of Ljubljana, Faculty of Civil and Geodetic Engineering (ULFGG), Slovenia

Albania Geological Survey / The Geotechnical Society of Bosnia and Herzegovina / Center for Scientific Support in Disasters – Federal University of Parana, Brazil/ Geological Survey of Canada / University of Alberta, Canada / Northeast Forestry University, Institute of Cold Regions Science and Engineering, China / China University of Geosciences / Chinese Academy of Sciences, Institute of Mountain Hazards and Environment / Tongji University, College of Surveying and Geo-Informatics, China / The Hong Kong University of Science and Technology, China / Shanghai Jiao Tong University, China / The University of Hong Kong, China / Universidad Nacional de Colombia / Croatian Landslide Group (Faculty of Civil Engineering, University of Rijeka and Faculty of Mining, Geology and Petroleum Engineering, University of Zagreb) / City of Zagreb, Emergency Management Office, Croatia / Charles University, Faculty of Science, Czech Republic / Institute of Rock Structure and Mechanics, Department of Engineering Geology, Czech Republic / Brown Coal Research Institute, Czech Republic / Cairo University, Egypt / Technische Universitat Darmstadt, Institute and Laboratory of Geotechnics, Germany / National Environmental Agency, Department of Geology, Georgia / Universidad Nacional Autonoma de Honduras (UNAH), Honduras / Amrita Vishwa Vidyapeetham, Amrita University / Vellore Institute of Technology, India / National Institute of Disaster Management, India / Agency for Meteorology, Climatology, and Geophysics of the Republic of Indonesia (BMKG Indonesia) / University of Gadjah Mada, Center for Disaster Mitigation and Technological Innovation (GAMA-InaTEK), Indonesia / Parahyangan Catholic University, Indonesia / Building & Housing Research Center, Iran / Italian Institute for Environmental Protection and Research (ISPRA) - Dept. Geological Survey, Italy / University of Calabria, DIMES, CAMILAB, Italy / Istituto de Ricerca per la Protezione Idrogeologica (IRPI), CNR, Italy / DIA– Universita degli Studi di Parma, Italy / University of Torino, Dept of Earth Science , Italy / Centro di Ricerca CERI - Sapienza Università di Roma, Italy / Kyoto University, Disaster Prevention Research Institute, Japan / Japan Landslide Society / Korean Society of Forest Engineering / National Institute of Forest Science, Korea / Korea Infrastructure Safety & Technology Corporation / Korea Institute of Civil Engineering and Building Technology / Slope Engineering Branch, Public Works Department of Malaysia / Institute of Geography, National Autonomous University of Mexico (UNAM) / International Centre for Integrated Mountain Development (ICIMOD), Nepal / University of Nigeria, Department of Geology, Nigeria / Moscow State University, Department of Engineering and Ecological Geology, Russia / JSC "Hydroproject Institute", Russia / University of Belgrade, Faculty of Mining and Geology, Serbia / Comenius University, Faculty of Natural Sciences, Department of Engineering Geology, Slovakia / Geological Survey of Slovenia / University of Ljubljana, Faculty of Natural Sciences and Engineering (ULNTF), Slovenia / Central Engineering Consultancy Bureau (CECB), Sri Lanka / National Building Research Organization, Sri Lanka / Landslide group in National Central University from Graduate Institute of Applied Geology, Department of Civil Engineering, Center for Environmental Studies, Chinese Taipei / National Taiwan University, Department of Civil Engineering, Chinese Taipei / Asian Disaster Preparedness Center, Thailand / Ministry of Agriculture and Cooperative, Land Development Department, Thailand / Institute of Telecommunication and Global Information Space, Ukraine / California State University, Fullerton & Tribhuvan University, Institute of Engineering, USA & Nepal / Institute of Transport Science and Technology, Vietnam / Vietnam Institute of Geosciences and Mineral Resources (VIGMR).

ICL Associates

State Key Laboratory of Geohazard Prevention and Geoenvironment Protection (Chengdu University of Technology), China / Czech Geological Survey, Czech Republic / Department of Earth and Environmental Sciences, University Aldo Moro, Bari, Italy / Department of Sciences and Technologies, University of Sannio, Italy / Department of Earth and Environmental Sciences – University of Pavia, Italy / Geotechnical Engineering Group (GEG), University of Salerno, Italy / Niigata University, Research Institute for Natural Hazards and Disaster Recovery, Japan / Ehime University Center for Disaster Management Informatics Research, Japan / Tian-Shan Geological Society, Kyrgyzstan / Institute of Environmental Geoscience RAS (IEG RAS), Russia / Russian State Geological Prospecting University n.a. Sergo Ordzhonikidze (MGRI-RSGPU) / TEMPOS, environmental civil engineering Ltd., Slovenia / Institute of Earth Sciences – Faculty of Geoscience and Environment, University of Lausanne, Switzerland / Middle East Technical University (METU), Turkey / North Dakota State University, USA

ICL Secretariat:

Secretary General: Kyoji Sassa
International Consortium on Landslides, 138-1 Tanaka Asukai-cho, Sakyo-ku, Kyoto 606-8226, Japan
Web: http://icl.iplhq.org/, E-mail: secretariat@iclhq.org
Tel: +81-75-723-0640, Fax: +81-75-950-0910

© Springer Nature Switzerland AG 2021

N. Casagli et al. (eds.), *Understanding and Reducing Landslide Disaster Risk*,
ICL Contribution to Landslide Disaster Risk Reduction,
https://doi.org/10.1007/978-3-030-60311-3